T0133701

ERGONOMICS IN DESIGN

METHODS & TECHNIQUES

Human Factors and Ergonomics Series

PUBLISHED TITLES (CONTINUED)

ERGONOMICS IN DESIGN METHODS & TECHNIQUES

edited by
MARCELO M. SOARES
FRANCISCO REBELO

CRC Press
Taylor & Francis Group
Boca Raton London New York

CRC Press is an imprint of the
Taylor & Francis Group, an **informa** business

CRC Press
Taylor & Francis Group
6000 Broken Sound Parkway NW, Suite 300
Boca Raton, FL 33487-2742

Printed on acid-free paper
Version Date: 20160523

International Standard Book Number-13: 978-1-4987-6070-6 (Hardback)

Visit the Taylor & Francis Web site at
http://www.taylorandfrancis.com

and the CRC Press Web site at
http://www.crcpress.com

Printed and bound in the United States of America by Publishers Graphics,
LLC on sustainably sourced paper.

Contents

SECTION I   Theoretical Issues

SECTION II   Human Characteristics in Design

SECTION III   Methodological Issues

SECTION IV Design Development

Preface

People currently deal with various entities (hardware, software, buildings, spaces, communities, and other people), to meet specific goals while going about their everyday activities in work and leisure environments. Such entities may be considered as parts of a system (Chapanis, 1996; Wilson, 2014). These systems have become more and more complex and incorporate functions that hitherto had never been considered such as automation, use in virtual environments, connectivity, personalization, mobility, and friendliness. This interaction, no matter how simple or complex, must meet the expectations and needs of users and workers in a healthy safe, efficient, and enjoyable manner. Moreover, this ought to be considered as a way to improve the performance of the system and in turn be able to achieve higher productivity in all areas of our lives.

Although these principles are basic ones, it is not always the case that the interaction between human beings and the system takes place appropriately from the point of view of ergonomics. According to Karwowski et al. (2011), applying ergonomic principles and knowledge in the design process helps to meet users' expectations, thereby reducing the risks of malfunction or potential failures, which can lead to accidents and health problems and can also lead to a product or a system being more widely acceptable.

According to Karwowski et al. (op. cit.), this is not always the case nor is the designing of products or systems that meet human needs an easy task. Ergonomics seeks to optimize the relationship between systems and humans. Good design can be regarded as that which considers users' abilities and limitations and their needs and expectations, when developing a product, in order to make aspects of its aesthetics, functionality, and manufacture compatible with its usability and users' experience. According to these authors, design methods which involve applying the principles and the knowledge of human factors and ergonomics (HF/E) contribute to achieving these characteristics, thus reducing the risk of malfunction or failure and the potential for accidents, thereby contributing to well-being and good health and to the design being acceptable and of use, all of which at the same time reduces the total costs during the life cycle of the product and the system.

Karwowski (2005) states that HF/E are presented as a single discipline that focuses on the nature of the interactions between human beings and artifacts, based on a unified perspective of science, engineering, design, technology, and management of systems that are compatible with human beings. Thus, the discipline of HF/E can promote a holistic view, an approach centered on the human being, which considers the physical, cognitive, social, organizational, environmental, and other factors related to design. These factors should be considered in order to meet users' diverse preferences including such different factors as age, gender, and health issues.

Soares (2012) argues that the paradigm of human-centered design as applied to products and systems is to improve levels of user satisfaction and efficiency of use; to increase comfort; and to assure safety in normal use, as well as in the foreseeable misuse of a product or a system. Achieving these design goals is a major challenge. The use of appropriate design methods contributes to improving the critical features of products and workplaces, such as the ease with which they can be used, their functions learned, and their efficiency, comfort, safety, and adaptability taken advantage of, such that all of these features meet users' needs and contribute to user satisfaction. Rebelo et al. (2012) argue that when developing a product, user experience must be part of the user-centered design methodology. In this context, a new goal of ergonomics for design emerges, namely, to provide users with good experiences of using the product, which materialize in positive emotions. This is also a goal of design, in this context. Therefore, both disciplines are consonant with each other, thereby contributing to users' delight with the product and their environment.

We are pleased to introduce the book *Ergonomics in Design: Methods and Techniques*. New technologies are studied in various chapters of this book. Thus, its aim is to provide a comprehensive review of the state of the art of current ergonomic methods and techniques, which are being applied to products, machinery, equipment, workstations, and systems while taking new technologies and their applications into consideration.

This book is divided into four sections. Section I deals with theoretical aspects. Investigating what kind of information can be transferred with haptic feedback and how this should be designed is the subject of Chapter 1. Making a comparative analysis of the (analytical) theoretical method of design with the experimental method of design, used in HF/E, is also part of this section. Other aspects tackled in this section include an analysis of how inclusive design should help people and communities prepare themselves for the future as well as improve their present quality of life. It does so by presenting a study on the physical, physiological, and psychological effects of colors in their interaction with human beings in terms of their clothes; how corporate visual identity design acquires special relevance because this is one way to make complex information more accessible, understandable, and easy to memorize; a study of cultural usability and its importance for different ethnic groups that use information technology on a daily basis; an analysis of signposting systems by using ergonomic principles, standards, and codes; the analysis of the relationship between ergonomics and sustainability in the built environment; and a study on virtual reality applied to digital games.

Section II deals with the theme of human characteristics in design. Chapter 10 presents an ergonomic analysis applied to the design of homes for the elderly and how this can influence their quality of life. Chapter 11 analyzes the use of psychophysical data by ergonomics and evaluates the individual perception of effort in tasks that simulate measuring torque and pulling strength. Chapter 12 presents a method for evaluating the comfort of work equipment, with reference to a digital hand model, and takes into consideration models that involve discomfort that can be used to guide the design of handles. Finally, an ergonomic analysis applied to school furniture used in classrooms in Portugal is presented.

Section III is related to methodological issues. In this section are covered topics such as to generate user-oriented design solutions for ship bridges which support the existing activities of professional workers and surprise the users with innovativeness by offering new possibilities; to apply colors as an ergonomic factor to urban furniture, bearing in mind that pertinent color application enhance its use; to design the reference form enabling the user to shift his viewpoint from the main eyesight to the ancillary information presented by the monocular see-through head-mounted display (HMD) within a short team; to combine H/FE and design thinking that might shift decision-makers' ideas toward ergonomics; to evaluate usability issues regarding an experimental virtual environment (VE) prototype that was specifically designed to conduct ergonomic and design studies with older populations/users; to promote an independent way for elderly people with low vision conditions to act autonomously in hospital environments using a sensorimotor wayfinding system; and to evaluate perspectives on seniors as computer-users in different contexts of healthcare services.

Section IV analyzes design development. First, there is research that investigates human interaction with postural behaviors during sleep. Then several haptic devices are outlined and the potential benefits and application of this technology on the usability evaluation of product design in a virtual environment is demonstrated. The next chapters of this section respectively address the effect of packages' shape on hazard-related perceptions (i.e., levels of content hazardousness and awareness of consequences); the issues involved in defining the requirements for designing the Knowledge-Based Doctrine Tool (KBDT) which supports the cooperative and participatory processes of knowledge elicitation that are intuitive and easy to implement; anthropometric design that deals with the problem of the safety and comfort of passengers in city buses; a study to understand if the differences between two economic and social different countries may interfere with ergonomic and human factors involved in the school supplies transportation task; an analysis of the inclusive service design approach, guided by the human factors perspective, in order to contribute to evaluating and designing a more inclusive service bus used by younger and older people; the enhanced sustainability embodying human factors in building design; an analysis of five different tops for frozen shoulder users; and finally the application of neuroscience in design and human-system interactions.

We hope this book will be useful for a large number of professionals, students, and practical users wishing to incorporate the methods and ergonomics techniques presented here in product design, jobs, and systems in various applications. We also hope that this book will contribute to ensuring high-quality standards of HF/E applications and to promote HF/E research excellence at universities and other organizations (Dul et al., 2012).

Marcelo M. Soares
Recife, Brazil

Francisco Rebelo
Lisboa, Portugal

REFERENCES

Chapanis, A. 1996. *Human Factors in Systems Engineering.* New York, NY: John Wiley & Sons.
Dul, J. et al. 2012. A strategy for human factors/ergonomics: Developing the discipline and profession. *Ergonomics* 55(4), 377–395.
Karwowski, W. 2005. Ergonomics and human factors: The paradigms for science, engineering, design, technology, and management of human-compatible systems. *Ergonomics* 48(5), 436–463.
Karwowski, W., Soares, M.M., Stanton, N. 2011. *Human Factors and Ergonomics in Consumer Product Design: Methods and Techniques.* Boca Raton, FL: CRC Press.
Rebelo, F., Noriega, P., Duarte, E., Soares, M. 2012. Using virtual reality to access user experience. *Human Factors* 54, 964–982.
Soares, M.M. 2012. Translating user needs into product design for the disabled: An ergonomic approach. *Theoretical Issues in Ergonomics Science* 13(1), 92–120.
Wilson, J. 2014. Fundamentals of systems ergonomics/human factors. *Applied Ergonomics* 45, 5–13.

Editors

Marcelo M. Soares is currently a full professor in the Department of Design at the Federal University of Pernambuco, Brazil. He was an invited lecturer at the University of Guadalaraja, Mexico and the Indian Institute of Technology Kharagpur, Kharagpur, India. He was also an invited lecturer at the University of Central Florida, Florida and the Technical University of Lisbon, Portugal. He earned an MS (1990) in production engineering from the Federal University of Rio de Janeiro, Brazil. He earned a PhD at the Loughborough University in England (1998). Dr. Soares is a certified professional ergonomist from the Brazilian Ergonomics Association, of which he was president for 7 years. Dr. Soares has provided leadership in Human Factors and Ergonomics in Latin America and in the world as a member of the Executive Committee of the International Ergonomics Association. He was also appointed as the chairman of IEA 2012 (the Triennial Congresses of the International Ergonomics Association), which was held in Recife, Brazil (2012). His research, teaching, and consulting activities focus on manufacturing ergonomics, usability, product design, information ergonomics, and applications of virtual reality, thermography, and neurosciences in products and systems. Professor Soares currently serves on the editorial board of *Theoretical Issues in Ergonomics Science, Human Factors and Ergonomics in Manufacturing*, and several publications in Brazil. He has more than 180 papers published in congresses worldwide and several books and book chapters. He has undertaken research and consultancy work for several companies in Brazil. Professor Soares is the coeditor of the *Handbook of Human Factors and Ergonomics in Consumer Product Design: Methods and Techniques* and the *Handbook of Human Factors and Ergonomics in Consumer Product Design: Uses and Applications*.

Francisco Rebelo graduated in ergonomics and earned his PhD in ergonomics in 1996. He is currently professor of product ergonomics, design of systems, usability and game design, at University of Lisbon (ULisboa). He is also involved in entrepreneurship activities at that university. He is the director of the Ergonomics Laboratory at ULisboa, mainly involved in the modeling of virtual environments to study and optimize human interaction in ergonomic design studies.

Francisco Rebelo's research focuses on interaction design with emerging technologies, creating human-centered positive experiences that can enhance user safety and health and systems efficiency.

Contributors

Carlos Aceves-González
Loughborough Design School
Loughborough University
Loughborough, United Kingdom

Tareq Ahram
Institute for Advanced Systems
    Engineering
University of Central Florida
Orlando, Florida

Audrey Abi Akle
ESTIA Recherche
ESTIA
Bidart, France

Pedro Arezes
School of Engineering
University of Minho
Guimarães, Portugal

Erminia Attaianese
Department of Architecture
University of Naples Federico II
Naples, Italy

Hande Ayanoğlu
UNIDCOM
IADE—Creative University
Lisboa, Portugal

Rafaela Q. Barros
Department of Design
Federal University of Pernambuco
Recife, Brazil

Gregory Bedny
Evolute
New Jersey

Inna Bedny
Evolute
New Jersey

Klaus Bengler
Lehrstuhl für Ergonomie
Technische Universität München
München, Germany

Miguel de Aboim Borges
CIAUD Research Center in Architecture,
    Urbanism and Design
University of Lisbon
Lisboa, Portugal

Katarzyna Borgieł
Laboratoire de l'Intégration du Matériau au
    Système (IMS)
University of Bordeaux
Bordeaux, France

Breno Carvalho
Centre for Social Sciences
Catholic University of Pernambuco
Recife, Brazil

Sharon Cook
Loughborough Design School
Loughborough University
Loughborough, United Kingdom

Fernando Moreira da Silva
CIAUD Research Center in Architecture,
    Urbanism and Design
University of Lisbon
Lisboa, Portugal

Ricardo Dagge
CIAUD Research Center in Architecture,
    Urbanism and Design
University of Lisbon
Lisboa, Portugal

Gustavo Desouzart
Health Research Unity
Polytechnic Institute of Leiria
Leiria, Portugal

Emília Duarte
UNIDCOM
IADE—Creative University
Lisboa, Portugal

Marcello Silva e Santos
Department of Industrial Engineering
UNIFOA—University Center of Volta
    Redonda
Volta Redonda, Brazil

Christianne Falcão
Centre for Sciences and Technology
Catholic University of Pernambuco
Recife, Brazil

Ernesto Filgueiras
Department of Communication and Arts
University of Beira Interior
Covilhã, Portugal

Yusuke Fukuda
Mechanical System Development
    Department 1
Brother Industries, Ltd.
Nagoya, Japan

Margarida Gamito
CIAUD Research Center in Architecture,
    Urbanism and Design
University of Lisbon
Lisboa, Portugal

Joaquim Góis
Department of Mining Engineering
University of Porto
Porto, Portugal

Alexander González
School of Architecture
Universidad Pontificia Bolivariana de
    Medellín
Medellín, Colombia

Peter Gust
Institute for Engineering Design
University of Wuppertal
Wuppertal, Germany

Eija Kaasinen
VTT Technical Research Centre of
    Finland Ltd
Espoo, Finland

Hannu Karvonen
VTT Technical Research Centre of
    Finland Ltd
Espoo, Finland

Waldemar Karwowski
Institute for Advanced Systems Engineering
Department of Industrial Engineering and
    Management Systems
University of Central Florida
Orlando, Florida

Anthony Lins
Centre for Social Sciences
Catholic University of Pernambuco
Recife, Brazil

Cristina C. Lucio
Department of Design and Fashion
Maringa State University
Maringa, Brazil

Thomas Maier
Institute for Engineering Design and
    Industrial Design
Research and Teaching Department
    Industrial Design Engineering
University of Stuttgart
Stuttgart, Germany

Petri Mannonen
Department of Computer Science and
    Engineering
Aalto University
Espoo, Finland

Rui Matos
Superior School of Education and Social
    Sciences
Polytechnic Institute of Leiria
Leiria, Portugal

Andrew May
Loughborough Design School
Loughborough University
Loughborough, United Kingdom

Filipe Melo
Laboratory of Motor Behavior
University of Lisbon
Lisboa, Portugal

Christophe Merlo
Laboratoire de l'Intégration du Matériau au
    Système (IMS)
University of Bordeaux
Bordeaux, France

Stéphanie Minel
ESTIA Recherche
ESTIA
Bidart, France

Cristina Olaverri Monreal
Lehrstuhl für Ergonomie
Technische Universität München
München, Germany

Miwa Nakanishi
Mechanical System Development
  Department 1
Brother Industries, Ltd.
Nagoya, Japan

Andre Neves
Department of Design
Federal University of Pernambuco
Recife, Brazil

João Neves
CIAUD Research Center in Architecture,
  Urbanism and Design
University of Lisbon
Lisboa, Portugal

Paulo Noriega
CIAUD Research Center in Architecture,
  Urbanism and Design
University of Lisbon
Lisboa, Portugal

Isabel L. Nunes
UNIDEMI
Universidade Nova de Lisboa
Caparica, Portugal

Maria Lúcia L. R. Okimoto
Department of Mechanical Engineering
Federal University of Parana
Curitiba, Brazil

Marie Monique Bruère Paiva
Post Graduate Program of Design
Federal University of Pernambuco
Recife, Brazil

Luis Carlos Paschoarelli
Laboratory of Ergonomics and Interfaces
Paulista State University
Bauru, Brazil

Cristina Pinheiro
UNIDCOM-IADE Research Unit in Design
  and Communication
CIAUD Research Center in Architecture,
  Urbanism and Design
University of Lisbon
Lisboa, Portugal

Maria Eugénia Pinho
Faculty of Engineering
University of Porto
Porto, Portugal

Nara Raquel Silva Porto
Post Graduate Program of Ergonomics
Federal University of Pernambuco
Recife, Brazil

Daniel Raposo
Escola Superior de Artes Aplicadas
Instituto Politécnico de Castelo Branco
Castelo Branco, Portugal

Bruno M. Razza
Technological Center
Maringa State University
Maringa, Brazil

Francisco Rebelo
ErgoLab—FMH
ULisboa and CIAUD FA
University of Lisbon
Cruz Quebrada, Portugal

Lara Reis
Ergonomics Laboratory
University of Libon
Lisboa, Portugal

Thaisa Francis César Sampaio Sarmento
Post Graduate Program of Design
Federal University of Pernambuco
Recife, Brazil

Letícia Schiehll
CIAUD Research Center in Architecture,
  Urbanism and Design
University of Lisbon
Lisboa, Portugal

Christian Schulz
Institute for Engineering Design and Industrial
    Design
Research and Teaching Department Industrial
    Design Engineering
University of Stuttgart
Stuttgart, Germany

Danilo Silva
Laboratory of Ergonomics and Interfaces
Department of Electric Engineering
Paulista State University
Bauru, Brazil

Inês Simões
CIAUD Research Center in Architecture,
    Urbanism and Design
University of Lisbon
Lisboa, Portugal

Mário Simões-Marques
CINAV—Portuguese Navy
Almada, Portugal

Gabriel Soares
Centre for Social Sciences
Catholic University of Pernambuco
Recife, Brazil

Marcelo M. Soares
Post Graduate Program of Design
and
Post Graduate Program of Ergonomics
Federal University of Pernambuco
Recife, Brazil

Svetozar Sofijanic
College of Vocational Studies
Belgrade Polytechnic
Belgrade, Serbia

Luís Sousa
Department of Civil Engineering
University of Porto
Porto, Portugal

Evica Stojiljkovic
Faculty of Occupational Safety
University of Nis
Nis, Serbia

Júlia Teles
CIPER—Interdisciplinary Center for the
    Study of Human Performance
University of Lisbon
Cruz-Quebrada, Portugal

Takahiro Uchiyama
Department of Administration Engineering
Keio University
Yokohama, Japan

José Alfredo C. Ulson
Department of Electric Engineering
Paulista State University
Bauru, Brazil

Aydin Ünlü
Institute for Engineering Design
University of Wuppertal
Wuppertal, Germany

Vilma Villarouco
Post Graduate Program of Design
and
Post Graduate Program of Ergonomics
Federal University of Pernambuco
Recife, Brazil

Mikael Wahlström
VTT Technical Research Centre of Finland Ltd
Espoo, Finland

Julie A. Waldron
Human Factors Research Group
The University of Nottingham
Nottingham, United Kingdom

Johann Winterholler
Institute for Engineering Design and
    Industrial Design
Research and Teaching Department Industrial
    Design Engineering
University of Stuttgart
Stuttgart, Germany

Daigoro Yokoyama
Department of Administration Engineering
Keio University
Yokohama, Japan

Aleksandar Zunjic
Faculty of Mechanical Engineering
University of Belgrade
Belgrade, Serbia

Section I

Theoretical Issues

1 Multimodal Information Transfer by Means of Adaptive Controlling Torques during Primary and Secondary Task

Johann Winterholler, Christian Schulz, and Thomas Maier

CONTENTS

1.1 INTRODUCTION

The technical progress of human–machine interfaces leads to an increase of functions and, associated with this, to a growing number of control elements and audiovisual displays (Winterholler 2009; Hampel and Maier 2011). Furthermore, users expect and request new control concepts, for example, touch operation while driving a car. As a result a reduced usability and an overload of the human perception channels can be noticed. In particular, the visual perception channel is overloaded, which might lead to operating errors.

An approach to optimize this is to use adaptive control elements, which are characterized by their ability to vary and adapt their gestalt (structure, shape) depending on the context of the human–machine interaction (Petrov 2012; Janny et al. 2013). In combination with a haptic feedback, an adaptive control element can be adapted to different tasks. For example, it can also be used to transmit information haptically so that the visual perception channel of the user is relieved in situations of complex information input. The aim of the current research is to investigate which kind of information can be transferred and how the haptic feedback should be designed.

1.2 BASICS, STATE OF TECHNOLOGY AND RESEARCH

In the following some basics are described. To understand the mentioned problem it is necessary to have a look at human–machine interaction. After a theoretical overview, this topic is explored using the example of the automobile. Afterward multimodality of the topic will be discussed. Then important components of human–machine interaction such as parameters of control elements or structure of displays will be treated. Finally, this systematic approach allows us to narrow down and describe the parameters necessary for the investigations.

1.2.1 Human–Machine Interaction

In the literature there are various terminologies for human–machine interaction. Typical synonyms are human–machine interface, human–machine communication, or human–machine interaction design. According to standard (DIN EN 60073) human–machine interaction is a part of a (human–machine) system that is provided as a direct communication medium between user and system and which allows the controlling and monitoring of the system. Standard (DIN EN 894-1) describes the human–machine system as a closed control system in which the user gets information from a product by perceiving and recognizing the information. The perception takes place (visually, acoustically, somatosensorically, and sometimes in olfactory or gustatory manner) via the human senses and the corresponding receptors. After the user has processed the information, he interacts with the product. This interaction, which is called behavior or reaction, can be a verbal statement or a direct visual or kinesthetic operation and use. Figure 1.1 shows the described process according to Seeger (2005) and Maier and Schmid (2014). As shown a human–machine interaction consists of a display

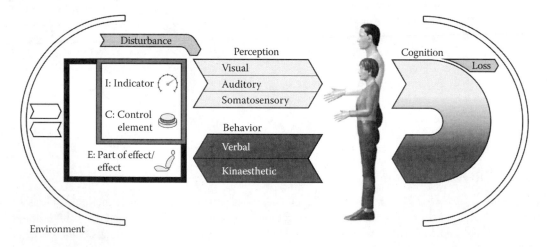

FIGURE 1.1 Process of human–machine interaction. (Adapted from Maier, T. and M. Schmid. 2014. Interface-Design. Manuskript zur Vorlesung. Manuscript, Stuttgart, Universität, Institut für Konstruktionstechnik und Technisches Design. Germany based on Seeger, H., *Design technischer Produkte, Produktprogramme und -systeme. [Industrial Design Engineering]*, Springer-Verlag, Berlin, 2005.)

and a control element. The task of the display is to transfer information (visually, acoustically, or haptically) to the user. In return the user can influence the presented information by using the control element. That means human–machine interaction is a continuous control loop.

In practice, there are different human–machine systems. The range is wide and extends, for example, from control and communication systems for any class of vehicles, to different kinds of robotic, medical, and computer-based systems (Johannsen 1993). Although these systems cover a wide spectrum they all have one thing in common. It is the interaction between human and machine in order to fulfill a task at its best. That means the task has to be fulfilled correctly, safely, and within an appropriate time.

1.2.2 HUMAN–MACHINE INTERACTION ON THE EXAMPLE OF AUTOMOBILE

A typical example of human–machine interaction is the operation of an automobile. In this case there are tasks which have to be monitored and operated by the user continuously. According to the literature such tasks are called primary tasks (Bubb 2001). That means the primary task for the user while driving a car is to navigate, maneuver, operate, and stabilize the car. To solve the primary task the user needs a steering wheel, foot pedals, and a gear selector lever. All other control elements are not necessary for the primary task. But as our own investigations show (Winterholler 2009) there are now up to a hundred or even more control elements in a car cockpit. Many control elements belong to secondary tasks (Bubb 2001) such as the operation of the light switch, direction indicator, horn, or assistance systems like active cruise control, lane change control, or speed control. Much more control elements belong to tertiary tasks (Bubb 2001) such as the adjustment and selection of infotainment system, navigation system, seats, or air conditioning. Thereby, most of the control elements are not related to the primary driving task. Rather they only serve for the satisfaction of comfort, entertainment, or information needs. Figure 1.2 shows the development of control elements and functions in cars since 1886 (Mercedes-Benz) and 1974 (Volkswagen).

It is noticeable that the number of control elements and functions was equal and clear in former times (compare cars of Mercedes-Benz from 1886 to 1927). With the development of the radio and air conditioning the number of control elements and especially the number of functions increased. Suddenly, there were control elements that had two or even more functions. In consequence, the operation and interaction became extensive and complicated. Drivers had to handle a large number of control elements and to learn a lot of functions.

To reduce the gradually increasing number of control elements some car manufacturers developed a central control element at the beginning of this century. It is situated in the center console between the front seats and can be pressed, turned, and sometimes swiveled, or pushed to the side. It is used in combination with a display that is positioned near the driver's field of view on the dashboard (see Figure 1.3).

Although the central control element can be operated "blindly" and the display is positioned near the driver's field of view there is a (visual) distraction from the primary task during the operation. The reason is, on the one hand, the large number of functions that can be operated with the central control element. According to McCann (2002) there are over 700 vehicle functions. On the other, whenever the driver sets a value or selects an option by using the central control element he has to take his eyes off the road, because value adjustments or selections must be monitored and controlled visually. Even if the distraction only lasts for a very short period of time, this time could be a life-saver in a critical situation. For example, a speed of 30 km/h (18.6 miles/h) and a distraction of 1 s mean that the car drives 8 m (26.3 ft) uncontrolled.

In the near future this effect will probably increase. Users expect and demand new assistance and infotainment systems, for example. According to an online survey, which was conducted at the Institute of Ergonomics of the Technical University of Darmstadt, many drivers are interested in controlling, viewing, and using their smartphone contents via the car infotainment system (Franz et al. 2014). That means new services and applications may become an important sales pitch for

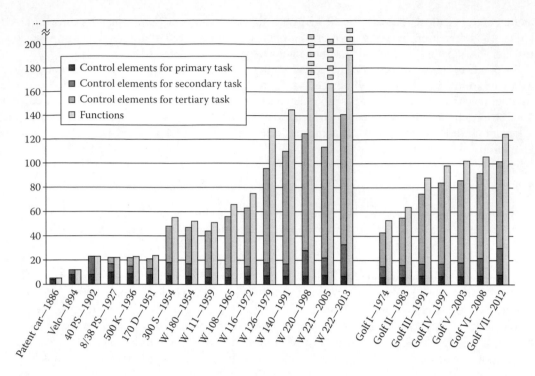

FIGURE 1.2 Development of control elements and functions in cars of Mercedes-Benz and Volkswagen. (Adapted from Winterholler, J. 2009. Entstehung und Entwicklung von Stellteilen im Pkw. Stuttgart, Universität, Institut für Konstruktionstechnik und Technisches Design. Germany. Unpublished student research project.)

passenger cars in the near future (Toll Collect 2013; Franz et al. 2014). Consequently this also means a further increase of functions and control elements, which have to be handled by drivers, also in critical situations. So car manufacturers have to develop new and better control concepts in order to reduce driver workload and increase safety.

It seems that car manufacturers have recognized the trend and the disadvantage mentioned. In recent years new and modern control concepts have been developed. For example, touchpad technology was integrated into the central control element, which allows a "blind" entering of letters, numbers, and symbols by using natural handwriting. That means the driver does not have to take his

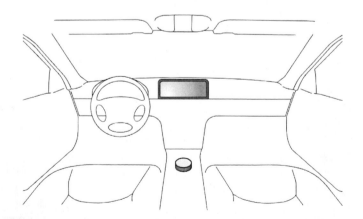

FIGURE 1.3 Central control element with display in a car.

eyes off the road when entering a navigation destination or a telephone number. A brief glance after the input is sufficient so that the distraction of the primary task can be reduced.

Another very interesting improvement is a central control element that blocks at the beginning and end of a list or menu bar. This means that certain pieces of information are transferred through the haptic perception channel. But not every car manufacturer offers this feature.

These described new developments by car manufacturers are moving in the right direction. But there is much more potential to reduce operation time, visual distraction, and cognitive stress of a driver by increased use of the haptic perception channel. The approach of the current research is to investigate how haptic feedback of a rotary control element should be designed in order to use it for information transfer. There is much more information than just the beginning or the end of a list that can be transmitted through the haptic perception channel.

1.2.3 MULTIMODALITY

As Figure 1.1 shows, one or more sensory channels (modalities) are involved in the perception process. A transmission and perception of stimuli through different modalities is called multimodality. According to Schomaker et al. (1995, p. 6) "multimodality is restricted to those interactions which comprise more than one modality on either the input or the output side of the loop and the use of more than one device on either side."

There are several investigations, which document that a multimodal information transfer can reduce operation errors and increase operation time and safety (Akamatsu 1994; Göbel et al. 1995; Dennerlein et al. 2000; Debus et al. 2002; Grane and Bengtsson 2005; Timpe 2006; Pfeffer and Maier 2012; Zühlke 2012). Pfeffer and Maier, for example, verified that a multimodal information transfer can increase the efficiency and reduce the workload by analyzing the heart rate variability (spectral analysis) and a NASA-TLX questionnaire (2012). But the results depend on the tasks and an approach like a multimodal information transfer has to be examined case by case. Because the human perception is restricted the transferable amount of information is limited (Luczak 1998). For example, there are also investigations, which show that users can be impaired and overburdened by multimodal information transfer (Cockburn and Brewster 2005).

1.2.4 PARAMETERS OF ROTARY CONTROL ELEMENTS

Figure 1.4 shows fundamental parameters of a rotary control element. They all have an impact on the availability, speed, and accuracy of operation (Luczak 1998). For some of them there are a lot of concrete recommendations in the literature (Neudörfer 1981; Bullinger 1994; DIN EN 894 1-2; DIN EN 894 3-4; Schmidtke and Jastrzebska-Fraczek 2013). The shape, for example, influences the kind of grip (contact actuation, power-grip, or clench actuation). If a rotary control element is operated, for example, by three fingers (power-grip) the diameter should be larger than 30 mm (1.18 in) and the height more than 25 mm (0.98 in).

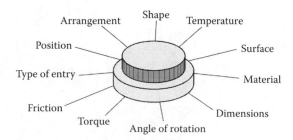

FIGURE 1.4   Parameters of a rotary control element.

In an electro-mechanical system each of these parameters can be structured or modified in such a way that it can be used for the haptic coding and transmission of information. For example, there are investigations regarding a rotary control element, which can vary and adapt its gestalt (structure and shape) depending on the context of the human–machine interaction. According to Petrov such a control element would be called adaptive control element (Petrov 2012). In his research he varied individual parameters of a gestalt and investigated whether people recognize the variation or not. Sendler (2008) even constructed a control element which can vary its gestalt depending on the context. Further examples can be found in Doerrer (2003) or Michelitsch et al. (2004).

Janny et al. (2014) describe variable properties (e.g., elasticity, coefficient of friction, profile depth, and width of profile) of the parameter surface and a test body with which different material properties can be tested. The other parameters can be designed in a similar way so that a variation and adaptation of their typical material properties is possible depending on the context.

1.2.5 STRUCTURE AND DIFFERENT KINDS OF DISPLAYS

Displays are required to present information during human–machine interaction. According to standard (DIN EN 894-2) there are respectively visual, acoustic, and tactile haptic displays. Furthermore, the standard distinguishes between digital display, analog display, and alphanumeric display.

Bullinger (1994) describes four general kinds of visual display forms (difference display, setpoint display, range display, and actual value display), which can be designed as an analog and digital display. With the exception of the difference display all mentioned displays consist of a scale and a pointer (see Figure 1.5), which depend on the form of the display (round or linear). In addition, every display has a symbol in the form of a numeral, letter, or abstract sign (DIN 43790).

Modern digital displays also include graphic content and often a computerized menu. But basically, they just use "beautifully presented" scales with symbols. This means that the display can usually be reduced to a scale with symbols.

While visual and acoustic displays are already well studied and well-known in practice, there are hardly any concrete findings regarding haptic displays (Winterholler et al. 2014a). However, such displays could support other types of displays and relieve overloaded perception channels (Bullinger 1994).

According to Hampel (2011), a haptic display is an element that transfers information by active touching and moving and that requires the use of muscular strength. As mentioned, such an element can be a control element that adapts and varies its gestalt (structure and shape) or operating force depending on the context of the human–machine interaction.

1.2.6 NARROWING OF PARAMETERS

The question is which visual content of a display can be transferred through the haptic perception channel using a rotary control element. It is very difficult to transfer certain graphical contents

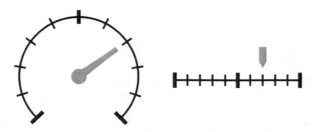

FIGURE 1.5 Round and linear display with scale and pointer. (Adapted from Bullinger, H.-J., *Ergonomie. Produkt- und Arbeitsplatzgestaltung*, Teubner-Verlag, Stuttgart, 1994.)

such as the image of a car in the setup menu or the course of a road in the navigation menu through the haptic perception channel. But it is possible to transfer certain conditions or changes of a scale haptically. Research (Anguelov 2009; Hampel 2011) and our own investigations (Winterholler et al. 2013, 2014a,b) show that rotary control elements can be used, for example, to haptically transfer the center of a list or scale, preferred values, changes of grouped zones, or the beginning and end of a list. Furthermore, it is possible to transfer increasing or decreasing values of a list by a rotary control element (Winterholler et al. 2014a). But which of the parameters mentioned above are suitable for this purpose? The easiest way is to use the parameters that affect the operation of a normal rotary control element directly anyway, which are controlling torque and angle of rotation.

1.2.7　Controlling Torques of Rotary Control Elements

Controlling torques of rotary control elements can be divided into different types of entry (Hampel 2011). For example there is a mono-stable, a continuous, and a discrete value input. As Figure 1.6 shows the main difference between them is the number of detents. Hampel describes further differences and can be consulted for more information (Hampel 2011).

This scientific work investigates the discrete value input. The type of entry can, for example, also be divided into a rising and falling saw tooth shape as well as into a sinusoidal shape. The parameters that can be varied are amplitude and rotary angle in each case (Hampel 2011; Winterholler et al. 2014a,b).

Scientific investigations (Anguelov 2009; Hampel 2011; Winterholler et al. 2013) and practice (Reisinger 2009) show that the falling saw-tooth shape is preferred in terms of best comfort and precision impression. This can be explained by the typical characteristics of this function. The steep rising angle of the rest position ensures high stability. As a result, this causes an impression of high comfort and operating quality.

1.3　METHOD

This work describes a research study, which investigated the use of variable controlling torques for information transfer. On the basis of the scientific findings by Hampel (2011) and Winterholler et al. (2013, 2014a,b), the information transfer through the haptic perception channel during a primary task was investigated. At first, preliminary investigations were done in order to prove the purpose.

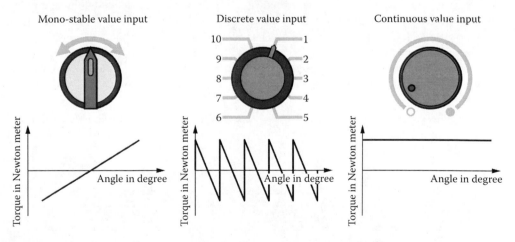

FIGURE 1.6　Different kinds of entry of controlling torques of rotary control elements. (Adapted from Hampel, T. 2011. Untersuchungen und Gestaltungshinweise für adaptive multifunktionale Stellteile mit aktiver haptischer Rückmeldung. PhD dissertation, Stuttgart, Universität, Institut für Konstruktionstechnik und Technisches Design. Germany.)

Subsequently, main investigations were carried out with more participants. In both research studies a torque test bench was used, which was developed by Hampel (2011). The diverse functions were created with MATLAB® Simulink and the control surface with dSpace ControlDesk.

1.3.1 Experimental Setup of the Preliminary Investigation

This research study investigated whether information transfer through the haptic perception channel is basically possible while a primary task is done. Further, this study investigated whether the operation time can be reduced by the haptic perception channel in addition to the visual perception channel for information transfer. That means the user would get distracted less from the primary task and thus could better react in critical or difficult situations. The hypothesis of this research study was that the results such as operation time, reaction time, task fulfillment, and number of errors are better when information is being transmitted multimodally (visual and haptic).

To investigate this hypothesis an experimental setup was necessary. First of all, the primary task was defined. The aim was to design a task that requires the attention of the user all the time. For this a software was programmed in which a rectangle appeared at regular time intervals at different positions (see Figure 1.7). After each position change the rectangle's color changed randomly between red, green, blue, or orange. The position of the rectangles also changed randomly. To simulate different degrees of complexity the time in which the rectangles were visible could be programmed freely. For a task with a low complexity the rectangle was visible for the time of 3 s and for a task with a higher complexity for the time of 2 s. Through this software the permanent observation of an environment and the reaction to changes are simulated.

In addition to the primary task a secondary task was defined based on the scientific findings by Hampel (2011) and Winterholler et al. (2013, 2014a,b). The aim of the secondary task was to set a defined value along a 10-step scale using the rotary control element of the test bench. All in all three different secondary tasks were defined, where the principle was always the same. Each of these tasks had to be done without a haptic feedback (visual), with a haptic feedback, and combined (visual and haptic). Figure 1.8 shows the experimental setup.

The aim of the experiment "center mark" was to set up the value along a 10-step scale from two or three (the starting point varied) to seven. At first, the value had to be set up without a haptic feedback. The subjects only saw a 10-step scale with an indicator and had to set up the value while they did the primary task. After that the task was repeated but now the value had to be adjusted purely haptically. For the haptic feedback, a single amplitude change at a defined point at the torque function was selected. Starting from a rotary angle of 24° and a torque of 0.08 N m the amplitude was increased at the defined point to 0.15 N m. Hampel (2011) and Winterholler et al. (2013, 2014a) showed that such a function can be used in order to find a special point, in this case the center of a scale. The subjects knew that the amplitude change simulates the center of the scale. That means they only had to indicate the amplitude change and then set up the value by turning the rotatory

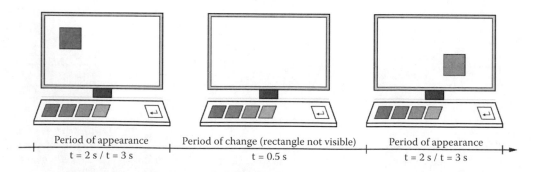

FIGURE 1.7 Primary task during preliminary investigations.

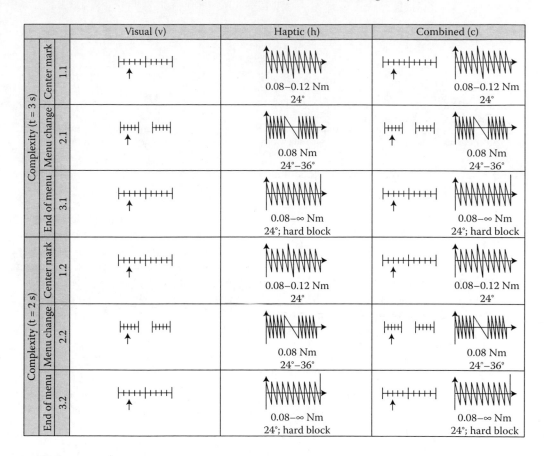

FIGURE 1.8 Theoretical experimental setup of the preliminary investigation.

control element two detents forward. In the third situation the subjects had the opportunity to set up the value both visually and haptically (see Figure 1.8). It was left to the subjects if they set the value with the help of the visual control or with the help of the haptic feedback.

In the second experiment ("menu change") a change between two menus was simulated. Therefore, the scale and the haptic feedback were adapted (see Figure 1.8). The haptic feedback was a change at the rotary angle at a defined point. Based on the results of Winterholler et al. (2013, 2014a), the rotary angle changed from 24° to 36°. The torque of the falling saw-tooth shape was set up to 0.08 Nm.

The aim of the third experiment ("end of menu") was, similar to the first experiment, to adjust a value from two or three (the starting point varied) to seven. The difference was the haptic feedback. Instead of the amplitude change at a defined point a hard block at the end of the torque function was defined. Again a torque of 0.08 Nm and a rotary angle of 24° were set up. The aim was to recognize the end of the scale (value 10) with the help of the block and then to adjust the value 7 by turning the rotary control element three detents back.

In total 18 experiments were carried out in a period of one week with a total of 10 subjects (5 male and 5 female). All experiments took about 30 min and were carried out under laboratory conditions. The average age of the subjects who were primarily students was 26 years (4.0 years standard deviation). The average body size of the subjects was 175.6 cm (6.0 cm standard deviation). Figure 1.9 shows the real experimental setup.

The test procedure always began with the investigator describing the whole investigation. After that the torque functions and the special changes (amplitude change, rotary change, and block) of

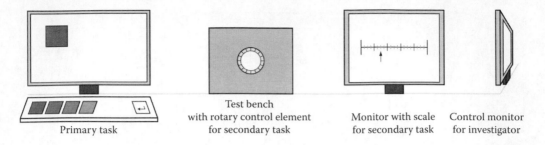

Primary task   Test bench   Monitor with scale   Control monitor
                     with rotary control element   for secondary task   for investigator
                         for secondary task

FIGURE 1.9   Real experimental setup of the preliminary investigation.

the functions were explained theoretically. Then the subjects had the opportunity to extensively test the functions at the test bench. Afterward the primary task was explained with the request to perform the task during the experiments as well as possible and without errors. To compare and analyze the reaction times, a reference measurement of each subject was done. For this purpose the subjects had to do the primary task (without any secondary task) for a time of 2 min. This provided comparative values, which could be compared with the measured values. In a following test exercise, which took about 10 min, the subjects had enough time to test the experimental setup. After this teach-in phase, the 18 experiments were carried out in a random order.

1.3.2 Results of the Preliminary Investigation

Figure 1.10 shows the results (operation time, reaction time, task fulfillment of secondary task, and number of errors during primary task) of the preliminary investigations. The diagrams are scaled equally to those of the main investigations so that they can be compared more easily.

1.3.2.1 Operation Time

It is noticeable that the average operation time (marked as a dot) is always higher in tasks, which were done purely haptically. It is also visible that these tasks have the largest outliers and that the subject partially needed 20 or more seconds to fulfill the tasks.

The remarkable and positive fact is that the average operation time of combined tasks (visual and haptic) is in some experiments (e.g., in the experiments "menu change") better than the average operation time of purely visual tasks. The subjects were partially over 0.5 s faster when they used a combined feedback.

The results of the "end of menu" experiment are interesting. Theoretically, the operation time should be much higher when the subjects use a haptic feedback for adjustment. Finally, the operating displacement is longer than in the purely visual setting. The operation time is higher, as expected, for tasks done with just a haptic feedback. But for tasks done in combined manner, the operation time is equal to the operation time of tasks done visually. A reason could be that a combined feedback provides a more secure feeling. Therefore, the subjects can concentrate on the primary task because they do not have to avert their glance. And this leads to an equal operation time.

1.3.2.2 Reaction Time

Also interesting are the results of the reaction times. In all cases the mean reaction time is higher than at the reference measurement (0.66 s). This leads to the conclusion that the subjects are more burdened due to a simultaneous processing of multiple tasks (primary and secondary tasks). The highest mean reaction times have tasks that were done purely haptically. Better are the results of tasks that were done with a visual or combined feedback. But it is not possible to say that the reaction time of tasks which were done, for example, with a combined feedback, is better. In some cases the reaction time of visually fulfilled tasks is better and vice versa. Noticeable is that the mean reaction time of tasks with high complexity (t = 2 s) in general is lower than the one of tasks with lower complexity (t = 3 s).

1.3.2.3   Task Fulfillment of Secondary Task

Figure 1.10 also shows the task fulfillment of the secondary task. As expected, the tasks, which were done with visual control, were always 100% fulfilled. The situation is similar in tasks, which were done combined (visually and haptically). With one exception, the tasks were also always 100% fulfilled. Tasks that were done purely with haptic feedback (without a visual control) have partially insufficient results. The results in these cases are bad both in the experiment "center mark" and "menu change." The fact that the results are comparable between experiments with low complexity (t = 3 s) and higher complexity (t = 2 s) is interesting. It seems that this kind of complexity does not have an effect on the results, especially, when the tasks are done visually or combined.

1.3.2.4   Number of Errors during Primary Task

Finally, Figure 1.10 shows the number of errors of all subjects during the primary task while doing the secondary task. The most errors were made in the experiment "menu change" with low complexity (about 35% of the errors) and high complexity (about 20% of the errors). The fewest errors were done in the experiment "end of menu" with high complexity. Two things are noticeable. There were no haptically fulfilled tasks without any errors. And again the results of tasks with higher complexity (t = 2 s) are better.

All in all the results show on the one hand that a purely haptical information transfer is not recommendable because the efficiency (operation time, reaction time, task fulfillment, and number of errors) is bad. On the other, the results show that the efficiency of a combined information transfer (visual and haptic) is basically not worser than the efficiency of a visual information transfer. But all in all, it is difficult to say that the above-mentioned hypothesis was definitely confirmed. For this the differences of the results are too low and have to be proved with more subjects.

1.3.3   EXPERIMENTAL SETUP OF THE MAIN INVESTIGATION

Similar to the preliminary investigations a primary task was defined. The aim was again to create a task, which requires the attention of the user all the time. As Figure 1.11 shows this time arrows were used because it is more intuitive for the user to push the corresponding button on the keyboard than to push color-coded buttons while observing the items. The time in which the arrows were visible was set up to 2 s (high complexity) for all experiments. The reason is that the results at the preliminary investigations were better when tasks were fulfilled with a higher complexity (t = 2 s). Furthermore, the selected time corresponds more to reality.

In addition to the primary task, again secondary tasks were defined with the aim to set a defined value along a 10-step scale using the rotary control element of the test bench. All in all three different secondary tasks ("center mark," "menu change," and "end of menu") were defined. Figure 1.12 shows the complete experimental setup of the main investigation including the varied and examined parameters.

The procedure was equal to the preliminary investigation. Each of these tasks had to be done without haptic feedback (visual), with pure haptic feedback (haptic), and combined (visual and haptic). The difference was that more parameter combinations of the above-mentioned controlling torques were investigated.

The investigation involved 20 participants (10 male, 10 female, primarily students) with an average age of 24.7 years (1.9 years standard deviation) and an average body size of 174.4 cm (8.3 cm standard deviation). The participants were confronted with the three main experiments ("center mark," "menu change," and "end of menu") within a time period of about 45 min.

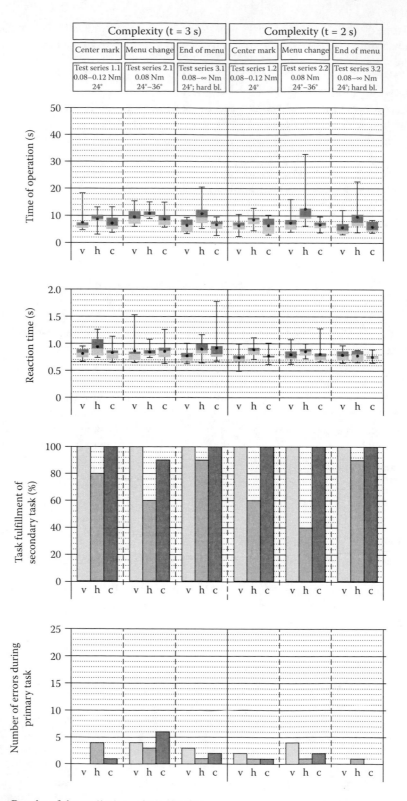

FIGURE 1.10   Results of the preliminary investigation.

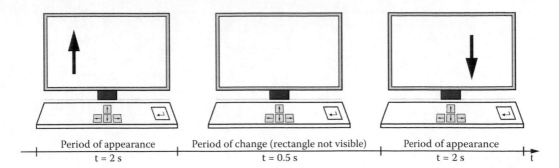

FIGURE 1.11 Primary task during main investigation.

As Figure 1.12 shows the experiments "center mark" and "menu change" included four individual experiments and the experiment "end of menu" two individual experiments. Combined with the three different kinds of secondary tasks the participants had to accomplish a total of 30 individual experiments. Both the main experiments and the individual experiments were carried out in a random order.

Since subjects other than those in the preliminary investigation participated in this investigation, the test procedure again began with a detailed description of the whole investigation. That means the torque functions and the special changes (amplitude change, rotary change, and block) of the functions were explained theoretically. Then the subjects had the opportunity to extensively test the functions at the test bench. Afterward the primary task was explained and a reference measurement of the reaction time was done. To avoid learning effects there was a teach-in phase, which took about 10 min. During this time the subjects had the opportunity to test the whole experimental setup (primary and secondary tasks) and to ask questions.

1.3.4 Results of the Main Investigation

Figure 1.13 shows the results (operation time, reaction time, task fulfillment of secondary task, and number of errors during primary task) of the main investigation.

1.3.4.1 Operation Time

The results of the total operation time while accomplishing primary and secondary tasks are very interesting. In general, the operation time of tasks with just a haptic feedback is highest (worst) at all test series of the experiments "center mark" and "menu change." It is also visible that these tasks have the largest outliers and that some subjects needed 35 or more seconds to fulfill the tasks. This effect has already been observed in the preliminary investigations.

In the experiments "end of menu" tasks with visual feedback have the highest (worst) and tasks with combined feedback the lowest (best) operation time. The large difference at test series 3.1 and 3.2 between tasks with combined and visual feedback is remarkable. The tasks were solved on average over 2 s faster when combined feedback was used. That means a block at the beginning or end of a scale does considerably support the user.

In the experiment "menu change" the mean operation time is lower (better) at all tasks with just a visual feedback is noticeable. This is interesting because in the preliminary investigation the mean operation time of tasks with a combined feedback was always lower. A reason could be the changed parameter settings. Both the amplitude and the rotary angle change as the defined point were increased (compare parameter settings at Figures 1.8 and 1.12).

In the experiment "center mark" the mean operation time is in two of four cases better at tasks with a combined feedback (see test series 1.1 and 1.2 at Figure 1.13). It seems that a too high amplitude change leads to worse results regarding tasks with combined feedback.

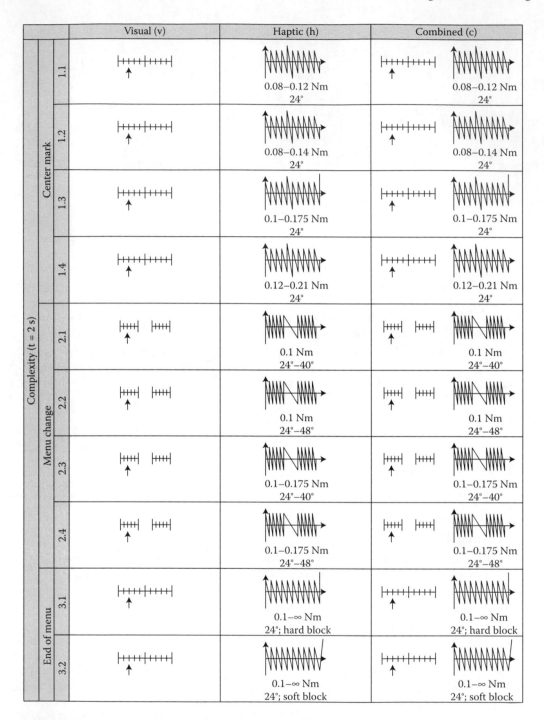

FIGURE 1.12   Experimental setup of the main investigation.

1.3.4.2 Reaction Time

Similar to the preliminary investigation, the mean reaction time of each subject is higher than the reference mean reaction time. With one exception tasks, which were done purely haptically, have the highest mean reaction times. As the literature says, users are more burdened in such kinds of tasks although there is no visual distraction.

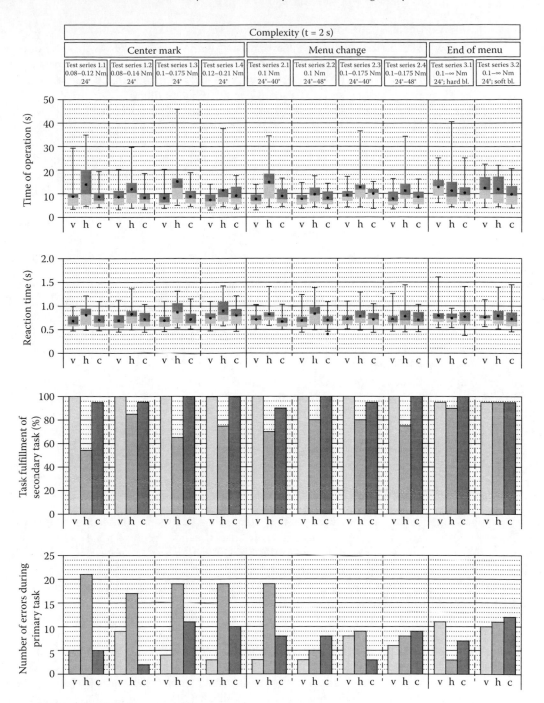

FIGURE 1.13 Results of the main investigation.

In the experiments "center mark" the mean reaction time is least at tasks that were done with just a visual feedback. The situation is different, however, in the experiments "menu change" and "end of menu." With one exception (test series 2.2) in these cases, the mean reaction time is lower at tasks that were done with a combined feedback in comparison to tasks that were done with visual feedback. At the test series 3.1 (experiment "end of menu") the task with purely haptic feedback even have the lowest mean reaction time compared to the task with visually combined feedback.

1.3.4.3   Task Fulfillment of Secondary Task

As the results show tasks carried out during the experiments with visual control were 100% fulfilled nearly always (8 of 10 tasks were 100% fulfilled). One subject was careless and made mistakes in both "end of menu" experiments. A bit worse are the results of experiments, which were done combined (visual and haptic). In this case just 5 of 10 tasks were 100% fulfilled.

It is noticeable that three of the six mistakes were done by one subject, which can be caused by different factors, for example, carelessness, less haptical sense, or overload and excessive demand of the perception channel. Another reason can be improper parameter settings of the controlling torques. But exactly that is what was supposed to be investigated: parameter settings of controlling torques, which achieve the best results at tasks with combined feedback. Experiments, which have been done with purely haptic feedback, have generally worse results. The task fulfillment is better than 80% in just three cases.

1.3.4.4   Number of Errors during Primary Task

The worse results of experiments that have been done with purely haptic feedback, during the experiment "center mark," are remarkable. But also the other results of tasks with haptic feedback are not very positive. With one exception, there are no cases, which have less than five errors.

It is noticeable that experiments conducted with combined feedback (visual and haptic) partially have a lower number of errors (e.g., in test series 1.2, 2.3, and 3.1) than experiments conducted with purely visual feedback. But in most cases the number of errors is lower at tasks that were done with just a visual feedback. That means not each investigated parameter combination is suitable for a combined information transfer.

But all in all, multimodal information transfer can reduce the number of errors during human–machine interaction depending on the parameters. It has to be emphasized that these observations are specific to the participants. Over 50% of all errors were done by five participants. The analysis of the individual data of the participants shows that neither gender nor hobbies such as playing an instrument or doing sports have an influence on the results. So the individual characteristics of each person such as perception, alertness, or emotional mood seem to be crucial.

1.4   CONCLUSION AND OUTLOOK

The investigations show that the haptic perception channel can be used in combination with visual displays to transfer information. As the results show special haptic feedbacks such as amplitude changes, rotary changes, or blocks at the end of the torque function are recognized. Especially, the results of the "end of menu" experiment in the main investigation are very positive. At the test series 3.1 the task with combined feedback achieved best results at all measured variables (operation time, reaction time, task fulfillment, and number of errors).

Unfortunately, there are no test series in the "center mark" and "menu change" experiments that achieved best results in all measured variables. But there are some parameter combinations, which have the potential to be better than just a visual feedback, for example, test series 1.2, 2.3, and 2.4. But this statement must be backed up by further investigation in order to show that a combined transmission of information to the user, for example, via a central control element, leads to a reduction of operation time, reaction time, number of errors, and a better or equal task fulfillment.

In a next step, the investigation must be carried out with more subjects. The results should be regarded as a preliminary investigation, which should be pursued and intensified. In this context, the investigated torque functions should be adapted. Some of the subjects indicated that they did not perceive the haptic feedbacks clearly while completing the primary task. A reason could be the distraction and complexity of the primary task, which must be taken into account in the determination of the torque function parameters.

All in all, this approach allows innovative interfaces so that the usability of products can be improved and the operational safety can be increased.

REFERENCES

Akamatsu, M. and S. Sato. 1994. A multi-modal mouse with tactile and force feedback. In *International Journal of Human-Computer Studies*, eds. E. Motta and S. Wiedenbeck, 443–453. Duluth, MN.

Anguelov, N. 2009. Haptische und akustische Kenngrößen zur Objektivierung und Optimierung der Wertanmutung. PhD Dissertation, Technische Universität Dresden, Dresden, Germany.

Bubb, H. 2001. Haptik im Kraftfahrzeug. In *Kraftfahrzeugführung*, eds. T. Jürgensohn, K.-P. Timpe, 156. Berlin: Springer-Verlag.

Bullinger, H.-J. 1994. *Ergonomie. Produkt- und Arbeitsplatzgestaltung*. Stuttgart, Germany: Teubner-Verlag.

Cockburn, A. and S. Brewster. 2005. Multimodal feedback for the acquisition of small targets. *Ergonomics*, 48(9), 1129–1150.

Debus, T., Becker, T., and P. Dupont. 2002. Multichannel vibrotactile display for sensory substitution during teleoperation. In *Telemanipulator and Telepresence Technologies VIII. Proceedings*, ed. M. Stein, 42–49. Bellingham, Wash.: Society of Photo-Optical Instrumentation Engineers (SPIE).

Dennerlein, T., Martin, D., and C. Hasser. 2000. Force-feedback improves performance for steering and combined steering-targeting tasks. In *Conference on Human Factors in Computing Systems, Proceedings of the SIGCHI Conference on Human Factors in Computing Systems*, ed. T. Turner, 423–429. New York: ACM Press.

Doerrer, C. 2003. Entwurf eines elektromechanischen Systems für flexibel konfigurierbare Eingabefelder mit haptischer Rückmeldung. PhD dissertation, Technische Universität Darmstadt, Darmstadt, Germany.

Franz, B., Zöller, I., Kauer, M., Schulz, L., Abendroth, B., and R. Bruder. 2014. Being always on in vehicles—The use of apps while driving bears risks. In *Proceedings of the 5th International Conference on Applied Human Factors and Ergonomics*, eds. T. Ahram, W. Karwowski, T. Marek, 2792–2800. Krakow, Poland.

Göbel, M., Luczak, H., Springer, J., Hedicke, V., and M. Rötting. 1995. Tactile feedback applied to computer mice. *International Journal of Human–Computer-Interaction*, 7(1), 1–24.

Grane, C. and P. Bengtsson. 2005. Menu selection with a rotary device founded on haptic and/or graphic information. *Proceedings of the First Joint Eurohaptics Conference and Symposium on Haptic Interfaces for Virtual Environment and Teleoperator Systems*, 475–476. IEEE Computer Society, Washington, DC, USA.

Hampel, T. 2011. Untersuchungen und Gestaltungshinweise für adaptive multifunktionale Stellteile mit aktiver haptischer Rückmeldung. PhD dissertation, Stuttgart, Universität, Institut für Konstruktionstechnik und Technisches Design. Germany.

Hampel, T. and T. Maier. 2011. Active rotary knobs with adaptive torque feedback—Conflict of goals between positioning accuracy and difference threshold. *11th Stuttgart International Symposium Automotive and Engine Technology*, Stuttgart, pp. 437–452.

Janny, B., Haug, M., and T. Maier. 2014. Optimierung gestalterischer Faktoren für die altersgerechte Mensch-Produkt-Schnittstelle durch Greifkraftmessung. In *Entwerfen Entwickeln Erleben 2014—Beiträge zum Technischen Design*, eds. K. Krzywinski, M. Linke, C. Wölfel, G. Kranke, 279–290. Dresden: TUD Press.

Janny, B., Winterholler, J., Petrov, A., and T. Maier. 2013. Adaptive control elements for navigation systems. In: *Posters, Part I, HCII 2013*, CCIS 373, ed. C. Stephanidis, 473–477. Berlin, Heidelberg: Springer-Verlag.

Johannsen, G. 1993. *Mensch-Maschine-Systeme*. Berlin, Heidelberg: Springer-Verlag.

Luczak, H. 1998. *Arbeitswissenschaft*. Berlin: Springer-Verlag.

Maier, T. and M. Schmid. 2014. Interface-Design. Manuskript zur Vorlesung. Manuscript, Stuttgart, Universität, Institut für Konstruktionstechnik und Technisches Design, Germany.

McCann, K. 2002. Bringing haptic technology to automobiles. *AutoTechnology*, 2(5), 38.

Michelitsch, G., Williams, M., Osen, M., Jimenez, B., and S. Rapp. 2004. Haptic chameleon: A new concept of shape-changing user interface controls with force feedback. *Conference on Human Factors in Computing Systems*, Vienna, pp. 1305–1308.

Neudörfer, A. 1981. *Anzeiger und Bedienteile. Gesetzmässigkeiten und systematische Lösungssammlungen*. Düsseldorf: VDI-Verlag.

Petrov, A. 2012. Usability-Optimierung durch adaptive Bediensysteme. PhD dissertation, Stuttgart, Universität, Institut für Konstruktionstechnik und Technisches Design, Dissertation, Germany.

Pfeffer, S. and T. Maier. 2012. Beanspruchungsmessung beim Einsatz taktiler Displays—eine Abschätzung der informatorischen Belastung. 58. Kongress der Gesellschaft für Arbeitswissenschaft, Kassel, pp. 443–446.

Reisinger, J. 2009. Parametrisierung der Haptik von handbetätigten Stellteilen. PhD dissertation, München, Technische Universität München, Lehrstuhl für Ergonomie, Germany.

Schmidtke, H. and I. Jastrzebska-Fraczek. 2013. Ergonomie. *Daten zur Systemgestaltung und Begriffsbestimmungen.* München: Hanser-Verlag.

Schomaker, L., Nijtmans, J., Camurri, A., F. Lavagetto et al. 1995. A taxonomy of multimodal interaction in the human information processing system. A report of the Esprit Project 8579 MIAMI, Nijmegen University, NICI.

Seeger, H. 2005. *Design technischer Produkte, Produktprogramme und -systeme [Industrial Design Engineering].* Berlin: Springer-Verlag.

Sendler, J. 2008. Entwicklung und Gestaltung variabler Bedienelemente für ein Bedien- und Anzeigesystem im Fahrzeug. PhD dissertation, Technische Universität Darmstadt, Darmstadt, Germany.

Standard, DIN 43790. 1991. Basic principles for the design of line scales and pointers.

Standard, DIN EN 60073. 2003. Basic and safety principles for man–machine interface, marking and identification—Coding principles far indicators and actuators (IEG 60073:2002); German version EN 60073:2002.

Standard, DIN EN 894-1. 2009. Safety of machinery—Ergonomics requirements for the design of displays and control actuators—Part 1: General principles for human interactions with displays and control actuators; German version EN 894-1:1997+A1:2008.

Standard, DIN EN 894-2. 2009. Safety of machinery—Ergonomics requirements for the design of displays and control actuators—Part 2: Displays; German version EN 894-2:1997+A1:2008.

Standard, DIN EN 894-3. 2010. Safety of machinery—Ergonomics requirements for the design of displays and control actuators—Part 3: Control actuators; German version EN 894-3:2000+A1:2008.

Standard, DIN EN 894-4. 2010. Safety of machinery—Ergonomics requirements for the design of displays and control actuators—Part 4: Location and arrangement of displays and control actuators; German version EN 894-4:2010.

Timpe, K.-P. 2006. Mensch-Maschine-Interaktion bei der Fahrzeugführung. In *Handbuch der Mess- und Automatisierungstechnik im Automobil*, eds. H.-J. Gevatter and U. Grünhaupt, 3–8. Berlin, Heidelberg: Springer-Verlag.

Toll Collect GmbH. 2013. *LKW-Maut in Deutschland. Nutzerinformationen.* Berlin: Im Auftrag des Bundesamt für Güterverkehr.

Winterholler, J. 2009. Entstehung und Entwicklung von Stellteilen im Pkw. Stuttgart, Universität, Institut für Konstruktionstechnik und Technisches Design. Germany. Unpublished student research project.

Winterholler, J., Böhle, J., and T. Maier. 2013. Informationskodierung mittels variabler Stellmomente im nutzerzentrieren Design. In *Stuttgarter Symposium für Produktentwicklung*, eds. D. Spath, B. Bertsche, H. Binz. Stuttgart.

Winterholler, J., Böhle, J., Chmara, K., and T. Maier. 2014a. Information coding by means of adaptive controlling torques. In *Human Interface and the Management of Information. Information and Knowledge Design and Evaluation, Part I, LNCS 8521*, ed. S. Yamamoto, 271–280. Heidelberg: Springer-Verlag.

Winterholler, J., Chmara, K., Schulz, C., and T. Maier. 2014b. Use of adaptive controlling torques with the aim to transfer information by the haptic perception channel. In *Proceedings of the 5th International Conference on Applied Human Factors and Ergonomics*, eds. T. Ahram, W. Karwowski, T. Marek, 1640–1648. Krakow, Poland.

Zühlke, D. 2012. *Nutzergerechte Entwicklung von Mensch-Maschine-Systemen [Useware-Engineering für technische Systeme].* Heidelberg: Springer-Verlag.

2 Time Structure Analysis in Ergonomic Design with Cognitive and Manual Components of Work

Gregory Bedny, Waldemar Karwowski, and Inna Bedny

CONTENTS

2.1 INTRODUCTION

Design is one of the most important issues in ergonomics (Karwowski, 2005). Ergonomists use the term "design" in many different ways. Consequently, there are many definitions of ergonomic design. The term "design" emerged from engineering. The purpose of design is creation of new products, software, manufacturing goods, technological processes, etc. In the last decade, the term "design" has been used not only in reference to material systems, but also to the design of algorithms of human activity during task performance and related technological processes. Design cannot be reduced to experimental procedures. The early stages of design include analytical procedures. In this chapter design is referred to as "creation and description of ideal models of inexistent artificial objects, in accordance with previously set properties and characteristics, with the ultimate purpose of materializing these objects." Hence, in this work, we perform a comparative analysis of the theoretical (analytical) method of design that has been developed within the framework of the systemic-structural activity theory (SSAT) with experimental methods of design currently used in ergonomics. In ergonomics analytical procedures of design are most efficiently utilized when studying the anthropometric characteristics of the work place. Such procedures are also are used in biomechanical analysis of human movements. However, there is no efficient analytical method for studying cognitive components of work outside of SSAT. Moreover, cognitive components of human activity are usually combined with manual components which make analytical procedures of ergonomic design even more complex. The specificity of design lies in the fact that developing the model of an object being design, the object itself does not yet exist as a materialized system. The design process can be viewed in terms of stages of sequential refinement of design models. At the initial stages a designer has only the ideal image or mental model of the object being designed in her/his mind. During the subsequent stages, the conceptual or mental model of the object is externally described using symbols and signs. This makes the model available to the other specialists involved in the design process.

There are three basic types of models: conceptual (mental), symbolic (verbal, graphical, mathematical), and physical. Any model is only an approximation of a real object or event. The real object or event has many features and properties. So, it is almost impossible to include all of them in any one model. This is why the design process requires the creation of a number of interdependent models of the same object. The first stage of ergonomic design, similarly to engineering design, is largely an analytical process for which empirical methodology (e.g., observation and questionnaires) is not useful. Usually physical models and the related empirical methodology should be used during the final stage of the design, to verify existing symbolic models. Typically, the conceptual models are immediately embodied in physical models of equipment or software that are then are used for the experiment. Usually, an ergonomist uses observation and experiments and does not utilize design models of human activity in finding solutions. This is due to the fact that in ergonomic design there are no principles for developing conceptual models of human activity that include cognitive and behavioral components. However, experiments should be supplemental tools in design. There is another issue in ergonomic design. Specialists in cognitive psychology reduce design problems to creation of mentalistically orientated modeling methods. There are two kinds of mental models during task performance. One of them describes the information-processing system of brain, rather than real activity during interaction with equipment. Such models sometimes are erroneously considered as design models. The real design models describe the structure of activity during task performance. It is important to understand that any changes in the configuration of equipment or software in a probabilistic manner also changes the structure of activity during task performance. The analysis of relationship between structure of activity and configuration of equipment or characteristics of the interface is the basic principle of design in SSAT.

Design models should be described in a standardized manner. Examples of such standardized models in engineering are drawings. Human activity structure during task performance also should be described in a standardized manner. In order to develop the new system it is important to find out a list of tasks performed by workers or users, discover their logical organization, and describe the structure of activity during the performance of various tasks. At the next step structure of activity during task performance is compared with equipment configuration or computer interface and then the design solution can be developed. Description of activity structure or creation of activity models is important for discovering the most efficient methods of task performance when the same equipment is used (Bedny and Meister, 1997; Bedny and Karwowski, 2007). Development of such models is important for reliability and safety analysis (Bedny et al., 2010; Bedny and Harris, 2013). Such models are also useful in analyzing dynamics of activity structure evolvement during the training process. Analytical models are important for describing the historical evolution of equipment design. In order to meet new requirements it is often necessary to track all modifications of not only material components of a system but also modifications in a structure of activity. Utilizing Vygotsky's (1978) terminology, we describe the genesis of human activity during interaction with new equipment or software.

Contemporary work tends to be highly automated. People are often responsible for monitoring the state of the automated systems. The significance of cognitive aspects of human work has increased tremendously. In abnormal situations an operator has to intervene in the functioning of an automated system and perform complex problem solving tasks that include manual components. Therefore, manual control is also important for automatic systems. Providing efficient transition from automatic to manual control is a critical factor in contemporary task analysis and design. Total elimination of manual control in automatic systems has a negative effect on the operator's performance because it leads to monotony which is particularly important to avoid in the case of vigilant tasks. In many production operations manual components are the main part of the work process. So, motor components of activity will always be important in human work. In modern production conditions the nature of motor actions has changed and heavy physical work is now significantly reduced. Motor actions cannot be considered as automated reactions triggered by external stimuli or selected from memory as ready responses. They are goal directed elements of human activity.

The initiation of a conscious goal (goal acceptance or goal formulation) constitutes the starting point of an action. An action is finished when the actual result of the action is evaluated in relation to the goal of this action. A motor action consists of motions that are elements of this action. A motor action is a self-regulative system. Cognitive actions can be described similarly. Thus, activity in general is a complex structure that unfolds in time and functions according to the principles of activity self-regulation. This makes activity a very flexible and adaptive system. The time structure of activity is a critically important concept in ergonomic design. Analysis of the time structure of activity that includes cognitive and motor components is an important step in ergonomic design. It is reasonable to ask what the relationship is between ergonomic design and time study. At first glance, these are two independent areas. However, in reality they are closely interrelated. Evaluation of ergonomic design always involves analysis of human activity or behavior which is a process. More precisely activity is a complex structure that consists of various elements that unfold over time. Our research has shown that the structure of the activity is the most complex when its cognitive and motor components are combined during task performance. Activity is a multidimensional system that should be described by various models. Description of time structure of activity is a critical stage of task analysis and design because it depicts the way activity as a process unfolds in time. In this work we also will show that analytical and experimental procedures should be integrated. The combination of analytical method with experimental procedures allows the development of a unified theory of ergonomic design of work which includes the complex combination of cognitive and manual components.

2.2   ALGORITHMIC TASK ANALYSIS VERSUS CONSTRAINT-BASED APPROACH

Currently, there are no effective methods of task analysis that can be efficiently used in the study of variable human activity. So, the assumption is that there are unpredictable external disturbances acting on the system and therefore there is no single right way of getting the task done. Human behavior is dynamic, requiring workers to adapt to moment-by-moment changes in the context.

Hence, the conclusion was that the principle of discovering "one best way of task performance" is basically incorrect. Hence there is the necessity to use a constraint-based approach to task analysis. The basis of this principle is an assertion that the subjects can perform the task utilizing any chosen method within the specified constraints. Vicente (1999) distinguished two approaches to task analysis. One is known as instruction-based and the other one is the constraint-based approach to task analysis. The instruction-based approach strictly determines all required procedures of task performance and is considered by this scientist as being not very efficient in contemporary task analysis. Vicente suggested resolving conflict between these approaches by introducing the concept of *constraints* that specifies *what should not be done by a performer*. According to the second approach subjects independently decide how to perform the task within the existing constraints. However, any design implies some constraints, even when there is one best way of performance, such as time constraints, safety constraints, etc. Moreover, a performer can utilize multiple methods of task performance inside existing constraints. Some of them can be efficient and others are inefficient. Introduction of constraints can be done utilizing different parameters. Constraints exist in the traditional engineering design of equipment and software. Even in the presence of the same constraints there is still a possibility of utilizing different versions of design solutions. Some of them are more efficient than others. Similarly, when performing a task with the same constraints a subject can use efficient and inefficient methods of task performance. The formalized stage of ergonomic design involves the creation of models of human activity during task performance. Such models should describe the structure of activity during task performance in a standardized manner and at the next stage this structure should be compared with a configuration of designed software or equipment. Based on such comparison the optimal equipment or interface design solution can be found. Therefore, the statement "a worker decides how to achieve the goal" (Vicente, 1999, p. 70) within existing constraint assumes that there is no design in ergonomics.

According to SSAT any design solution has to take into account constraint-based principles. For this purpose it is necessary to identify the most effective strategies to accomplish a particular task in specified constraints conditions. In SSAT qualitative systemic analysis of activity strategies during task performance are performed based on analysis of the mechanisms of self-regulation (Bedny et al., 2015). Outside of SSAT self-regulation is viewed as a homeostatic process. Such understanding of self-regulation is too mechanistic and cannot describe the flexible strategies of human performance. In SSAT self-regulation is a goal directed process. Thanks to self-regulation of activity subjects create a conscious goal and develop strategies to achieve this goal. Therefore, the description of these strategies during task performance is an important stage of task analysis especially for computer-based tasks. Flexible human activity can be described in a formalized manner by utilizing the algorithmic description of task performance.

The algorithmic analysis of activity is a particularly powerful method of the morphological approach that consists of subdividing activity into qualitatively distinct psychological units and determining the logic of their sequential organization. Each member of the human activity algorithm consists of tightly interdependent homogeneous actions (only motor, only perceptual, or only decision-making actions, etc.) that are integrated by a higher order goal into a holistic system. A member of such an algorithm is perceived by a subject as a component of his/her activity, which has a logical completeness. Usually, the number of actions in one member of an algorithm is restricted by the capacity of short-term memory. While motor actions can be performed simultaneously, mental or cognitive actions are usually performed sequentially. Cognitive actions can be combined with motor actions according to that described in the SSAT rules. Members of an algorithm called "operators" and "logical conditions" are units of activity analysis. Operators represent actions that transform objects, energy, and information. For example, we can describe operators that are implicated in receiving information, analysis of a situation and its comprehension, shifting gears, levers, etc. Logical conditions are members of an algorithm that include decision-making process and determine the logic of selecting the next operator. Each member of an algorithm is designated by a special symbol. Operators can be designated by the symbol O and logical conditions by the symbol l. If decision making is performed based on information extracted from memory the symbol l^μ is used. Symbol μ designates memory function that complicates decision making.

Symbol "l" that depicts a logical condition always has an arrow to the right of it with a number on top that reflects the number of this logical condition. For example, the logical condition l_1 is associated with a number on top of the arrow $\uparrow^1$. An arrow with the same number but in reversed direction is placed in front of a corresponding member of the algorithm to which the arrow refers, $\downarrow^1$. Thus the syntax of the system is based on a semantic denotation of a system of arrows and superscripted numbers. An upward pointing arrow of the logical state of the simple logical condition "l" when, "l" = 1, requires skipping all following members of the algorithm until the next appearance of the superscripted number with a downward arrow (e.g., $\downarrow^1$) meaning that the operation with the downward arrow with the same superscripted number in front of it is the next to be executed. If a logical condition equals 0, then the next member of the algorithm is activated. An algorithm that includes a logical condition with two outputs 0 or 1 is a deterministic algorithm.

A complex logical condition has multiple outputs. For example, $l_1 \uparrow^{1(1-6)}$ indicates that this is the first complicated logical condition that has six possible outputs: $\uparrow^{1(1)}$, $\uparrow^{1(2)}$, $\uparrow^{1(3)}$, ..., $\uparrow^{1(6)}$. The combined probability of these outputs should be equal to 1. Each output has its own probability. Arrows after logical conditions demonstrate transition from one member of an algorithm to another ($\uparrow^1 \downarrow^1$) meaning that the logical condition according to the output addressed from the upward to the downward arrow is associated with the particular member of an algorithm. Therefore, a human algorithm can be deterministic as well as probabilistic (Bedny, 1987). A deterministic algorithm has logical conditions with only two probabilities 0 or 1. A probabilistic algorithm has more than two outputs with various probabilities or two outputs which can have any value between 0 and 1.

All operators that are involved in receiving information are categorized as afferent operators, and are depicted with the superscript α, as O^α. If an operator is involved in extracting information

TABLE 2.1
Algorithmic Description of Task "Bypassing the Car in Front"

Member of Algorithm	Description of Algorithm Member
$\overset{3}{\downarrow}\overset{2}{\downarrow}\overset{1}{\downarrow}O_{11}^{\varepsilon}$	Continue driving
$O_2^{\alpha th}$	Mental order or command "to bypass the vehicle ahead of my car"
$O_3^{\alpha th}$	Look at the speedometer and evaluate speed
$l_1\overset{1}{\uparrow}$	If speed makes it possible bypass (Yes) go to O_4^{α}. If no go to O_1^{ε}
O_4^{α}	Look forward
O_5^{th}	Position of cars in front allows bypassing?
$l_2\overset{\overset{2}{\underset{2}{}}}{\uparrow}$	If "No" continue driving (go to O_1^{ε}); if "Yes" go to $O_6^{\mu\alpha}$
$O_6^{\mu\alpha}$	Keep information in memory about position of car in front and look backward
$O_7^{\mu th}$	Position of cars behind also allows bypassing?
$l_3^{\mu}\overset{3}{\uparrow}$	If "No" continue driving (go to O_1^{ε}); if "Yes" go to O_8^{α}
O_8^{ε}	Perform bypassing (turn wheel of car right and then left and go ahead)

from long-term memory, the symbol μ is used, as O^{μ}. The symbol $O^{\mu w}$ is associated with keeping information in working memory, and the symbol O^{ε} is associated with executive components of activity, such as the movement of a gear. Operators with the symbol O^{ε} depict efferent operators. The above description shows that O^{ε} cannot include any cognitive actions. Similarly, O^{α} can include only perceptual actions. If an operator is involved in extracting information from long-term memory (only mnemonic actions), the symbol μ is used such as O^{μ}. Sometimes after receiving information (performance of O^{α}) it is impossible to use this information immediately. As a result, a performer keeps this information in working memory and therefore symbol $O^{\mu w}$ is used. This symbol describes an element of activity that is involved in keeping information in working memory.

Thinking actions often can be performed based on externally provided information (e.g., mental manipulation of externally presented data), or with reliance on the information held by or retrieved from memory (manipulation of data in memory), or thinking actions requiring keeping intermittent data in memory. In this case, we describe thinking operators as $O^{\alpha th}$ or $O^{\mu th}$ (α means that thinking operator is performed based on external, for instance, visual information, and μ means that such operator requires complicated manipulation in memory). Such a symbolic description is used when visual information or information from memory is a critical factor for the performance of the considered members of an algorithm. Due to the limited scope of this chapter, we cannot examine the method of algorithmic description of activity in detail. Consequently, we consider a hypothetical example. The task is "A driver bypasses a car in front of his/her car." This is a real scenario that every driver is familiar with (see Table 2.1).

The algorithm should be read from top to bottom. In the left column there is a symbolic description of a member of an algorithm in a standardized form which is an example of psychological units of analysis because they have clearly defined psychological characteristics.

2.3 ACTIONS AS MAIN UNITS OF ANALYSIS IN ALGORITHMIC TASK DESCRIPTION

The term "activity" refers to a logically ordered system of motor and cognitive actions (Bedny, 2015). This system forms the structure of activity. An action is a relatively bounded element of activity, which fulfills an intermediate conscious subgoal of activity. The performance of all required actions leads to the achievement of the task's goal. The term "action" in activity theory is its main building block that is understood as an element of activity. The concepts of cognitive and behavioral actions are the basis for creating algorithmic models of activity during task performance. We will consider them here in abbreviated manner. Outside of SSAT the concepts of activity and action are often used interchangeably while in SSAT they are two different concepts. An action can be defined as a discrete element of activity that is directed to achieve its conscious goal. Actions consist of operations (psychological operations). Operations that are components of particular actions determine the method of performing such actions. Achievement of the action's goal and assessment of its result is the end point of such action that separates it from the next action. Cognition is not just a system of cognitive processes. It also is a system of cognitive actions and operations. Standardized description of cognitive and behavioral actions is necessary for the description of activity structure and particularly for design purposes. Motor and cognitive actions should be considered as complex self-regulative systems. Motor actions consist of motions that can be viewed as psychological operations when analyzing motor components of activity. Such units of analysis as mental and motor actions are basic elements of the structure of activity during task performance. Usage of these elements allows the designer to substitute the experimental domain with the domain of analytical description making experimental methods supplementary in the ergonomic design process. The analytical comparison of the structure of activity and the configuration of the equipment or interface becomes the central component of ergonomic design. A number of features of cognitive actions have certain analogy with those of motor actions. They are goal directed, have a beginning and an end, function according to the self-regulation principle, and so on.

Motor actions presuppose the existence of material objects with which a subject interacts to transform these objects according to the actions' goals. Cognitive actions transform information. More precisely, they manipulate not material objects but operative units of information (OUI) or operative units of activity. These units of information perform functions that are similar to the ones material objects have for motor actions. Such internalized operational units of cognitive actions should be regarded as internal mental tools of activity. Operational units of activity are semantically holistic entities that are formed during acquisition of a specific activity. A person can mentally manipulate images, extracting units of information from memory, mentally manipulate them while thinking, even without an external representation of data.

One way of describing cognitive actions is by utilizing technological terms or terms that describe some task elements associated with a considered action. Taking a reading from a pointer or a digital display are examples of perceptual actions that are described based on technological principles (technological units of analysis). Depending on the distance of observation, illumination, and constructive features of a display, the content of a mental operation and time of a performed action can vary. Such description as "taking a reading from a pointer on a display" does not describe clearly what action is performed by a subject because reading conditions can vary. If we use such description as "simultaneous perceptual action" with duration of 0.30 s we can really understand what action is performed by a subject. This is an example of perceptual actions that is described based on psychological principles (psychological units of analysis). In a similar way a "simple decision-making action at the sensory-perceptual level" with duration of 0.35 s is another example of a psychological unit of analysis.

If we know the duration of such actions, we can understand a worker's actions from the psychological point of view. A combination of technological and psychological units of analysis is specifically useful for describing the structure of activity during task performance. Extraction of cognitive actions during task performance might be a complex task. We recommend the use of the new eye movement analysis method developed in SSAT (Bedny et al., 2012). This method is not covered here. Cognitive actions sometimes have very short durations and it is often not easy to extract mental operations out of the content of cognitive actions. Therefore, in our further discussion, we offer a standardized description of holistic cognitive actions.

In SSAT we identify three basic approaches to studying work activity:

Parametric approach is concentrated on the study of various parameters of activity that are treated as relatively independent. For example, experimental methods in cognitive psychology are the parametric method of study.

Functional analysis is concentrated on the study of the most preferable strategies of task performance. Analysis of strategies of activity is based on functional models of self-regulation of activity. The main unit of analysis is the functional mechanism or functional block.

Morphological analysis studies the structure of activity during task performance. The main units of this analysis are cognitive and behavioral actions and operations. Functional and morphological analyses are basic methods of studying work activity in SSAT. In this chapter we briefly describe morphological analysis. Below we present examples of the description of cognitive actions.

1. Direct connection actions unfold without distinctly differentiated steps and require a low level of attention. They can be further distinguished as sensory actions (detection of noise); decision about a signal at a threshold level; obtaining information about distinct features of objects such as color, shape, sound, etc.
2. Simultaneous perceptual actions such as identification of clearly distinguished stimuli that are well known to a subject and require only an immediate recognition, perception of qualities of objects or events (recognition of a familiar picture).
3. Mnemonic (memory) actions: memorization of units of information, recollection of names and events, etc. Direct connection mnemonic actions include involuntary memorization without significant mental effort.
4. Imaginative actions: manipulation of images based on perceptual processes and simple memory operations (mentally rotating a visual image of an object from one position to another according to a specific goal).
5. Decision-making actions at a sensory-perceptual level: operating with sensory-perceptual data like decision making that requires selecting from at least two alternatives (detecting a signal and deciding to which category it belongs out of several possible categories).

There are also two other groups of mental actions. For example, the second group is mental transformational actions. Such actions as successive perceptual actions; explorative-thinking actions; thinking actions of categorization; logical thinking actions; decision-making actions at verbal thinking level; recoding actions, etc. can be related to this group of actions (Bedny, 2015).

It is possible to combine experimental and analytical procedures during systemic-structural task analysis. After receiving initial experimental data a practitioner can develop theoretical models of activity during task performance and then switch to analytical procedures of task analysis. Visual information often plays a critical role in task performance. Hence, experimental task analysis can be conducted by utilizing eye movement data. SSAT utilizes a new method of eye movement interpretation. In cognitive task analysis practitioners utilize the cumulative scan path of the entire task utilizing such data as scan path length (in pixels), the total number of fixations, cumulative dwell time, or average fixation times in respective areas of interest, etc. Such a method is not effective for analysis of cognitive actions. In SSAT the cumulative scan path is divided into small fragments that are associated with one or several images of eye movements. These images allowed tracing eye movements from one element of the visual field to another and extracting various types of cognitive actions with high precision. Cognitive actions are described and classified based on standardized principles developed in SSAT (Bedny et al., 2012).

Let us consider the method developed in SSAT to describe motor actions. Outside of activity theory there is no clear understanding of the concept of motor actions. A motor action consists of motor operations (motions) that are integrated by a goal of this action. A motor action is a self-regulated element of a motor component of activity. In contrast in engineering psychology and ergonomics the term motor response is used as a synonymous of motor action. Such methods of motor actions' description can be utilized only when an operator reacts to isolate signals, using discrete actions in highly predictable situations. According to SSAT motions are elements of motor actions (psychological operations of motor actions). The MTM-1 system utilizes a standardized method of motion description. Under a standardized motion we understand a single motion of a hand, a leg, a body, or a wrist that has a definite purpose and falls under the rules of a standardized description. We utilize the MTM-1 system for descriptions of motor actions due to the fact that it has a standardized method of motion description. This is a totally new method of using the MTM-1 system in ergonomic studies. According to SSAT a standardized motor action is a complex of standardized motions (usually no less than two or three motions) performed by a subject that is unified by a single goal of an action and a constant set of objects and work tools. The amount of motions in one motor action is restricted by the short-term memory capacity (usually no more than 4–5 motions).

The MTM-1 system ignores the concept of motor action and the factor of flexibility of activity during tasks performance. Therefore in our approach the MTM-1 system is combined with algorithmic analysis of activity and standardized motions are considered as elements of motor actions. This allows us to describe a flexible time structure of human activity that includes cognitive and motor components.

Technological and psychological units of analysis can also be utilized for description of motor actions. Psychological units of analysis describe motor elements of activity in a standardized manner which allows unified and unambiguous interpretation of motor components of activity. Technological units of analysis simply describe some elements of task using a common language. "Take part" and "press a button" are examples of the description of motor actions using technological units of analysis. It is not a very precise description of motor action that covers a variety of motor actions from a psychological point of view. For example, distance to an object and the direction of action can vary when performing the action "take part." This action can be performed automatically or under precise conscious control. Therefore, the action "take part" in various conditions can require totally different elements of activity. Ignoring the difference between technological and psychological units of analysis can result in inadequate task analysis and make it impossible to design a model of activity. The action "take part" can be more precisely described if we transfer technological units of analysis into psychological ones. By using MTM-1 we can describe the motor action "take part" utilizing psychological units of analysis. For example, this action can be described as "R40D + G1C1." According to MTM-1 this motor action includes two standardized motions and this description means "reach under careful control at 40 cm distance, and grasp a cylindrical object with diameter of more than 12 mm with some interference while grasping." In this example, one can see that a motor action comprises of two motions (psychological operations) that are integrated into a motor action by the conscious goal of this action (grasp object). This description shows that an action should be performed carefully and concentration of attention during performance of this action is required. According to MTM-1 the performance time of motion R40D = 0.66 s and G1C1 = 0.26 s. Therefore, the performance time of this motor action is 0.92 s. Performance time of the real execution of this motor action might vary around this time. This motor action requires the third level of attention concentration and therefore according to SSAT should be related to the third category of complexity. However, under stressful conditions or in the presence of contradictory information such action can be moved to the fourth category of complexity.

Let us consider another example. Suppose, a worker performs the following motor action: "Move an arm and grasp a lever." This is a description of a motor action in technological terms. A goal of this motor action is to grasp a given lever's handle for further use. This is a technological unit of analysis. Using the MTM-1 system we can describe this motor action as "R40B + G1A." In MTM-1 R40D means "reach an object with average level of attentional concentration and grasp it when the low level of control is required." According to MTM-1 performance time of motion R40A = 0.40 s and G1A = 0.07 s. Performance time of this motor action is 0.47 s. This motor action requires the second level of attention concentration and therefore according to SSAT should be related to the second category of complexity.

When we compare action "take part" and "Move an arm and grasp a lever" we can see some similarity in the performance of the actions. The purpose of the actions is similar. However, from the psychological point of view the second action with the same distance of movement requires less control and concentration of attention. Technological units of analysis provide a very ambiguous description of human activity. For example, instead of "take part" we can write "move an arm and grasp a part" for description of the same action. In SSAT the lowest category of complexity for cognitive actions is the third category. As an example, a simultaneous perceptual action with duration of 0.3 s according to those developed in SSAT rules is related to the third category of complexity. Hence, the first and the second category of complexity can be assigned only for motor actions. Time structure activity analysis is important for task complexity evaluation.

MTM-1 system does not use the concept of motor action and there is no distinction between a conscious goal and a purpose there. However, the purpose is not always conscious. Therefore, a subject can be unaware of how he/she performs separate motions. So, such terminology as "motion and time study" is not adequate from the SSAT perspective. Motor components of activity include hierarchical units of analysis such as motor actions and motions (motor operations in activity theory). One can clearly describe a motor action only if she/he defines standardized motions imbedded in motor actions. Motor actions that are performed by multiple parts of the body cannot be integrated into one action. For example, two motions simultaneously performed by the left and right hand cannot be considered as one motor action. Finally, we need to point out that a specialist in the MTM-1 system begins task analysis by dividing the activity during task performance into discrete motions.

In contrast to the traditional method of using MTM-1, where specialists immediately divide a task into its constituent motions, in SSAT we utilize the following steps:

1. Conduct qualitative task analysis and discover preferable strategies of task performance
2. Describe task algorithmically
3. Describe cognitive and motor actions in each member of an algorithm
4. Describe a list of motions in each motor action by utilizing the MTM-1 system and determine their duration

The separate stage involves analysis of cognitive actions and their relationship to motor actions. In SSAT the first stage is qualitative analysis and then motor components of a task are divided into actions and then each action is, in turn, divided into operations (motions). Combination of MTM-1 with algorithmic analysis gives us an opportunity to describe very flexible components of motor activity. This becomes possible when we determine the relationship between cognitive and motor actions.

2.4 DESCRIPTION OF ACTIVITY TIME STRUCTURE

Any task that is performed in a human–machine or human–computer system requires some time for its performance. In work activity, the time of task performance emerges as one of the most important criteria of work productivity and efficiency. Time not only reflects the distinguishing features of external behavior but also the specifics of the internal cognitive processes. For example, chronometrical studies play an important role in cognitive psychology (Sternberg, 1969). Although cognitive psychology makes use of time as performance measure, the concept of "time structure of activity" is not recognized in ergonomics. For example, performance time is used in ergonomics to evaluate the workload. In a simplified manner this method can be presented as an evaluation of the ratio of the task performance time to the total available time. This measure is utilized for evaluation of demands imposed on the operator. However, this method is not sufficient for evaluating cognitive demands imposed by the task that depend not only on the index considered above but also on complexity of the task. The more complex the task is the more cognitive efforts it requires. The time occupied by performing the task can be shorter and at the same time cognitive demands can be higher. Without activity time structure analyses we cannot evaluate task complexity (Bedny, 2015).

Activity is a structurally organized system that consists of interdependent elements and unfolds in time as a process. Therefore, the concept of the activity time structure is critically important in task analysis and ergonomic design. In ergonomics and industrial engineering, time structure of activity is practically an unknown concept and it is misinterpreted in psychology. Some psychologists explain the time structure of activity during the skill acquisition process by assuming that, at the initial stage of skill acquisition a trainee cannot perform actions simultaneously or one action right after another which causes pauses between motor and cognitive components (Hacker, 1985). Such pauses are considered as empty time intervals. However, such "empty" time intervals are not really empty. An operator continues processing information during such pauses. Without developing a time structure of activity we cannot perform quantitative assessment of task complexity due to the fact that activity is a process and we have to evaluate the complexity of this process. At this stage of analysis all activity elements are translated into temporal data that demonstrate duration of standardized elements of activity. Technological units of analysis should be transformed into psychological units of analysis. Description of activity time structure is important for designing tools, equipment, and for HCI. The main idea is that changes in equipment configuration and computer interface probabilistically change the time structure of activity. A specialist can evaluate and change equipment characteristics based on time structure analysis. The time structure of activity helps a specialist evaluating efficiency of performance of production operations, thus it can be used in the evaluation of safety and training. The time structure of activity cannot be developed until we determine strategies of task performance and a possibility of performing activity elements simultaneously or sequentially. Before developing the time structure of activity it is necessary to describe each task algorithmically.

The following are the stages of time structure development:

1. Determine a content of activity with a required level of decomposition for defining its elements (psychological units of analysis)
2. Determine duration of elements while considering their influence on each other
3. Define distribution of activity elements over time, taking into account their sequential and simultaneous performance
4. Specify the preferable strategy of activity performance and its influence on the duration of separate elements and of the whole activity
5. Determine the logic and probability of transition from one temporal substructure to another
6. Calculate duration and variability of activity during a task performance
7. Define how strategies of activity change during skill acquisition, and estimate which elements are intermediate and which ones are final in the time structure

The critically important step in the development of the activity time structure is determining strategies of task performance. For example, in a dangerous situation, when actions have a high level of significance, an operator performs them sequentially, even if they are simple. However, in a normal situation, when the consequences of errors are not severe, the same simple actions will be performed simultaneously. Strategies of activity also depend on the logical components of a work process and the complexity of separate elements of activity. SSAT offers rules that determine the possibility of combining various components of activity, specifically: (1) to combine motor components of activity; (2) to combine cognitive components; (3) to combine motor and cognitive components of activity.

In SSAT such possibility depends on a level of concentration of attention during the performance of such elements (Bedny, 2015). At the final stage of a design process analytical models are tested experimentally and some corrections are possible. These are common steps not only for ergonomic but also for engineering design. There is a probabilistic relationship between these models and real performance. The more times a subject performs the same tasks the more closely this subject's activity approaches developed models. We consider as an example the task that includes a combination of cognitive and motor actions.

To analyze this task we have conducted a laboratory study utilizing a control board especially developed for this purpose. There was a panel for the participants on one side of this control board, and a panel for the experimenter on the other side.

The experimenter's panel allowed setting the program that would present various versions of the task to the participant. The duration of the performance of the presented versions of the task has been registered by the timer with precision of 0.01 s. The participant's panel had (1) a signaling bulb that displays either number 1 or 2; (2) a pointer indicator; (3) a digital indicator that showed a number from 1 to 10; (4) a green signaling bulb; and (5) a red signaling bulb. All indicators were located on a vertical panel slanted according to ergonomic standards. The controls were located under these indicators on another horizontal panel. All the indicators and controls were situated in order from left to right and had linear organization. The controls were as follows: (6) a four-position switch; (7) a hinged lever that could be moved to four perpendicular positions (up, down, left, and right); this lever had a button on the top of the lever's handle, which could be pressed by the thumb. A hinged lever (7) could be moved only after the button had been depressed. The next control was a 10-position switch (8). The panel also had a red button (9) and a green button (10). In summary, there were five indicators and five controls. Each control was located under the corresponding instrument. Consequently, eye movement from instrument to instrument and movement of the right hand had linear organization. We do not describe our experimental procedure in detail and simply present below the time structure of activity during task performance in tabular form. During experimental trials various versions of task have been presented to the subjects. We present the time structure of activity during performance of the first version of task performance in tabular form. Table 2.2 depicts the time structure of activity during task performance in tabular form.

When there are complex combinations of activity elements (activity elements are performed simultaneously) the most informative is a graphical form of time structure description. This method of presentation is usually done after a tabular form is developed. Figure 2.1 depicts the graphical model of the activity time structure for the first version of task performance and Table 2.2 presents it in tabular form.

In the graphical model of time structure of activity in Figure 2.1 individual elements of activity are presented by horizontal lines. The elements are specified by symbols above the segments. Segment under EF designates the duration of such mental action. Similarly, other segments designate the duration of various elements of activity. According to introduced rules, EF can be further divided into perceptual and decision-making mental operations in order to distinguish an afferent member of the algorithm (perceptual component of activity) from a logical condition (decision-making component of activity at sensory-perceptual level). In more complicated situations the duration of decision-making actions can be evaluated experimentally or required data can be taken from other sources. For example, O_6^α and l_3 are also involved in the decision-making process at the sensory-perceptual level. We have determined the duration of these elements using data from a *Handbook of Engineering Psychology* (Myasnikov and Petrov, Eds., 1976) and divided them into two mental operations. The elements of activity that are overlapped by other longer elements are designated by a dashed line. For example, O_{10}^α and l_5 are overlapped by O_{11}^ε. So, for O_{10}^α and l_5 we did not assign performance time when we calculated the duration of the whole task. One can make a conclusion about the possibility to perform actions simultaneously or sequentially not only based on analyses of separate actions or operations, but also based on the analysis of possible strategies of task performance. For example, if a performer is very skilled and/or consequences of wrong actions are not significant, then actions can be performed simultaneously. If actions are not automated and errors are undesirable, they should be performed sequentially.

TABLE 2.2

Time Structure of Activity during Performance of Task According to the First Version of Algorithm

Members of Algorithm (Psychological Units of Analysis)	Description of Elements of Tasks (Technological Units of Analysis)	Description of Elements of Activity (Psychological Units of Analysis)	Time (s)
O_1^α	Look at first digital indicator	Simultaneous perceptual operation	0.15
l_1	If the number 1 is lit, turn switch to the left (perform $_1O_2^\varepsilon$); if the number 2 is lit turn switch to the right (perform $_2O_2^\varepsilon$)	Simultaneous perceptual operation	0.15
$_1O_2^\varepsilon$ or $_2O_2^\varepsilon$	Move two-positioned switch 6 to the right or move the switch 6 to the left	M2.5A	0.14
O_3^α	Determine that digital indicator 3 or signal bulbs 4 or 5 are not on	Simultaneous perceptual operation	0.15
l_2	Decide to move an arm to the hinged lever 7 and press button 8 (perform O_4^ε)	Decision-making operation at a sensory-perceptual level	0.15
O_4^ε	Move right arm to the four-position hinged lever 7, grasp the handle and press button 8 with the thumb	RL1 + R13A + AP2 G1A	1.15
$O_5^{\alpha w}$	Wait for 3 s	Waiting time	3.00
O_6^α	Determine the pointer's position on the pointer indicator 2	Simultaneous perceptual operation	0.15
l_3	Decide how to move hinged lever 7 (if the pointer's position is 1, perform $_1O_7^\alpha$; if…2 perform $_2O_7^\varepsilon$…., if 4 perform $_4O_7^\varepsilon$)	Decision-making operation at a sensory-perceptual level	0.15
$_1O_7^\varepsilon$ … $_4O_7^\varepsilon$	Move the four-position hinged lever 7 to the position that corresponds to the number of pointer indicator	M5B	0.27
O_8^α	Determine that digital indicator 3 demonstrate number 5	Simultaneous perceptual operation	0.15
l_4	Decide to move multi-positional switch to position 5 (if the digital indicator 3 displays number 5 perform $_1O_5^\varepsilon$)	Decision-making operation at a sensory-perceptual level	0.15
$_5O_9^\varepsilon$	Turn multi-positioning switch 9 to the required position 5	RL1 + R13A + G1A + T150S	1.12
O_{10}^α	Determine that bulb 5 (green) is turned on	Simultaneous perceptual operation	Overlapped by motor activity (0.15)
l_5	Decide to press the green button 11 (if the green bulb 5 is on [$l_5 = 1$] perform O_{11}^ε).	Decision-making operation at a sensory-perceptual level	Overlapped by motor activity (0.15)
O_{11}^ε	Move an arm to the green button 11 and press it	RL1 + R26B + G5 + AP2	1.46
Total	Work time 3.73 s; waiting time 3 s		6.73

Some symbols in Figure 2.1 require additional explanation. M2.5A means "move object against stop" when distance is 2.5 cm. Letters P and D above a segment mean "perception" and "decision making." RL1 means "normal release performed by opening fingers." R13A means "reach an object in a fixed location, when distance is 13 cm." AP2 designates "apply pressure with effort less than 15 kg." G1A means "easily grasped." This element is overlapping with AP2. The letters WP mean

Members of algorithm (psychological units of analysis)	Graphical description of elements of activity (psychological units of analysis)
O_1^α l_1	P. DM
$_1O_2^\varepsilon$ or $_2O_2^\varepsilon$	M2.5A
O_3^α l_2	P. DM
O_4^ε	RL1 R13A AP2 G1A
O_6^α	WP
O_6^α l_3	P. DM
$_1O_7^\varepsilon \dots {}_4O_7^\varepsilon$	M5B
O_8^α l_4	P. DM
$_5O_9^\varepsilon$	RL1 R13A G1A T15OS
O_{10}^α l_5	RL1 R26B P. DM
O_{11}^ε	P. DM RL1 R26B G5 + AP2

FIGURE 2.1 Graphical presentation of time structure of a task performance utilizing experimental control board (the first version of the algorithm).

"waiting period." M5B designates "move an object 5 cm to an approximate location (requires an average level of concentration of attention)." T180S designates "turn 180 degrees when weight factor is small (from 0 to 1 kg)." Other elements of activity are designated similarly.

Let us consider units of analysis that are used for the algorithmic description of a task and temporal analysis of activity. The first two members of the algorithm (O_1^α and l_1) are the result of artificially dividing the element EF into two separate mental operations that are related to different members of the algorithm. This is done to distinguish between members of the algorithm that are associated with decision making at a sensory-perceptual level and the members of the algorithm that are comprised of simultaneous perceptual actions (operation).

In all other situations we also divide tasks into separate members of the algorithm according to recommendations described in SSAT. Usually a member of an algorithm includes 1–4 actions that are integrated by a high-order goal which is a goal of a member of an algorithm (a goal that should be achieved during performance of a particular member of an algorithm). For example, a member O_2^ε contains one motor action "move an arm to a lever, grasp it and simultaneously press a button with a thumb." This action, in turn, is comprised of the following motor operations: (motions) "move arm," "grasp a handle," and "press a button with a thumb." All these operations are integrated by a goal of a motor action. Other members of the algorithm are described similarly. In this example we covered only one strategy of task performance. Therefore, logical conditions did not include associated arrows and transitions to corresponding members of the algorithm.

Let us consider another example of time structure analysis. Here we consider a fragment of activity time structure during the performance of a manual production operation in laboratory conditions. This helps us to understand basic principles of time structure development. We present the complex time structure of an activity in which cognitive components are combined with motor components. Each element of the activity is relatively simple. However, the logical structure of the activity is sufficiently complex. The task requires installing 30 pins into the holes according to specific rules. Some pins had a flute. When subjects grasp pins with the flute from the box, they use the following rules:

1. If a fluted pin is picked up by a subject's left hand, it should be placed in the hole so that the flute is inside the hole.
2. If a fluted pin is picked up by a subject's right hand, it should be placed in the hole so that the flute should be above the hole.

Graphically, the time structure of activity during performance of this fragment of task when subject grasps two fluted pins is depicted in Figure 2.2.

In Figure 2.2, RH means right hand, LH means left hand, and MP means mental processes (dashed line). P means the probability of appearance of these elements of activity during task

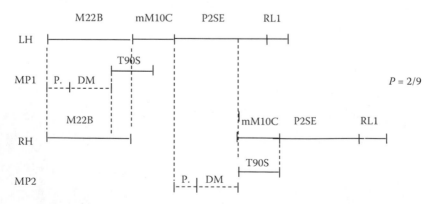

FIGURE 2.2 Graphical model of activity time structure during installation of pins "take two pins and install them into the holes."

performance ($P = 2/9$). As can be seen, some elements of activity are performed simultaneously and some in sequence. The following is a verbal description of these elements:

1. MP1—Receive information about the pin's flute in the left hand and decide to turn it into the required position. It consists of two mental operations: simultaneous perception (P) and simultaneous decision making (DM). P + DM should be considered as one cognitive action.
2. Left hand performs uninterrupted motor action which includes the following motions (according to the system MTM-1): M22B—move pin with left hand into the approximate position (22 sm); m10C—without interruption move pin into exact position (10 sm).
3. Simultaneously while moving the pins, turn pin in the left hand 90° so that the flute is inside the hole after installation.

Let us consider how the subject performs the cognitive action about the pin's flute in the right hand and then performs the motor action with the right hand according to the obtained information. The motor action performed by the right hand includes exactly the same motions. However, the strategy of this action performance is different. According to rules in SSAT two cognitive actions cannot be performed simultaneously. The minimal level of attention concentration for performing the simplest cognitive action is the same as the highest level of attention concentration for the motor action (the third category of complexity). Motion mMC also requires the third level of attention concentration (the third level of complexity). Therefore, cognitive action "receiving information and decision making" can be performed only after completion of mM10C performed by the left hand. Thus, after simultaneous performance of M22B by both hands a subject interrupts a motor action by the right hand until motion mM10C is competed by the left hand. After that a subject performs decision making (MP2) for the right hand followed by the right hand motor action mM10C. This is the most preferable strategy for performing the second cognitive action and the motor action associated with it by the right hand.

We described the strategies of performance of two cognitive and motor actions by the right and left hands. If a subject performs these actions several times his real performance approaches this model.

The material presented here clearly shows that the activity is a process that has a complex structure that evolves in time. Its time structure has various cognitive and motor components. If we know the duration of the individual elements, the probability of their occurrence, and their nature of being combined in time, it is possible to calculate the complexity of the analyzed activity structure. However, in this work we do not consider the method of complexity measurement. Here, we only want to point out that the preceding time structure description of task performance can help predict strategies that are used by a subject during task performance.

It is interesting to compare the time structure of activity in Figure 2.2 with a timeline chart which is utilized in the timeline analysis of the same task (see Figure 2.3). This method is widely used in contemporary ergonomics (Kirwan and Ainsworth, 1992).

If we compare the time structure of an activity depicted in Figure 2.2 and timeline chart in Figure 2.3 we can see that they are totally different methods of describing the temporal parameters of the subject's performance. In SSAT psychological units of analysis are utilized for description of the time structure of activity. When developing the timeline charts specialists utilize technological units of analyses.

2.5 CONCLUSIONS

The time structure of activity during task performance is an unknown concept outside of SSAT. Rather than considering separate parametric characteristics of activity, such as time of task performance, reaction time, reserve time, or conduct time line analysis (a rather crude method of time

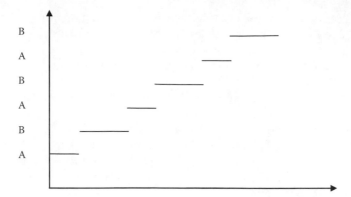

FIGURE 2.3   Timeline chart of the task "take two pins and install them into the holes three times" wherein vertical axis A indicates move two hands and grasp them; B indicates install the pins into the holes; horizontal axis indicates performance time in seconds.

measurement of task performance) we suggest using the concept of activity time structure. In order to develop the time structure of activity it is necessary to bridge from technological units of analysis to psychological units of analysis. Technological units of analysis describe the time structure of activity in technological terms or in a common language. Such units just describe each stage of task performance. Psychological units of analysis, on the other hand, describe typical elements of activity in a standardized manner. If we know the duration of separate elements of activity, rules of their performance in sequence or simultaneously, the logic of transition from one element to another, and probability of their performance, we can develop the time structure of activity.

One of the critical stages of time structure development is the algorithmic description of task performance. A human algorithm should be distinguished from a mathematical or computer algorithm. The main units of analysis in a human algorithm are cognitive and behavioral actions. Algorithmic analysis of task performance divides human activity not only into actions, but also into algorithmic elements or members of an algorithm such as operators, and logical conditions. Each member of an algorithm can be considered as a subsystem of activity which has a higher order goal that integrates the same type of actions into such a subsystem. A logical condition is a decision-making process that determines which member of an algorithm and therefore which actions are selected. An algorithm can be probabilistic as well as deterministic. Deterministic algorithms have logical conditions with only two outputs, 0 or 1. If an operator's activity is multivariant, it can be described by a probabilistic algorithm where logical conditions might have more than two outputs and vary from 0 to 1. Hence logical conditions can transfer activity flow from one member of the algorithm to another with various probabilities. Each member of an algorithm has a special symbolic designation that reflects the psychological meaning of this member of the algorithm.

The algorithmic analysis of human activity eliminates contradiction between the so-called "the best way of performance" and "the constraint-based approach." Actions comprise of smaller units-operations. For motor actions such units are motions which make it useful to utilize the MTM-1 system for description of motor components of activity. In SSAT the MTM-1 system is used in a totally new fashion. Motions are considered to be components of motor actions which in turn are components of a member of an algorithm. Combining the MTM-1 system with algorithmic analysis allows us to described very flexible human activity. Usually, the most complex time structure of activity is encountered when the activity is a combination of relatively simple cognitive actions and motor actions of various complexities due to the fact that cognitive actions cannot be performed simultaneously and should be performed in sequence. This is why we paid special attention to such types of activity time structure where motor components are combined with cognitive components.

Building activity time structure is an efficient tool for assessment of the efficiency of human performance. Combination of activity elements and therefore the time structure of activity is determined by equipment configuration. Changes in equipment or software design lead to changes in the time structure of activity. If during interaction with equipment the time structure of activity is very complex this means that a performance method should be changed or that the equipment or interface is not designed efficiently. Hence, the efficiency of ergonomic design and efficiency of task performance can be evaluated based on analyzing the activity time structure. The developed method can be used as a purely analytical tool or in combination with simplified experimental procedures. The design should not be reduced to purely experimental methods as it is done in cognitive psychology. In contemporary human–machine and human–computer systems activity during task performance is flexible and therefore only after the time structure of activity is developed, the total time of task performance can be estimated.

REFERENCES

Bedny, G. Z. 1987. *The Psychological Foundations of Analyzing and Designing Work Processes.* Higher Education Publishers, Kiev.

Bedny, G. Z. 2015. *Application of Systemic-Structural Activity Theory to Design and Training.* CRC, Taylor & Francis, Boca Raton, London, New York.

Bedny, G. Z. and Harris, S. H. 2013. Safety and reliability analysis methods based on systemic-structural activity theory. *Journal of Risk and Reliability* 227, 549–557.

Bedny, G. and Meister, D. 1997. *The Russian Theory of Activity: Current Application to Design and Learning.* Lawrence Erlbaum Associates, Mahwah, New Jersey.

Bedny, G. Z. and Karwowski, W. 2007. *A Systemic-Structural Theory of Activity. Application to Human Performance and Work Design.* CRC, Taylor & Francis, Boca Raton, London, New York.

Bedny, I. S., Karwowski, W., and Bedny, G. Z. 2010. A method of human reliability assessment based on systemic-structural activity theory. *International Journal of Human-Computer Interaction* 26, 377–402.

Bedny, G. Z., Karwowski, W., and Bedny, I. S. 2012. Complexity evaluation of computer-based tasks. *International Journal of Human-Computer Interaction* 28, 236–257.

Bedny, G. Z., Karwowski, W., and Bedny, I. S. 2015. *Applying Systemic-Structural Activity Theory to Design of Human Computer Interaction system.* CRC, Taylor & Francis, Boca Raton, London, New York.

Hacker, W. 1985. Work Psychology and Engineering Psychology. Psychological Structure and Regulation of Work Activity. Manufacturing Publishers, Moscow (Translation from German).

Karwowski, W. 2005. Ergonomics and human factors: The paradigms for science, engineering, design, technology, and management of human-compatible systems. *Ergonomics* 48(5), 436–463.

Kirwan, B. and Ainsworth, L. K. (Eds.). 1992. *A Guide to Task Analysis.* Taylor & Francis, London, Washington, DC.

Myasnikov, V. A. and Petrov, V. P. (Eds.). 1976. *Handbook of Engineering Psychology,* Manufacturing Publishers, Leningrad, Russia.

Sternberg, S. 1969. Memory-scanning, mental processes revealed by reaction-time experiments. *American Scientist* 57, 421–457.

Vicente, K. J. 1999. *Cognitive Work Analysis. Toward Safe, Productive, and Healthy Computer-Based Work.* Lawrence Erlbaum Associates, Mahwah, New Jersey.

Vygotsky, L. S. 1978. *Mind in Society. The Development of Higher Psychological Processes,* Harvard University Press, Cambridge, Massachusetts.

3 From Vision Science to Design Practice

Cristina Pinheiro and Fernando Moreira da Silva

CONTENTS

3.1 INTRODUCTION

The aging process is characterized by progressive and multiple acquisitions of some deficiencies, predominantly related to vision, hearing, dexterity, mobility, and cognition, which can lead to high levels of disability and dependency. This process includes physical degeneration, with reduction of the overall physical condition, reduction of agility, impaired vision and loss of hearing, memory faculties, and sense of direction. Deficits lead to changes in sensory perception and a decreased sense of well-being, which often involves strong feelings of insecurity (Meerwein et al., 2007). Knowledge about vision, color vision, perception, color interaction, chromatic contrasts, should always be present in the design process. The same way, inclusive design should help people and communities to prepare themselves for the future, as well as should improve the quality of life in the present.

Designers and architects need to create environments that are easier on everyone's eyes with or without visual impairments. Visual design objects and environments should be completely usable by older people, should be easier to read, and should meet the visual needs of this growing population.

Although the anatomical and physiological processes of aging are distinct from aging eye diseases, the vision changes they produce may be similar. Knowledge about these problems is essential to understand the mechanisms underlying age-related changes in visual function.

Information about neuroanatomical changes in the visual system helps guide the development of strategies for compensating age-related deficits in visually guided skills (Schieber, 1994).

With the aging process, the gradual decline in the functioning of vision will affect the performance of most daily visual tasks. Color contrasts and color combinations, with proper lighting, can make all the difference in the case of people whose vision is impaired as a result of aging, reducing risks and increasing safety. The challenge will be to create objects and environments that can help compensate for the most common types of vision loss, improving the remaining vision with lighting and proper use of contrasts and colors. Design that is more visually accessible—and not just visually appealing—can change lives for the better (Evamy and Roberts, 2004, p. 5).

3.2 VISION MECHANISM

The mechanism of vision and color vision involves a complex system composed by the eyes and the brain. Vision is perhaps the most important human sense, and also the most complex from the physiological point of view, and there is still a lot to discover. Although color vision is a widely studied subject, there are still aspects from processing visual information in the brain which remains partially unknown. Perception is a sensation perceived by the individual through the senses and combined with an interpretation performed in the brain.

The retinal cells, when excited by light rays generate electrical impulses that are conducted through the optic nerve to the brain. When the electrical impulses reach the visual cortex they are decoded and recombined to create the image. The retinal receptors (nerve cells specialized to transform optical signals into electrical signals), when excited by light rays, generate electrical impulses that are carried through the optic nerve to the brain to be interpreted. The brain gives meaning to these signals, extracting useful information from a biological point of view and turning it into conscious information. "The building of the external world in our minds requires the detection of physical energy from the environment and codification in neuronal signals, as well as the selection, organization and interpretation of our feelings. The first one is a process traditionally named as *sensation* and the second, a process traditionally named *perception*" (Durão, 2006, p. 156).

3.2.1 DISTINCTION BETWEEN LOW VISION AND BLINDNESS

The term "visual impairment" can be used for all situations, from low vision (partial impairment) to blindness (total impairment), but indicates that the classification is based on a measurement of the visual function (visual acuity, visual field, etc.), rather than an assessment of functional vision, for example, the ability to read newspapers.

There is no universal consensus on low vision and blindness definitions. In a broader sense, *low vision* can be defined as any visual impairment that cannot be corrected medically, surgically, or with conventional glasses (*The Eye Digest*, April 26, 2007). Someone with low vision has a severe reduction of visual acuity or contrast sensibility, a significantly blocked field of vision or all three conditions.

"Normal sight is 'the ability to see comfortably what is around us, whether far away or near, with or without glasses'" (Canadian Ophthalmological Society, apud Evamy and Roberts, 2004, p. 37). This *normal vision* is often referred to as 20/20 vision. This means that a person can see clearly at 20 ft what should normally be seen at that distance.

The vision between 20/60 and 20/190 is described as low vision, and means that a person can see at 20 ft what could be seen for a person with normal vision at 60–190 ft. This is already considered *partial sight*.

People are classified as blind, despite being able to see something, when vision is at 20/200 (Evamy and Roberts, 2004). If a person sees 20/200, the smallest letter which can be read at 20 ft, could be seen by a normal eye at a distance of 200 ft. This is the Snellen acuity test (Watt, 2003). Fractions, 20/20, 20/30, etc. are measures of vision sharpness, related to the ability to identify small prints, with high contrast, at a specified distance. Blindness is a condition with loss of abilities to perform all visual tasks such as reading, face recognition, cooking, etc.

August Colenbrander, from Smith-Kettlewell Eye Research Institute and California Pacific Medical Center, explains that contrary to what is commonly thought, 20/20 is not actually normal vision or the average, much less the perfect acuity. Snellen set a standard reference. The normal visual acuity in healthy adults is one or two better rows. The average acuity in a sample of population does not fall to the level of 20/20 until the age of 60 or 70 years. This explains the presence of two smaller lines down 20/20—20/15 and 20/10 (Strouse, 2002).

Amblyopia, also known as low vision means a reduced visual capacity—whatever the origin—that does not improve using optical correction. There is a low visual acuity either of organic origin (organic amblyopia—with injuries to the eyeball or the optic tracts), whether functional (functional amblyopia—without organic damage).

Between 1/10 vision and total blindness there is a continuous line, where we can distinguish:

a. Light perception—Distinction between light and dark
b. Luminous projection—Distinction of light and the place from which it emanates
c. Shape perception—Vision of fingers up to 1 m
d. Perception of forms and colors—Vision of fingers at 2.5 m

3.2.2 Low Vision Symptoms

- Difficulties for recognizing faces from family or friends
- Difficulties in reading—printed material looks distorted or incomplete
- Difficulties to see objects and potential hazards such as stairs, walkways, walls, rough surfaces, and furniture
- Difficulties to perform tasks requiring good near vision, such as reading, cooking, sewing, fixing things
- Difficulties to choose and match colors of clothing
- Difficulties to perform tasks because light levels appear to be weaker than they used to be
- Difficulties to read traffic signs, public transport signs, shop names, or other information in urban environment like signage

Some of the most common causes of low vision are cataracts, macular degeneration (AMD), diabetic retinopathy, glaucoma, hemianopia, retinitis pigmentosa, resulting among others, in problems or discomfort with glare and bright lights.

The functional evaluation of vision is an important step to help improve the quality of life. Vision is one of the primary senses and serious or complete loss of sight also has a major impact on a person's ability to communicate effectively and function independently.

3.3 COLOR VISION AND COLOR PERCEPTION

Color vision involves a complex mechanism composed of eyes and brain. The capacity to distinguish different colors is due to a physiological process that occurs in the eyes, when the light reaches the retina, but it is subordinated to a whole of neural processes in the cortex. A substantial quantity of visual information treatment occurs in the eye, but the fact is that an even larger quantity occurs after the nervous signals have left the eye. It is in the brain that the visual stimuli are elaborated and associated in underlying structures to basic mental functioning. Although much information about the external reality comes from the sense of vision, the answer to its profitability is in the brain.

The phenomenon of color vision includes the ability not only to discriminate between different colors, but to respond to them as a means of conveying information, stimulating emotions and practising deception (Lancaster, 1996, p. 112).

So color is a sensory perception, a specific visual sensation produced by visible radiation that is interpreted by the brain. "Color is not the property of objects, spaces, or surfaces, it is the sensation caused by certain qualities of light that the eye recognizes and the brain interprets" (Mahnke, 1996, p. 2).

When light rays pass through the eye, several structures help to refract the light to focus appropriately. The first of these structures is the cornea and the next is the aqueous humor in the chamber behind the cornea which helps to feed it and maintain its curvature.

The light continues on its way, passing through the pupil (whose size is adjusted by the iris) that controls the amount of light. Then, the light rays reach the lens, a transparent disk suspended in muscles that change this form. This adjustment process, known as accommodation, helps to project light correctly on the retina, improving visual acuity. After passing through the lens, the light moves through a large chamber filled with a clear and viscous substance called the vitreous body.

The retina is composed of three cell layers, one of which contains the receptors, rods, and cones. *Cones* are receptors sensitive to color, located preferably in the fovea, and are responsible for color vision. They are of three types—*cones S, M*, and *L*, depending on their sensitivity: S cones are sensitive to blue—receptors sensitive to light of short wavelength; M cones are sensitive to green—receptors sensitive to light of medium wavelength; L cones are sensitive to red—receptors sensitive to light of long wavelength.

Rods are colorblind and converge on the retina's bipolar cells. As a result, they have poor acuity but they support our perception in low lighting levels (*scotopic vision*). Cones, sensitive to color information, are responsible for acuity and vision in daylight conditions (*photopic vision*). The posterior pole of the retina is constituted by *macula lutea*, region of the retina responsible for central vision, where most of the cones are located.

The retina is the inner layer of the eye, and the endpoint of the light movement through the eye. The *blind spot* is the point on the retina where the optic nerve fibers are joined and blood vessels enter/exit the eye and does not contain cones or rods.

The brain performs the task of cortical color processing in a very complex way. The elementary units of the brain are the neurons. The neural system underlying the vision begins in the retinas, but the region responsible for vision lies in the visual cortex in the occipital region of the brain.

The images are organized in such a way that information which either one or both eyes see in the left field of vision travels to the left half of the brain, and information which either eye sees in the right field of vision travels to the right side of the brain (Figure 3.1).

The brain is divided in two halves and form two complex hemispheres that are connected by the *corpus callosum* and the optic chiasm, the region where the optic tracts cross to the opposite hemisphere. Before reaching the cortex, the retinal potential actions are diverted through an area of the brain called the *lateral geniculate nucleus* (LGN), which transmits a huge amount of visual information to the primary visual cortex.

At the optic chiasm half of the axons of ganglion cells remain on one side, while the other half crosses to the opposite hemisphere of the brain and reaches two groups of cells—LGN located deep in the brain. The LGN send their fibers to the striate cortex. After synapsing at the LGN, the visual tract continues on the back to the primary visual cortex (V1) located at the back of the brain within the occipital lobe. Within V1 there is a distinct band (striation). This is also referred to as "striate cortex," with other cortical visual regions referred to collectively as "extra striate cortex." It is at this stage that color processing becomes much more complicated.

All the process of transmission, decoding, and interpretation of information occurs in an almost instantaneous time period. Some ganglion cells are particularly sensitive to movement and contrast, while others are more sensitive to form and detail, and others transmit information about color.

In the area of visual cortex, the brain reconstructs the electrical impulses transforming into images, performing continuously every day without failure; however, many aspects of this operation and capabilities still remain unknown. The conscious visual perception depends on this area of the brain called the *primary visual cortex*, also known as the V1 area, area 17 of Brodmann, or striate cortex. Other visual areas are continuously being discovered.

The area called 17 not only performs one visual task, but several: the blobs in the visual cortex, rich in cytochrome oxidase are specialized in receiving information about color, while interblob

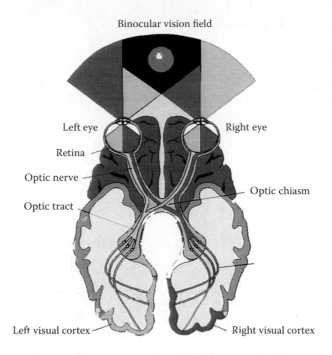

Binocular vision field

Left eye

Retina

Optic nerve

Optic tract

Right eye

Optic chiasm

Left visual cortex

Right visual cortex

FIGURE 3.1 Binocular vision and optic tracts. (Modified from Fehrman, K. and Fehrman, C. 2004. *Color: The Secret Influence* (2nd ed.). New Jersey: Pearson Prentice Hall.)

areas separate at least two streams of information, describing movement, shape, and depth (Zeman, 2007). The V4 area is essential to see colors and the V5 to perceive movement.

Neuroscience is continuously discovering correlations between the visual experience and the cortical activity. Zeki (2006) considers the visual brain not as an organ that passively narrates the external events of our world, but is actively involved in producing what we see. The brain can "read" the activation of the three types of cones to create the perception of multiple colors. Color vision has low resolution, while high resolution depends on the brightness differences.

Perception of color is an overall perception of all wavelengths of light that compose the visual scene. Color is examined in V4, dynamic form in V3, and movement in the cortex MT. The areas beyond V1 and V2 are more specialized to process the different attributes of visual scenes, such as color, shape, or movement (Zeki, 1978; De Yoe and Van Essen, 1985; Shipp and Zeki, 1985; Livingstone and Hubel, 1988, apud Bartels and Zeki, 1998).

In addition to V1, there are two general streams processing information, one for movement and location and the other for color and form, which are known as the *ventral and dorsal streams*, respectively. The ventral stream ends in the temporal lobe, the dorsal stream in the parietal lobe. It is the ventral stream that is more involved in the perception of color and form (Wagner and Kline, 2000).

The global color analysis appears to occur in cortical area V4. Unlike the cells in V1 and V2, V4 cells respond only to a narrow band of wavelengths. In V4 there is a direct correlation between the perceived color and the wavelength, made possible by a global analysis of information from neighboring cells. Damages in V4 can harm and even eliminate the ability to see or imagine color. Clinical evidence shows that damage restricted to a processing system, leads to a no perception (agnosia) of the attribute to which this system is specialized, but does not lead to a global blindness.

About the color processing system, we know that some individuals with lesions in the V4 area (Bartels and Zeki, 1998) become afflicted with *achromatopsia* and only are able to see the world in shades of gray. They may, however, distinguish between different wavelengths of light, albeit with higher thresholds, but are unable to give them color names (Vaina, 1994; W. Fries and Zeki, unpublished results, apud Bartels and Zeki, 1998).

3.4 DEFECTIVE COLOR VISION

The normal trichromat is the individual who has a normal perception of all colors. Color vision considered normal is trichromatic, but there may be deficiencies in the eye level and at the level of sensitivity to colors. People with proper functioning of all cones and a normal color vision are able to see all the mixtures and subtle differences between colors, using cones (retinal receptors) sensitive to the three wavelengths of light—red, green, and blue.

Discromatopsia or colorblindness is an inherited disorder in color vision and is linked with the X chromosome, but the woman transmitting this color vision anomaly may or may not be color blind. The different anomalous conditions are protanomaly, which is a reduced sensitivity to red light, deuteranomaly which is a reduced sensitivity to green light and the most common form of color blindness, and tritanomaly which is a reduced sensitivity to blue light and is extremely rare.

Trichromat—Sees using all three color receptors (red/green/blue); *anomalous trichromat*—reception of one pigment misaligned (anomalous).

Dichromat—Only two of the three visual pigments exist—red, green, or blue is missing. *Protanopia*—abnormal red cones. *Deuteranopia*—abnormal green cones. *Tritanopia*—abnormal blue cones. *Monochromatism*—abnormal cones.

Protanopia affects one in every hundred men, two cones in operation, the absence of the L cone, without distinguishing between red and green. There are also those less sensitive to light in the spectrum above orange. The brightness of the red, orange, and yellow is much reduced compared to normal. *Protanopes* are more likely to confuse black with many shades of red, dark brown with dark green, dark orange, and dark red; some blues with some reds, purples, and dark pinks; mid-greens with some oranges.

Deuteranopia affects one in a hundred men, two cones functioning, no cone M, without distinguishing between red and green. The *deuteranope* suffers from the same problems of color discrimination as the *protanope*, but without the abnormal darkening, and is unable to discriminate colors in the green–yellow–red segment of the spectrum. *Deuteranopes* are more likely to confuse mid-reds with mid-greens; blue-greens with gray and mid-pinks; bright greens with yellows; pale pinks with light gray; mid-reds with mid-brown; light blues with lilac.

Tritanopia—Two cones functioning, no cone S, without distinction between blue and yellow.

People affected by tritanopia are dichromats and only the sensitivity of cones to long and medium wavelengths are present, resulting in the inability to see colors in the blue–yellow range. The most common color confusions for tritanopes are light blues with grays, dark purples with black, mid-greens with blues, and oranges with reds.

The monochromacy of L, M, or S cones describes another condition where only one cone type works, and therefore, only a single color is perceived. People with monochromatic vision can see during the day but cannot distinguish between hues.

Achromatopsia is extremely rare occurring only in approximately 1 person in 33,000 and its symptoms can make life very difficult. In achromatopsia (sometimes called *rod monochromacy*), there is an absence of cones and poor visual acuity. Because an achromatic vision depends entirely on rods, it works better in low light levels, and in peripheral vision. Usually someone with achromatopsia will need to wear dark glasses indoors in normal light conditions. Achromatic congenital people perceive black, white, and shades of gray, but do not understand the concept of color. People with cerebral achromatopsia see in shades of gray, but because they already had color vision, can perceive the absence of color.

3.5 AGED VISION

3.5.1 ELDERLY COLOR PERCEPTION

The aging process is unique to each person, and despite the fact that we are not able to accurately predict the specific effects of this process, most individuals will present changes in sensory organs as they age. It is through the senses that we establish communication links with the outside world. With the gradual reduction of these sensory abilities, communication can be modified, affecting the way in which the elderly receive and respond to all this information from the surrounding environment (Larsen et al., 1997).

Structural changes in the eye cause age-related functional disorders. But if vision is processed in the brain and brain functions also decrease, the processing of visual information as well as interpretation capabilities also take place more slowly. Within the aging process, visual functions gradually decline, changes occur in the eye, retina, and visual nervous system, and the quality of vision worsens; these changes in vision are normal and are not diseases.

The most common problems of aged vision are: Loss of focusing capability presbyopia, senile miosis, decreased visual field and depth perception; decreased visual acuity; loss of central vision; difficulties with light–dark adaptation; increased sensitivity to glare, dazzle with the brilliance; and loss of contrast sensitivity, reduced ability to discriminate colors. Some of the most important changes in the aging eye include these anatomical changes:

- *Reduction of pupil size*—The maximum diameter begins decreasing and becomes smaller with the aging process (*senile miosis*). Under conditions of dim illumination, the resting diameter of the pupil falls from approximately 7 mm at age 20, to around 4 mm at age 80 (Lowenfeld, 1979, apud Schieber, 1994). The area of the pupil controls the amount of light that can reach the retina. With age, the extent to which the pupil dilates also decreases, and because of the smaller pupil size, older eyes receive less light at the retina; most older adults spend more time adjusting to changes in levels of illumination, and have greater difficulty seeing in dim light. A 60-year-old person receives only about 40% of the same amount of available light received by a person of 20 years of age.
- *Loss of focusing capability*—Presbyopia, incapacity to see near objects correctly, can be corrected with eyeglasses. It is one of the most common degenerative changes associated with aging. It is not a disease, it is the natural evolution of vision, whose symptoms begin at 40–50 years of age.
- *Decreased visual field and depth perception.*
- *Difficulties with light–dark adaptation*—Most older adults take more time adjusting to changes in light levels and have more difficulty to see in low light conditions. It has been estimated that for the same level of light, a retina of a 60-year-old person receives only one-third of the light received at age 20 (Figueiro, 2001). Even if there is enough time for adjusting the eye to darkness, at 60 years of age we need 10 times more light than at age 20, to detect the presence of light, and at age 80 we need 100 times more than at age 20 to detect that light. This demonstrates how strong age-related deficits in light sensitivity are. However, in normal lighting conditions, at age 40 we need twice the light that is required at age 20, and 3 times more light at 60 years.
- *Increased sensitivity to glare, dazzle with the brilliance*—Glare also negatively affects the perception of contrast. Older adults feel visual discomfort under bright light conditions, or at night with oncoming headlights.
- *Loss of contrast sensitivity*—Means the need of sharper contrasts and sharper edges to discriminate between objects; it is true that older people need contrasts 3 times and a half higher than individuals of 20–30 years.

Owsley and colleagues (1983, apud Connolly, 1998) determined functions of contrast sensitivity of 91 persons aged between 20 and 80 years. The resultsshowed that from the age 40, the contrast sensitivity of higher spatial frequency starts to decrease, and at the age of 80 it is reduced to 83%. The function of contrast sensitivity for older observers is affected by low light conditions (Owsley et al., 1983, apud Connolly, 1998).

This loss of contrast sensitivity means that strong contrasts and sharper contours are required to distinguish objects.

• *Decreased visual acuity; loss of central vision*—Loss of clarity in the visual field means that the affected individuals will see the world as a damaged image with unclear objects; as the viewing conditions get worse, the perceived images will become progressively more blurred and indistinct (Barker et al., 1995).

Along with the gradual reduction of the amount of light that reaches an aged retina, the opacity of the ocular media also increases. The cornea remains clear but becomes thicker and more likely to scatter light. The lens becomes more dense, more yellow, and less elastic.

With the *increasing opacity of the lens*, accommodation is reduced. *Accommodation* is defined as the ability to focus, that is, the ability of the lens to change thickness and be able to focus at short distances.

Partial sight, aging, and congenital disabilities of color vision, produce changes in perception that reduce visual effectiveness of certain color combinations.

Shinomori (2005) states that although color appearance changes little with age, the ability to discriminate colors is significantly reduced in certain conditions. Because the yellowing of the lens causes a selective absorption of short wavelength light, the ability to discriminate colors is reduced mostly on blues and greens. The color blue may appear dark and hard to distinguish from green, because the lens absorbs blue light selectively.

The yellowing of the lens, the progressive opacity, and loss of transparency are the most important causes of decreased visual performance of the aging eye. Consequently, the ability to distinguish some color contrasts decrease, and depth perception also decreases, affecting the perception of differences between figure and ground, three dimensionality, and also the perception of violet, blue, and green becomes more difficult (Meerwein et al., 2007). The color blue may look dark and hard to distinguish from green because the elderly yellow lens absorbs selectively blue light; violet, blue, and green are perceived as paler; therefore, when using these colors, it is better to choose them in a stronger hue. The blue/green area of the color spectrum becomes harder to distinguish than the red/yellow end. Most older individuals have more difficulty seeing in low light conditions. It is believed (Fairchild, 2005) that the yellowing of the lens is responsible for this effect. The aged lens causes glare by scattering light, and absorbs and scatters shorter wavelength light (blue and violet). As the lens hardens, the level of this absorption and scattering increases. In other words, the lens becomes more and more yellow with age.

With aging, other problems may interfere with vision and the ability to perceive colors by older adults: cataracts, age-related macular degeneration (AMD), glaucoma, diabetic retinopathy, and *retinitis pigmentosa*. As a consequence, color vision, night vision, visual quality, and light entering the eye decrease in this group of people.

Cataracts—Are cloudy areas in the lens, causing decreased visual acuity, blurred vision, glare and reduced color perception, fragmented and unfocused vision. People with cataracts also have a decreased depth perception, making it difficult to discern whether the objects are placed in the background or foreground. "The haze element of the crystalline lens greatly affects color perception of objects in the way that all the colors are desaturated. This point has been often neglected and only the color element was treated. We like to emphasize that more attention should be paid to the haze element when we investigate the color vision of elderly people" (Ikeda, 2009).

The perception of color improves after cataract surgery and most patients notice a greater brilliance in colors of the blue area of the spectrum (Marmor, 2007).

According to statistics on elderly people, 70% of their eyes develop cataract when they are between 65 and 69 years old, but the percentage increases to 100 at the age of 90 years old. This implies that to know about the vision of elderly people is to know about the vision of the cataract eye (Ikeda, 2009).

After surgical removal of the cataract, the stimulus of cone S greatly increases, usually resulting in a large change in color appearance. It takes about 3 months to stabilize the visual system, which exceeds the duration of other types of chromatic adaptation, according to a survey of the Department of Ophthalmology & Vision Science at the University of California Davis.

In a study of Obama et al. in Panasonic (Ikeda, 2009) it was found that after surgery, people stated that the fog disappeared, and they could see people's faces clearly.

AMD (age-related macular degeneration)—This condition also known as *age-related maculopathy* (ARM) is associated with aging and gradually destroys the macula, a cluster of light-sensitive cells in the central part of the retina rich in cones, where the visual acuity and color vision are the best.

The macula includes the fovea, which allows a sharp central vision and allows seeing of the details. Central vision is needed to see objects clearly and for everyday tasks such as reading and driving. About 23% of people over the age of 65 show some form of macular degeneration.

People often report that the objects in their central field of vision becomes distorted, changing shape, size, or color, and appear to move or even disappear. Visual acuity sometimes may fall to less than 20/400. The rest of the retina has no problem, so the peripheral vision remains normal.

Glaucoma—This is the name for a group of conditions where the optic nerve is damaged by the high pressure within the eyeball. It is often referred to as "the sneak thief of vision" because shows no symptoms.

Certain body tissues within the eye are fed by the liquid called the aqueous humor. The eye pressure is "normal" when the amount of produced fluid balances the amount drained. The eye needs a certain pressure to keep shape and size. However, with increasing fluid pressure, retinal nutrition is precluded, neural cells die, leading to so-called "tunnel vision." It happens very slowly, over months and sometimes years.

If glaucoma is detected on time it may be treated effectively; if not, the damage might progress, causing a loss of peripheral vision and may eventually lead to complete loss of vision.

As stated by Moreira da Silva (2011), Lakowski and Drance (1979, apud Moreira da Silva, 2011) found that a large number of patients with ocular hypertension (OH) showed acquired color vision losses. These losses were particularly in the blue–green part of the spectrum, the called tritan defects (Kelly, 1993, apud Moreira da Silva, 2011). They seemed to precede nerve fiber bundle defects in the visual field (Drance et al., 1981, apud Moreira da Silva, 2011). This loss of chromatic sensitivity in the area of short waves of the spectrum in people with glaucoma was confirmed by others.

Corneal edema refers to a problem of the corneal tissue, which can cause scars on the inner layer. A person with *corneal edema* is likely to see a color rainbow or halos around bright light sources, especially at night, when the brightness is more evident.

Diabetic retinopathy, one of the leading causes of blindness related to aging, involves dilatation of blood vessels and in a minority of cases the abnormal growth of new blood vessels (i.e., neo vascularization).

These new vessels are fragile and might spill blood into the vitreous humor, reducing the light that reaches the retina and cause blurred vision. They may also fail to provide the oxygen needed to meet the metabolic needs of the photoreceptors. The death of cones in this area can damage acuity and color vision. Diabetes can damage these blood vessels and can also cause cataracts and double vision.

Optic neuritis is an inflammation of the optic nerve which can result in blurred vision and distortion or lack of color vision. Although the cause of *optic neuritis* is unknown, it is believed that it starts with plaque formation around the optic nerve myelin sheath. People suffering from *optic neuritis* reported increased sensitivity to light, pain with eye movement, scotomas, and loss of color vision.

Consequences of visual aging problems:

- Decreased visual field and visual acuity
- Loss of central vision and contrast sensitivity
- Reduced ability to discriminate colors particularly the blue and green area of color spectrum
- Difficulties with light–dark adaptation

- Increased sensitivity to glare
- Dazzle with brilliance
- Decrease in contrast sensitivity and depth perception
- Difficulties in reading and focusing on near objects
- Difficulty to perform visual tasks under dim lighting or changing levels of illumination

Some research from Reading University has shown that although there is some knowledge in medical areas about how the perception of color is affected by eye problems such as macular degeneration, cataracts, glaucoma, retinitis pigmentosa (affecting about 80% of the population in the United Kingdom), very little of this knowledge has been associated with the decisions taken within the building construction and architecture (Bright et al., n.d.).

Findings from *Project Rainbow* from the same university (1997) suggest that visually impaired people can determine color differences but there are areas where difficulties exist, mainly in differentiating blue-toned green from green-toned blue of similar lightness and chroma. Due to declining vision following old age, very dark colors should not be placed next to each other since they seem to be difficult to distinguish, and the same goes for light colors. Since color preferences remain more or less stable throughout life and since color and color design are highly appreciated among most people it is indicated that the color scheme ought to take a greater advantage of this than is common today (Wijk, 2012).

In the process of aging the most disturbing aspect is the feeling that nothing happens without effort or at least as fast as before. The older see elements almost as well as ever seen, but the organization of perception as a whole takes longer and requires more attention. Having to focus more, and perceive a complex situation is a slower and more difficult task.

3.6   INTO THE DESIGN PRACTICE

3.6.1   Light, Color, and Contrasts

Light is essential for the perception of color. The visual senses of human beings work with three dimensions in order to perceive space. Light and color interaction is the base for understanding space and the person's innate or learned experiences are important when trying to interpret the surroundings and its spatial properties. Light and color are important parts within the physical environment and can be used to support this highly frail group of people (Nordin, 2012).

> Understanding how the relationships between colors of a chosen palette will affect the final outcome of an overall composition is integral to mastering the use of color (Moreira da Silva, 2011, p. 45).

In optimal conditions of light, normal vision is good. Under dim lighting or with changing levels of illumination, some individuals have difficulty in performing visual tasks. Increasing ambient light levels to about 50% more than is comfortable for the young, will increase vision. Aged and visually impaired people and others who are not recognized as visually impaired may be unable to perceive some or all colors; however, they can perceive light and dark, and since this is also a feature of colored surfaces their appearance can be influenced by the nature of the lighting condition, that can significantly affect the way color contrasts of an interior environment are perceived.

As stated before, partial sight, aging, and congenital disabilities of color vision produce changes in perception that reduce the visual effectiveness of certain color combinations.

Two colors, which contrast strongly for someone with normal vision, may be more difficult to distinguish for someone with a vision problem. Most older individuals with loss of contrast sensitivity have more difficulty to see in dim light conditions, and they need sharper contrasts and sharper edges to discriminate between objects. It is important to use color contrasts to increase visibility, and a conscious color scheme to make the environment attractive and safe for the elderly.

Wijk (2012) proposed a more frequent use of contrasting colors in order to accomplish visual distinction in the environment, to support depth and spatial perception, and to simplify object recognition. In addition, neutral colors and lack of contrast minimize attention, contrary to strong color cues.

Different visual limitations mean different capabilities to distinguish color. Many people with decreased visual abilities are able to distinguish more readily between colors of different lightness than in the same color hue. According to Mollerup (2005), when considering color differences, usually one thinks about hue differences: green, yellow, etc. However, for individuals with visual limitation, the brightness of a color is a more relevant parameter. To *wayfinders* with visual problems, it may be easier to distinguish between a light blue and a dark blue than distinguish between a blue and a red with the same brightness. The most important color contrast for visually impaired people is the contrast between color lightness, defined as color reflectance. To this people it is more important and effective than contrasts of hue or contrasts of saturation.

Designers can help to compensate the deficits of elderly people by making colors differ more dramatically on the basis of all three characteristics—hue, lightness, and saturation. "Two colors that contrast sharply to someone with normal vision may be far less distinguishable to someone with a visual disorder. It is important to appreciate that it is the contrast of the colors one against another that makes them more or less discernible rather than the individual colors themselves" (Arditi, 1999).

Findings from researchers of Reading University suggest that visually disabled people can determine color differences, but there are areas where difficulties exist, mainly in differentiating blues from greens of similar lightness and chroma. We need to compensate for the most common types of vision loss, improving the remaining vision with proper use of contrasts and colors, balancing but improving the contrast between figure and ground, avoiding some opposite or complementary colors that are very difficult to read.

3.7   GRAPHIC DESIGN AND PRINTING GUIDELINES

Good legibility helps all users, but for people with low vision the issue is crucial for reading text.
Here are some guidelines for visual communication project and printed material:

- It is important to have enough contrast between the background and the text: the human eye requires contrasts for visibility and legibility
- The more an object contrasts with its surrounds, the more visible it becomes
- When color value is too close between text and background colors, it can create legibility problems
- Too much contrast or the use of complementary colors takes the idea of contrast too far: colors will appear to vibrate and will create legibility problems
- Colors that are close in value tend to blur together, and their borders melt
- Black on red should be avoided when designing printed material
- Black text on a dark blue background or black type on a red background are hard to read. Yellow jumps out at the reader on a blue background
- White or yellow type on black or a dark color is more legible
- Yellow text on a red background is difficult to read
- Patterned backgrounds or an image in the background of text reduces its legibility
- The most common forms of color blindness are associated with the inability to discriminate red and green wavelengths, so it is not effective to use red on green
- Bright colors can produce glare, which might distract the user and cause the eyes to become tired
- A clear open typeface (font) should be used for text
- The characters must be of good proportions with clear character shapes. Avoid the use of too ornamented letters and with a complex design
- Small type and very bold type tend to blur for some people, reducing legibility

- Paper with a reflecting surface is not good for legibility
- Avoid shades of blue, green, and violet for conveying information since they are problematic
- Use no more than five colors when coding information
- Be sure the elements have a contrasting color value unless you want the elements to just blur together
- The use of too many colors might increase the chance of including a low contrast combination in the display
- When using colors, one must have in mind that older individuals have a harder time distinguishing between colors in the cooler range—blues and greens particularly
- Color is not appropriate as the sole differentiating feature between different elements—they should vary in other design features as well
- Varying the value of colors (the lightness or darkness) by at least two levels will enable most people to differentiate between the colors

People with macular degeneration, the most common cause of decreased vision among older people, need more space between letters and between lines, to distinguish individual type characters and follow the shape of the text from one line to another. Many people with low vision also have difficulty with glossy and very reflective papers that cause excessive glare.

Words or sentences which have no space between letters are not effective; and there are difficulties if there is no proper line spacing too. We should be careful choosing type fonts, because details may help to perceive differences and avoid confusion between type characters (Figure 3.2).

Among all the visual design elements, graphics and texts are inseparable from color performance. Aged people perceive less contrast between colors and perceived brightness is also different from people with normal vision. When selecting pairs of colors for typographic fonts and background, the contrasts of light against dark is preferable to emphasize the differences in hue or chromaticity.

In other words, the lettering is perceived faster and more legible when there are substantial differences in brightness between letter colors and background colors. Thus, colors should be manipulated to maximize contrasts and facilitate perception. If we increase the contrast of brightness in a color design project, it will increase its visibility.

Visocky O'Grady (n.d.) recommends accentuating the difference in 70% of brightness value between letters and background. Also Osborne (2001) refers to the same idea. The influence of contrast in reading is important not only because text of a wide range of contrasts is encountered in the environment but also because many ocular conditions lower the effective contrast of the reading stimulus.

Most studies of the role of contrast in reading, however, have treated only the luminance dimension. In general, reading is found to be fastest when the luminance difference between text and background is maximal (Moreira da Silva, 2011).

FIGURE 3.2 Details that help to perceive differences and avoid confusion between characters. (Modified from (Herrmann, R. 2011, apud Pinheiro 2012) Pinheiro, C. 2012. *Visual Communication and Inclusive Design-Colour, Legibility and Aged Vision*. Unpublished PhD Thesis. Technical University of Lisbon.)

In addition to biological changes, differences in cognitive ability are also closely associated with aging. Heyl and Wahl (2003, apud Pettigrew, 2004) postulate that the aging of the central nervous system results in the deterioration of the neural pathways to the brain, causing slower cognitive processes, including processing of visual stimuli.

The ability of attention also decreases with age, meaning that cognitive processing becomes more demanding. It becomes more difficult to memorize new information, especially if it conflicts with information previously learned; nevertheless older people are able to conveniently process information if they are given enough time. Changes in visual acuity and cognitive processing related with age result in a need to modify presentation of texts to maximize understanding by older people. The typographic characters should be slightly larger but not too large (Braus 1995, apud Pettigrew, 2004). We cannot separate the chromatic relations from other aspects that contribute to effective communication, such as typographic composition, text formats, shapes, proportions, and scale of all elements that constitute the graphic design object. If one of the issues fails, the readability and the legibility get compromised. Printings in matte paper instead of glossy, presented in environments with minimal glare, will improve readability and legibility (Spotts and Schewe, 1989, apud Pettigrew, 2004; Braus 1995).

3.8   INTERIOR AND EXTERIOR ENVIRONMENTS

Visually comfortable environments, where *seeing well* is easy and comfortable and is one effortless task, helps preserve the sense of competence and independence of older people, improving their quality of life.

Good lighting can make the difference between seeing and not seeing for older adults with poor vision and between comfort and discomfort. Designers, caregivers, allied medical professionals, and other service providers can improve quality and well-being of older people by recommending good lighting to mitigate some of the common problems associated with aging eyes (Figueiro, 2001).

- Light colors on the walls and ceilings reflect more light in an indoor space, which is often useful. Doors, floors, and furniture should have darker tones to contrast with the walls.
- The use of plain colors and matte finishes help prevent dazzle, reflections, and glare.
- Colors in contrasting shades are recommended to highlight furniture, equipment, and potentially dangerous objects and situations. A good color contrast can help in locating emergency exits. Special attention should be given to the location of mirrors to avoid confusion.

It is also suggested that color could be used to attract the attention of cues in the environment of the elderly. For example, it is well known that orientation in the urban environment can be facilitated by wayfinding cues, symbols, and proper lighting to enhance visibility (Ulrich, 2006, apud Wijk, 2012).

In exterior environments, the characteristics of ambient light are much more difficult, if not impossible, to control; so hanging posters and outdoors in public spaces should focus on the best visibility conditions, avoiding reflections, or undesirable glare. In the case of advertising panels (mupis) with proper lighting, night visibility conditions are more guaranteed because the light box makes them stand out from the surrounding environment.

3.8.1   ABOUT LIGHTING REQUIREMENTS

Generally speaking and also referring to interior spaces, lighting design should consider these aspects:

- Should be strong but balanced. Light levels should be increased, focused toward the task performance areas that require well-seen small details, and adding devices or lamps of greater intensity near the location of these tasks. Even though older people become more sensitive to light, very high brightness levels can be uncomfortable or even incapacitating.

- Reducing glare and reflections. Avoid direct view of lamps, using interior blinds or curtains to minimize glare from the windows; the shiny surfaces of objects can produce many reflections, which are more keenly felt when the luminous object is seen against a dark background; a source of bright light that is well protected from sight can provide good lighting in a room and minimize situations of intense glare.

Examples include indirect lighting located in architectural features such as crown moldings or niches, or luminaires having protection elements such as grids or deflectors.

Lighting design projects in interior spaces should also consider the following:

- Be uniform, minimizing the dark areas within an interior space. A general light or "ambient light" should ensure that there are no dark areas. Interiors should be sufficiently lit to allow good visibility so that people can move without risk.
- Improve perception of color, using lamps with good color-rendering properties to help perceive them well. For good color, use a bulb or tube, with a color temperature in the range of 2700–3500 K and CRI (color-rendering index) of at least 80. The CRI characterizes the way the light produced by the lamp makes the objects seem "natural." The color temperature of a lamp and the CRI, as well as the level of light affect the color reproduction; the better the quality of lighting on a task, the more "natural" colors will look.
- The large range of artificial light sources available have individual color-rendering properties. Their selection should be based on the need for color recognition in the interior as well as energy efficiency. When lamps with improved color rendering are used, then all observers, including visually impaired people, will be better able to perceive contrast differences.

The illumination level throughout a building during the day and night should remain relatively constant. Special attention to light position should be given, avoiding lights at eye level, because they can be blinding, even painful for people with specific ophthalmological problems.

Extra light should be used to accentuate stairs, handrails, signage, lighting, and important spots like phones; on the other hand, lighting levels should be uniform throughout the building.

Always avoid creating situations of glare and brightness by adjusting the angle of lights, to direct the beam of light out of the line of sight.

Artificial lighting should be located so that shadows or silhouettes are avoided. Consider the same idea for bright light sources and windows, because sharply angled lights can produce relatively large and strong shadows.

Matte finishes instead of glossy surfaces on walls, doors, furniture, handrails, and floors can also help prevent glare and reflections. For example, if there is a window at the end of the corridor, the sunlight can make the floors look shiny and even wet. Glare can be mistaken for water on the floor that the person tries to step around.

Since many elderly lose some color sensitivity, good color-rendering lamps may enhance the color discrimination that remains. Incandescent lamps, including halogen, render colors very well. Many types of fluorescent lamps render colors nearly as well as incandescent lamps, and have much longer lives.

General and spot lighting are necessary for the development of some visual tasks. Some light sources can be distracting for visually impaired people. This can make the identification of visual clues difficult. Strongly directional daylight from windows or panels can also cause problems. Colors in contrasting shades are recommended to highlight furniture, equipment, and potentially dangerous objects and situations.

Stairs can be dangerous without color contrasts. Small items need a bigger color difference from their surroundings to differentiate them. Shiny surface finishes are confusing and should be avoided.

Large amounts of light reflected from surfaces will cause problems, reducing contrast and increasing dazzle with brilliance. Some lighting and color conditions produce glare that may create an uncomfortable environment, and some visual discomfort. Polished concrete floors can be perceived as wet, which may create an uncomfortable feeling of insecurity. Color contrasts increase visibility and good lighting conditions makes the environment attractive and safe for the elderly.

Visual communication design objects, interior spaces, urban environments, products, signage, and all kinds of visual information will be effective and easier to read with correct lighting. Bringing knowledge to projectual practice, applying principles of visual ergonomics, will help people to improve their quality of life, moving safely in urban environments, living comfortably in interior spaces, and reading visual printed information with minimum effort.

3.9 CONCLUSIONS

The effective communication, legibility, readability, and visibility of text will depend not only on color combinations, but on the interaction of many other factors such as shape and design of typefaces, type size, the x-height, spaces between letters (kerning), words (tracking), lines (leading), colors, and contrasts between text and background, page layout, form and weight of text, avoiding confusion between letters and numbers, reading distance, lighting conditions, and surface of printing paper.

As we were studying legibility concepts and color contrasts, we always aimed to contribute with some principles to projectual practice in other areas. When including people who might normally be ignored in the design process, interior spaces design objects, urban environments, products, signage, and all kinds of visual information will be effective and easier to read, not only for visually impaired people but also for all of us.

These areas of knowledge will improve the design process and contribute to an inclusive and efficient design practice. So, bringing this knowledge to projectual practice, that is, applying the principles of visual ergonomics, will help people to improve their quality of life, move safely in urban environments, live comfortably in interior spaces, and read all visual information with minimum effort.

REFERENCES

Arditi, A. 1999. *Effective Color Contrast: Designing for People with Partial Sight and Congenital Color Deficiencies.* New York: Lighthouse International.

Barker, P., Barrick, J., Wilson, R. 1995. *Building Sight.* London: HMSO and RNIB (Royal National Institute for the Blind).

Bartels, A., Zeki, S. 1998. The theory of multistage integration. *Proceedings of the Royal Society B, London,* 265, 2327–2332.

Bright, K., Cook, G., Harris, J. (n.d). *Color Selection and the Visually Impaired—A Design Guide for Building Refurbishment. Project Rainbow.* http://www.reading.ac.uk/kqFINCH/nhe/research/rainbow/rainbow.htm (accessed October 14, 2011).

Connolly, K. 1998. *Legibility and Readability of Small Print: Effects of Font, Observer Age and Spatial Vision.* www.psych.ucalgary.ca/PACE/VA-Lab/gkconnol/Introduction.html (accessed March 9, 2007).

Durão, J. 2006. Funções Perceptuais para o Design de Espaço (Perceptual functions for space design). Caleidoscópio. *Revista de Comunicação e Cultura,* 7, 156.

Evamy, M., Roberts, L. 2004. *In Sight—A Guide to Design with Low Vision in Mind.* Switzerland: Rotovision.

Fairchild, M. 2005. *Color Appearance Models, Color Vision* (pdf). England: John Wiley & Sons Ltd.

Fehrman, K., Fehrman, C. 2004. *Color: The Secret Influence* (2nd ed.). New Jersey: Pearson Prentice Hall.

Figueiro, M. 2001. *Lighting the Way: A Key to Independence.* Rensselaer Polytechnic Institute. http://www.lrc.rpi.edu/programs/lightHealth/AARP/pdf/AARPbook1.pdf (accessed April 6, 2009).

Herrmann, R. 2011. How do we read words and how should we set them? http://opentype.info/blog/2011/06/14/how-do-we-read-words-and-how-should-we-set-them (accessed June 30, 2011).

Ikeda, M. 2009. A neglected but an important color perception of elderly people. *Proceedings of AIC Color Congress*, Sydney, Australia.

Lancaster, M. 1996. *Colorscape*. London: Academy Editions.

Larsen, P., Hazen, S., Martin, J. 1997. Assessment and management of sensory loss in elderly patients. *AORN Journal*, 65(2), 432–437.

Mahnke, F. 1996. *Color, Environment & Human Response*. New York: John Wiley & Sons.

Marmor, M. 2007. Normal age-related vision changes and their effects on vision. Lighthouse International: www.lighthouse.org/educationservices/professionaleducation/patient-management/managing-the-patient-with-low-vision/normal/ (accessed June 1, 2007).

Meerwein, G., Rodeck, B., Mahnke, F. 2007. *Color Communication in Architectural Space*. Basel: Birkhäuser.

Mollerup, P. 2005. *Wayshowing*. Baden: Lars Müller Publishers.

Moreira da Silva, F. 2011. Color and Inclusivity—A Visual Communication design Project with older people. CIAUD, Faculdade de Arquitetura, Lisboa.

Nordin, S. 2012. Professional's thoughts on light and color in nursing home facilities, in Matusiak, B., Fridell Anter, K. (Eds.). *Nordic Light and Color*. NTNU-The Faculty of Architecture and Fine Art, p. 139. Norway: Trondheim.

Osborne, H. 2001. In other words…communicating across a life span…universal design in print and web-based communication. *Boston Globe's on Call Magazine*, January.

Pettigrew, S. 2004. Creating text for older audiences. *M/C: A Journal of Media and Culture*, 7. http://journal.media-culture.org.au/0401/010-pettigrew.php (accessed November 21 2007).

Pinheiro, C. 2012. *Visual Communication and Inclusive Design-Colour, Legibility and Aged Vision*. Unpublished PhD Thesis. Technical University of Lisbon.

Schieber, F. 1994. Recent developments in vision, aging and driving: 1988–1994. *Technical Report*. http://citeseerx.ist.psu.edu/viewdoc/download?doi=10.1.1.117.4645&rep=rep1&type=pdf.

Shinomori, K. 2005. Ageing effects on color vision—Changed and unchanged perceptions. *AIC Color 05—10th Congress of the International Color Association*. Granada, Spain, May 8–13, 2005.

Story, M., Mueller, J., Mace, R. 1998. The universal design file: Designing for people of all ages and abilities. http://designdev.ncsu.edu/openjournal/index.php/redlab/article/view/102 (accessed December 5, 2008).

Strouse, W. 2002. *How Visual Acuity Is Measured*. http://www.mdsupport.org/library/acuity.html (accessed November 8, 2010).

Visocky O'Grady, J. and K. Visocky O'Grady. (n.d). *Access Ability. A Practical Handbook on Accessible Graphic Design*. http://www.rgd-accessibledesign.com/ (accessed October 15, 2011).

Wagner, B., Kline, D. 2000. *The Sensorineural Basis of the Color Experience*. University of Calgary. http://psych.ucalgary.ca/PACE/VA-Lab/Brian/neuralbases.htm.

Watt, W. 2003. *How Visual Acuity Is Measured*. http://lowvision.preventblindness.org/eye-conditions/how-visual-acuity-is-measured/. (accessed november 11, 2010).

Wijk, H. 2012. Evidence-based health care design—How can it be measured? Aspects of color and light, in Matusiak, B., Fridell Anter, K. (Eds.). *Nordic Light and Color*. NTNU-The Faculty of Architecture and Fine Art, pp. 67–72. Norway: Trondheim.

Zeki, S. 2006. The visual brain as a cognitive system. *Lecture Series*. http://vision.berkeley.edu/VSP/content/program_info/courses_zeki_lectures.html (accessed March 6, 2009).

Zeman, A. 2007. The consciousness of sight. http://www.bmj.com/content/317/7174/1696.full (accessed March 26, 2007).

4 Bridging Fashion Design and Color Effects
The ColorErg

Fernando Moreira da Silva

CONTENTS

4.1 INTRODUCTION

Color is a property and the language of form (Mahnke and Mahnke 1993, p. 62).

Nature provided us with color to fuel our body and spirit. Color enriches our entire system, giving us vital energy which is an essential part of life.

Observing nature, color often appears as a means of defense and conservation, as with plants and poisonous animals whose showy color warns potential predators, or even a function of attraction that allows the reproduction of plants and the mating calls of certain animals. The human being, in its evolution, inherited psychological and physiological reactions to color, and although they cannot be controlled or explained objectively, make color a necessary medium for the information, communication, and understanding of the environment. In this respect, Michael Lancaster (1996, p. 8) states that color has the function to attract attention, convey information, add emotions, and stimulate illusions (Gamito and Silva 2009).

As highly colored beings, our forms are made of vibrant and ever-changing colors and the human being answers colors actively or passively in everything we do. Light waves affect us in every minute of our lives and enter our energy system, whether we are awake, sleeping, visual, or blind. Our growth, blood pressure, pulse, temperature, muscle activity, immune system, etc., are all affected by light rays. The colored rays affect not only our bodies but also our emotions, dispositions, and mental faculties.

We all have a personal relationship with color. Often we give ourselves an instinctive color treatment, just by choosing clothing of a particular color, or putting around us certain colors in our homes or gardens. Most of our reactions are, however, unaware and it is only when we begin to use color in an informed way that we can take advantage of this extraordinary life force in order to improve the quality of life and our well-being. Humans are not the only ones to be affected by color. In the animal and vegetable world, color may mean survival or extinction. Color is used to attract, camouflage, as a sign of danger, or to send sexual signals. Color is instinctive to life and it is as important for our species as it is for the animal and plant kingdoms. There are many ways to introduce color in our system. Through the understanding of physiological and psychological color effects, we can select the best colors for our clothes, and our home and work environments. We can learn to use color to take advantage of it:

- Through the power of color in the food we eat and drink
- Through color techniques as therapy to heal, maintain health, or alleviate many aches and imbalances (not a new idea, since there are many ancient systems of healing by color developed throughout the world)

Of course you cannot forget color symbolic, religious, and cultural power. Always used in religious rituals, color is the language of the soul. There is a color language we have lost and have to learn again, which brings benefits to our health, happiness, and perhaps even for the survival of our planet. There are very few studies on the subject, and even these are merely descriptive of a period or a time. Until now, there has never been an attempt to understand the global phenomenon of color. Does the one who determines trends, who designs and manufactures fashion have proper training in color and its potential (for good and for bad)? Is it that when people possess more knowledge about the use of color and the effects it produces in every one of us, we will wear and use fashion differently?

ColorErg is a research project which aims to continue a former study on the physical, physiological, and psychological effects of color, in their interaction with human beings through clothing. Since this is closely linked to fashion, the study will also address fashion as a variable. The main objectives focus on the acquisition of scientific knowledge in the area that can serve as a projective tool for fashion designers, as well as to contribute, through the dissemination of its results, as a reference to the use of color to users in general.

A user-centered research methodology has been used, with participatory design, using survey methodologies, direct observation, and active research, supported by mechanical means, in the laboratory.

4.2   COLOR SYMBOLOGY

Juan Eduardo Cirlot (1958) in the *Dictionary of Traditional Symbols* presents a summary of various symbolic images of color in human communication—"the color symbology is the most universally known and used consciently, in liturgy, in heraldry, in alchemy, in art and in literature (Cirlot, p. 167)." He also quotes as examples: "the generic division established by the optical and experimental psychology in two groups of colors, the hot and spare ones, which correspond to processes of assimilation, activity and intensity (red, orange, yellow and, by extension, the white), and the cold and re-entrant ones which correspond to processes of disassimilation, passivity, and debilitation (blue, indigo, violet and, by extension, black), having green as an intermediary hue of transition and of communication between the two groups" (...) The co-ordination of colors with psychical functions, changes from one culture to another, from society to society and also from individual to individual. However, as general rule:

- Blue, the color of space and the light sky, is the color of thought
- Yellow, the color of the sun which comes from afar, comes through the darkness as a messenger of light and disappears again into the dusk, is the color of intuition, i.e., of a function which illuminates the origins and tendencies of events
- Red, the color of the palpitating blood and of the fire, is the color of life and fiery senses

- Green, the color of the directly perceptive vegetations, is the color which represents the perceptive functions... "The positive color and the negative color: Frequently in symbols, the opposition of black and white, as being positive and negative, either as a simultaneous polarity or as a successive and alternated mutation" (Cirlot, p. 125 and 143).

The concept of symbolism and color language may be extended to literature, fashion, folklore, and ethnography.

Symbolism represents the associations and impressions together, consecrated by the traditions transmitted through the centuries, by means of civilizations and religions. As represented in the color experience pyramid (Mahnke 1996), there are color symbolisms that belong to the inheritance of humanity and are the same to most cultures; and other ones, less lasting, that are the characteristics of some cultures and specific groups.

Wolfgang Goethe (1989) interpreted the colors according to expressions and symbolisms creating a language which, according to his view, should be applied to the use of color:

- "Red—indicates force (power): the higher manifestation of color; it expresses the ideal satisfaction.
- Green—indicates weakness: the color of simplicity; it expresses real satisfaction."

Between green and red, which are set as poles, he defines two groups of variations:

- "Hot colors, indicating: active, agile, vigorous, ambitious
  - Yellow/red: the vigorous, the convulsive
  - Red/yellow: the splendid, the pleasant, the happy, the lively
  - Yellow: light
- Cold and dark colors, indicating: passive, turbulent, soft and distant
  - Blue/red: the turbulent
  - Red/blue: the vigour, the joy, the gloomy, the shady, the dichotomy open/closed
  - Blue: the cold
- Hot colors give grace and enchantment, cold colors are an incentive to sternness and dignity."

4.3 COLOR CODE

Code is a system of principles which grants a certain value to certain signals. Value is mentioned and not signification, in order to give a more general character, because signification is only used in respect to the communication between human beings; however, a communication system can be operated between machines. The relationship between machines is done by codes of values because they don't understand the meaning of signals. In human communication, the receiver has a voluntary act of comparison between message and code and decodifies the message (Eco 1967, p. 72).

The *message* one deals with is a *visual one* and the *signs* must correspond to one's own perceptive necessities, which belong to human beings. Therefore, the system of principles must grant a certain value to certain signals to become a code and this certain value has a significative basis.

In terms of color, particularly of *color/space language*, the organization of codes obeys general rules which are strictly connected to the possibility of manipulation of the language of color, to the boundaries and differential thresholds of perceptive capacity in general, be that of signals, be it of the visual field, and to the categories of signification which are embraced by the language of color, as analyzed before, according to *color/space signs* which are the *communication of color*, and those which communicate by color, or with characters of simple signs, of superimposed signs, or of supersigns.

Therefore, the chromatic codes are organized as having the repertoires of *color/space signals* as a basis, programmed according to *categories and shade classes*, and they are dependent on the range of *repertoires of signs* of the individuals who participate in the *communicative process*, in consideration of the environment and common experience context.

The *primary and principle codes of color/space* are organized according to solicitations of types of harmonic shades, extension of boundaries, and levels of differentiation between shades.

By these codes, which are applicable to the *programming of languages* to all project fields, one can define the chromatic harmonization types which, from a group of *color/space signs* constitute a new *sign indicator* of a certain communicative situation which is characterized as a message by the tonal quality and by the quantity of visual vibrations between the shades that it transmits.

Having as a base the *formation of sign repertoires*, the codes of superimposed *color/space signs*, for designation of communicational intentions of something more than *color/space*, are defined according to the solicitation of the context.

These codes follow the common *communicational structure of the languages* being applied to them, beyond the chromatic reference given by the previous code, the qualification established by the communicative structure.

Selections of unities of interest in the codification are applied to the *sign repertoires* and for the *whole* a *system of rules* for the *message transmission* is established.

4.4 COLOR BIORHYTHMS

One can also associate colors to the seasons. There are personalities more connected to a particular season, in detriment of others.

If we see the case of the traditional Chinese medicine that associates each of the seasons to a color and an element and these, in turn, are associated with different body parts and organs.

Even if we have our favorite colors, there are also many colors to which we are attached, depending on our mood and emotions. These changes in our color preferences can give a valuable indication of the changes in our energy levels, disease, and areas of our lives where it is developing an imbalance that needs attention.

Our preference colors and their changes may be associated with health changes or changes in life (as is the case of the holidays).

Knowledge of color and its potentialities can lead us to try to find balance, including colors for protection.

There is also the case of colors that displease us. We may be concerned about the fact of looking at some colors and feel disgust. These can reveal hidden diseases or areas of our own image that need more attention.

The colors that displease us can lead us to determine our own vulnerabilities. For example, if we do not like pink, that might mean we hate being in situations where we feel in dependence on others; if brown displeases us, we may feel resentful of others.

If we want to create balance and harmony in our lives, we need to appreciate the importance of giving natural rhythm to our life. The realization that life is not constant and our activity and energy levels change through the seasons, helps us to be gentle with ourselves and develop a sensitivity to nature and, once again, to the nature cycles.

The colors we chose to use and are there around us tell us if there is an imbalanced area that needs our attention and to be cured. The ideal for a balance in our biorhythm is to build our own biorhythm colors chart. If we find that certain colors are missing, it is possible we feel the lack of their energy in our own system.

4.5 COLOR NEUROSCIENCE

An understanding of the brain and nervous system functions is important in the overall understanding of color. It is intended to underline the involved key process within sight and brain. Neurology is the study of the nervous system, particularly with regard to humans. This field is meaningful for the study of color since it allows an understanding (as far as medical science allows) of the process involved between the arrival of the light wave and the physical reactions which result inside the human body.

The eye structure is a complex one. In the spirit of this document, the function of each element cannot be described in detail, but a brief reference will be made. The cornea (positioned in front of the eye) allows light to enter the structure. The size of the pupil is defined by the extension of the iris, which determines the amount of light entering the eye. The lens refracts the light, which passes through the vitreous humor before focusing on the retina.

The retina structure itself is very complex, and acts as the function of the first level of light interpretation. The most important receptors are the rods and cones, which respond to noncolor and color, respectively. The central fovea is the most sensitive part of the retina where the concentration of cones is maximum. The optic disc (or white spot) is the area where the blood vessels enter the sight, and nerves are in and out of the eye as the optic nerve. This region has no rods or cones and is often referred as the blind spot.

The rods are also responsible for adapting the light and dark by the process of a change in the production of the pigment protein "rhodopsin."

The cones need a relatively bright light to operate, so with the decreasing light it also reduces the color of perceived objects. The cones host the visual pigment "iodopsin," which contains the protein opsin. There are three types of opsin-color responses: blue, red, and green, each of which is sensitive only to the appropriate light spectrum.

As the color of a particular wavelength reaches the retina, all photoreceptors are able to respond do it. The combination of the receptors that respond allows the brain to interpret the exact color of light. For example, if light with a wavelength of 600 nm reaches the retina, 30% of the green photoreceptors respond in conjunction with 55% or 64% of the red receivers. From this information, the brain will calculate the color in question is orange. The reason why there are two possible red responses is due to the fact that 60% of the population has the amino acid at position 180 of the opsin protein, whereas the rest have alanine in that position. Therefore, although all of them describe red and its variations the same way, it is likely that there might be a difference in color perception.

From the retina, visual information passes to the optic nerve, joining in the optic chiasma. Here, the images are organized in a way that the information that one or both eyes see in the left side of the sight is directed to the left half of the brain, and vice versa.

With the help of both eyes focusing on the same object, a human being can realize the depth and the distance, while each eye produces a slightly different image of the same object. The visual cortex of the brain completes the task of organizing visual information, which began in the retina. As previously mentioned, the processes in the brain are poorly understood, although the brief description above reflects current medical studies.

Studies in cats have been carried out (Graham 1997) in order to determine the role of the various areas of the brain (just a little research has been done in human beings). Five areas were identified within the cat brain which control the perception of shape, motion, color, shape and movement coordination, and movement and color coordination. The perception of form is therefore very important for cats, with movement and color respectively in second and third places.

Once the visual information has reached the human brain, it is interpreted through the various elements present in this structure. The cerebral cortex contains 90% of all nerve cells, and it receives and interprets the sensory impulses. The brain contains the voluntary and conscious process, with the right-hand side controlling the imagination and the left-hand side controlling the logic. The thalamus deciphers the sight, hearing, taste, and touch; the hypothalamus regulates blood pressure and body temperature, translating the emotions into physical responses (of physiological type). The pituitary gland is the largest endocrine gland of the body, controlling all other glands; the cerebellum is responsible for muscle coordination; and the straight formation regulates emotion. All these glands are interconnected and all are related to the response to visual stimuli.

4.6 COLOR PHYSIOLOGY

The context used in this document to "color physiology" is the following one: the effect and the extension the brain stimulation (involved in the process described above) has on other human body systems.

According to Mahnke (1996, p. 136), the environment may be responsible for the introduction of "nervousness, headaches, lack of concentration, inefficiency, poor provisions, visual disturbances, anxiety and stress." All of these are physiological responses caused by a large number of stimuli; the effect that color in particular has on the human being has been investigated for many years. Freiling (1990) points out the largest comebacks in the light of a particular color: red color is devastating; blood pressure is inconsistent, evidencing a pulse increase, it induces dryness in the throat and headaches. The subjects tend to move away from the source. The yellow light leads the subject to the tendency to move away from the source, and a subjective sensation in the hands. "Yellow causes tension but at the same time releases and activates motor activity"; violet-blue light (and blue) "... leads to calm and to concentration capacity"; the green light "...has a similar effect to the stimulus of light balancing heterogeneous trends" (Freiling 1990, pp. 122–125).

The following investigation was conducted by Birren (1982). Although not specifically related to the behavior of humans, some of the findings are particularly relevant. Different colored lights affect the rate of increase of various plants, albeit stimulating hue varies with the species. The behavior of insects, fishes, reptiles, and birds is affected by the variation of wavelengths, including the end of the spectrum, tending to ultraviolet light. The hue of light has a profound effect on birth ratio of various animals, for example, mice living under a full light spectrum reproduce at sex ratio of 50:50; mice living under blue light produce 30% of males and 70% of females, while those living under pink light produce 70% of males and 30% females. Surprisingly, these ratios are also reflected in blind mice under similar lighting.

Birren also mentioned (1982) that studies carried out by Krakov, Allen, and Schwartz in 1942 revealed that loud noises, strong smells, and tastes assets increases human sensitivity to green and reduces sensitivity to red. Blue light decreases the activity and the crying of babies (which have no cultural experience of blue).

Goldstein (1942) suggests that time is over stimulated to red light and substimulated under green light, although not all investigations have confirmed this interpretation.

In studies specifically directed to the approach of the physiological effects to the direct exposure of colored light, according to Birren (1982), the researcher Metzger (unknown date) used light sources directed only to one eye. The effect was that the muscles, only in the same side of the body, responded to the stimulus. Metzger also found changes "...in the superficial and deep sensations, both demonstrating a deep dependence on the optical stimulus" (p. 78). The subjects were asked to stay standing with arms stretched horizontally forward, placing the light source directed only to one eye. When the colored light was used, the red led the arm of the illuminated side of the body to move out. The green light caused the arm to move in.

Ellinger (1963) developed similar experiments directing light to only one side of the face and neck of subjects. When the red light was used, the subjects with arms stretched likewise drove to the light source, verifying the removal with the use of blue light. According to Ellinger (1963, p. 235), "This reaction also occurs when the eyes are closed."

Birren (1982, p. 129) concluded: "It seems clear that humans, like all other living beings react to radiation ... We can therefore generalized that color affects muscle tension, cortical activation (brain waves), heart rate, breathing, and other functions of the autonomous nervous systems—and certainly this brings defined emotional and aesthetic reactions, tastes, pleasant and unpleasant associations." Once again Birren (1982) quotes Brown (unknown date), who basically says: "The skin sees in Technicolor." It is "... also a good detector and seems to reflect the direction in which the brain neurons process color information. The experiments demonstrating body reactions to color argue that colors lead to emotional states, which are specific to different shades" (p. 126 and 157).

There are two phenomena that are fully recognized in the neurological world, which are "the constancy of color" and "the observation of the negative." The constancy of color is the name given to the tendency of the eye to view objects with the same form under different light conditions. For example, a white sheet of paper appears white under a red light or a bright blue one (although different if viewed simultaneously under the two sources separately).

The observation of the negative is the name given to the eye's ability to, after a prolonged exposure to a specific hue, create an image of its complementary color (or opposite). For example, if the eye constantly observes a blue object, and suddenly the eye sees a white area, a pink–orange version of the blue object on a white surface will appear, for a while. Some surgeons have used this principle to prevent this green observation purposes after lengthy operations involving the sight of blood (red). All operating theaters are now painted green. Experiments involving hypnotized subjects have confirmed the existence of this phenomenon of observation of the negative. According to Birren, Walls in 1942 asked hypnotized subjects to focus on not existing colors: "The subjects observed complementary colors, although their retinas had not been stimulated" (Birren 1978, p. 38). Surprisingly, "these were people who, in the state of waking, unaware the existence of negative observation of nowhere, but it should be expected the existence of complementarity to a particular stimulus" (Birren 1978, p. 57).

4.7 COLOR PERCEPTION

Color perception includes all color characteristics and proprieties, and the different color effects, which are essential to a good comprehension of color and an adequate color application.

Colors have their own characteristics, or attributes, which allow an objective description, and contribute to a good perception and identification. The color characteristics usually considered, are hue, luminosity (value), and saturation (chroma). To these three fundamental attributes, we may aid temperature. Hue is a universal variable present in all colors (Munsell 1976), and is defined by its wavelength that places it on the visible spectrum. In common language, it is confused with the color's name.

Primary hues activate only one eye sensibility. However, as the other sensibilities are activated in higher or less intensity, there appear other hues, each of them pure, without mixing with white, black, or gray, and with their own wavelengths.

Luminosity or value is the lightness graduation of a hue, it is the distinction between any color and a lighter or darker one (Munsell 1976). It may be changed by addition of white or black.

By the perception laws, when all the eye sensibilities are activated similarly and simultaneously, white corresponds to the higher luminosity, and black corresponds to the lesser sensibility . On the neutral luminosity axis, going from white to black through the grays, the three sensibilities are activated in various degrees, but no one is dominant.

Chroma or saturation defines the state of purity of a color. All pure colors have their saturation at 100% corresponding to their wavelength. When saturation is nonexistent, the color is achromatic and belongs to the neutral luminosity axis.

By the perception laws, to brighten up a color it is necessary to partly activate the sensibilities that do not belong to this color, although, when the color becomes lighter, it also becomes less saturated. To darken a color, its sensibility must be less activated, and the color loses luminosity and saturation.

Color temperature may be considered in comparison with other colors or by its effects. However, it can be also considered a color characteristic having a correspondence to its wavelength. Color measurement by electronic sensors shows that surfaces covered with a range of reds, orange, or yellow are warmer than surfaces covered with blues or greens. The temperature decreasing scale of the principal colors are: red, orange, yellow, blue, cyan, magenta, and white. Dark colors are warmer than lighter ones, absorbing more incident light. So, within a hue, temperature may change with its luminosity. In a composition or environment, the human eye is spontaneously attracted by warm colors, and need more concentration to perceive cold colors. Pigment mixage implies mixing their temperatures. Also, the relative color temperature makes that a color may seem more or less warm in comparison with other colors.

It is common to represent color through a tri-dimensional form, known as a "color solid" and adopted by several authors, in chromatic systems used in arts and in sciences, following the fundamental relation between the three original dimensions. Judging color in spatial issues and when comparing it with the perception of space regarding the light, one positions in first place the attribute of luminosity or value. Nevertheless this characteristic, or attribute, in terms of chromatic definition evolves into an integrated dimension to the attribute of hue, which in terms of an informative meaning of color is the one that characterizes it the most. So, in the tri-dimensional conception, when getting together the three attributes, the following relationship is established:

$$\text{Hue} \frac{\text{Luminosity (or value)}}{\text{Saturation (or chroma)}}$$

So, combined stimuli of luminosity and saturation belong to a certain hue in the perceptive chromatic whole. So, Moreira da Silva (1999) proved that there is a tri-dimensional relation of color/space, which forms a unity, and is represented by the following expression:

Being: h = hue; v = value; c = chroma; C = color; S = space C/S = (h v/c)/S

This relationship becomes the perceptive unity which is the basis for the other possible formations in the color/space relationship.

Newton established the concept of color complementarity, but it was Ewald Hering's opponent color theory that related this concept with the vision mechanism.

The perception of a color hue and luminosity is activated by the reaction of the eye sensibilities to the emission of certain wavelengths. The wavelengths that would activate the eye sensibilities, not yet activated, constitute the complementary color. Therefore, the complementary color is the same as light rays absorbed, and the addition of the two colors rebuilds the white light. Complementarity to white is black, and primary colors are complementary to secondary colors. However, the totality of complementary colors is not limited to these colors and every color from the visible spectrum has its complementary one, its pair to the white color reconstruction.

After image is a complementarity relationship provoked by eye tiredness and by visual memory. This phenomenon consists in, after observing one color insistently, seeing the complementary color over a white background. This reaction is more immediate when the observed color has great intensity, and it happens because the cones sensible to this color become saturated and, when the eye moves to the white background, are temporarily impeached and can only activate the complementary sensibilities.

The complementary after image is usually called negative after image. Nevertheless, after image can be positive when it happens under the stimuli of an intensive light, and is similar to the observed color. Every hue forms an after image with its complementary hue, and the after image of white is black.

The after image effect can be neutralized by a complementary background color, it can also be diminished or neutralized, applying a white or black contour. Successive contrast is the reaction to after image provoked by colors, observed one after the other. When the eye is moving rapidly over a colored surface, it observes the color on which it focuses and also, the after image from the previously observed one. This phenomenon happens because the eye, trying to activate all the sensibilities, tints the adjacent colors with the observed complementary hue.

The successive contrast is provoked only by hues, independent of their luminosity. If two colors with a great luminosity contrast are placed side by side, when they separate the lighter color darkens and the darker one lightens.

Simultaneous contrast is the result of an after image which alters the appearance of a color, by the influence of the adjacent colors, the influenced color having a minor area compared to the influent color. This reaction occurs on saturated, or less saturated hues, and neutral colors, and may affect an object shadow.

Neutral colors are the most affected by the simultaneous contrast, and in consequence, a gray background makes other colors look more intense. Being a form/ground contrast, this can happen in relation with all color characteristics.

The shadow color of the object depends on the incident light. Every colored light creates a shadow in its complementary color, which is not the effect of simultaneous contrast, but real color. White light, the day light, activates every eye sensibility and its shadow is black. However, the colors of the shadows provoked by various colored lights correspond to additive color mixtures. On the other hand, in the absence of day light, the resultant colored light shadows are different. One colored light has a black shadow, because it is the absence of light. When the object is illuminated by two complementary colors, each color provokes a complementary shadow and where the shadows overlap the color black is seen, while the mixed light is the result of the additive synthesis of the two colors. With analog lights, each light projects a shadow on the other incident color, the overlapped shadow will still be black, and the mixed light will be the additive synthesis of the two colors. The results of the three primary colors projection are three primary colors shadows, one overlapped black shadow, and a white light mixture, because the additive synthesis of the three light beams reconstitutes the sunlight.

4.8 COMMUNICATIONAL STRUCTURAL RELATIONSHIPS WITH COLOR

From the world of perception one goes to the world of *communication*. In order to accomplish an informative act it is not enough to be in touch with the surrounding environment through sensorial reactions. At that point perception becomes the channel through which the *signals* are transmitted and transformed into *signs* which are more than simple responses to physical stimuli; they are the basis of meaning, having in one unit *the signals' physical support and the references to the objects and images which are connected to them.*

> Signal is the physical concreteness of a message (Cherry 1957, p. 21).

The significative contents result from an action between: a physical support, or information support; an idea or thought indicating a meaning and giving sense to something physical; and a behavior between thought and physical support, creating a significative reality.

Summarizing, this *triad of communicative elements* define what in the *science of communication* is called a *sign* (Cherry 1957; Pignatari 1976). "A *sign* is a stimulus—i.e. a sensitive substance—whose mental image is, in our mind, associated to another stimulus, bringing the other about for a communication to take place" (Guiraud 1973, p. 112).

$$\text{Significant} + \text{Signification} = Sign$$

Sign is, therefore, comprised by a significant and by a meaning. The plan of significations constitutes the plan of expression, and the plan of meanings constitutes the one of content. The signification can be conceived as a process, as an act which joins together the significant and signification, the result of the act is, consequently, the *sign*.

This way, the signification is not a "*thing*," but the psychic representation of the "*thing*" (Afonso 1983). *Sign* is used to transmit information, to tell someone something that another one knows and what one wants the others to know, too. It comprises the following communicational process: origin—sender—channel—message—receiver (Eco 1967).

Sign, is a result between a significant, a physical instrument which carries a meaning, and the meaning itself which gives it a communicative sense.

In *color/space language*, the physical structure composed by chromatic signals supports a significative charge in accordance to the color itself. The *sign language* derives from the relationship between *structure and meaning* (Pignatari 1976). A *sign* being a basic element of the significant organization, its interpretation is based in the unity definition, that is, the *sign* is the *unity of signification*. The *group of signs* articulated among themselves creates the language, a sign means that the message be invoked.

According to Rudolf Arnheim (1971, p. 53), things and events would not be worth anything if there was no information: events simply cannot occupy the mind only as sensorial reflexes "without the information of what is happening in space and time, the brain cannot act," they are the natural relationships between perception and thought.

One talks about *information* in a general sense. However, in a strict sense, according to the *science of communication*, "information theory represents a method of computing the units, the transmissible and transmitted signals," not interfering in the field of significance: "it doesn't represent a method of computing the significants units, semiotics being the study of *meanings*" (Eco 1967, p. 134).

> That color can play an enhancing role in the communication of visual messages is readily apparent.
> It can serve to attract attention, as to increase the readership of an object (Tannembaum 1966, p. 38).

As this study, in the search for equilibrium between the *sensible knowledge* and the *scientific knowledge*, the message structural level does not restrict itself to definitions of physical computations of the *color/space unity*, in order to have a perceptive psycho-physiological relationship. It is essential to establish a link between this level and the other one, in which the computation of unities is of a subjective nature; the judgment of expression and contents goes beyond the simple quantification of information, which only has a meaning after a semantic–aesthetic qualification. Therefore, there must exist an equilibrium between the information transmitted by a group of computable signals and the meaning given by judgment of sensible values, so that the *language of color/space unity* can effectively constitute a *communicative basis* for the *transmitter/receiver* relationship.

As previously mentioned, in terms of *color/space*, there is a *visual communication* because it relates to specific perceptive channels for the chromatic sensations of sight.

Conjugated with *space*, *color*, even if physically present in all kinds of perception inherent to the sensation of space, participates in them only in the theoretical sense of the meaning, because it is not susceptible to information content through other sensorial channels. Consequently, *space*, *characteristically defined by color*, belongs to the *visual communicative structure*. Therefore, all characteristics which are peculiar to it in the structural formation, have correspondence in the visual meaningful formation.

According to the analysis of Abraham Moles (1971, p. 28) in his studies about art and computers, "a piece correspond to a great number of communication systems, which in principle can be separated objectively by an observer, and even subjectively by the receiver if he is paying attention." He quotes as an example: "In a real message between human beings, the receiver and the sender distinguish spontaneously a hierarchy of levels correspondent to repertoires of different signs. Therefore: spots of light, alphabet letters, words, expressions, constitute syntactical elements which correspond in the written language to overlaid levels. The signs of a certain level join themselves in a stereotyped way to become the supersigns, for example, the words which are elementary signs in a following superior level."

But in visual communicative reality, we never see the form *separated from the color nor the color separated from the* form. Only in special conditions could one visualize a mental image (and not the real one), with both languages separated (Moles 1971, p. 57).

However, in its concept, there exists the distinction, and in practical terms, one talks about form, triangular, square, circular, spherical, cubic form, etc., and one talks in color, blue, red, brown, gray, etc., a *sign association* happens naturally. The form has stronger links with reasoning, the reason to relate configured structures: there is a mental construction in geometrical language. *Color* has stronger links with *emotion*, the feeling to relate light and shade: there is a mental harmony in chromatic language. Both of them complement each other by reason and emotion, because one constructs the other, one brings the other into harmony, in a sensible whole which can be translated as a unified language of form and color, thereby, a group of unities: *color/space—signs*.

Kandinsky (1975) gives each form a specific color, which would permit an interpretation of superposition of sign repertoires, like Abraham Moles (1971) proposes. One can find the following relations of *form* and *color* given by Wassily Kandinsky (1975): "Correlations Line—Plan—Color."

acute angle ↔ triangle ↔ yellow

right angle ↔ square ↔ red

obtuse angle ↔ circle ↔ blue

After analyzing the correlations given by Kandinsky, Faber Birren (1961, pp. 164–167) explains: "*Red* suggests a cubic form, because it is hot, dry and opaque, qualitatively. It is heavy, solid and substantial and offers a strong visual attraction. Because it is very well focused by the eye it suits structured plans and rigid angles. *Yellow*, implies a triangle form or a pyramid form with its vertices turned to the bottom. It is the color of bigger visibility in the spectre and, qualitatively, it is acute, angular and crisp. It is lighter than substance because it doesn't have much weight; it is more space than volume. *Blue* implies a form of a circle or of a sphere. It is cold, wet, transparent, atmospherical. It is the one with less focus in the sight and, normally, it produces a smudged image, particularly distant. Because it is extense it doesn't give harsh details." The same author illustrates the correlations with secondary colors, the ones obtained by the mixture of two primary colors (in filters and pigments); sequentially it sets up the correlations according to the same scheme of primary colors and with the following interpretation:

right angle ↔ rectangle ↔ orange

acute angle ↔ hexagonal ↔ green

obtuse angle ↔ ellipse ↔ violet and purple

For these correlations, Faber Birren (1969), makes intermediary interpretations, in accordance with component colors, that is, between red and yellow for orange, between blue and yellow for green, and between blue and red for purple.

In conclusion, transcribing words from Goethe (1989, p. 154), "since color occupies a very important place in the series of elementary phenomena designated to it in a more complete variety, filling in like the limited circle does, we shouldn't be surprised to verify that its effects are always determined and significative, and are promptly associated to the mind's emotions. We shouldn't be surprised in observing that when presented by itself, acting on the mind, the combinations which creates harmonies and disharmonies, impresses us even in its more elementary characters, without a relation with nature, form or the object whose appearance is deposited on its surfaces."

4.9   COLOR AND ERGONOMICS

Color has always been a concern of every civilization, from the more remote and primitive till the more developed and actual ones. Therefore, the comprehension of the color phenomena has been a research objective for philosophers, theorists, artists, and scientists.

Color can greatly influence your ergonomics and your fashion design project when it is applied appropriately. However, it can really mess things up when it is not used in support of our cultural and psychological understanding of color. But when you get it right it can be a game changer.

Michel Pastoureau (1997) defends that nowadays colors can only be understood when related with the colors from the past, with which they are in continuity or rupture. Far more than the color use, humans have tried to control and explain the color phenomenon, ever since prehistoric times.

Although the idea of "color" may seem a simple concept, it conjures up very different ideas for each of us. To the physicist, color is determined by the wavelength of light. To the physiologist and psychologist, our perception of color involves neural responses in the eye and the brain, and is subject to the limitations of our nervous system (Lamb and Bourriau 1999, p. 1).

Color and light [...] have great impact on our psychological reactions and physiological well-being. Research has proven that light and color affect the human organism on both a visual and non-visual basis. It is no longer valid to assume that the "only" significant role of light and color is to provide adequate illumination and a pleasant visual environment (Mahnke and Mahnke 1996, p. 3).

Frieling points out the major responses to light of a certain color: the color red is devastating; blood pressure is inconsistent, there is a higher heartbeat, induces dryness in the throat and headaches. The subjects have a tendency to move away from the source. Yellow light leads the subject to the tendency to move away from the source, and a subjective sensation in hands. "Yellow causes tension at the same time releases and activates the motor activity." The violet-blue (and blue) light "... leads to calmness and to the ability to concentrate." The green light "... has a similar effect to the stimulus of the light scale heterogeneous tendencies" (Frieling 1990, pp. 25–26).

Cheskin (1947), from the Color Research Institute of America, conducted tests involving four different interior spaces, each of which of a single color (red, yellow, blue, and green), including table, chair, and typing machine equally colorful, for subjects using the machine. The results were as follows:

- Red room: increased blood pressure and pulse, overstimulation, difficulty of working
- Blue room: blood pressure and pulse rate decreased, the activity decreased
- Yellow room without any effect on blood pressure or pulse rate
- Green room: produced monotony. No other effect registered

Other research carried out by Mahnke (1996) concludes that the ultraviolet light may have a wide range of physiological effects such as "a decrease in pulse rate, a dropped in blood pressure, skin temperature changes in metabolism and a reduction in reaction time, an improvement in health and resistance to certain types of infections."

The principles of physiology and psychology are fully interconnected. Both are controlled by neurological processes, although the exact nature of the process that controls the psychological reaction is effectively unknown. Graham (1997) presented a thread that outlines the path between a biological cell and psychology, in order to overcome the existing hole in medical theory.

4.10   COLOR PSYCHOLOGICAL ASSOCIATIONS

Research on the psychological aspects of color is difficult for the mere reason that human emotions are none too stable and the psychic make up of human beings varies from person to person (Wright 1998, p. 28), p. 39.

Even without detailed knowledge about the psychological effects of color, it is well known that it can affect our moods, our disposition. There are some colors that are exciting and inspiring, just as there are others that are depressed. How often we use expressions like "I feel blue," "green with envy," "red with anger," without thinking about the meaning that lies behind each of these words. Our feelings and emotions are directly affected by the hormonal balance or imbalance in our body. Once this is affected by the colors, it has, of course, an indelible influence on our feelings and dispositions.

The psychological effects of color have been extensively investigated through the use of colorful cards, instead of using painted compartments. As Sivik (1997, p. 17) concluded, "... there is considerable disparity between the conclusions drawn by cards or buildings." Hesselgren (1987, p. 39) states: "Most of the tests on the preference of colors now indicate the effect of the color contrast between color under investigation and its background (usually white)."

It is believed that most of color effect is based on associations of education from childhood, such as the sky and sea are blue, the grass is green, etc. The principles are applied in architecture due to taste associations, which can be translated into benefits to the overall effect of the building. The second way of experiment involves the projection of light of a particular color directly onto the retina of a subject so as to compare the physiological reactions (Ellinger 1963).

Some colors can calm us, while others can stimulate our mental activity. Through the process of restoring balance of color energy directed to the pituitary, it is possible to restore metabolic and emotional balance. This can relieve stress, tension, anxiety, or depression. Certain colors can help us deal with our feelings of loneliness, frustration, or pain. The use of color to modify emotional energy also results in the change of perception of the world and our way of living. Once color is directly linked to the subconscious, we can use it to diagnose and treat a problem at a deep level.

So far, we have developed some studies based on relevant literature data and experiments with users, crossed with scientific research conducted in several countries, which have already led to some psychological associations to some of the most used colors.

We also added a survey by inquiry, questionnaire based, and involving 623 people from Portugal and the UK, both genders, of ages between 23 and 72. Here, we present some of the results, for example:

RED— vital, powerful, ambitious, hot, intimate, sensual, determined, friendly, brave, antidepressant, furious, impatient, and angry

ORANGE—hot, insurance, creative, stimulant, entertaining, cheerful, with humor, independent, and antidepressant

YELLOW—cheerful, light, bright, mentally stimulating, logical, smart, orderly, optimistic, clear thinking, and fearless

GREEN—harmonious, mental and physical relaxing, peaceful, natural, refreshing, calm, sincere, insurance, free, generous, restrained, and personal domain

TURQUOISE—refreshing, cool, calm mentally, youth, power concentration, control, communication power, and confidence

BLUE—cold, clear, relaxing, mentally calm, brings peace, tranquility, wisdom, spacious like the sea or the sky, sensitive, hopeful, faithful, believer, flexible, and quiet

INDIGO/VIOLET/PURPLE—dramatic, spiritual, creative, intuitive, media, mystical, inspiring beauty and art, protective, and clean

MAGENTA—support, natural, kind, considerate, condescending, and compassionate

WHITE—peaceful, cathartic, cold, and isolated

BLACK—comforting, mysterious, female, protective, and restrictive

GRAY—independent, secure, tab, lonely, and self-critical

SILVER—changeable, harmonious, feminine, and sensitive

GOLD—desire, understandable, powerful, high ideals, and abundance

BROWN—support, mentally mean, land, and seclusion

4.11 OUR PERSONAL COLORS

A group of scientists from the USA have recently discovered that there is a connection between children's color of the eyes and their personalities: children with dark eyes are more gregarious and extroverted than the clear-eyed.

We all tend to have one or several colors that best express our personality and make us feel comfortable and give extra vitality and inspiration. A person's color is the one that captures the essence of that person's entire personality. It will not necessarily give her any extra energy but, if the person is already feeling good, it will create the maximum impact, projecting the full force of her personality (Wright 1998).

Many people are led to change their own colors through hair coloring or the use of colored lenses. They are subconsciously driven to the personality usually associated with certain colors they think they are missing. Looking at the colors that people use most often, we get to know a bit more about them. There are several types of psychological tests that lead to a higher analytical knowledge of our personal colors. One of the most widely known and applied is the *Max Luscher color test* which was developed in order to be used by psychiatrists, psychologists, and physicians in order to provide them with relevant information about a particular person by knowing his or her color choice.

The *Luscher color test* was devised by psychologist Max Luscher in 1969. Its effectiveness has been known in advertising and industry (automotive and fashion) for years. It is uncanny what this test can consistently reveal. What is far more revealing is our unique living relationship with color, which is revealed over time and in a variety of contexts under many influences. http://www.john-paulcaponigro.com/blog/136/the-luscher-color-test-online/, accessed in October 12, 2012.

This personal knowledge about the meaning of colors and what they reveal about our personality, can lead to a better use of color in fashion (Rossignol 2001).

We can synthesize the essential data collected and interpreted, by stating that if we prefer to wear red, we are impulsive, excitable, and energetic; there is a demonstration of ambition and a taste for seeing things happen quickly; we like to be the best in everything we do; we can be a little insensitive to others' feelings, since we like to be the center of attention; red means vital force, with our nervous activity pushing us to achieve results and succeed; the habit of wearing red may indicate that we attach great importance to sexual desire and eroticism—this energy may be best used in the form of creative force, leadership, as well as development and expansion; or that we are bold and outgoing, but we tend to get angry and bad-tempered if we cannot make our way.

The preference to use orange means that we are competent, action-oriented, and impatient. We are also independent, organized, and self-motivated. Orange is the color of creativity and being practical. Orange means that our energy levels are high and are sometimes restless; we have a strong will, and we tend to be active and competitive. We are also excitable and can seek to dominate others. Bright orange and burnt orange can make us feel frustrated or blocked. We should try to use the orange peach directing our energy to others in a careful manner.

If our preference is yellow, we have an interesting and stimulating personality; we like to be active and involved in what is happening—alive and vital, can cope well with life's challenges; it represents spontaneity and communication; we are active, ambitious, willing, and researchers; there is a desire and hope of greater happiness, which implies a conflict in which it is necessary to free ourselves; it pushes us in search of the new, the modern, the developed, and the informal.

The use of green is an indicator that we are cautious people and we are led to not easily trust others, in addition to being observers of life, but do not want to involve ourselves more than necessary. It gets us quite a calm life. We are benevolent, humanistic, and self-oriented. If we use the blue–green is because we need a peaceful environment, wishing to free ourselves from stress, and move away from conflicts and disagreements. We have difficulty controlling the situation and the problems resulting therefrom, proceeding cautiously. We have a feeling of speed and keen eye for detail. We use pale yellow with green, in order to help us to share and to develop an optimistic attitude.

The use of light blue indicates that we are creative, perceptive, and sensitive, we have a good imagination and a practical way of approaching life. Our approach can be analytical and we are advised to use our knowledge to solve problems. We like to do things at our own pace and not be rushed. It is a sign that we need a safe and peaceful environment.

We can also apply this to other color dark blue: we are intelligent, self-confident, and have a great depth of feeling; we feel a responsibility for others and like to make decisions; we need tranquility around us and we must be surrounded by attention, tenderness, affection, and love; noisy people disturb us; we may suffer from mental stress, leading us to the lack of action and relaxation.

The use of the *violet* color means that we have a sensitive, compassionate nature and can easily impose on us. We must have attention with regard to the choice of friends as sensitive as us. To be happy, we must work where we are needed. We should try to use purple or magenta, which is a color with more red. This leads to increasing our self-confidence and developing of our vulnerability protection.

If our preference is the color purple, this is an indicator that we are very intuitive and we have deep feelings and great aspirations. We are always interested in the best, including friends. Lesser people do not enter our schemes, except only when they are needed. We should try not to be arrogant, giving more time and attention to listen to others. The purple color can make us feel tax systems, rules and regulations accredited by others. We have to be sensitive to our own bodily and spiritual needs.

The use of gray indicates that we are very individualistic. Many people may get the impression that we are self-sufficient, since we present a great self-control and prefer to remain without getting involved. We tend to isolate ourselves, which can lead to loneliness. We can be passive when we feel stressed and overwhelmed. We need rest, relaxation, and freedom from everyday stress. Perhaps, we need a break by introducing ocean blue or green. People who use gray are usually those who think and can be good critics. Those who lack judgment and struggle to form opinions should use gray.

Brown clothes suggest that we are honest people, grounded, and like a structured and sustainable lifestyle. We are lovers of the best things life can offer, lovers of good food, drink, and company, and sex. Do not forget that brown is the color of mother earth. This is a protective color; can accumulate emotion or a secret that leads us to put ourselves in our "shell" and be afraid of the outside world, although we feel protected by the use of brown or brownish colors. There is a desire to be emotionally safe and accepted by the outside world. We need to understand our own value and to leave out the "smallness" of mind.

If our preference is white, as this color contains all the other colors of the spectrum, it shows that we have a positive, well balanced, and optimistic personality; we possess high values and we must be open-minded and communicative. On the contrary, if we prefer to use black, it shows that we have strong will power, we are disciplined and always opinionated; it also demonstrates that we have an organized and independent character, being sometimes too stubborn and independent; we have a lack of confidence in ourselves and have an innate ability to lead life with efficiency; this color represents renunciation and those who choose to wear it constantly want to renounce everything as a stubborn protest; however, the black used on occasions demonstrates that we have control on ourselves in order to communicate an authoritarian image (Chiazzari 1998).

4.12 FASHION AND DRESSING FOR HEALTH

The colors we use can provide protection against many physical ailments as well as can give us emotional inspiration. Obviously, clothes also protect us from the elements and are a form of personal expression, protect us from the world around us and affect the way we feel and think. The fabrics act as color filters, enabling certain chromatic waves to pass through our skin. This means, for example, in the case of a green fabric, the natural light which passes through the fabric absorbs green vibrations and green passes into our system.

White clothes allow all wavelengths of light to pass and thereby to "feed" us with equal amounts of different colored energies. On the contrary, black takes away all the light from us and this is the reason why black clothes often form a layer of protection around us. The use of black often draws energy and can have a negative effect on our health, since the body needs light energy in order to function normally as a living organism.

For many reasons, there is a strong relationship between the colors of our clothes and the effect that their energies have on us. The colors we use can give us protection against many physical problems and also provide emotional support and inspiration.

We must choose colors depending on the harmonization with our color-season type and related with

- Our energy levels of the moment
- Our physical health status
- Our state of mind

The state of the art correlating color and clothes, or fashion, is still not very significant. So, several studies using quasi-experimental methodology were implemented. We worked with 12 different sample groups, each of them composed of 10 women of ages between 21 and 58. During the course of two and a half years, we developed experiments in UK and Portugal, which results, crossed with the literature evidence, led us to important conclusions, yet not totally conclusive.

These are the main results achieved:

RED
  To wear. When we need to raise the morale, or we feel tired and lethargic, or we need to encourage physical exercise and a competitive spirit. Red helps positive progress and success, and to put our plans into action. We use red when we want to feel sexy and alive.
  To avoid. When we get tired easily, or when we suffer from constipation, myalgia, mononucleosis, or chronic fatigue syndrome. Do not use it if we have a colored physique, high blood pressure, or if we anger easily. Also avoid if we are nervous or tense.

PINK
  To wear. If we are caring persons, lovers, and we need to be compassionate and friendly. Rose encourages the self-esteem. Some shades of pink to help loving exchanges. Salmon and deep pink surround us with love. Shocking pink stimulates our most sensual pleasures.
  To avoid. If we are emotionally immature, if we need the figure of a father or a mother, or if we are too dependent on others. We may feel that we are too sensitive to external dependencies.

ORANGE
  To wear. If we suffer from depression, or we need to bring joy and light to our lives. The orange relieves the serious air of thought and promotes a smile. Encourages independence of mind and self-motivation as well as helps to release creativity and negative emotions related to a poor self-image. Orange is good for increasing adrenaline, and pain, especially in the neck, arms, and limbs.
  To avoid. If we feel confused, frustrated, or claustrophobic. Orange is not advisable if we feel sick or nauseous, it will make us feel worse.

YELLOW
  To wear. When we want to be alone, so that we become detached or impartial. Yellow promotes rational thinking and reasoning, and can improve memory. It aids communication, sharing, and self-expression. Yellow is a good color for accessories. Golden yellow promotes shine and vitality, as well as the ability to meet new challenges. Yellow adds value and is a confidence builder.
  To avoid. If we are predisposed to criticism and suffer a loss of anchorage and stability. Yellow can encourage selfishness and is not advised if we give too much importance to material wealth. Yellow can cause irritability and a feeling of nausea. In this case, we introduce the gold instead of yellow, which will encourage us to find our inner desires.

GREEN
  To wear. If we are hyperactive but we have difficulty making decisions or have a clear judgment. Apple green promotes health, happiness, and innovation. Green grass produces in us the effect to understand and help others, as well as encourage abundance in our lives. Blue green promotes optimism and faith in ourselves and in others.
  To avoid. If we need to have action. Green gives space and time when we do not make decisions, but can promote the hassle, repression and inactivity.

BLUE
  To wear. If we need peace and relaxation, and if we want to have an empty mind, when we suffer from mental fatigue. Blue helps to restore self-confidence, independence,

and responsibility for others. Blue brings insight and wisdom, encourages decision making, and helps us to connect to our intuitive sense. Also aids communication and strengthens our powers of speech.

To avoid. If we feel depressed, since blue can make us feel worse. If we need to recharge power or if we are nervous, since blue does not give us this help. If we are totally relying on our mental faculties to achieve solutions, perhaps we need to stay more in touch with our emotions. In this case, we must introduce some oranges in our wardrobe.

PURPLE/VIOLET

To wear. If we want peace and love without anxiety, and authority without requirement. Violet fabrics make the peace and calm necessary to meditation and prayer, making us aware of our sensibilities. The use of purple helps us to open the mind to higher forces, by building a channel for creative energy.

To avoid. If we do not like silence or possess feelings of invasion of our own privacy. We should not use it if we feel oversensitive and we need to be socially acceptable, or if we feel taxed above the rules and regulations. We should try to use magenta, since the red rays help build our self-esteem and promote action.

LILAC/MAUVE/LAVENDER

To wear. If we want to promote sensitivity, a more peaceful and sweet nature. Mauve induces a feeling of reserve, although it is good for others. It promotes listening to our more intuitive side, so that we can use these insights to help others. We must use mauve whenever we need time for relaxation and meditation.

To avoid. If we suffer from a lack of freedom, or are surrounded by people who are not sensitive to our feelings. If we need the sensitive support of others, we must use pink more.

WHITE

To wear. If we need a greater inner knowledge. The use of white makes us more open minded, clear, and receptive to new ideas and plans without action. White gives us time to stop and think, to reflect without having to make a decision.

To avoid. If we are feeling lonely, isolated, or if we need to participate, act, or make decisions. When we are ready, we can use the yellow to promote communication and division. Pink can bring sensitivity to our lives. Blue can bring understanding and orange can promote action.

BLACK

To wear. If we are self-sufficient in authority and control, and we are in a position to protect. If we need to seek the advice or ideas of others, black protects us. We have to be in total control of ourselves, so that we can communicate an image of authority. We should always use black with touches of another strong color, in order to counterbalance any negative effects.

To avoid. If we are depressed or desperate due to lack of self-recognition and self-denial. Black rejects the help of others and promotes isolation.

4.13 CONCLUSIONS

ColorErg has already achieved important results, which go far from the literature review on the research subject because of the project's empirical phase. We are implementing a mixed research methodology based on user-centered design and participatory design, using survey methods, as well as direct observation, active research, supported by mechanical means, and quasi-experiments with sample groups of women.

Through recent researches, we know which areas of the brain are activated, as well as can measure the behavior of humans in what concerns color issues. So, an experiment with users is under development in order to check brain reactions to the different color dimensions, comparing the results with those obtained by the other previously used methods.

The main objectives focus on the acquisition of scientific knowledge in the area that can serve as a projective tool for fashion designers, as well as to contribute, through the dissemination of its results, as a reference to the use of color to users in general.

As expected results, we intend to achieve

- Systematization of scientific knowledge reusable by all within the interaction color/user, through clothing
- Guidelines for the use and application of color in clothing design projects, in fashion design

REFERENCES

Afonso, N. 1983. *Le sens de l'Art*. Lisboa: Imprensa Nacional-Casa da Moeda.
Arnheim, R. 1971. *El Pensamiento Visual*. Buenos Aires: Eudeba.
Birren, F. 1961. *Color Psychology and Color Theory*. New York: University Books.
Birren, F. 1969. *A Basic Treatise on the Color System of Wilhelm Ostwald–The Color Primer*. New York: Van Nostrand Reihold Co.
Birren, F. 1978. *Color and Human Response*. New York: Van Nostrand Reinhold Co.
Birren, F. 1982. *Light, Color and Environment*. New York: Van Nostrand Reinhold CO.
Cherry, C. 1957. *On Human Communication*. Massachusetts: The MIT.
Cheskin, L. 1947. *Colors: What They Can Do for You*. New York: Liveright Pub. Corp.
Chiazzari, S. 1998. *The Complete Book of Color*. London: Element.
Cirlot, J. E. 1958. *Dictionary of Traditional Symbols*. Barcelona: Ed. Luis Miracle.
Eco, U. 1967. *Estética e Teoria dell' Informazione*. Milano: Ed. Valentino Bompiani.
Ellinger, R. G. 1963. *Color Structure and Design*. New York: Van Nostrand Reinhold Co.
Frieling, H. 1990. *The Color Mirror: The Quicktest for Character Diagnosis with the Colors of the "Frieling-Test."* Berlin: Musterschmidt.
Gamito, M. and Silva, F. M. 2009. Cor no Mobiliário Urbano: um factor de Inclusividade, Orientação e Identificação (*Color on Urban Furniture: A factor of Inclusivity, Orientation and Identification*). In *5° CIPED (5th CIPED)*, Bauru: UNESP.
Goethe, J. 1989. *Theory of Colors*. Cambridge, MA: MIT Press.
Goldstein, K. 1942. Some experimental observations concerning the influence of colours on the function of the organism. *Occupational Therapy and Rehabilitation* 21, 147–151.
Graham, H. 1997. *Discover Color Therapy*. New York: Ulysses Press.
Guiraud, P. 1973. *A Semiologia*. Lisboa: Presença.
Hesselgren, S. 1987. *Hesselgren's Color-Atlas: Color Manual*. New York: T. Palmer.
http://www.johnpaulcaponigro.com/blog/136/the-luscher-color-test-online/ (accessed October 12, 2012).
Kandinsky, V. 1975. *Curso da Bauhaus*. Lisboa: Edições.
Lamb, T. and Bourriau, J. (eds.). 1999. *Color: Art & Science*. UK: Cambridge University Press (1st edition 1995).
Lancaster, M. 1996. *Colorscape*. London: Academy Editions.
Mahnke, F. 1996. *Color, Environment, and Human Response*. New York: John Wiley & Sons, Inc.
Mahnke, F. and Mahnke, R. 1993. *Color and Light in Man-Made Environments*. New York: Van Nostrand Reinhold.
Moles, A. 1971. *Art et ordinateur*, Paris: Casterman.
Moreira da Silva, F. 1999. *Colour/Space: Its quality management in architecture*. Unpublished PhD thesis, University of Salford.
Munsell, A. H. 1976. *The Munsell Book of Color; Munsell's Color Notations*. New York: MacBeth Division of the Koolmorgan Corp.
Pastoureau, M. 1997. *Black: The History of a Color*. USA: Princeton University Press.
Pignatari, D. 1976. *Informação, Linguagem e Comunicação*. S. Paulo: Ed. Perspectiva.
Rossignol, M. C. 2001. *O poder das cores: como influenciam o seu dia-a-dia (The Power of Colors: how they influence your daily life)*. Cascais: Pergaminho.
Sivik, L. 1997. *Generality Aspects of Color Naming and Color Meaning*. Department of Psychology, Göteborg: Göteborg University.
Tannembaum, P. H. 1966. Colour as a Code for Connotative Communication. *The Penrose Annual*. 59, pp. 115–120. London: Herbert Spencer.
Wright, A. 1998. *The Beginner's Guide to Color Psychology*. London: Color Affects Ltd.

5 Communication Efficiency and Inclusiveness in the Corporate Visual Identity

Daniel Raposo and Fernando Moreira da Silva

CONTENTS

5.1 INTRODUCTION

Much of the literature devoted to design sees man as the center of all design activities. From this perspective, design starts to meet a human need, contributing to the improvement of life quality in an ecological and sustainable way, while conciliating with commercial or market issues. However, the history of communication design and other visual arts is marked by stylistic variations from which images and brands representing corporations cannot be dissociated.

As happens in other areas, communication design can have a contribution to users' daily lives. The graphic designer should shape messages that belong to others to communicate them to specific audiences.

A theoretical body of work comprising the draft corporate visual identity as the consequence of a purpose and a pursuit of communication effectiveness is presented here. The literature review is focused on the design of the corporate visual identity signs, where, identified, the main arguments of drawing—which contribute to the effectiveness of visual identity communication and brand marks drawing—get special relevance as a way of making complex information more accessible, understandable, and easy to memorize.

Visual identity is presented as an artificial interface to mediate communication and interaction between two groups of people: those within the company and all stakeholders.

The graphic signs drawing process is discussed in view of its need to ensure the identification, differentiation, and articulation between the denotative and connotative meaning, according to human perception and understanding.

It analyzes several ways of creating semantic emphasis to enhance the sense of the message, influencing brand perceptions. These associations also influence the memorization, facilitate the recognition, and contribute to a significant visual experience.

5.2 PURPOSE OF THE CORPORATE VISUAL IDENTITY

It is a common error to mistake concepts such as identity and corporate image. For a matter of rigor, in this case, we understand that corporate identity is a set of intangible attributes assumed by the organization as their own, which constitute the "discourse of identity," and develops within organizations, as with individuals. It is the information used in the visual messages.

Corporate identity results from the corporate culture, which, according to Tajada (2008), is the dominant set of beliefs and values of the organization, corporate philosophy, standards, and characteristics of the working group's habits, traditions, and behaviors. But to Villafañe (1993), the definition of corporate culture is more evident; the author believes that it is the process of social construction of self-identity, that is, the appropriation of meaning (or new meanings). For the author, the identity comes from the history of the company, the business plan, and the corporate culture.

The concept of corporate culture can be subdivided into realistic, idealistic, and ambitious plans, which are related to the mission and conditions of identity and is a natural consequence from acting and communication between groups of people over time in a work context (Tajada, 2008).

In short, the corporate identity results from a set of not necessarily similar visions, in which each social subject is aware of what it is, a notion of how it intends to be seen. It is an ideological view that results from what the organization wants (on a real level), which are its immediate and/or projectual perspectives (Chaves, 1988).

Constituting itself as a way to think and act in a group, the corporate culture contributes to the formation of identity, that is, "… to the organization or parts of it have the sense of being as a consistent and specific being, that assumes his history and place in society" (Kapferer, 1991, pp. 30–31).

Regarding the corporate image/branding image, it is the result of different mental perceptions from stakeholders, about a specific reality, or the result of the brand communications, the level of satisfaction to what is offered, fulfillment of expectations, and experiences of different audiences.

The corporate image relates to an analysis by the audiences, the result of all the data obtained, and concerns the organization (which can cause different interpretations or images).

As seen, the perception of corporate visual identity is the result of a syntaxis, a set of connotations seen in one or more moments and gathered as one in the mind of one or a group of persons in such a way and with little awareness that contributes to the corporate image.

The vast majority of authors (e.g., Chaves, 1988; Costa, 2001, 2008a) who have written on this subject believe that the mere existence of a company is enough to create its corporate image, but the real problem resides in how to obtain the desirable one. From this perspective, when a company exists, it becomes publicly evaluated. The same authors emphasize the importance of a program to obtain the desired corporate image, which is the host of designer decisions during the design of the corporate visual identity system, eventually subdivided (adapted in content and codes) in different ways to communicate to different audiences.

In this sense, the corporate identity is a set of attributes assumed as their own by the organization, which constitutes the "identity discourse." It is developed inside the organization, as with an individual. Corporate identity is a complex picture, a set of visions that is not necessarily similar. Each social subject has a notion of what it is, and the way it pretends to be seen.

The corporate image is the ultimate goal of the communication design project, but does not refer to design or to graphic images, but first to the mental image that the audience made from a company or organization (Costa, 2004).

The designer's work affects the strategic choices and, particularly, the need to define, plan, and materialize the corporate identity through signs (only visual elements) in what is called by corporate visual identity. Referring to this principle, Zimmermann (1993, p. 11) understands that, "through the symbol or logotype, colors, typography: or the connection between a multiplicity of visual relationships among all these basic elements of Visual Identity, a company publicly shows its image and, simultaneously, his being."

We understand that the system of visual identity is the combination and relationship of the various signs with different natures and individual meanings to form a unique semantic meaning (Costa, 2004). Heskett (2005, p. 145) explains, "A system can be viewed as a set of interrelated, interacting, or interdependent elements forming, which can be considered as forming a collective entity."

The various signs of identity implicitly or explicitly carry the same brand sense, then by recognition, recall, and association allow you to create one global meaning. The identity system aims to obtain a pretended corporate image, as a result of an effect in a chain of actions that occurred in the social imaginary (Costa, 2009).

It is important to note that not all aspects of the identity system are visual, such as sound environments or the pronunciation of the brand name; the olfactory mark; the design and quality of products and services; interior design; uniforms; internal communication; communication and external relations; web and management 2.0.

Thereby, the corporate visual identity role includes all messages organized in a system according to a program designed to create a distinctive style and is capable of generating a corporate image that complies with the corporate identity (Costa, 2004).

This transformation of the corporate identity into the corporate image seems to have to overcome various obstacles to work. The program itself, or desire previously established, from which the strategy to guide the corporate visual identity results, complies both securities or corporate interests as well as market demands: "The company should aspire to conquer the leadership in the broadcast message but should not be forgotten the demand from the public neither the response of recipients corporate message" (Sanchis, 2005, p. 23). On the other hand, in the case of corporate visual identity, a receiver gets messages in different times and contexts, which makes the decoding depend first on its ability to interconnect the meanings properly, and second on the communicative quality of the developed identity system.

As Munari (1979) understands, once all receivers recreate the received messages, there is always a difference between the communication of the corporate visual identity and the corporate image, that is, between the message sent and its processing or its integration by the receiver. If this recoding of received messages is combined with some inertia in the system of corporate visual identity, it is possible to have an imbalance instead of the intended corporate image (Chaves, 1988).

Although the description seems to be a bit short, Llorens (1999) says that the programs of corporate visual identity are developed in two main stages: strategic and creative. In the strategic phase, values are identified as well as the identity of the company, continuing with the delimitation of the strategic axes to follow. The creative stage consists in coding visually the corporate values, according to the goal or strategic purposes (activity from which result several graphic objects that visually and publicly (re)present the company).

In the course of his work, the communication designer is faced with different needs and requests from the company. The designer is presented as the mediator that optimizes and adds value, in a humanization process that tries to ensure that the corporate visual identity fits the corporate program, and to the cultural profile of the audience, this is, "The designer encodes visual messages by translating the needs of the sender into images and content that connect with the receiver" (Hembree, 2011, p. 14).

In this regard, says Férnandez Iñurritegui (2007), the communication designer is to interpret and encode corporate message using graphic signs culturally common to the issuer (company) and receiver (audience).

Following the premises of communication design, the designer is the interface optimizer, the mediator between the corporate messages from the company to the audience. The designer's concerns about the corporate visual identity project are selecting and manipulating the intersubjective codes required for proper decoding in a given context. The corporate visual identity project refers to a system of integrated graphic signs that gain a new semantic meaning when drawn together. From the design's perspective, the expression of the signs of corporate identity determines the graphic-semantic positioning of the brand, according to notions of value (Raposo, 2014b).

Designing a brand is not a task left to chance. Brands have to ensure distinction, differentiation, and recognition. As stated by Zimmermann (1998, p. 84), as opposed to being fashionable, it is to share the same language with a group, "to have style it is to be unique, to be different." In the scope of corporate visual identity, the concept of style refers to "a quality or a characteristic way of expressing itself" (Schmitt and Simonson, 1998, p. 111), and with this way of becoming public it could use all the expressions of communication design.

In this sense, before starting the project of corporate visual identity, which is defined as a desirable corporate image, there is a process referred to variously as brand picture or brand personality.

Aaker (1997) proposed the "Dimensions of Brand Personality" as "the set of human characteristics associated with a brand" caused by direct or indirect contact between the audience and the brand (Plummer, 1985, quoted by Aaker, 1997, p. 347).

Kapferer (2012) considers that the definition of brand personality by Aaker (1997) is too broad to include any intangible attribute, as intellectual abilities, gender, and social class, and maintains the model of the "Brand Identity Prism (Kapferer, 1992)," which is based on knowledge of psychology, as well as on the design project.

In his proposal, Kapferer (2012) presents six dimensions that shape the personality and brand identity: (1) physical (objective characteristics, symbols, and attributes); (2) personality (subjective characteristics, character, and attitude); (3) culture (set of values that define the context in which the brand grows); (4) relation (beliefs and associations, and how they will interconnect with the public); (5) reflection of the public (creating value through stereotypes and aspirations of the public); and (6) self-image (the way the public sees itself and how this relates to the brand perception).

According to Azoulay and Kapferer (2003), the brand personality is a concept that consists in assigning a set of human characteristics to the brand, such as values, age, emotions, behaviors, attitudes, and beliefs. The brand personality fosters lasting relationships between the issuer and the audience, because it relies on the communication of identity on a symbolic level recognizable by the public, which seeks to identify objects with archetype, related with styles of life (Martins, 1999).

According to Zimmermann (1998, p. 84) in opposition to being trendy, which is about sharing a group language, to be similar to others, "having a style is to be unique, being different." In the visual identity program, "when it comes to style, we speak of a quality or a characteristic way, a specific way of expressing herself" (Schmitt and Simonson, 1998, p. 111), and by this way to create a public expression that should be coherent with the corporate behavior in general.

Consequently, as noted, the designer deals with the task of optimizing a message that does not belong to him and to seek its maximum efficiency. That is why personal style and trends can interfere with the semantic efficiency of communication design objects, specifically when they are reduced to these concepts.

For the same reasons, Davis (2005) refers that defining the brand style is an important decision that should capture the spirit of corporate identity, values, and personality. Consider that style is like the clothes of the company, it is used depending on the context in which it will be used, and depending on the personality of anybody who wears it. This idea is reinforced by Schmitt and Simonson (1998, pp. 111–112) for whom "styles perform several and important functions for the companies. They contribute to build the visibility and brand reputation; intellectual and emotional associations; distinguish products and services among others; setting relations of affinity; help distinguish varieties within product lines, adjust the marketing mix to different target markets."

Over the years a series of studies have emerged that aim to understand how the creation of meaning works in visual communication, which has often led to semiology, or to semiotics. Chaves (2003, p. 123) presents a possible explanation, considering that "harassed by the self phobia to the randomness, when the functional or technological factors will be absent, they will seek for explanations in other fields, such as, for example, the semantic. They will cling to the semiological science as a lifesaver to come as sort of sign technology."

We know that from the existing signs, graphic design uses symbols symbolically that are established and that vary with the culture and context of use. Resnick (2003, p. 123) considers that "symbolism is the term used to describe the art or practice using symbols. A symbol is a thing standing for or representing something else, especially a material thing taken to represent an immaterial or abstract concept."

Although it is common to find studies that do it, Costa (2008b) argues that semiotics is insufficient to explain the processes of graphic design, as it is a branch of study from linguistics, based in orality and in the written language, and not on images—"The language of images does not speak a purely intuitive lexicon. However, photography is a language without code." Costa (2008b, p. 70) supports its assertion explaining that there is an absence of a limited repertoire of signs known prior by the sender and recipient in images, and able to combine among them in the discourse to create meaning. The opposite happens with text, whose repertoires are the letters, the alphabet, words, and the grammatical rules. Images do not have such a limited repertoire of signs and are based on the representation or display of scenes or objects that constitute the instantaneous and overall structure of speech (Costa, 2008b).

The use of semiotics to understand the functioning of communication design should be limited to the common aspects between written and visual language, such as syntactic (the graphic shape of the sign, its denotation within a system or code), the semantic (symbolic value, the sign expressiveness, and connotation), and the pragmatic (legibility, contrast, differentiation, flexibility in use, perception, and comprehension of the sign in a given context). In the communication process, there is necessarily more than one person, at least the transmitter (which induces) and the receiver (which is induced and deduces). Although in both cases the deduction is conditioned by culture, it does not induce (Costa, 2008b).

This view is shared by Frascara (2008), who says that semiotics sits on an inflexible logic to human and cognitive behavior; for example, when it comes to a specific audience, or like rhetoric, because it only affects in the exposure of meanings, it lacks the data that we can find in sociology, psychology, or marketing: "Rhetoric and semiotics help, but are insufficient when it comes to building real and specific answers to real and specific audiences, relating to real and specific messages dealing with real and specific problems" (Frascara, 2008, p. 95). We know that everything has meaning and that not everything that conveys meaning communicates, that it always lacks the sense that allows decoding (Costa, 2011, p. 52).

Following the same line of thought, Acaso (2006, p. 25) writes that the visual language "in particular has little to do with the rest of the languages we know, since both the writing and the verbal are subject to specific rules, fully structured and defined." Also, the same author (ibid, pp. 27–28) clarifies that "The visual language (…) is the oldest semistructured communication system we know (…) the one that has the most universal character." For the author, the feature that more distinguishes the visual language from others is its resemblance to reality, and the many ways of representing itself and because their signs do not have a univocal meaning.

According to Dondis (1976, p. 25), there is a basic and common perceptual visual system to all human beings, who experience variations, for example, by culture. The author believes that it will never be possible to establish a precise system to the visual language as exists in the written language; it would be a necessary storage structure for encoding and decoding, a structure with "a logic that visual literacy is unable to reach." That is, the visual language does not have a signs repertoire or a universal and unique code. The visual language is composed of basic elements such as color, shape, letters, graphisms, proportions, textures, tones, images, and rhythms, each one with its own meaning and possibility to change or to be added to the other graphic signs to form a tone of voice or connotation (Bonnici, 2000, p. 76).

Research on the visual language, grounded in semiotics, can be dangerous because, as says Cloutier (1975, p. 103), "we must avoid extrapolating too systematic and the structural linguistic analysis can't be fully applied to the study of languages without language, those who do not have a precise code." But, Smith et al. (2005, p. xiii) go further and state that "in visual communication, however, there is no unifying theory, nor should there be, because the area represents the intersection of thought from many diverse traditions."

The different models of communication developed by various authors refer to a need to share signs allowing a common code, but this is rarely comprehensive, that is, "...this ideal situation, of complete congruence between the stock of signs on a coincidences field only exists in artificial languages" (Frascara, 2008, p. 96). By this way is understood that in the full role of signs, from sender and receiver, the communication process is only possible by a number of more or less shared signs, more or less understood by the same way (intersubjective signs to transmitter and receiver). These considerations are not incompatible with the existence of an effective communication design method, they just show the complexity of the process and the importance of the designer as author, as a mediator, or as an agent in a society that communicates.

Frascara (2008, p. 27) says that "design is an intellectual, cultural and social activity: the technological aspect belongs to a dependent hierarchy" (the author refers to the production and distribution). The same direction is pointed by Providência (2003, pp. 197–198), clarifying that "we understand that in design—drawing an artifact for cultural interaction—the drawing is one of the stages in the design process, the result of a desire that precedes its purpose, which is revealed as a technical thing but in its genesis, is poetic (...). The author, moved by the desire (feeling of absence, desiderium) intentionally builds a substitute (purpose) that fills the empty space of that desire. The desire creates the design, serving a finality (...) the author, to respond to the order (purpose) may assign a metaphor value, shaping his poetry."

That is why the designer must not assume his personal style in his work, since the code belongs to the sender and to the receiver—"Design must solve a problem and disappear in its solution, should not be the protagonist of the object. An object-design is a way to an end, because it has excessive details. And design, as Papanek says, should never be an end in itself" (quoted by Zimmermann, 2003, p. 70). Munari (2001, pp. 49–53) makes it very clear, referring that "... unlike the artist and stylist, the designer does not have a personal style to which can appeal to formally resolve his problems. What the true designer produces doesn't have aesthetic features allowing to characterize him."

To Kroehl (1997, p. 18), communication involves encoding in which a complex reality is simplified and transformed into messages appropriated to the context and culture, and again enlarged by the decoding process. This process to transform complex data into common information is the communication goal, and graphic design is the first way to grant its efficiency, being a true cultural interface.

5.3 BRAND MARKS: THE SUMMIT OF CORPORATE VISUAL IDENTITY

Often, the concepts of brand and mark are misunderstood. Throughout human history, the brand has received different purposes and its definition was modified into a series of graphic and intangible meanings that are the synthesis of what is a certain reality. On the other hand, the brand refers just to a graphic sign used to evoke something. The mark is the sign itself, while the brand is what is made present by invocation.

It is also important to distinguish the brand from the product (the reality behind the brand which gives the semantic direction) and from the brand image, which is broader and includes corporate reputation and mental images in audience minds (Raposo, 2012).

Brands are concepts that are developed artificially to look like a natural phenomenon. The brand's success occurs when it effectively becomes a natural phenomenon; that is, when the audiences appropriate it in a consistent manner with the corporate identity and the strategic corporate goals. That is why brands are developed in accordance with extended audience maps, seeking to correspond to expectations and desires compatible with its positioning. For this reason, the effectiveness of brand communication depends on its ability to become common and accessible to broad audiences in terms of age and culture, whether or not experiencing some degree of disability.

Depending on the desired name and positioning, the brand mark is developed to grant uniformity and distinction to all graphic demonstrations of the organization. The brand mark is the graphic sign used as a signature by the organization or brand to identify, distinguish, and relate the different media of visual communication. This graphic sign can be a logotype or a symbol, or even be both as a single element.

The brand mark fulfills the function of invoking something, to differentiate and identify, functioning as the signature that ensures the authorship or ownership. For this reason, after the name, the brand mark is the identity sign most used in visual communication, which is why it acts as a container of meanings. That is, beyond the denotative and connotative meanings assigned during its design, brand marks will get others, which are assigned by audiences depending on their personal experiences with the product, service, and communication. Thus, the brand mark acts as a brand ambassador, the most obvious sign of the corporate visual identity system, which it also integrates (Chaves and Belluccia, 2003).

Often it is considered that the brand is the only way to differentiate companies, products, or services with similar characteristics. In this regard, Beltrán (2014) states that such claims are a fallacy, because what happens is that the brand plays the role to distinguish features which previously were not so evident. Regarding the corporate identity, Beltrán (2014) considers that it is possible to establish a hierarchy between what he calls primary realities (those that are more evident, distinctive, and specific, but less frequent) and the secondary, or tertiary (which are less differentiated, less known, and more frequent), but all having content that may be relevant or negative for the brand (particularly when they are auditory, audible, or visual).

Kapferer (1991, p. 53) understands that the logo and the symbol "provide information about the personality and brand culture," and that its purpose is to ensure that the brand is recognizable through them.

The design of a program's visual identity is more than a cosmetic process. And for a similar reason, the designer should be aware that designing a new brand has different requirements than redesign. Change is possible, but it requires a more rigorous coordination during its publication process, and also a bigger effort in its resignification, because there is always resistance to change. In the case of the design of a new sign, one must create the context and disseminate narrative using specific graphics and articulation (Gernsheimer, 2008).

When it comes to a design project, without a starting point, there may be problems in terms of corporate visual identity. Immediately in the research phase, there may be lack of information about the corporate identity, the mission and functionality of the company, but also on its market and operational strategy. In some cases, the designer develops projects that do not correspond to the reality of company objectives; also some cases occur where the designer-style overrides the corporate identity or is limited just to follow trends, or when the CEO imposes a graphic solution, yielding differentiation problems.

In terms of redesign, there are several levels of change that go from a total rupture of the symbols and codes to a restyling or facelift, a formal amendment without changing the symbols. Cases of redesign occur when the company has reputation problems and wants a new repositioning; when the company's operations are expanded to other sectors or activities; when the corporate visual identity is obsolete or incompatible with current culture; when the designer rushes to undervalue the work of others to replace it with its own; when a new CEO decides to show his leadership, forcing a change in visual identity.

The design is a profound change in terms of denotative and connotative meanings. In turn, the redesign focuses particularly on denotative meanings.

According to Costa (2011), on visual identity projects, designers and clients tend to value the symbols more than the logotypes. However, symbols and logotypes are just two different ways to solve the need of a graphic identity sign, and we have a large number of successful and reputed logotypes that can demonstrate their efficiency. In general, the symbol is evident and has a more arbitrary nature than has the logotype. However, we forget that, in addition to its denotative value, the logotype is also connotative and its graphical expressiveness depends on the interest of the program (as we can see by comparing the logotype of Siemens with Coca-Cola).

Beltrán (2014, p. 29) argues that the effectiveness of corporate visual identity conforms to communication needs and the spread of the attributes of the reality in question, it is still essential that it "is not arbitrary, but associated with the representing reality or realities; and consequently all communications must be accurate and appropriate, in accordance with the objectives."

According to Rand (1985), the corporate visual identity is the design of the solution to a communication problem, which is why the designer's work is to seek simple solutions, subtle and attractive, to capture the eye's attention and contain surprise elements. In this sense, brand marks play a particularly important role as a global synthesis.

The brand graphic must be compatible, in semantic terms and graphics, with the corporate identity, which, in turn, corresponds to the corporate identity and the strategic objectives of the company, serving to enhance the qualities of a reality. In this sense, Beltrán (2014, p. 30) states that visually nothing should be left to chance in the brand mark, so "its appearance, its size, its color, its texture, its position must be subordinated to its function."

The effectiveness of the message is not guaranteed when the information arrives at the receiver; the correct decoding also depends on its strength to stand out, and the interest with which it is received, and from its attributed value—the message is filtered. The filter is symbolic and cognitive; Neumeier (2006, p. 34) believes that "the differentiation happens by the way the human cognitive system works. Our brain acts as a filter to protect us from a vast amount of irrelevant information that surrounds us daily." For the author, the visual cognition requires that human factors are considered (perceptual, visual, cultural, or symbolic), but also others such as aesthetics, which help to create differentiation and interest. It is important to clarify that "aesthetics is not necessarily associated with enjoyment, but more correctly with experience, which is one of feeling" (Jamieson, 2007, p. 92).

Designing a message and transmitting it properly requires that the designer knows the visual codes shared between issuer and receiver, and especially how to combine those signs, enabling them to create interest and condition the behaviors. Wheeler (2003, p. 20) says that "the design must be appropriate to the company, its target market, and the business sector in which it operates." Also, Costa (1980, p. 23) states, "The question is therefore to establishing an optimal coordination and coherence among all manifestations of the company, which has certainly an effect on the reputation of the business and its quality, i.e. its image."

In the context of brand design, drawing is not an end in itself, it is how to shape the message and the communication program. And "communication begins with the perception. Every perception is an act of finding meaning (…)" (Frascara, 2006, pp. 69–70). This quest for meaning leads to a general idea, a set of symbols, or attributes mentally assigned to the entity, the Corporate Image (Tajada, 2008).

According to Villafañe (1999, p. 68), "a visual identity program is a series of core elements regulated by a combinatorial code that sets the program itself." The elements of this repertoire are the brand marks (symbol, logo, monogram, etc.) and the identity communication media system, that is, the name, colors, graphisms, corporate typography, the layout, and its semantic articulation to create a specific style that will be applied to numerous types of objects (Chaves and Belluccia, 2003; Wheeler, 2003).

That is why Providência (2003, p. 201) states that "designers are interpreters of the world: and its artifacts are suprafunctional objects, that unlike engineering objects, they often present an ulterior motive or an 'artistic' value; but, on the other hand, and in these cases antagonistic to art, they don't abdicate to their integration into the mundanity of everyday and domestic things."

Villafañe (1999) points out that during the design of a visual identity program, the designer must ensure that it complies with four principles:

1. The need for the visual identity to be a synthesis of corporate identity, projected globally according to their reality and emphasizing the positive attributes, but without lying
2. The visual identity highlights the strengths of the project or the business strategy
3. The semantic consistency between the behavior and corporate culture with the visual identity and the direction of the communication
4. The integration of the plan of visual identity in the overall strategy of the company and financial plan according to the proposed corporate image

The graphical representation of one concept or object can assume different styles or tones (rigorous, realistic, simple, deviant, expressive, synthetic, etc.) and with them determine the semantic content. In this case, the graphic expression affects the semantic meaning. So, a coherent graphic style contributes to position the sign or can be used to give emphasis to specific corporate values and most important personality characteristics (Chaves and Belluccia, 2003).

To adapt to different audiences, it is possible that the corporate visual identity take styles/themes/tone of voice, which are customized characteristically to communicate the corporate identity according to a specific audiences' culture or requirement. This is the kind of tone or visual rhetoric adopted for the corporate communication and corporate visual identity system.

Visual identity styles can be organized into two opposing main ranges: the informality (visual dynamism, formal and chromatic contrasts, irregularities, open or unfinished effects, lush or trendy), and structural (proportioned, compactness, pregnant, regular, symmetrical, balanced, simple, contrasts, closed, fewer colored, and enduring). Yet there are a variety of possibilities and style intersections which can be used (Villafañe, 1999).

The notion of corporate credibility works as with people; that is, as consumers we prefer to interact with brands that seem the most trusted, which we regard as more professional, more competent, and understood in the subject—"The logo should serve as the credible voice of the company's graphics program. But, once again, just as in the case of a person, the logo must be a believable representation of the business it symbolizes to be effective" (Haig and Harper, 1997, p. 26).

Agreeing with the same idea, Doyle and Bottomley (2006, p. 115) hold opinions stating that "when people encounter a new brand, they necessarily rely heavily on what the brand is trying to signal about itself. One way a brand can do this is through the lettering it adopts."

The topic or the tone is the kind of narrative, which is selected to provide a specific visual identity style. According to Schmitt and Simonson (1998, p. 153), the topic or tone of voice is a "mental anchor point and, by specific reference," used to express characteristics of a company or brand to the public or to a segment of the audience of that corporation.

Writes Frascara (2006, pp. 23–31), "design is to coordinate a long list of human and technical factors, the invisible to visible, and to communicate" (...) "the designer essentially designs an event, an act in which the public interacts with the design and communication occurs. The purpose of the visual communication designer, then, is the design of communicational situations." In communication design, the messages are set (encoded) in accordance to a program of an emitter in order to be easily and correctly received (decoded) by the recipient, persuading his actions without harm to him.

We may conclude that the visual identity design acts as one of the major means of materialization and coding of corporate values. The designer has the task of interpreting and meeting corporate personality to give it a strategic direction through a global graphic language. To do this, the designer must know each corporation well to adjust the program to communication needs, because, as explained by Chaves and Belluccia (2003, p. 48), although each organization is unique, "Few organizations can reduce their communication to a single language: different themes and different audiences require dividing the corporate discourse in several rhetorics."

5.3.1 BRAND MARKS AS CULTURAL INTERFACES

Azoulay and Kapferer (2003) state that the brand personality is a concept that consists in assigning a set of human characteristics to the brand, such as values, age, emotions, behaviors, attitudes, and beliefs. The brand personality fosters lasting relationships between the issuer and the audience because it relies on the communication of identity on a symbolic level recognizable by the public, which seeks to identify objects with the archetypal, related with styles of life. To be really effective, all these complex and intangible dimensions need to become visible.

A communication system comprises selecting the graphic signs that are more appropriate and possible to coordinate to express a certain global meaning. In these communication processes there is an hierarchy order in which the graphic signs fulfill specific identification and differentiation functions, like the brand marks and colors, whereas the secondary ones are complementary and reinforce, clarify, or support the style, such as graphisms, texts, formats, or textures (Rand, 1993; Chaves and Belluccia, 2003).

Smith et al. (2005, p. 48) argued, "The eyes are, in fact, extensions of the brain into the environment. The last and most sophisticated of our senses to evolve, our eyes send more data more quickly and efficiently through the nervous system than any other sense." In this way, the eye is responsible for capturing the data to be perceived, that is, "perception, the process by which we derive meaning through experience, is a dynamic, interactive system that utilizes built-in genetic programming to synthetize sensory input, memory, and individual needs."

Therefore, and assuming he knows the current and common culture in a given context and time, the designer can consciously direct the semantic meaning of graphic signs to awake certain concepts compatible with the object and be seen as strategic in a specific market (Raposo, 2008).

In terms of drawing, the brand marks share many perceptive requirements with pictograms, but they are always a conventional result, are more emotional, and have a persuasive character.

In general, we can consider that brand marks require readability and contrast, that is, reading and readability of the name and the symbol (re)presented. But a brand mark also requires odd aesthetic qualities, whereas here the aesthetic function is to create differentiation, recognition, identity, and memorability (Heskett, 2005). Even when brand Mmrks are logotypes, they are words designed to be read and, above, to be viewed.

Perceived by the eye, the brand marks contain isolated meanings, which are decoded and expanded, when associated with other related graphics signs. The organizations become represented, identified, or recognizable by differentiated visual styles charged with meaning.

Parramón (1991) said that brand marks must be legible, memorable, graphically unique (original and different from all the others), and expressively related to the concept it represents. As discussed, the designer should clarify the concept of visual identity through a graphic style, shapes, or colors, which is possible because the "human being thinks visually. The images act directly on the perception of the brain, impressing first for be analyzed later, the opposite of what happens with the words" (Strunck, 2007, p. 52).

The way we memorize graphic signs is not always conscious, but apparently they settle in the audience's mind and they can be used for design purposes. Rögener et al. (1995, p. 14) say, "The subconscious reacts in a large variety of ways to even the most minimal stimuli. Sensory impressions are connected to symbols—that is to say, coded—and stored in the subconscious as experience. Symbols can activate any experiences at any given time and bring them forward to the conscious level."

Rand (1985, p. 7) says that "because graphic design, in the end, deals with the spectator, and because it is the goal of the designer to be persuasive or at least informative, it follows that the designer's problems are twofold: to anticipate the spectator's reaction and to meet his own aesthetic needs. (...) It is in symbolic, visual terms that the designer ultimately realizes his perceptions and experiences; and it is in a world of symbols that man lives. The symbol is thus the common language between artist and spectator."

From a theoretical point of view, any image, object, or concept may be the starting point for designing a brand mark, but, in practice, their semantics efficacy depends on a program and on a certain design in a cultural context and use. To this purpose, Rand (1985, p. 48) states that "visual statements such as illustrations which do not involve aesthetic judgements and which are merely literal descriptions of reality can be neither intellectualy stimulating nor visually distinctive. But the same token, the indiscriminate use of typefaces, geometric patterns, and abstract shapes (hand or computer generated) is self-defeating when they function merely as vehicle for self-expression"; the brand mark must be designed to be distinctive in an environment of use.

Formal synthesis is desirable for a brand mark, so that it contains only the necessary data to make it recognizable, contrasting, legible, artificial, individual, and memorable, as the eye prefers simplicity. Various laws of Gestalt theory teach that the eye pursues formal simplicity (Costa, 2011).

Associating these premises, the brand mark's graphical synthesis contributes to differentiation, memory, and fascination, especially when the design includes the exaggeration of attributes considered as most relevant and unique to materialize the visual identity program (Ramachandran and William, 1999; Strunck, 2007).

Finally, by excluding details, the drawing provided to brand marks is a new formal synthesis allowing them to be more flexible to a variety of media and in different sizes (Strunck, 2007).

The signs of identity are based on real objects or concepts, but some of their features are omitted, while others are exaggerated to graphically express and connote specific meanings. The brand marks design process can ensure the shape synthesis by accentuation or flatness of characteristics that best promote the recognition of the object or concept.

However, in the last few years, there have been many identity projects using three-dimensional or iconographic, or descriptive and realistic brand marks. Healey (2012, p. 12) states that it was the way designers found to offer something new to their customers or to follow trends, and he writes that "the logo also needs to be updated with the expectations of an increasingly sophisticated audience."

On the other hand, Rand (1985) believes that there are many complex symbols and images, or even objects, that have been transformed into symbols of high efficiency, as a result of their use in a systematic, coherent, and articulated way. Still, brand marks require simplicity that can be demonstrated in a simple blur test where its formal structure and key profile should resist. Besides, "a trademark, which is subjected to an infinite number of uses, abuses, and variations, whether for competitive purposes or for reasons of 'self-expression,' cannot survive unless it is designed with utmost simplicity and restraint—keeping in mind that seldom is a trademark favored with more than a glance. Simplicity implies not only an aesthetic ideal, but a meaningful idea, either of content, or form, that can be easily recalled" (Rand, 1985, p. 34).

The need to ensure that the brand mark has a recognizable structure and a profile compatible with a specific meaning is located in the mental repertoire of the receiver. According to Joly (2008, p. 20), "The mental image is distinguished from mental scheme, which combines enough and necessary visual traits to recognize a drawing or a necessary visual form. This is a perceptual object model, from a formal structure that we have internalized and associate with an object, which can be evoked by some minimum visual features." This is to say that mental images are a more complex and intersubjective phenomena, because they are a specific form of internal representation, with an associative cognitive, prolonged when compared with other similar perception forms. Carrieras and Codina (1992, p. 52) explained that "the mental image is obtained according to an amodal perceptual process. The term 'amodal' has been established following several studies made on congenitally blind people, who proved that a mental image is not uniquely based on visual perception."

From the perspective of psychology, the theory of geons (geometric ions) presented by Irving Biederman (1987), and according to which there are at least 36 geometric components made up with three-dimensional shapes (such as cylinders, cones, pyramids, etc.) stored in our mind as structural descriptions, that is, with these forms all objects can be perceptually decomposed, identifying the most shared structures between the object and geometric figures.

In this sense, Rand (1993, p. 58) highlights that drawing a brand mark with a complex shape or trying to express what the company does with an illustration "will only make identification more difficult and the 'message' more obscure. A logo, primarily, says who, not what, and that is its function. Its effectiveness depends on distinctiveness, visibility, adaptability, memorability, universality, and timeleness." So, the subject matter in the symbolic origin of the brand mark depends on the brand personality, the corporate program, the audience culture, the competitors and market rules, the media opinion, and in the society interests.

However, brand marks "effectively distill a great deal into a concise symbol that is ideally attractive, cohesive, conceptual, distinctive, enduring, legible, memorable, relevant, sophisticated, and versatile: the ten characteristics of a great logo," argues Gernsheimer (2008, p. 19). In this way brand marks must have fascination power created by the use of surprising features; a clear visual hierarchy in its components and a semantic reinforcement; the different elements must form a structured group, the symmetry and the use of stylistic features related to the subject, contributing to build attractiveness; nevertheless, meeting the cultural codes of an era and time is recommendable and should not follow trends or be ephemeral; readability, comprehension, and contrast positively help in the process of memorization; the sign must be flexible and have a clear structure to ensure the recognition, regardless its size or medium in which it has been used (Gernsheimer, 2008; Hardy, 2011).

Besides the shape, color also plays an important role in the recognition process, and particularly to create secondary meanings and for brand mark memorization. According to Wheeler (2003), the sequence of cognitions starts by acknowledging the shape, and after comes the color meanings and secondary associations, and then only after the content decoding. This reveals the importance of color to corporate visual identity.

The subject of color is vast and complex. As a matter of rigor we will focus on aspects concerning corporate visual identity. As Mollerup (1987) says, the color may even be part of the verbal identity (name) like "Yellow Cab Corporation" or "Yellow Pages." Or, in this case, although not present in graphical form, color is invoked by the word.

The selection of a color for an identity project may be related to the corporate culture, with a symbolic reality, psychology, economics, or technical limitations, to the need to distinguish themselves from competitors or with the intention to join an existing code in an activity sector, also with aesthetic reasons from designers or their customers, fashions, or epochs, among others (Wheeler, 2003).

Moreira da Silva (2013, p. 80) studied color as an inclusiveness factor and states that selection of the color may occur "on a number of factors: for their emotional appeal, their implied meaning, their match to a certain printed brochure, or because the designer (or the client) likes a certain colour."

Heller (2005, p. 18) states that "there is no color without meaning," but stresses that its effect is determined by its context of use and perception. The approach of Fraser and Banks (2004) is consistent and added that the meaning of the color depends on the culture of the perceiver.

Color can be used to evoke emotions, associations, differentiate, or increase the relationship between brand and subbrands—"While some colour is used to unify and identify, other colours may be used to clarify a brand architecture, through differentiation products or business lines" (Wheeler, 2003, p. 128).

There appears to be a relationship in some sectors between the corporate color and the represented product or market where it operates, such are the color brown for sausages or the color green for pharmacies in Portugal. Through associations it is possible that a certain color helps to identify new attributes in a brand, for example, wealth, feminine or masculine, freshness, elegance, nature or technology, energy, and so on. The color in the corporate visual identity is strongly connected with the cultural context and the market in which the corporation operates. O'Connor (2011) refers to "environmental color mapping" as the process by which you select and identify the characteristics of a color in a given context.

To ensure that the colors are properly perceived also for individuals with low vision, particularly for the elderly, Moreira da Silva (2013, p. 164) considers the need of ensuring the contrast, the size and space for eye comfort, but also that "older individuals have a harder time distinguishing between colours in the cooler range—blues and greens particularly. Also, some individuals are colorblind and find it difficult to distinguish between red and green. Therefore, colour is probably not appropriate as the sole differentiating feature between different elements—they should vary in others features as well. Varying the value of colours (the lightness or darkness) by at least two levels enable most people to differentiate between the colours."

Good design should include a considered relationship between the shapes and colors based on a criteria and code that have to remain consistent in all the corporate visual identity systems (Martins and Moreira da Silva, 2012).

5.3.2 Brand Marks Denotative and Connotative Meaning

The correct understanding of the brand mark depends on the semantic level resulting from an inter-subjective redundancy between the denotative and connotative meaning.

Denotation refers to the meaning base and concrete, while the connotation corresponds to subjective, figurative, or symbolic meanings that work by association and are beyond the denotative meaning.

Blanchard (2003, p. 36) states that "the connotation is an extension of the meaning by the receiver, by interpreting the context created by the sender, in accordance to its own culture, allowing him to perceive what has not been mentioned in words, through secondary associations."

Mollerup (1987) refers to this principle when he argues that brand marks produce different types and meaning levels, induced by the graphic expression of the sign; that is, from the different levels of meaning that depend on the connotation, resulting from visual style. A metaphorical brand mark that reinforces graphically what it stands for has more meaning levels.

The connotation results from secondary associations created by the graphic symbol in a given culture or context or by the graphic expressiveness. In terms of graphic expression, we refer to the basic elements of any visual message: point, line, direction, tone, color, texture, scale and proportion, movement, space, reality, and appearance (Dondis, 1976).

We know that graphic signs can take different levels of iconicity, depending on their formal relationship with reality between two extremes (ranging from hyperrealism to the schematic representation). The scale of iconicity by Morris and Hamilton (1965) has had many developments promoted by various authors, which will not be included in this discussion. Costa (1990, adapted from his 1989 proposal) explains that the connotative value increases proportionally to the schematic level of the graphic signs, but also its need to be fixed by agreement. But there are many possible ways to draw a schematic sign. Therefore, just not the same object, but also the same structural shape can take different styles or graphic expressions and each one will tend to create distinct connotations. This is different levels of iconicity, from iconography through the geometric synthesis into abstraction, correspond to distinct secondary meanings.

Referring to the design of graphic signs, based on objects or concepts, as proposed by Resnick (2003, p. 123), the meaning of brand marks can be divided into different types of connotation as analogies, metaphors, or puns:

> Analogy, the term for a description derived from a process of reasoning from a parallel or similar case, explaining what unlike things share in common.
>
> Metaphor is a figure of speech in which one thing is compared to another to suggest a likeness or analogy between them.
>
> A pun is the humorous use of a word or image to suggest alternative meaning, a play on words with more than one meaning.

In a similar way, Villafañe (1999, pp. 89–90) also established a set of denotative and connotative possibilities that Brand Marks can assume in function of their iconicity and expressiveness:

1. Analogical association (creating a relationship, a description of a concept which becomes similar)
2. Allegorical association (recognizable elements of reality combined in an original or unusual way)
3. Logical association (follows a pattern of signs already in place)
4. Emblematic association (the appropriation of external positive values)
5. Symbolic association (adding emotional content)
6. Conventional association (no attempts to highlight any particular attribute, is especially for identification)

In a similar way, Oejo (2000, p. 170) also refers that brand marks can empower secondary associations, and presents these categories:

1. Analogy (the graphical representation resembles the object)
2. Allegory (combination of recognizable elements)
3. Logic (descriptive of the activity or business)
4. Flagship (heraldry or other institutionalized meanings)

5. Symbolic (use of ideological elements), convention (elements whose meaning is agreed)
6. Contiguity (elements that are as a whole)

From the perspective of neuroscience, the authors Ramachandran and William (1999) present a subset of eight principles or laws underlying all the diverse manifestations of human artistic experience, that they divide into

1. Peak shift (fascinating power created by the shape)
2. Isolation and allocating attention (emphasis or isolation of individual components to give them more visibility)
3. Perceptual grouping and binding (ability to distinguish figure and ground perceptually grouped in an environment)
4. Contrast (distance or approach between the shape and the environment)
5. Perceptual problem solving (the shape contains an element of surprise, which is not entirely obvious or common, which affects its capacity of fascination)
6. Abhorrence of unique vantage points (particular and unique point of view from which one looks and designs the sign)
7. Metaphors (use of metaphors or graphic analogies)
8. Symmetry (aesthetic notion of well-being and equilibrium)

Talking about graphic signs perception, Spiekermann and Ginger (2003, p. 39) explain that all observers formulate an opinion or mental idea of the message based on the first look, even if just lasts a split second. "In other words, an overall impression is created in our minds before we even read the first word. It's similar to the way we respond to a person's presence before we know anything about him or her, and then later find it difficult to revise our first impression."

However, as we have noted, the language of symbols is only one possibility to create the identity system. It was also noted that the creation of connotations begins with issues of expression and graphic style. "Unlike logos or brand names, typography may not trigger the usual suspicion or defence mechanism in the consumer. In other words, typography contains a subtle message or soft power, operating in the realm of the subconscious. (...) We see that the font not only carries information or rational meaning, but send other, subtler messages by way of font shape's characteristics. This, clearly, is where identity surfaces—through the spirit of the letter" (Spiekermann and Ginger, 2003, p. 45). It should be noted that some graphic signs, such as typography, do have a great emotional weight that goes far beyond the merely denotative meaning, as its double nature of being a symbolic and graphic sign (Montesinos and Hurtuna, 2004).

5.4 CONCLUSIONS

Following the premises of graphic design, the designer is the indirect mediator of the corporate message. The designer focuses on the corporate visual identity project, selecting and manipulating the intersubjective codes required for proper decoding in a given context.

The visual identity project refers to a system of integrated brand marks that gain a new semantic meaning when drawn to express certain global meaning.

From the perspective of design, brand marks are identity signs that influence the embodiment of the graphic-semantic positioning in the visual identity project, but there is no data on the graphic capabilities of the letter in influencing the definition of corporate image.

When the designer develops projects disregarding corporate reality or his intention focuses on his personal style, rather than design, he develops styling. It is a deformation of the semantic meaning of corporate visual identity and interference in its effectiveness, since it forgets the corporate image to be created.

As observed, the efficiency of brand marks depends on its adjustments to human factors to promote correct understanding, such as cultural codes and perception requirements.

From different perspectives and fields of knowledge, we observed how authors consider that the efficiency of brand marks to be recognizable relies on its need to have a well-defined structure and profile compatible with a specific mental scheme.

On the other hand, the graphic expression or style given to this same brand mark can increase the number of secondary associations as a result of the comparison with mental images.

The different possibilities to draw connotations are approaches to establish connections between sender and receiver, creating the audience's interest by giving them something more fascinating and unreal. To do this, designers must consider the different levels on the scale of iconicity as well as a better graphic style.

Visual identity is the leading factor during the formation of the corporate image, that is, the set of perceptions and mental images created by the public for positioning a company compared to others.

The complexity of the visual language allows the designer to develop a vast number of meanings to be used in different cultural and market contexts. But when there is a program to accomplish, it is fundamental to seek the appropriate signs, and to coordinate them in accordance to the corporate purpose and human factors of the audience. Therefore, it is not possible to use graphic signs randomly.

The meaning of signs is mutable because meanings belong to people, not the graphic shapes. But graphisms express certain types of denotations and connotations able to articulate the mental images of that which live in individuals. The corporate image results from this mediated social process in which corporate identity is the message content, that is, of the visual identity.

REFERENCES

Aaker, A. L. 1997. Dimensions of brand personality. *Journal of Marketing Research*. 34 (3): 347–356.
Acaso, M. 2006. *El lenguaje Visual*. Barcelona: Ediciones Paidós Ibérica, S.A.
Azoulay, A. and Kapferer, J. N. 2003. Do brand personality scales really measure brand personality? *Journal of Brand Management*. 11 (2): 143–155.
Beltrán, F. 2014. *Félix Beltrán Diseñador Gráfico*. Valencia: Ediciones Vuelta del Ruiseñor.
Biederman, I. 1987. Recognition-by-components: A theory of human image understanding. *Venue: Psychological Review*. 94 (2): 115–147.
Blanchard, G. 2003. *Aide au choix de la typo-graphie*. Cours supérieur. Reillanne: Atelier Perrousseaux.
Bonnici, P. 2000. *Linguagem Visual: O misterioso meio de comunicação*. Lisboa: Destarte.
Carrieras, M. and Codina, B. 1992. Spatial cognition of blind and sighted: Visual and amodal hypothesis. *European Bulletin of Cognitive Psychology*. 12 (1): 51–78.
Chaves, N. 1988. *La Imagen Corporativa Teoria e metodología de la identificación institucional*. 1st ed. Barcelona: Editorial Gustavo Gili, S.A.
Chaves, N. 2003. El diseño ni arte ni parte. In *Arte¿?Diseño: Nuevos capítulos en una polémica que viene de lejos*, ed. A. Calvera, pp. 119–138. Barcelona: Editorial Gustavo Gili.
Chaves, N. and Belluccia, R. 2003. *La marca Corporativa: Gestión y Diseño de símbolos y logotipos*. Buenos Aires: Paidós.
Cloutier, J. 1975. *A era do Emerec ou a Comunicação Audio-scripto-visual na hora dos self-media*. Lisboa: ITE.
Costa, J. 1980. *La puesta a punto de La Imagen de empresa en las entidade financieras*. Documentos del CIAC—Imagen de Empresa—Comunicación: Mass media, diseño, casos práticos. Barcelona: CIAC.
Costa, J. 1990. Los recursos combinatorios del grafismo functional. In *Grafismo Funcional. Enciclopedia del diseño*, eds. A. Moles and L. Janiszewski, pp. 121–154. Barcelona: Ediciones CEAC.
Costa, J. 2001. *Imagen Corporativa en el siglo XXI*. 1th ed. Buenos Aires: La Crujía Ediciones.
Costa, J. 2004. *La Imagen de Marca*. Barcelona: Paidós Diseño.
Costa, J. 2008a. *En torno a los 60 años de la Ciencias de las Comunicaciones*. Lição magistral. Barcelona: Universitat Abat Oliba CEU.
Costa, J. 2008b. La fotografía creativa: las tecnologías de la ficción; el color en la fotografía. *Las variables creativas; el universo de la imagen*. México: Editorial Trillas.

Costa, J. 2009. Diseño y Creación de Conceptos y Empresariales Innovadores. In *Cuadernos de Diseño. Diseño, innovación, empresa*, ed. F. Jarauta, pp. 149–166. Barcelona: Instituto Europeo di Design.

Costa, J. 2011. *Design para os Olhos: Marca, Cor, Identidade, Sinalética*. Lisboa: Dinalivro.

Davis, M. 2005. *More than a Name: An Introduction to Branding*. London: Ava Academia.

Dondis, D. 1976. *La Sintaxis de la Imagen*. Barcelona: Editorial Gustavo Gili.

Doyle, J. R. and Bottomley, P. A. 2006. Dressed for the occasion: Font-product congruity in the perception of logotype. *Journal of Consumer Psychology*. Elsevier/North-Holland, Amsterdam 2 (16): 112–123.

Férnandez Iñurritegui, L. 2007. *El discurso del Diseño: Los Signos de Identidad como textos visuales*. 1th ed. Bilbao: Universidad del País Vasco/Gobierno Vasco.

Frascara, J. 2006. *El Diseño de Comunicacion*. Buenos Aires: Ediciones Infinito.

Frascara, J. 2008. *Diseño gráfico para la gente: Comunicaciones de masa y cambio social*. Buenos Aires: Ediciones Infinito.

Fraser, T. and Banks, A. 2004. *Designer's Color Manual: The Complete Guide to Color Theory and Application*. San Francisco: Chronicle Books.

Gernsheimer, J. 2008. *Designing Logos: The Process of Creating Symbols that Endure*. New York: Allworth Press.

Haig, W. L. and Harper, L. 1997. *The Power of Logos: How to Create Effective Company Logos*. New York: International Thomson Publishing Company.

Hardy, G. 2011. *Smashing Logo Design*. Chichester: John Wiley & Sons, Ltd.

Healey, M. 2012. *Design de Logótipos*. São Paulo: Rosari.

Heller, E. 2005. *Psicología del color: Cómo actúan los colores sobre los sentimientos y la razón*. Barcelona: Gustavo Gili, S.A.

Hembree, R. 2011. *The Complete Graphic Designer: A Guide to Understanding Graphics and Visual Communication*. Massachusetts: Rockport Publishers.

Heskett, J. 2005. *El diseño en la vida cotidiana*. 1st ed. Barcelona: Editorial Gustavo Gili.

Jamieson, H. 2007. *Visual Communication: More than Meets the Eye*. Bristol: Intellect Books.

Joly, M. 2008. *Introdução à análise da imagem*. 12th ed. São Paulo: Papirus Editora.

Kapferer, J. N. 1991. *Marcas: Capital da Empresa*. Lisboa: Edições Cetop.

Kapferer, J. N. 1992. Strategic brand management: New approaches to creating and evaluating brand equity. New York: Free Press and London: Kogan Page.

Kapferer, J. N. 2012 *The New Strategic Brand Management Advanced Insights and Strategic Thinking*. 5th ed. London: Kogan Page.

Kroehl, H. 1997. *Communication Design*. Zürich: ABC 2000.

Llorens, C. 1999. *Identidad Corporativa e imagen de marca*. IPMARK—Información de Publicidad y Marketing, Barcelona.

Martins, J. 1999. A Natureza Emocional da Marca: como encontrar a imagem que fortalece sua marca. São Paulo: Negócio Editora.

Martins, D. R. and Moreira da Silva, F. 2012. Perceptive and ergonomics concerns in corporate visual identity. In *Advances in Usability Evaluation Part II*, eds. F. Rebelo and Marcelo M. Soares, pp. 614–623. Boca Raton, Florida, EUA: CRC Press—Taylor & Francis Group.

Mollerup, P. 1987. *The Corporate Design Programme*. European/EEC Design Editions. Copenhagen: Danish Design Council.

Montesinos, J. L. M. and Hurtuna, M. M. 2004. Manual de tipografía: del plomo a la era digital. Valencia: Campgràfic Editors.

Moreira da Silva, F. 2013. *Colour and Inclusivity: A Visual Communication Design Project with Older People*. Lisbon: Caleidoscópio.

Morris, C. and Hamilton, D. J. 1965. Aesthetics, signs, and icons. *Philosophy and Phenomenological Research*. 25 (3): 356–364.

Munari, B. 1979. *Design e Comunicação Visual*. Lisboa: Edições 70.

Munari, B. 2001. *Artista e Designer*. Lisboa: Edições 70.

Neumeier, M. 2006. *The Brand Gap: How to Bridge the Distance between Business Strategy and Design*. Berkeley: AIGA.

O'Connor, Z. 2011. Logo colour and differentiation: A new application of environmental colour mapping. *Color Research & Application*. Wiley Periodicals, Hoboken, New Jersey. 31 (1): 55–56.

Oejo, E. 2000. Firmas Maestras: La Imagen Gráfica de La Marca. In *La fuerza de la publicidad: Saber hacer buena publicidad. Saber administrar su fuerza*, ed. M. Moliné, pp. 168–175. Aravanca: McGraw-Hill, Universidad Antonio de Nebrija.

Parramón, J. M. 1991. *Lettering & Logotypes*. New York: Watson-Guptill Publications.

Plummer, J. T. 1985. Brand personality: A strategic concept for multinational advertising. *Paper presented to the AMA Winter Marketing Educator's Conference*, Young and Rubican, New York.

Providência, F. 2003. Algo más que una hélice. In *Arte¿?Diseño: Nuevos capítulos en una polémica que viene de lejos*, ed. A. Calvera, pp.195–214. Barcelona: Editorial Gustavo Gili.

Ramachandran, V. S. and William, H. 1999. The science of art: A neurological theory of aesthetic experience. *Journal of Consciousness Studies*. 6 (6–7): 15–51.

Rand, P. 1985. *A Designer's Art*. New Haven: Yale University Press.

Rand, P. 1993. *Design, Form and Chaos*. New Haven: Yale University Press.

Raposo, D. 2008. *Design de Identidade e Imagem Corporativa: Branding, história da marca, gestão de marca, identidade visual corporative*. Castelo Branco: Edições IPCB.

Raposo, D. 2012. La letra como signo de identidad visual corporative: Codificación y decodificacón del sistema de identidad. PhD thesis, Faculdade de Arquitetura da Universidade de Lisboa, Portugal.

Raposo, D. 2014a. Dynamics on brand design and communication. In *What's On: Cumulus Spring Conference*, eds. J. Sampaio and T. Franqueira, pp. 67–77. Aveiro: Aveiro University. Retrieved from http://cumulusaveiro2014.web.ua.pt/wp-content/uploads/2014/12/CAV14_00.pdf.

Raposo, D. 2014b. The graphic speech of the brand. In *Proceedings of 9th Conference of the ICDHS— International Committee for Design History and Design Studies—Under the Theme Tradition, Transition, Trajectories: Major or Minor Influences?* eds. H. Barbosa and A. Calvera, pp. 445–450. Aveiro: Aveiro University.

Resnick, E. 2003. *Design for Communication: Conceptual Graphic Design Basics*. New Jersey: John Wiley & Sons, Inc.

Rögener, S., Jan Pool, A., and Packhäuser, U. 1995. *Branding with Type: How Type Sells*. California: Adobe Press.

Sanchis, J. L. 2005. *Comunicar con êxito: Teoria y pratica de la comunicación*. Barcelona: Ediciones Gestión 2000.com.

Schmitt, B. and Simonson, A. 1998. *Marketing y estética: la gestión estratégica de la marca, la identidad y la imagen*. Bilbao: Ediciones Deusto.

Smith, K. L., Moriarty, S., Kenney, K., and Barbatsis, G. 2005. *Handbook of Visual Communication: Theory, Methods, and Media*. 1th ed. London: Routledge.

Spiekermann, E. and Ginger, E. M. 2003. *Stop Stealing Sheep & Find Out How Type Works*. 2nd ed. California: Peachpit Press.

Strunck, G. 2007. *Como criar identidades visuais para marcas de sucesso*. 3rd ed. Rio de Janeiro: Rio Books.

Tajada, L. Á. S. d. 2008. *La auditoria de la imagen de empresa: métodos y técnicas de estudio de la imagen*. Madrid: Editorial Sintesis.

Villafañe, J. 1993. *Imagen positiva: Gestión estratégica de la imagen de las empresas*. Madrid: Ed. Pirámide.

Villafañe, J. 1999. *La gestión profesional de la imagen corporativa*. Madrid: Ed. Pirámide.

Wheeler, A. 2003. *Designing Brand Identity: A Complete Guide to Creating, Building, and Maintaining Strong Brands*. New Jersey: John Wiley & Sons, Inc.

Zimmermann, Y. 1993. *Zimmermann Asociados*. Barcelona: Gustavo Gili, S.A.

Zimmermann, Y. 1998. *Del Diseño*. Barcelona: Editorial Gustavo Gili.

Zimmermann, Y. 2003. El arte es arte, el diseño es diseño. In *Arte¿?Diseño: Nuevos capítulos en una polémica que viene de lejos*, ed. A. Calvera, pp. 57–74. Barcelona: Editorial Gustavo Gili.

6 Aspects of Usability Assessment in the Multicultural Approach

Maria Lúcia L. R. Okimoto, Klaus Bengler, and Cristina Olaverri Monreal

CONTENTS

6.1 INTRODUCTION

Culture plays an increasing role in interaction, acceptance, and learnability, especially of digital products. The trend is mainly in the rise of digital products in several applications that require input from the user. In order for products to be successfully marketed, product designs should accommodate users' cultural differences. Considering these aspects, various authors pointed out the need for studies in cultural usability (Chua et al., 2005; Evers and Day, 1997; Guan et al., 2006; Marcus et al., 2003; Nisbett and Masuda, 2003; Ono, 2006). The importance of cultural usability is growing with the increasing numbers of different national and ethnic groups that use information technology on a daily basis. Confirming the importance of culture on the concepts is given. In the ease operation of everyday products, the cultural factor is described as an important established element (DIN/ISO 20282-1 – BSI/ISO 20282-1:2006, 2006).

6.2 THEORIES ON THE SUBJECT

6.2.1 CULTURE CONCEPTS

First, it is important to clarify how the term "culture" is understood for research in a usability context. Culture is defined as a phenomenon which is essentially dynamic and intimately linked to the process of social and economic development of a society (Ono, 2006). According to another viewpoint, researchers believe that there is a causal chain running from social structure to social practice to attention and perception to cognition. This concept is being applied within the Nisbett's theory, based on logics versus dialectics and a cognitive perception (Chua et al., 2005; Nisbett and Masuda, 2003). Cultural models of use (CM-U) theory as opposed to psychophysiological approaches were proposed centered on social-cognitive approaches to usability by the authors in Clemmensen (2009), Frandsen-Thorlacius et al. (2009), Shi and Clemmensen (2007). Hofstede's cultural model (Barber and Badre, 1998) can help identify some main elements for the structural analysis of a cultural context through the parameters: PD—power distance, CI—collectivism × individualism, FM—feminine × masculine, UA—uncertainty × avoidance, and CO—confusion × orientation. Elements of this model can also be used separately, for example, the power distance was used to evaluate the cultural effect on structured interviews (Vatrapu and Pérez-Quiñones, 2006). Another method used is the culturability inspection method (CIM), in which identifying cultural markers are applied to the summative usability of software (Nisbett and Masuda, 2003). Cultural diversity in industrial design has been identified by symbolical, practical, and technical products requirements (Ono, 2006). Another conceptual model proposed is based on design preferences and interface acceptance: "Modified Technology Acceptance Model" (Shi and Clemmensen, 2007).

6.2.2 CULTURE AND COMPLEX SYSTEM

Culture is understood as a complex concept that can be both a structure and a process (Evers and Day, 1997). From this perspective, we can consider that usability, in a cultural context, is derived from a complex system. As a result, these cultural variables should be treated as elements of a complex system (Bertalanffy, 1976). In a complex system, understood according to the general system theory (GST), the user (as a living organism) is in constant transformation in the universe, changing and altering the environment and itself (Bertalanffy, 1976). We can observe various elements concerning a complex system (Barber and Badre, 1998; Clemmensen, 2009; Shi and Clemmensen, 2007) using previous culture and usability studies. The most important elements used to describe this complex system are cognition and perception. Cognition and perception vary between different cultures, and cultural practices encourage and sustain certain kinds of cognitive processes, which then perpetuate cultural practices. Following GST (Bertalanffy, 1976), it is correct to say that a user's cognitive process can, at the same time, perpetuate and turn continuously. When seeing culture and usability as a complex system, we need to look for all elements, as GST is a general science of wholeness.

First, we need to discuss the following question: Which variables are relevant to understanding the cultural aspects relative to usability context? To answer this question, we have collected associated studies on culture and usability that have been published in scientific congresses or journals in the past 15 years. Our goal is to provide selected usability studies associated with cultures from different languages, especially those cases applied to product design. In the next step, variables and methods previously used to assess culture and usability context were identified. After reading the articles, we characterized the usability research into practical elements in order to classify them in a certain design group with suitable characteristics. We consider two types of usability evaluation, summative and formative (Tullis and Albert, 2008). The formative usability test is used to identify or to diagnose the conceptual design (project phases), while the summative usability test is applied to check the finished product or part of it (prototype or product in market).

Table 6.1 shows an overview of the cited studies, including main characteristics and possible applications of the method/technique to acquire knowledge about intercultural aspects on usability. The concepts of three studies are most appropriate for summative usability, another three for formative usability. Another four studies have concepts which can be applied to both formative and summative usability tests.

Identifying the type of knowledge involved in the variables is very helpful to executing an intercultural usability test. The knowledge theory divides knowledge into two separate parts: tacit and explicit knowledge, (Nonaka and Takeuchi, 1995). Tacit knowledge is described as acquired knowledge, for example, when someone takes a decision, and they cannot describe why and what they did. Contrarily, explicit knowledge can easily be written down and transferred to other persons. Table 6.2 shows grouped variables found in the literature and shown in Table 6.1. We define the

TABLE 6.1
Aspects on Cultural Usability Research

Author	Characteristics	Type Usability Test
Barber and Badre (1998)	Inspection/collect remote information	Summative usability
Ono (2006)	Inspection/interviews with industrial designer	Formative usability
Nisbett and Masuda (2003)	Theory/discussion—cultural differences in attention and perception	Summative usability
Chua et al. (2005)	Experimental	Summative usability/learnability
Clemmensen (2009)	Theoretical/conceptual/structural model-usability	Theoretical basis (formative and summative usability)
Tholacius et al. (2009)	Research perception/satisfaction questionnaire	Formative usability
Evers and Day (1997)	Perception of the design elements with usability elements	Formative and summative usability
Shi and Clemmensen (2007)	Subjective aspects preparation/definition of usability testing	Structural elements (summative/formative)
Vatrapu and Pérez-Quiñones (2006)	Experimental/subjective aspects involved in usability testing	Formative and summative
Olaverri Monreal et al. (2014)	Tools to implement cultural factors in the design	Formative usability/HTA

TABLE 6.2
Researchers' Variables Found on Culture and Usability Context

Explicit Knowledge	Tacit Knowledge
Cultural Marks Knowledge Metaphors; specific icons; specific colors; grouping; language; geography; orientation; sound; font; links; regions; shapes; architecture; cultural diversity in industrial design, in relation to symbolical, practical, and technical requirements of products/accuracy rates from the object-recognition phase *Task Elements* Data from analysis: tasks and instructions, number the usability problems found/performance, time of information's display, number of mouse moves or clicks, eye movement patterns, effectiveness	*Feelings Elements* Perception usability: visual appearance of a system; weight; frustration, fun, and usefulness of systems; and ease of use *Socio-Cognitive Process* Uncertainty/avoidance; need for significant others; parallel versus sequential actions; diffuse versus specific; particularism versus universalism; collectivism versus individualism; high context versus low context; transference; complex spatial area on the visual scene; focal object information × contextual information; attention to the background field × perceived affordances in the environment, aesthetics *Structural Elements/Usability Test* Overall relationship between user and evaluator in task analysis. Considerations about characteristic's evaluator. Evaluator's cultural background: foreign evaluator and local evaluator. The communication patterns of local pairs and distance pairs. Cultural profile of the interviewer.

elements of visible variables that can be easily represented in formal language as explicit knowledge in cultural usability studies. Table 6.1 identifies two variables that can be easily shared with explicit knowledge: cultural marks knowledge and task elements.

Tacit knowledge is better suited for three types of variables that are not observable in formal languages such as feeling elements, social-cognitive process, and structural elements of usability tests. Tacit and explicit variables can also be found together. These variables have not been discriminated by the authors, because the main objective of this study was to collect and characterize the cultural variables.

6.2.3 Considerations on Cultural Variables in Complex Systems

As evidenced in the literature studies it is a challenge to introduce usability parameters for products in the global market. When considering the wholeness of the culture in terms of a usability system to make decisions about a feedback loop in a design process, we look at it as a complex system. We need to consider too the distinct design phases (conception, development, and prototype) and finally the user contact phase, where we could include all information, perceptions, and knowledge about the product. These phases have different needs on information feedback loops. The process of developing innovative products require methods of formative usability mainly if it has demands in globalized markets, and the cultural elements can affect product usability directly or indirectly. The surveys analyzed contribute for formative usability in both phases of knowledge, for example, tacit knowledge (feelings elements in perception/satisfaction questionnaire) (Frandsen-Thorlacius et al., 2009); explicit knowledge "task elements" (Olaverri Monreal and Bengler, 2011); and explicit knowledge "cultural marks" (Ono, 2006). The classical frame of knowledge, the "iceberg," show us the top of the explicit knowledge and the bottom of the tacit knowledge, but culture is not static like an iceberg, it is dynamic and changes. The dynamic and the real complexity of the system can be observed in Table 6.2, where the variables are treated and grouped by similarity but the range of variables is huge. The cultural usability system includes variables such as feelings, socio-cognitive processes, and also objective and subjective structural elements on usability tests. This allowed us to identify the level of complexity on intercultural usability. We envision further studies with a larger number of surveys of other authors.

6.3 APPLICATIONS OF USABILITY IN DIFFERENT CULTURAL CONTEXTS

From a review of the literature, we conclude that different cultural variables such as color, sound, semantics, signs, reading direction, etc. can affect product uses (Okimoto et al., 2013). These elements can often form true popular stereotypes, but they are not explicit elements. Previous culture and usability studies have shown that through understanding the processes of cognition and perception, we can recognize cultural elements (Barber and Badre, 1998; Marcus et al. 2003; Nisbett and Masuda, 2003). In product design methods, color is understood within a universal meaning (Davies and Corbett, 1997). Even today, this vision is still a part of industrial design training in many countries. Contrasting with this theory, colors bring unique meanings to different cultures (Barber and Badre, 1998). According to these authors, the color red can represent happiness in China, anger and danger in Japan, death in Egypt, aristocracy in France, and danger and "stop" in the United States. For the authors, color is viewed as a cultural stereotype, but this is a global perception. It is also a representation of the user's memory associated with the color, but not necessarily a representation that can be applied to all interfacing elements. The question that arises in this issue concerns the global stereotype. Is color one of the components of the usability of a product? To extend this discussion of the meaning of color, we proposed an exploratory study involving the perception of color within a computer task in two different contexts, Brazil and Germany. In this study, we propose a comparison between possible elements of usability in task and color. The survey included elements such as effectiveness, efficiency, satisfaction, and dissatisfaction, which were associated with a color palette. Our challenge was to understand how these elements were perceived and interpreted within the two cultures we looked at. We present and discuss some results of this survey in the next section.

6.3.1 Description of the Survey

A total of 62 individuals (31 Brazilian and 31 German) participated in the study. Their ages varied from 21 to 61 years. The mean age and the standard deviation for the Brazilian group was 38.13 (10.35) and for the German participants 28.9 (10.01); 16 of the Brazilian and 15 of the German participants were female. The Brazilian participants were postgraduate students or researchers in industrial design or mechanical engineering. The German participants were postgraduate students or researchers in ergonomics. We developed a questionnaire in the language of each country. The questionnaire was sent, via email, to persons who had previously agreed to participate in the survey. The survey was conducted between the months of October and November 2012 in both countries. Alongside each question, we provided a palette of 48 colors to correlate colors with usability elements. Participants were requested to correlate past experiences of a typing task in the following situations: when the task is completed correctly; when the task cannot be completed; when the task is performed in less time than expected; when the task takes a long time, much more time than expected; when the task is easy to do; when the task is very laborious; when you are satisfied with the use of computers, notebooks, netbooks, tablets, etc.; and when you become dissatisfied with the use of computers, notebooks, netbooks, tablets, etc. The survey was divided into four positive questions and four negative questions regarding usability perceptions of the computer task. The positive perceptions were task completed correctly, performed in less time, easy task, and user satisfaction. The negative perceptions were task is not completed correctly, the task requires much time, difficult task, and dissatisfied. Neutral perceptions were not investigated in this study.

6.3.2 ANALYSIS RESULTS

Respondents in both cultural groups used the computer daily. Computer use time per day was almost the same for Brazilians and Germans, 6.5 and 7 h, respectively. Preferred types of equipment for the Brazilians were notebooks (90.3%), desktops (74.2%), and tablets (22.6%). The Germans preferred equipment was notebooks (93.5%), desktops (74.2%), and tablets (25.8%). Preferred equipment indicated for data entry by all subjects in Brazil was keyboard (96.8%), mouse (93.5%), and touchpad (74.2%). In Germany, the preferred equipment for data entry was mouse (96.8%), keyboard (90.3%), and touchpad (42%). The comparative usability perceptions of positive elements resulting from computer tasks are presented in Figures 6.1 through 6.4. Figure 6.1a and b show the results of perception when the task is completed correctly. The colors associated with a correct task for German

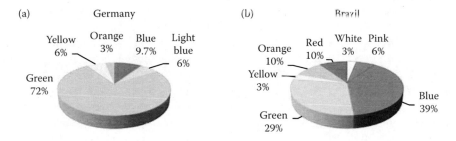

FIGURE 6.1 Colors associated with a correct task. (a) Germany and (b) Brazil.

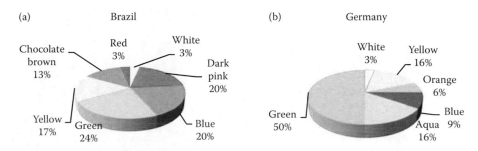

FIGURE 6.2 Color associated with task completed in little/less time. (a) Germany and (b) Brazil.

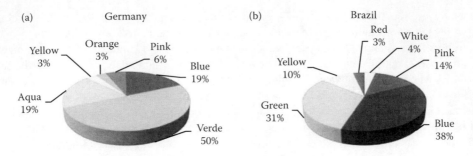

FIGURE 6.3   Color associated with an easy task. (a) Germany and (b) Brazil.

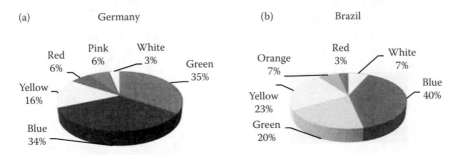

FIGURE 6.4   Color associated with a pleasurable computer task. (a) Germany and (b) Brazil.

subjects presented the following proportions: green 75%, blue 10%, light blue 6%, yellow 6%, and orange 3%. For Brazilian subjects, a correct task was associated as follows: blue 39%, green 29%, red 10%, orange 10%, pink 6%, yellow 3%, and white 3%.

Figure 6.2a and b presents the results for tasks performed in less time than expected. For German subjects, the associated colors were green 50%, aqua 16%, yellow 16%, blue 9%, orange 6%, and white 3%. And for Brazilian subjects, the colors associated with this result were green 24%, blue 20%, dark pink 20%, yellow 17%, chocolate brown 13%, red 3%, and white 3%.

Green was selected by 50% of German subjects in relation to an easy task, and for the remaining German subjects, preferences were distributed among the following colors, as shown in Figure 6.3a: aqua 19%, blue 19%, pink 6%, orange 3%, and yellow 3%. For 38% of Brazilian subjects, the easy task was more commonly correlated with blue, as shown in Figure 6.3b.

Figure 6.4 shows the results for the question about satisfaction with the use of computers, note-books, netbooks, tablets, etc. German subjects selected the following colors related to use satisfaction: green 35%, blue 34%, yellow 16%, red 6%, pink 6%, and white 3%, while Brazilian subjects expressed the following choices: blue 40%, yellow 23%, orange 7%, white 7%, and red 3%.

Perceptions of negative results related to a computer task are presented in Figures 6.5 through 6.8. The first negative question addresses feelings of inefficiency due to the perception of task incompleteness. A task that cannot be completed correctly is shown in Figure 6.5a and b, and the predominant color was red, for both groups. But German group agreement reached 78%, a high value compared with that of the Brazilian group, which was 26%. The following colors were also associated by the German group, though to a lesser degree: black 13%, orange 6%, and dark blue 3%. For the Brazilian group, there was a significant association with dark magenta, 23%.

The indicators of efficiency include task completion time and learning time. In this survey, lower efficiency is related to greater consumption of time, and is presented in Figure 6.6a and b. Consequently, the negative result for efficiency is understood here to be a completed task taking more time than initially expected. Lower efficiency is expressed through the following colors by German subjects: dark orange 31%, dark red 27%, violet 15%, magenta 12%, dark purple 11%, and

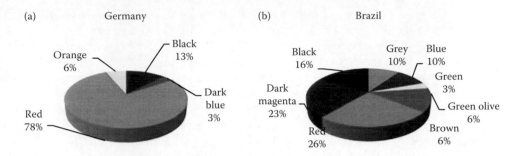

FIGURE 6.5   Color associated with a task that cannot be completed correctly. (a) Germany and (b) Brazil.

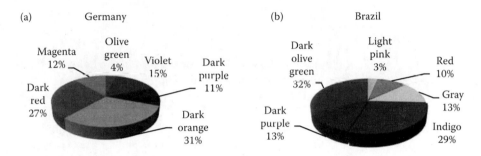

FIGURE 6.6   Color associated with completed task requiring more time. (a) Germany and (b) Brazil.

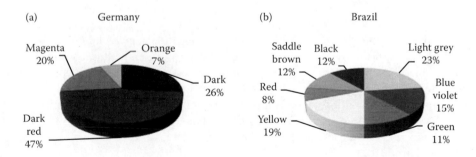

FIGURE 6.7   Color associated with a difficult task. (a) Germany and (b) Brazil.

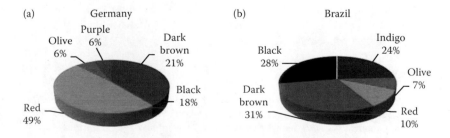

FIGURE 6.8   Color associated with a dissatisfying task. (a) Germany and (b) Brazil.

olive green 4%. Brazilian subjects selected shades of the following colors: dark olive green 32%, indigo 29%, dark purple 13%, gray 13%, red 10%, and light pink 3%.

The task's degree of difficulty could be evaluated in this study through a perceptual level. The results of the positive level, the easy task, were initially presented in Figure 6.3, with similar colors for both cultures (Germany and Brazil). In contrast, the color perceived for the negative difficulty level of the task presented a significant difference between the two cultures. The difficult task was associated with the following colors by German subjects: dark red 40%, black 23%, magenta 17%, brown 14%, and orange 6%. The Brazilian subjects selected mostly light tones: light gray 23%, yellow 19%, blue violet 15%, saddle brown 12%, black 12%, green 11%, and red 8%.

An unpleasant task is perceived with similar colors by both Germans and Brazilians, but the predominant colors are different for each group, as we can see in Figure 6.8a and b. For the German group, red is significant, with 49%, and other colors associated were dark brown 21%, dark gray 18%, olive green 6%, and purple 6%. For the Brazilian group, the associated color was dark brown with 31%, followed by black 28%, indigo 24%, red 10%, and olive green 7%.

6.4 CONSIDERATION

The results presented in this study show us the variation in perception in two cultural groups which use the computer daily. The most significant difference in input choice was related to the touchpad, which was preferred by 74.2% of the Brazilian group, higher than the German group's 42%. Comparative usability was perceived as consisting of positive and negative elements resulting from the computer tasks, according to the qualities of emotions elicited. We found significant differences for color perception in positive and negative emotions (success and failures in accomplishing the task). Subsequently, we sought to associate the emotions elicited by the task with the possible levels of usability generated by the typing task. In this study, we found a color variation differing from conventional stereotypes. We observe that the color ratings obtained from the German group are more homogeneous than those of the Brazilian group. Regarding the results for the positive level, the easy task was marked with similar colors by both cultures. In contrast, color perception of the negative emotion for a difficult task presented a significant difference between the two cultures. And in the Brazilian group, we found more preference for many colors for that same task. This fact was also reported by Davies and Corbett (1997) in a multicultural study. This author reported that Brazil has many sites that are particularly colorful, with no color being overly dominant. This is indicative of a cultural preference for many colors. In this study, the outsourcing of color as a component of usability shows us that the emotions elicited by the task reflect aspects of stereotypes developed by the culture to which a subject belongs. The topic of usability and culture is a relevant topic, and we suggest further studies to include color as an element that can change the usability of a product.

ACKNOWLEDGMENTS

This study was financially supported by the Brazilian National Council for Scientific and Technological Development-CNPq-Brazil, throughout the CsF-Science without Borders. We would also like to thank Univ.-Prof. Dr. phil. Klaus Bengler, from the Lehrstuhl für Ergonomie (LfE) at the Technische Universität München.

REFERENCES

Barber, W. and Badre, A. 1998. Culturability: The merger of culture and usability. In: *Proceedings of the 4th Conferences on Human Factors & the Web*, June 5, Basking, NJ, USA.
Bertalanffy, L. 1976. *General System Theory: Foundations, Development, Applications*. New York: George Braziller.

Clemmensen, T. 2009. Towards a theory of cultural usability: A comparison of ADA and CM-U theory. In: *Proceedings of the 1st International Conference on Human Centered Design: Held as Part of HCI International*, July 19–24, San Diego, California, pp. 416–425.

Chua, H., Boland, J., and Nisbett, R. 2005. Cultural variation in eye movements during scene perception. *Proceedings of the National Academy of Sciences of the United States of America*, 102 (35), 12,629–12,633.

Davies, I. R. and Corbett, G. G. 1997. A cross-cultural study of colour grouping: Evidence for weak linguistic relativity. *British Journal of Psychology*, 88 (3), 493–517.

DIN/ISO 20282-1 – BSI/ISO 20282-1:2006. 2006. Ease of operation of everyday products—Part 1: Design requirements for context of use and user characteristics.

Evers, V. and Day, D. 1997. The role of culture in interface acceptance. In: S. Howard, J. Hammond, and G. Lindegaard (Eds.). *Human-Computer Interaction INTERACT'97*, London: Chapman & Hall.

Frandsen-Thorlacius, O., Hornbæk, K., Hertzum, M., and Clemmensen, T. 2009. Non-universal usability? A survey of how usability is understood by Chinese and Danish users. In: *Proceedings of the SIGCHI Conference on Human Factors in Computing Systems*, ACM. New York, USA.

Guan, Z., Lee, S., Cuddih, E., and Ramey, J. 2006. The validity of the stimulated retrospective think-aloud method as measured by eye tracking. In: *Proceedings of the SIGCHI Conference on Human Factors in Computing Systems*, New York, USA, pp. 1253–1262.

Marcus, A., Baumgartner, V.-J, and Chen, E. 2003. User-interface design vs. culture. *Proceedings, International Conference on Internationalization of Products and Services (IWIPS 2003)*, July, Berlin, Germany, pp. 67–78.

Nisbett, R. and Masuda, T. 2003. Culture and point of view. *Proceedings of the National Academy of Sciences of the United States of America*, 100, 11163–11175.

Nonaka, I. and Takeuchi, H. 1995. *The Knowledge-Creating Company*. New York: Oxford University Press.

Okimoto, M. L., Olaverri Monreal, C., and Bengler, K. 2013. Usability assessment in the multicultural approach. In: A. Marcus (ed.). *Design, User Experience, and Usability. Health, Learning, Playing, Cultural, and Cross-Cultural User Experience* (pp. 89–94) (*Second International Conference*, DUXU 2013, Held as Part of HCI International 2013, Las Vegas, NV, USA, July 21–26, Proceedings Vol. 8013). Springer: Berlin, July.

Olaverri Monreal, C. and Bengler, K. 2011. Impact of cultural diversity on the dialog design of driver information systems. In: *IEEE Intelligent Vehicles Symposium (IV'11)*, June 5–9, Baden-Baden, Germany, pp. 107–112.

Olaverri Monreal, C., Okimoto, M. L., and Bengler, K. 2014. *Multicultural Text Entry: A Usability Study. Design, User Experience, and Usability. Theories, Methods, and Tools for Designing the User Experience*. Springer International Publishing, Crete, Greece, pp. 329–339.

Ono, M. M. 2006. Cultural diversity as a strategic source for designing pleasurable and competitive products, within the globalisation context. *Journal of Design Research*, 5, 3.

Shi, Q. and Clemmensen, T. 2007. *Usability and Internationalization. HCI and Culture*. Berlin, Heidelberg: Springer.

Tullis, T. and Albert, W 2008. *Measuring the User Experience: Collecting, Analyzing, and Presenting Usability Metrics*. Burlington: Morgan Kaufman.

Vatrapu, R. and Pérez-Quiñones, M. A. 2006. Culture and usability evaluation: The effects of culture in structured interviews JUS. *Journal of Usability*, 1 (4), 156–170.

7 Ergonomics and Wayfinding Design
Principles, Standards, and Visual Code

João Neves and Fernando Moreira da Silva

CONTENTS

7.1 INTRODUCTION

Signage systems are closely linked to the security of persons and goods, as well as pressing issues such as ergonomics, accessibility, orientation, and mobility, among others. Signaling traffic, tourism, emergency signaling, railway signaling, air, maritime, etc. well reflect the role of signage/signalization in contributing to the safety and mobility of societies and the social role of design as a methodology or discipline that conceives orientation signs in space—the signage systems.

The potential benefits of information design for social development are also looking for new approaches and methodologies, for research in and by design in an attempt to maximize new solutions that improve the lives of societies. Solutions at the level of signage systems do not always consider the real needs of multiple users, that is, the level of design of these systems lack design projects truly centered on the users and that incorporate real solutions at technical, aesthetic, ergonomic, and also inclusive levels.

The development of signaling systems, which are considered spatial information systems, involves the vital function of transmitting information in order to guide and direct certain individuals or groups/classes of individuals. It is a communication process that uses artifacts designed for the transmission of clear and unambiguous information and guidance for citizens: wayfinding.

The largest influx of people to certain places of tourist interest, tourist facilities, airports, train stations, events, shopping areas, public services, etc. raised the need to guide these people in an unknown space and communicate basic messages in a language understood by a majority.

Generally, each signage or signaling project is developed in an isolated form, not contemplating any connection with the ruling systems, often ignoring the legislation, the agreed international systems, with the notable absence of strategies in the territories-country, and in various regions of the world. The own regulatory authorities of process often disclaim their role in the study, uniformity, standardization, monitoring and above all, lack of development of methodological processes capable of generating synergies and interorganizational projects.

At the level of development of specific signs to touristic information systems, the lack above all is that the various actors understand that projects can, and should be, user-centered and do not have only aesthetic or economic criteria. It matters also to apply models, methods, or methodologies that place the design professional in the center of the process, avoiding empirical processes, but also involving other professionals such as ergonomists, architects, and technicians from central and regional government, municipal technicians, consultants, companies, the scientific community, and other entities related to the wayfinding process.

It is fundamental to create mechanisms to develop complementary systems that do not require the development of new codes and that facilitate displacement and access to tourists and travelers, generating visual codes which complement the expressive limitations of the images and texts as code to use in touristic messages, incorporating a new universal and instantaneous sign language.

Sign systems for touristic information developed in several countries, generally, are not developed in a systematic way and it is unknown (in most cases) that an applied methodology to project development for touristic information considers this problem as the starting point of the investigation. To reach this objective, it is necessary to incorporate in the projects to signalize touristic information, the principles and methodologies that are transversal to all the steps involved in the process, sufficiently adaptable to various types of design and allowing the generation of more effective systems from the point of view of communication, functional aesthetics, and ergonomics.

7.2   USERS, MOBILITY, AND TERRITORY

The development of railway networks, the advent of the automobile industry, and the growth of the aerial fluxes, allied to a growing world scaled globalization, brought a greater individual social mobility coming from different regions and continents. Commerce, industry, leisure, and other activities caused the abolition of borders, whether they are physical, linguistic, or even cultural, in order to make the circulation of people and goods easier.

According to Costa (1989), social mobility assumes the fluxes of individual groups, from different geographical provenances and with different sociocultural characteristics, moving from one point to another based on very distinct reasons. This social dynamic implicates the circumstantial basics, which means that the passage through determined places is sporadic as a result of a naturally itinerant activity. Therefore, it generates new situations, morphological and organizational unawareness of these places, and consequentially, it presumes a high level of intelligibility or indetermination, which raises dilemmas and even risks in the individuals' actuation necessities.

Social mobility supposes a displacement from a place to another in a certain territory, and of course it is related with the accessibilities, especially in the way that reduced mobility, special educational needs, or human factors are a conditioning aspect in accessing or displacing in a predetermined space.

The greater influx of people to public places has defused the necessity of giving those people an orientation in an unknown place and to communicate basic messages through an understandable language. On the other hand, that mobility brought along traffic developments associated with a growing flux of individuals that displace themselves from a point to another. That displacement, often performed in unknown spaces, defused the need to learn new rules, which become normalized through signs that facilitate the access/circulation to/in determined places.

The orientation systems are constituted by multiple signs, which require a profound and systematic study of a code in which, by the quantity not always are their characteristics apprehended, sometimes causing disrespect and alienation regarding the delivered message. Signage and signalectics are constituted by multiple levels, categorized in different categories regarding their characteristics, constituted by signs or panels which transmit a visual message, thanks to their location, shape, color, type, and even through symbols and characters.

The growth in recent decades in tourism activities, coupled with an increasingly global world, provided a more or less general abolition of physical borders, linguistic, and even cultural, facilitating the free movement of people and goods, enhancing trade, industry, recreation, and other activities related to tourism.

The mobility and greater influx of people to certain places or attractions, raised the need to target these people in an unknown space and communicate basic messages in a universal language, expressed through images in order to facilitate understanding, and resulting in the reduction of written messages in any language. To Massironi (1983, p. 118), this type of images (pictograms) help in orientation in stations, airports, hotels, services, and are also currently found on maps, tourist guides, multimedia applications, among others, and for which the requirements of export markets and circulation cannot predict the use of one language or the confusion of many languages at once.

The displacement within the tourism activities is made many times in unknown places, raising the need to develop new rules, which will then be formalized through signs that facilitate access or movement to certain places.

The growth and evolution of cities, the complexity of transportation routes, trade relations, and communications become essential in the signaling environment, necessary for the safe use of urban facilities, providing business, and the exchange of knowledge and ideas (Velho, 2007, p. 12).

Tourism, as an activity that involves the displacement or mobility of visitors, generates several needs, whether at the level of tourism resources, services, or offerings. Tourist information makes a huge contribution to mobility, the quality of accommodation, and the provision of tourist services and, ultimately, to develop more inclusive places and territories. Tourist information is presented in various forms, from tourist maps, books, leaflets, guides, panels, advertising panels, multimedia applications, web sites, signage, and tourist signs, among other forms of communication.

7.3   SIGN AND SYSTEMS

Signs, as words, need to be understood, to be organized in a speech or text, else they can become not understandable as a whole. Therefore, sign relation in the same network must be understood so that one can comprehend their meaning. Blending in groups of related signs, having in mind its utilization rules, one is in the presence of a code. Otherwise, a code is a system of signs with relations and meanings (Raposo, 2008, p. 12).

In communication, according to Aicher and Krampen (1995, p. 9), there are elements from two main groups that interrelate those from a fundamental group of signs; and those from a crucial message group admitted from the signs. Code is what one calls the coordination of these two main groups.

It is the code that establishes that a certain sign has certain meaning. Meaning is not natural when one looks at a sign. The signs whose meaning are determined by a code require an apprenticeship of its meaning (Fidalgo, 2005).

Therefore if, by code we refer to a system of signs with relation and meaning, it is important to deepen the definition of the system, which can be grouped diversely, that is, building signage, companies, traffic signs, theme parks, organizations, etc.

Beni (2001) defines a system as the set of parts that interact in order to achieve a given goal, according to a plan or principle, logically ordered, and sufficiently coherent to describe and explain the functioning of the whole. For Britto (2006), systems are part of a whole, coordinated among themselves and function as organized structure. Heskett (2005, p. 145) defines a system as a set of interrelated elements, interacting or independent entities that form, or one may consider to form, a collective entity. The objective of a system is to provide clear information on the consequences of choosing a route or a particular direction, but leaving the users decide exactly where they want to go.

A system requires principles, rules, and procedures to ensure harmonious interaction and ordered in the interrelation of ideas with forms. This means having systematic qualities of thought, from which it implies methodical, logical, and certain procedures. The author also adds that the objective of a system is to provide clear information on the consequences of choosing a route or a certain direction, thus leaving the users to decide exactly where they want to go.

Each artifact unit (signal) helps to form a whole (the system), that is, the signs (artifacts built by man) are not individually designed, but take into account the collective entity which unites them. The signal (unit belonging to a whole) is then a physical object with different meanings and with unique features that make it, on one hand different from the rest, and on the other still related to the system. Being the sign an physical object, with its own image, and to which a signification is conventionally assigned, then we are faced with a sign (Neves, 2006, p. 178).

The signage systems are composed by independent elements—the signs—that convey certain information or the obligation of an action and interrelate with the function of communicating messages with meaning (through code). Signage systems, to communicate messages, involve the use of pictograms, which are not more than simplified figurative signs that represent things and objects from the surroundings (Costa, 1998, p. 219).

The pictographic systems discussed in this chapter are understood as signage elements, signals, or interrelated information, making use of simplified figurative signs that represent things and objects in the environment (pictograms). Simply put, the set of signage elements (signs, signals, or information) that relates to form a code and that involve the use of pictograms is understood by a pictographic system.

7.4   HUMAN FACTORS AND WAYFINDING

Considering wayfinding as the organization and communication of our dynamic relationship with space and environment, we are in the field of an important area for design, in architecture and ergonomics, which is not only limited to the design of systems, but all that pertains to human interaction with spaces (Arthur and Passini, 1992).

Thus, wayfinding design seeks to develop activity in the design of signage systems and of spatial information, which aims to guide and assist the task of accessibility to a particular space or territory. It is understood to be fundamentally a user-centered approach, which favors the ergonomic, anthropometric, and inclusiveness issues, making signage systems truly universal or rather, accessible to a majority.

Several definitions of the term wayfinding are known, as a methodology for organizing indicators to guide people to their destinations (Beneicke et al., 2003). Wayfinding can also be an orientation process that uses spatial and environmental (natural, urban, or built) information. Wayfinding can be considered as a method to provide consistent information, in a clear and obvious way, to guide a person to their destination. This information can include maps and signs, clear clues to the architecture and interiors design of facilities, or through the use of color pattern and texture. Advanced systems of wayfinding can also be effective information systems that support organizational identity and branding strategies (Hablamos Juntos, s.d.).

The organization and communication of our dynamic relationship with space and environment can be defined as wayfinding. Successful design to promote wayfinding, allows people to (1) determine their location within an environment, (2) determine their destination, and (3) develop a plan that takes them from their location to their destination. The design of wayfinding systems should include (a) identifying and marking spaces, (b) areas of agglomeration, and (c) binding and organizing spaces through architectural features and graphics (Center for Inclusive Design and Environmental Access, s.d.).

Wayfinding design is the creation of resources and spatial information systems and the environment, to guide and direct people and can be considered as the process of organizing spatial and environmental information to help users find their way. Wayfinding should not be considered a separate or different activity than the traditional "signage design," but rather a broader, more inclusive way of assessing all the environmental issues which affect our ability to find our way (Kelly Brandon Design, s.d.).

A wayfinding system includes brands, signs, maps, and directional devices that tell us where we are, where we want to go, and how to get there. An effective wayfinding system can add an important dimension to the image of a museum, a transit system, an airport, an office building, or an entire city. It can be designed as an auxiliary understanding that provides information and guidance to people in a manner that is clear, appropriate, and friendly to the user, in order to help find the way through, and out of an environment (Wyman, 2004).

All wayfinding projects have a common factor, whether they are large scale or small scale, long term or short term, to public spaces or commercial companies, for new visitors or staff, indoor or outdoor, pedestrians or car drivers. All wayfinding projects are intended to be used by people. This means that all wayfinding projects will have to take into account human perception and human psychology (Mijksenaar, 2009–2011).

Thus, it is important to understand that wayfinding systems are directly related with the individual and collective cognition of users, requiring decoding and adaptation of the mental processes inherent to the signic process, namely perception, memory, and reasoning. We are then faced with a communication process in the field of cognitive ergonomics.

7.5 PRINCIPLES AND CODE

7.5.1 PRINCIPLES

The sign systems for tourist information are not developed in a systematic way and it is not known (in most cases) that there is one applied methodology to project development for tourist information, in terms of graphic development, in management terms, and in planning, development, and application.

It is therefore necessary to incorporate in the projects for signaling tourist information, a methodology that covers all the steps and actors involved in the process and that is sufficiently adaptable to various types of projects and permits the generation of more effective systems from the point of view of communication, functional, ergonomic, and even aesthetic.

Considering this problem as the starting point for this essay, based on the research area of design, specifically graphic design/information, it was understood that this test could produce results in terms of defining the principles and standards that would support decision making of designers when developing pictograms and ideograms for application in systems of signs to tourist information, which would make them more inclusive and focused on the diverse needs of users.

This test is assumed as a reflective proposal about certain principles and norms that is believed may have implications in improving the graphical and functional quality of visual code and part of the interpretation of the results of the entire investigation. The principles applied to the development of systems of signs for tourist information are as follows:

a. *Principle of necessity*: Is it necessary to develop a new system or a new symbol? Prior to the development of a system of signs for tourist information, it must be evaluated if any system ever developed or related and sufficiently recognized exists, that fulfills the same function without the need of a new code.

b. *Principle of universality*: For whom is the signaling or symbol? Regarding the signaling systems for the tourism area the universe and not the sample should always be considered, that is, the target audience of the system should be considered as the entire audience from various sources in order to make the system recognized by the majority. If possible, pictograms and ideograms recognized worldwide should be used.

c. *Principle of participation*: Who develops the system or symbol? The participation of the various actors (customer, company, designers, and other professionals and above all the user) is essential to the quality of the system or symbol developed and it is necessary that they incorporate the decisions of each design stage, to conceive a more inclusive system, one closest to the user's needs and hence with greater differentiation potential, with perceived quality and satisfaction of all needs of the stakeholders and recipients of the developed system.

d. *Principle of usability*: Are the systems of signs or symbols easy to use? Systems of signs or symbols for tourist information should be developed taking into account the ease with which the user interacts with the signaling artifacts, in order to facilitate access and displacement in a given space or territory. In this sense many cycles (analysis/design/testing) should be performed and their results evaluated.

e. *Principle of simplicity*: Are the system of signs and their symbols clear? The complexity of the information provided by the signaling system or the symbols and/or the need for fast reading and decoding by the user of the signs, impose a graphic system, simple, clear, and without possible ambiguities. In this sense, graphic design plays a key role in the transmission of information, and should be based on elementary graphic structures, flat colors, typography with effective readability and legibility, and other factors that may contribute to a system that fulfills its function: to inform.

f. *Principle of uniformity*: Do the system of signs or their symbols have a common graphic language? The graphical choices made in the development of a particular pictogram that belongs to a class or system must regulate the design of other signs, that is, there must be a common principle to the graphic language, regulatory and uniform for all systems. The uniformity results from the similarity of constituents and homogenization of the parties to establish a common code, the facilitator of the communication process between the system of signs for tourist information and the user.

g. *Principle of normativity*: Are there standards or rules that constrain the development and application of symbols or signs of the system? The adoption of international symbols or systems implemented in another country can be beneficial for decoding and recognition of the code applied. However all the precepts and rules that might constrain the development and application of a particular symbol or system should be analyzed. Thus, there must be an analysis of the normative documents (sectorial regulators, regional, national, and international) that may affect or assist in the development of the graphic system.

h. *Principle of perceptibility*: Was the perception question considered in the development of the symbols or system? The question of perception as fundamental to the quality of the system must be considered in the development of pictograms and ideograms for incorporation of the signs into tourist information systems.

7.5.2 STANDARDS

It is important to understand, in relation to the development of the graphic system, that all systems are different, distinctive, unrepeatable, and with distinct audiences. It is therefore important to consider the issues related to the design of the visual code. In the following paper various contents and processes for application in the graphic development of the visual code are presented, which are intended to contribute to the graphical and functional quality of the system.

The technical report ISO/TR 7239: 1984 (E) provides procedures for the development and implementation principles of the symbols for public information, addressing the report in three major areas: procedures for the development or adoption of symbols; the criteria for visual design; the implementation process of the symbols for public information (ISO/TR 7239, 1984, pp. 4–16).

As for the procedures for the development or adoption of symbols for public information, the technical report indicates that one should first check if the referent is no longer standardized (ISO 7001, 2007), before developing a new symbol. If not, before proceeding to developing a symbol for a certain function, it must be clearly established whether the graphic symbol is really necessary. Establishing the necessity of the existence of a new symbol, the development of the symbol for public information must be based on the results obtained by the procedure proposed by the standard (ISO/TR 7239, 1984, p. 7).

Regarding the content of the normalized image, the ISO 7001 establishes three elements: (a) the contents of the default image, (b) function, and (c) field of application. Generally, the information contained in the technical report is aimed at design professionals in visual communication, which are to be interpret and applied after appropriate assessment of specific environments.

As for building a symbol, the report reflects on the different elements of the system: graphism, such as the application of grid construction, symbol proportions, symmetry of shapes, directional arrows, shapes and contours, and reduction scales. The report also presents a number of technical rules for defining the shape of symbols, as well as the angles of vision, important for the design and implementation of public information symbols, and also contains other rules relating to the viewing distances–symbol size–displacement. Also presented are other rules for the implementation of symbols such as orientation, distance, symbol–text–arrow interaction, tonal contrast, and relationship with the corporate image.

In other ways, the standard ISO 22727:2007 (E) indicates some guiding principles for the creation and design of public information symbols, these principles being organized into three parts: the creation process, function and meaning, and finally the design of graphic symbol (ISO 22727, 2007, pp. 2–24).

According to the document, before developing a new symbol for public information, you should check whether that symbol and not another is really necessary, one should also identify its precise meaning, and the need to create a new symbol by determination, or if the symbol with the meaning required is already given in ISO 7001. For the design of a new symbol for public information, one must consider the existing graphic elements with similar meanings that can be used, adapted, or combined to form the new symbol. An analysis of the expected characteristics of the new symbol users and the context of their use, must also be performed.

For the design of the new symbol a model (layout) should be used. After the conception of the new symbol, it is highly recommended to conduct an evaluation of the understanding of the symbol in the context in which the symbol is being used (according to ISO 9186-1 standard). If necessary, the public information symbol must be modified.

A meaning, a function, and an image content must be assigned to the symbol for public information. To this end, one must consider the category to which the symbol belongs (categorization). Each symbol must normally be used to transmit only one message and it should be placed in only a single category. A meaning and a function that must be unambiguous must be assigned.

According to the ISO 22727:2007 standard, the design of a graphic symbol should be understandable; easily be associated with the intended meaning; be based on objects, activities, etc., or a combination thereof, which will surely be identifiable by the users; be easily distinguishable from other graphic symbols; contain only those details that help in the comprehension; and maintain these characteristics when reduced to 25% of the size of the conception layout grid.

The graphic symbol must be designed within a grid. This should not extend beyond the margins of the grid, but should make full use of the area within the margins of the model. It should also be constituted by the fewest possible components while maintaining comprehensibility. Any existing standardized representations of components of the symbol should be used without modification. The letters, numbers, punctuation marks, mathematical symbols, and other characters should be used only as an element of a symbol for public information. The symbols for public information should only be denied for reasons of comfort or convenience. The denial bar and cross are usually in red.

It is recommended that the designer address certain issues during the creation and design of a graphic symbol for use in public information in the sense of solving an identified problem of public information: meaning; alternative meaning, unintentional meaning; function; necessity; existence of symbols for public information; graphic symbols and elements of graphic symbol existence; field of application; users; more details for specific audience; related meanings; negation, project review, test data.

7.5.3 Visual Code

It is understood, in the context of this study, that the principles describe the general characteristics of development of a visual code with perceived graphical and functional quality. The standards describe the content and processes for application in the graphic development of the visual code. Together, the principles and standards constitute the guidelines in developing the visual code.

It is considered, in the development of graphics system, that all systems are different, distinct, and unrepeatable and with distinct audiences. It is therefore important to consider the issues related to the design of the visual code.

Any image that competes to form a pictogram, tends to take on the characteristics and convey the sense of the total category of objects which belong to the object under examination (Massironi, 1983, p. 118). That is to say that an image to be represented by a pictogram tends to regulate the design of other pictograms that are contained in the same category. Ordinarily, the image of an object has the property to display that object in all its uniqueness, loaded with all attributes that characterize it as unique. In the pictograms, the figure "man" should serve to mean "all possible men."

If a photograph of a man was used as a signal, the image would be much closer to the real man than that outlined by the pictogram, but it would be much less useful. If each figure must serve to signify "the whole set of possible objects belonging to that class," the figure of which we speak should never foreshadow an object, but the whole class of those objects, that is, a concept (Massironi, 1983, pp. 119–120).

According to Moles and Janiszewski (1992, p. 47), there are criteria that characterize the different types of images, such as iconicity/abstraction; complexity/simplicity; normativity; universality; historicity; aesthetics or cognitive load; and fascination.

The requirement of transmitting information by pictograms obliges us to conceive concise, simple, and quickly understandable signs; elementary graphic structures must be sought, for it to do justice to a certain type of perception (Aicher and Krampen, 1995, p. 101). In general, the conceptual model (taking into account the design of pictograms) should present information in a manner that is more simple, clear, and without possible ambiguities (Mijksenaar, 2001, p. 25). The design has the unique ability to shape information by certain techniques (Mijksenaar, 2001, p. 25), such as emphasis or understanding; comparison or structuring; grouping or order, selection or omission; option for immediate or delayed recognition; and presentation in an interesting way.

The European Conference of Ministers of Transport in 1991, supplementary to the Signs and Road Signs Convention, established basic principles for tourist signs (Organização Mundial do Turismo [World Tourism Organization], 2003, pp. 47–48), principles such as the principle of safety; principle of proximity; and principle of specificity.

According to Maria Avillaneda (2006, p. 88) it is essential to define the bases of creating a set of signs or graphics for developing a signal system, because the strict observation of each normative basis will be reflected in the signaling system functionality. The following normative foundations are defined (Avillaneda, 2006, pp. 88–97.): coherence; logic; terminology; location; clarity and precision; color; design; flexibility; and universality.

REFERENCES

Aicher, Otl; Krampen, Martin. 2002. *Sistemas de signos en la comunicación visual*. 4th ed. México: Gustavo Gili.
Arthur, Paul; Passini, Romedi. 1992. *Wayfinding: People, Signs, and Architecture*. New York: McGraw-Hill.
Avillaneda, María del Rocío Sánchez. 2006. *Señalética: Coceptos e y fundamentos, una aplicación em bibliotecas*. 1st ed. Buenos Aires: Alfagrama. ISBN: 987-22074-5-3.

Beneicke, Alice; Biesek, Jack; Brandon, Kelley. 2003. *Wayfinding and Signage in Library Design*. *Libris Design Project* (pp. 1–20). Retrieved from http://librisdesign.org/docs/WayfindingSignage.pdf.

Beni, Mário Carlos. 2001. *Análise Estrutural do Turismo*. São Paulo: Editora SENAC.

Britto, Janaina. 2006. Sistema de sinalização turística—a importância da sinalização turística para o desenvolvimento sustentável do Turismo. In *Revista de Estudos Turísticos*. 24th ed. Retrieved from www.etur.com.br/revista/.

Center for Inclusive Design and Environmental Access. s. d. *Universal Design New York: 4.1c Wayfinding*. Retrieved from http://idea.ap.buffalo.edu/udny/Section4-1c.htm.

Costa, Joan. 1989. *Señalética*. 2nd ed. Barcelona: CEAC.

Costa, Joan. 1998. *La esquemática: Visualizar la información*. 1st ed. Barcelona: Paidós. ISBN: 84-493-0611-6.

Fidalgo, António. 2005. *Sinais e Signos: aproximação aos conceitos de signo e de semiótica. Universidade da Beira Interior*. Retrieved from http://bocc.ubi.pt/pag/fidalgo-sinais-signos.html.

Hablamos Juntos. s. d. *Universal Symbols for Health Care*. Retrieved from http://www.hablamosjuntos.org/signage/default.index.asp.

Heskett, John. 2005. *El diseño en la vida cotidiana*. 1st ed. Barcelona: Gustavo Gili. ISBN: 84-252-1981-7.

ISO 22727. 2007(E). *Graphical Symbols: Creation and Design of Public Information Symbols—Requirements*. 1st ed. Geneva: International Organization for Standardization.

ISO 7001:2007(E). *Graphical Symbols: Public Information Symbols*. 3rd ed. Geneva: International Organization for Standardization.

ISO/TR 7239:1984(E). *Development and Principles for Application of Public Information Symbols*. 1st ed. Geneva: International Organization for Standardization.

Kelly Brandon Design. s. d. *Wayfinding*. Retrieved from http://www.kellybrandondesign.com/IGDWayfinding.html.

Massironi, Manfredo. 1983. *Ver pelo desenho: aspectos técnicos, cognitivos, comunicativos*. 1st ed. Lisboa: Edições 70.

Mijksenaar, Paul. 2001. *Diseño de la información*. 1st ed. Mexico: Gustavo Gili. ISBN: 968-887-389-6.

Mijksenaar, Paul. 2009–2011. *What is Wayfinding?* Retrieved from http://www.mijksenaar.com/content/18-wayfinding.html.

Moles, Abraham; Janiszewski, Luc. 1992. *Grafismo Funcional*. 2nd ed. Barcelona: Ediciones CEAC.

Neves, João. 2006. *O sistema de sinalização vertical em Portugal*. (Master thesis). Retrieved from RIA Repositório Institucional—University of Aveiro (Accession No. http://hdl.handle.net/10773/4761).

Raposo, Daniel. 2008. *Design de identidade e imagem corporativa*. 1st ed. Castelo Branco: Edições IPCB. ISBN: 978-989-8196-07-1.

Velho, Ana Lucia de Oliveira Leite. 2007. *O Design de Sinalização no Brasil: a introdução de novos conceitos de 1970 a 2000*. Master Thesis in Art and Design, Rio de Janeiro: Pontifical Catholic University of Rio de Janeiro.

World Tourism Organization. 2003. *Sinais e símbolos turísticos: Guia ilustrado e descritivo*. 1st ed. São Paulo: Roca. ISBN: 85-7241-450-9.

Wyman, Lance. 2004. Wayfinding Systems. *Webesteem magazine*, nr. 9. Retrieved from http://art.webesteem.pl/9/wyman_en.php.

8 Boundaries of Human Factors and Sustainability in Architecture

Alexander González and Julie A. Waldron

CONTENTS

8.1 INTRODUCTION

Human development is now measured through environmental, social, economic, operative, and net energy consumption indicators (Auliciems and Szokolay 2007). Taking into account all these variables, it was found that cities and buildings are not sustainable. In 2012, it was estimated that energy consumption from nonrenewable resources in the building sector in the United States was 47%; and 74% of this energy was used merely to operate the buildings (Architecture2030 2011). Furthermore, at the end of the twentieth century, a fourth part of this energy was spent in thermal comfort supply (Auliciems and Szokolay 2007). On the other hand, architecture in countries with emerging economies has developed as an exercise for the formal and aesthetical creation far away from environmental and human well-being concerns (Gómez Azpeitia 2007). The isolated understanding of environmental, human, and aesthetic factors in architecture is introducing new challenges for the design of contemporary buildings.

The negative consequences of this model of urban and buildings development are visible in the Human Development Report 2011 (UNDP 2011) and the *Environmental Performance Index* (Hsu et al. 2014). The *Human Development Report* is a calculation which initially integrated variables such as health, education, and life expectancy. However, for the 2011 report the *United Nations Development Programme* decided to also include equity and sustainability in order to show the effect of environment over human development, especially for people with low incomes, which are the most vulnerable to changes in this variable. The report of 2010, added the *Multidimensional Poverty Index MPI* (UNDP 2010), by including the measure of three aspects which are classified as environmental deprivations: health, education, and living standards. According to the report, these aspects are measured through a deep examination of the "... environmental deprivations in access to modern cooking fuel, clean water and basic sanitation. These absolute deprivations, important in themselves, are major violations of human rights" (UNDP 2011, p. 5).

On a global scale, people with the lowest income are the most vulnerable to environmental degradation due to their unfit place of living and lack of resources. The challenge in this scenario is integrating systemic sustainability and human factors in architecture for the design of a built environment capable of connecting actual *milieu* necessities with future feasibility.

This chapter presents the classic topics of "human factors" and "sustainability" and highlights the importance of its linking within architecture in order to achieve a "sustainable well-being" in the built environment.

8.2 BUILT ENVIRONMENT

The traditional concept of a city understood as "The conglomerate of buildings and streets, under the administrative control of a city council, of which the population is mainly dedicated to non-rural activities" Real Academia Española. (n.d.), must be reinterpreted nowadays as *built environment* in order to take into account the concept of sustainability. This allows understanding the city beyond its infrastructure, as the interaction its inhabitants have with it in a daily basis, since *environment* includes not only physical elements but also social, economic, and political processes which are closely interrelated (Ramírez 2009). Therefore, the city as a *built environment* is the result of the related interactions of nature, buildings, and humans which works as a system that exchanges matter and energy in a sort of urban metabolism, fed by a high demand of natural resources, and finally producing a great amount of waste (Di Pace et al. 2004).

The *built environment* is then the subject matter of sustainability. It is there where the goals of sustained development, social equity, human safety, and environmental quality must be projected and executed. Sustainability in this scenario grants its inhabitants (with the highest benefit for neglected areas) common and individual well-being through healthier work and recreational spaces, against the intrusion that represents the different issues brought by big conglomerates, such as pollution, stress, and the lack of safety perception (Rondón Sandia 2009).

According to Torres (2005), the approach of sustainability in the built environment must be reinforced by the concept of *sustainable society*, by acknowledging the physical as well as the social and cultural dimensions of the human being. The sustainability of the built environment is therefore in the middle of the dichotomy between the physical and the metaphysical spectrum, creating a balance for the benefit of the society.

However, the negative issues arising from the cities should not be the starting point for the discussion of sustainability for the development of the built environment. In this sense the contribution of architects and urban planners are of significant value. As was emphasized by Jaime Lerner (2003), if architecture and urbanism determine the physical scenery and the priorities between the environment and human society through the built environment, but the latter is currently in crisis because of limited resources, the responsibility of architects and urban planners in causing this, and must be not lower than their commitment to solve it, "The city is not the problem, it is the answer."

Notwithstanding this, some positions have diminished the character of architecture as genesis of the built environment, favoring instead forms and aesthetics as an expression of the artistic nature of architecture. Even though historically architecture has been considered one of the six *fine arts*, in the book *Observations on the Feeling of the Beautiful and Sublime* Immanuel Kant (1960) established the primacy of function over aesthetics in architecture. This idea is reinforced, according with Cotofleac (2009), with the analysis of the terms "use" and "concept" to define architectural design, from Kant's perspective, which means that there is no architecture without a previous theoretical analysis.

Conversely, architecture cannot be understood merely from the rational thinking perspective as an epistemological process, since the competences of the architect are not necessarily oriented to solve instrumental issues (Schon and DeSanctis 1986), contrary to the facts in exact sciences, because human habitat and the built environment involve complex dynamics from the viewpoint of his psychology, sensorial, cultural, and social conditions amongst others.

8.3 HUMAN FACTORS AND ARCHITECTURE

The study of human life and the body dimensions has been always present in architecture; some examples of these applications are visible in the classic and modern architecture. For instance, in the first century BC, Vitruvius described the human dimensions as part of the building symmetry and composition in architecture (Vitruvius 1771). Later on, in modern times, Le Corbusier developed *The Modulor* (1968) which was a rigorous anthropometric system to measure and shape more efficient spaces, having the human body as the starting point.

Both renowned architects contributed to the understanding of the body dimensions in the space in order to provide answers to the specific necessities of their time. The *Vitruvian Man* created by Leonardo Da Vinci (Accademia, Gallerie dell' 1490) is a good example of a holistic approach to the understanding of the human body from different perspectives: anthropometry, physiology, and art; all this applied to architecture. In this piece of art, Da Vinci represented the proportions described by Vitruvius (1771) in the chapter, "Of the Composition and Symmetry of Temples." There, the structure of the body and its dimensional proportions were taken as a referent for the construction of temples in ancient Italy and Greece "Thus; all measures were derived from the members of the human body; and as the several members thereof have a proportional relation to the whole, so in the temple of the immortal Gods we should endeavour, that the parts of the work may be so regulated, that their proportions and symmetry, separately, and unitedly, may be conformable and correspondent" (Vitruvius 1771, p. 47).

Even though Vitruvius, Da Vinci, and Borelli performed their studies from the perspective of the anatomy and physiology of the body, their work is evidence of the application of human factors in other disciplines such as architecture or other sciences (Kroemer et al. 1994).

Moving onto the industrial period, the studies in human factors were focused on improving the efficiency and accuracy of human tasks. According to Kroemer et al. (1994), during the nineteenth century the studies registered are concerned with motion at work, work payment system, and industrial systems. After a while, the crisis of the World War I triggered other types of studies such as "the minimal nutrition required to perform certain activities, the consumption of energy while carrying out agricultural, industrial, military, and household tasks..." (Kroemer et al. 1994, p. 5). At this stage in the history of humanity, the study of human parameters was focused on providing answers to the ongoing crisis, to the point that its purpose was not well-being but survival; and its application to design was also determined by this.

Later, in the modern world, the interest to understand the social environment favored the reflection about how humans interact with its context, and provided the opportunity to observe new dimensions of the human being. This can be observed in the studies of anthropometry conducted in this period, which had a different interest than of those conducted 50 years before. For instance, the *National Health Survey 1962* (Panero and Zelnik 1979) contains data representing how the anthropometric dimensions of the body can change when a person has access to education. This new perspective is evidence of the growing interest of academic sectors in linking the human body with social factors such as the educational system.

In the contemporary development of architecture, the study of the human being needs to evolve and fulfill the current necessities of sustainable societies. As mentioned by Kathryn B. Janda (2011, p. 4) "Buildings don't use energy: people do"; and, as it was presented by the *Human Development Index* (UNDP 2011), the necessities of humans in the last decade are strongly related to sustainability issues, in the same level of importance of aspects such as health and equity.

In order to integrate sustainability and human factors in architecture, it is necessary to understand the most complex relations of human society to favor the creation of buildings that can work as "social objects" (Hillier 2007). In the book *Space is the Machine* (2007), Hillier described how the building is a physical and spatial transformation of the situations that existed before, and how this transformation of materials on human beings should add value to the new space. This statement is closely connected with the definition of "sustainable development" (Brundtland 1987, p. 24): "Humanity has the ability to make development sustainable to ensure that it meets the needs of the present without compromising the ability of future generations to meet their own needs." This idea of contributing to human development through a responsible transformation of the environment, constitutes the needed linking between sustainability and human factors in a "sustainable well-being."

8.4 SUSTAINABILITY AND ARCHITECTURE

From the beginning of the twenty-first century, the practice of architecture is surrounded and limited by difficult socioeconomic conditions such as economic recessions, social inequity, extreme poverty, environmental degradation, and climatic change, to mention a few. For this reason, the issue affecting the performance of architects regarding sustainable design requires a holistic approach, that is, taking into account the socioeconomic context and the history frequently forgotten by architects (Monedero 2002).

The crisis in architecture ending the twentieth century and starting the twenty first is described by Monedero (2002), who takes as a starting point the analysis of the professional circumstances of European architects around the year 2000, which can be extrapolated to its Latin-American partners: external circumstances derived from the socioeconomic context, and internal causes arising from fundamental inconsistencies in the structure of knowledge of the profession. Amongst the external causes is presented the "demographic explosion," the displacement of the economy to the services sector, the growth of the professional classes, the incidence of new technologies, and the crisis affecting stability and politics in several economic sectors. Regarding the internal causes, which are natural to the professional practice of architecture, a culture of architecture is presented characterized by a private language which is distant from the common understanding of the people and deprives it of its nature as "public service."

However, the paradigm of sustainable development, as a standard to understand the contemporary relationship between human society and its environment, must avoid the possible polarization of speech by using the word "sustainable" to define architecture (as an attempt to reorient the concept of this discipline as a novelty which integrates in the process of design aspects such as economy, society, and environment of the built environment) (Achkar 2005). This practice has been commonly made through the use of adjectives such as "green," "bioclimatic," "smart," or "accessible," which are oriented particularly to the construction market avid for certifications rather than sustainability (Soria López 2005).

The basis for the sustainable development and its relation with the management and design of the built environment, are connatural with the scientific, methodological, historical, and sociological bases of architecture, because they belong to their own epistemology.

Notwithstanding that, society and the guilds of architects argue lack of understanding and diffusion issues of the relationship between architecture and environment. These issues are notorious from the current circumstances regarding the social, environmental, and energy problems that cities are facing nowadays.

Therefore, architecture is not the one to be redefined as *sustainable*; the principles of architectural design must be strengthened, understanding the current and future needs of the people and society, against the circumstances brought by the environmental, social, and economic context. In this sense, it should be clarified that every process of design is the logical relationship among people, architecture, and place, as design principle. This argument was defined by Muntañola, under the hypothesis of "Architecture as place":

> The logic of the place coincides always, in general lines, with the paradigm that in every period the man has had about the interrelations between himself and the environment surrounding him (…) the logic of the place marks always the measure under which the humanity is capable of represent itself. And that way we start to be more close to the heart of Architecture as a place to live (Muntañola 1974, p. 32).

Nowadays, it is necessary to think about architecture and urbanism as a science supported by ecology, technology, and human sociology; all of them understood as the relationship among people, architecture, and environment. This is the starting point for the analysis of the sustainable development in a built environment.

The concept *pedagogy of the place* (Muntañola Thornberg 1998) is a methodological process to guide the development of architectural projects, which requires meeting three standards during the design stage:

1. To explore the real environment where the subject lives
2. To apply construction skills, criticizing the shape of the place and discovering its socio-physical structure
3. The communication between the physical and the social fact

Thinking about architecture from the point of view of the pedagogy of the place, enables the construction of a curriculum which favors sustainable development.

Therefore, if architecture is considered as the design of relations between the environment and the people in order to guarantee their well-being, safety, and functionality, then the purpose of architecture is to bring together human factors, environmental quality, and energy efficiency in the final result of a built environment (Salazar et al. 2006). In order to achieve this, it is necessary to integrate different disciplines that study such phenomena and apply them into construction techniques, waste and material management, and a better use of limited natural resources, taking into account the socioeconomic context of the place where the project is located.

8.5 HIDDEN BOUNDARIES OF HUMAN FACTORS AND SUSTAINABILITY IN THE BUILT ENVIRONMENT

In the book *Hidden Dimensions* (1990), Edward T. Hall described some interesting invisible relations existing between body and space. For him, language and culture constitute important influencers of behaviors and spatial relations. For instance, in the chapter "Language of Space" (Hall 1990) Hall describes how Americans can only differentiate between "snow" and "slush," while Eskimos can have different words for every stage of the snow that allows them to differentiate, understand, and locate themselves in the Artic. This understanding of reality in different contexts shows how perception relies on the language used and other multiple invisible factors, which are needed for adaptation to environmental conditions.

Just like Eskimos found hidden characteristics in the snow, nowadays the design in architecture needs to involve properties that are "not evident" at first to the eye, but that can be applied in order to solve contemporary problems. Is in this context where it may be argued that one of the primary challenges for sustainable design is understanding that nowadays architecture must be an intellectual exercise delimited by hidden boundaries: an inner frontier, responsible of guaranteeing habitability and well-being of the people, and an outer frontier composed of the social, economic, and environmental context surrounding all human beings. Both boundaries vary according to the specific conditions of a particular project, as regards its natural resources, the morphology of the land, and the socioeconomic and cultural context (Figure 8.1).

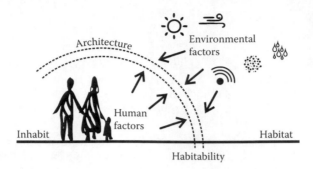

FIGURE 8.1 Representation of the relationship between environmental and human factors, with architecture as a boundary between them.

Hillier (2007) describes in his book how the act of drawing a physical boundary in the design process of a building represents a social decision. The hidden boundaries of sustainability and human factors in architecture are shaped by the relationship among humans, buildings, and environment; the tensions generated by them are translated in the final design in terms of appearance and function of the building. It may be equated to the skin of the built environment, which can be shaped by the tension of opposite factors; from inside out, the human factors demand for space, visual angles, comfort, etc. This generates a kind of pressure to the design of the building coming from inside of it. Opposed to it are the environmental and socioeconomic factors which depend on the context of the building: its physical location and the characteristics of the society and culture surrounding it. The latter generates an opposite pressure from the outside to the design of the building. Depending on the magnitude and importance of the issue to be solved, one of the factors will prevail over the other.

8.5.1 THE PERFECT BUILDING AND THE ACCEPTED BOUNDARY

Every design has the theoretical possibility of producing *the perfect building*, this *perfection* can be hypothetically achieved if all the forces shaping the boundaries respond to the most adequate solution for the issue presented, whether it is human, social, climatic, site, politics, health, efficiency, etc. However, the design is always influenced by external forces that change the magnitude of the vectors. This is the "real building," a built environment constrained and located within the accepted limits of its architecture, but not necessarily responding adequately or equally to all the phenomena affecting it (Figure 8.2).

8.6 HUMAN FACTORS, BUILDING, ENVIRONMENT, AND SUSTAINABLE INTERRELATIONSHIPS

This chapter presents the term "sustainability" as a transcendental part of contemporary architecture, but rejects the use of this term as a label or adjective to categorize the quality of the buildings. However, the role of sustainability goes beyond its individual value as it is also the concept that brings together the rest of the disciplines that study the built environment. Figure 8.3 presents the schema of the sustainable interrelationships: the circle represents the sustainability as unifier of the system; the triangle is composed by three nodes which are the human, the building, and the environment. This triad represents the definition of human factors from the architecture point of view; that is, "… the scientific discipline concerned with the understanding of interactions among humans and other elements of a system…" (Association, International Ergonomics 2015).

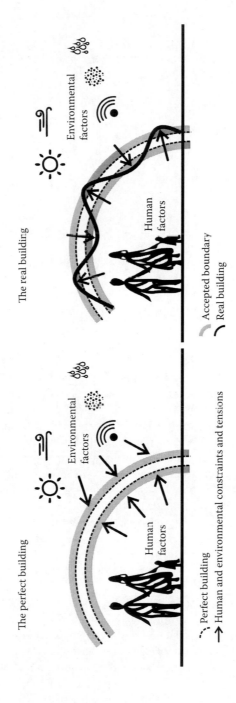

FIGURE 8.2 Perfect and real building and their boundaries.

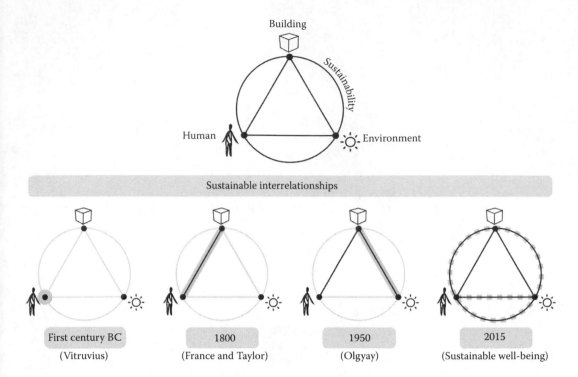

FIGURE 8.3 Sustainable interrelationships (top). The historical development of the interrelationships in architecture (bottom).

At the bottom of Figure 8.3, it can be seen how architecture has evolved to build the relationships between different elements of the system, according to the necessities of each period. First, the analysis of the built environment was focused on the human body as a parameter for design. At this point, the analysis is located exclusively in the human element of the scheme, as can be observed in the bottom left diagram. As it was mentioned before, Vitruvius' concerns were to understand the anthropometry, proportions, and biomechanics of the body to include them in the final design of the building. It is important to bear in mind that at that time architecture was based on aesthetic and religious concepts, and the preservation of natural resources were not such an issue as it is now. By the 1800s a connection was made between the human body and buildings/objects, and this is represented as a line of the equilateral triangle. At this stage and for the next century, humanity focused on enhancing the efficiency of the immediate interactions of the human body, such as the assembly lines of production promoted by Henry Ford, or the "space as a machine for living," as described by Le Corbusier (1968). The needs of that time were focused on increasing the capability of production and development of the cities. Therefore, sustainability and quality of life were not a primary concern.

In the mid-1900s, the interest in bioclimatic buildings started to grow. The book *Design with Climate* (Olgyay 1973) made reference to some statements about the influence of environmental factors for the design of buildings. This reflection started to pinpoint comfort and human well-being in the context of architecture. The necessity of including the environment and the human body into the design was a consequence of the climate change, the concerns about energy consumption, and demographic growth.

Today (2015) there is still an absence in the connections of the disciplines for the design of built environments: the relationship between the human inhabiting a building and the environment surrounding him. The importance of this lies in the need to add a human perspective to the environmental scale. This connection facilitates the equilibrium of human–building–environment in a symmetrical triangle.

Finally, sustainability acts as a joint of all the other elements: human factors, building, and environment. Therefore, it can only be said that a built environment is sustainable if the interrelationship of these three elements has been adequately made.

8.7   SUSTAINABILITY AS A CIRCUIT

Design as a process for an adequate built environment cannot be thought of as a standardized form to be used for every circumstance. It should be thought of as a planned process that provides specific solutions for particular issues arising from the place where the project is located. The starting point of this approach is considering the development of urban and architectural projects as a closed circuit and not linear as it was thought in the twentieth century.

The basis for the "analysis of the life cycle" in the construction industry applies to materials, processes, indicators, and management models, and allows to plan ahead aspects such as environmental impact, management of resources, efficiency in terms of ecology, and energy management. The main aspect of this exercise is time, and its phases are preoperative stage, design, construction, durability and operation, refurbishment, and deconstruction.

The cycle of sustainable management starts in a preoperative phase where the project is defined and its viability assessed from the legal, technical, economical, financial, environmental, and social point of view. Even though most of them are not within the architect's domain, it is important that he is aware of them as director of the design.

Some of the aspects considered in the preoperative phase are closely linked with the architectonic development of the project. For instance, the technical assessment refers to aspects such as the geological condition of the soil, accessibility, urban infrastructure, and connectivity with social services, all of which have direct influence over the final shape, its volume, and materials.

Regarding the environment, in the preoperative stage all the aspects that affect the habitability of the future building are analyzed, such as weather, pollution, and noise to mention a few. This determines the parameters in which the building can be inhabited guaranteeing comfort and well-being. It is also necessary to think of alternative sources of sustainable energy, as well as, the collection and use of rain and underground water.

One way of thinking about this process is to consider that technical feasability will have a high impact during the construction stage of the building, whilst environmental feasability will have it during its whole life (Figure 8.4).

8.8   CURRICULAR APPLICATION IN ARCHITECTURE SCHOOLS

The practical development of this integration of sustainability and human factors needs to be structured by an inter- and transdisciplinary work (Gonzalez and Waldron 2014). This collective effort must impact the different stages of architecture: formation of architects, professional practice, and standardization of design procedures; following the notion of *reciprocal reflexive practice* (González et al. 2011).

In the case of the Latin-American schools, this scenario represents a challenge and a need to integrate systemic sustainability in the curriculum for the education of architects, following the current trend of directing the training and practice of architects toward a reflexive and committed exercise with a sustainable development of human settlements (González Castaño and Trebilcock Kelly 2012).

The integration of human factors and ergonomics is not visible in the official curriculum of Latin-American schools of architecture, because the "user of a project" is studied merely as part

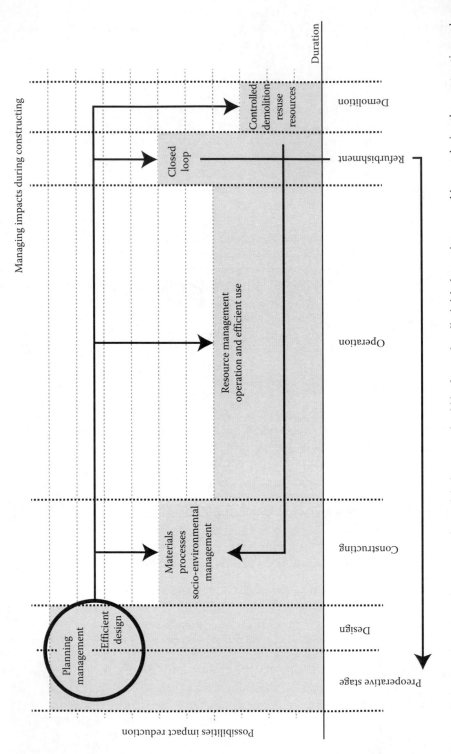

FIGURE 8.4 This figure shows how the preoperative and design stage should be focused to diminish the environmental impact during the construction and operation stages.

of the dimensions and activities that a building is going to contain. In some of the curricula there are courses related to accessible design for population with special needs, which are oriented to the relation between mobility and human mechanics. This can be considered as a chapter of human factors and ergonomics but it is not the total inclusion of this discipline in the curricula (González 2013).

In the meantime, the perception of sustainability in building design is interpreted as an obligation to reduce or restrict the design in order to diminish the environmental impact of a construction. This idea became a barrier for architects because they think it could affect their freedom and creativity for the design of buildings (Association Danish Architects 2013). The architects are not only forgetting the environmental and economic contexts, but also the human, cultural, and social aspects of a sustainable design; all of this related to the main environmental challenges and the real necessities of developing countries.

In a local example, the Law 435 in Colombia (Ley 435 de 1998) contains the legal framework for the practice of architecture and its related professions. This law expressly states that professionals in architecture must have a commitment to sustainability and human development, while undertaking urban and architectural design. Additionally, this regulation contains the Code of Ethics for the practice of architecture, which in its chapter II, article 16 states *"The Established Obligations with the City,"* as follows:

(...)
d) "To carefully study the environment that will be affected by every proposed task, assessing the environmental impact on ecosystems involved, built or natural, including socioeconomic context, selecting the best alternative to contribute for a healthy and sustainable environmental development, to achieve the best quality of life for the population;
e) To reject all kind of recommendations in works that represents reversible damages for humans and nature, in outdoors and indoors, assessing its environmental impact, in the short term as well as the long term."

In this legal context, the integration between ergonomics and systemic sustainability is shown as mandatory in the *Code of Ethics*, through the expressions sustainable development and quality of life quoted before. This statement about the relation between architecture, quality of life, and sustainable development and its imperativeness, conforms to the need of the complete development of the human kind.

8.8.1 Practical Foundation

The practical inter- and transdisciplinary effort to integrate human factors and sustainability in architecture contributes to a feedback design methodology where the consultants of a project participate during the design process in a reciprocal reflexive practice (González et al. 2011, Schon and DeSanctis 1986). In a conventional way, the development of a project is methodologically based on the knowledge of the design and the assessment made by external professionals or consultants.

However, each project has the possibility to present particular situations and conflicts that can exceed the effective practice of the design. In this stage, the professionals need to face their design process through a reflective practice or "reflection in action." This means, thinking of the object of design while it is been developed (Schon and DeSanctis 1986) (Figure 8.5).

The "reflection in action" is established as a theoretical foundation of the integration of sustainability and human factors, which outlines the difference with the development of conventional architecture. In consequence, designers and developers will have to start the discussion assuming a multidirectional work in order to contribute to the processes of other disciplines. In

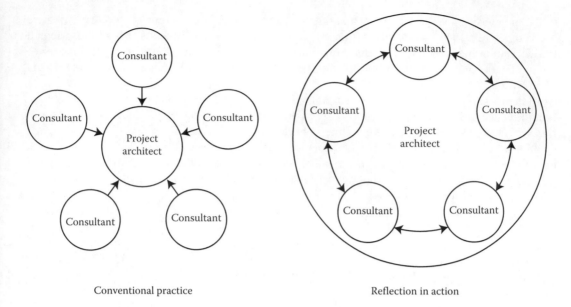

Conventional practice   Reflection in action

FIGURE 8.5   Schemes of the conventional development of a project and the proposed "reflection in action" practice.

addition, they have to guarantee environmental quality from the human factors and sustainable perspectives.

8.9   CONCLUSIONS

The hidden boundaries of sustainability and human factors in architecture are shaped by the relationship among humans, buildings/objects, and environment.

The parameters resulting from the relationship among humans, buildings/objects, and environment does not always respond to the needs of the project in terms of design, and this can affect the sustainability of the built environment.

Nowadays, the concept of well-being has to incorporate the concept of sustainability understood as the capability of built environments to connect actual environmental necessities with future feasibility. This is closely connected with the concept "sustainable development" (Brundtland 1987): "Humanity has the ability to make development sustainable to ensure that it meets the needs of the present without compromising the ability of future generations to meet their own needs."

"Sustainable" and "ergonomic" are not adjectives to define types of design nor are they labels for merchandising purposes. They are valuable elements of contemporary architecture, which aim to provide well-being to the inhabitants of the built environment.

Sustainability acts as a joint of all the other elements: human factors, building, and environment. Therefore, it can only be said that a built environment is sustainable if the interrelationship of these three elements has been adequately made.

Finally, the inclusion of sustainability and human factors in architecture has presented interesting growth during the last decade. In the statistics facilitated by the firm PVG Arquitectos SAS (2015), it can be observed how the number of architectural projects including environmental and human factors has grown three times in the last five years. Additionally, the graphic shows that the less consulted topic is *ergonomics–human Factors* and that the highest interventions are done in lighting and solar gaining, which have a high impact on the reduction of energy consumption of buildings (Figure 8.6).

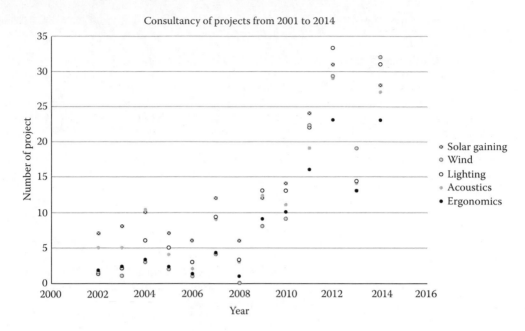

FIGURE 8.6 Chart of consulted projects between 2002 and 2014. (From PVG Arquitectos, SAS. 2015. Accessed April 02. http://pvgarquitectos.com/prehome.php.)

REFERENCES

Accademia, Gallerie dell'. 1490. Leonardo. L'Uomo Vitruviano fra Arte e Scienza (Leonardo. The Vitruvian Man between art and science). Accessed March 27. http://www.gallerieaccademia.org/wp/wp-content/gallery/leonardo-luomo-vitruviano-fra-arte-e-scienza/leonardo.png.

Achkar, M. 2005. Indicadores de sustentabilidad (Sustainability Indexes). In: Comisión Sectorial de Educación Permanente (Ed.), *Ordenamiento Ambiental del Territorio (Environmental planning of territory)* (pp. 55–70). Montevideo: DIRAC, Facultad de Ciencias. Universidad de la República. http://doi.org/10.1590/S1415-65552005000400014.

Architecture2030. 2011. http://architecture2030.org/the_problem/buildings_problem_why. Accessed March 25, 2015.

Association, Danish Architects'. 2013. Sustainable by design—The responsibility of the architect. Accessed March 27. https://arkitektforeningen.dk/artikel/nyheder/sustainable-by-design-the-responsibility-of-the-architect.

Association, International Ergonomics. 2015. What is ergonomics? Accessed April 02. http://www.iea.cc/whats/index.html.

Auliciems, A. and S. V. Szokolay. 2007. Thermal comfort. PLEA Notes. Passive and Low Energy Architecture International. Design Tools and Techniques, Note 3, 66. http://doi.org/10.1007/s00484-010-0393-2.

Brundtland, G. H. 1987. *Report of the World Commission on Environment and Development. "Our Common Future."* General Assembly United Nations.

Corbusier, L. 1968. *The Modulor: A Harmonious Measure to the Human Scale Universally Applicable to Architecture and Mechanics* [translated from the French by Peter de Francia and Anna]. Cambridge, Massachusetts: MIT Press, Texte imprimé

Cotofleac, V. 2009. Kant: Arquitectura y mímesis. *A Parte Rei: revista de filosofía* 63:1–9. http://dialnet.unirioja.es/servlet/articulo?codigo=3861990&orden=335304&info=link.

Di Pace, M., A. Crojethovich Martín, and A. C. Herrero. 2004. Ecología y ambiente (Ecology and Environment). In: M. Di Pace and H. Caride Bartrons (eds.), *Ecología de la ciudad (City Ecology)* (pp. 35–66). Buenos Aires: Universidad Nacional de General Sarmiento and Prometeo Libros.

Gómez Azpeitia, G. 2007. Los arquitectos y la investigación científica (Architects and scientific research). *Palapa. Revista de Investigación Científica En Arquitectura*, 2: 3.

González, A. 2013. *Integración Curricular de La Sostenibilidad En La Formación de Arquitectos En América Latina (Sustainability Curricular integration in architectural education in Latin America).* (Tesis doctoral, Universidad del Bío Bío) (Doctoral thesis)].

González, A., A. García, and J. Salazar. 2011. Práctica reflexiva recíproca para el diseño ambiental del espacio público (Reflexive Reciprocal Practice for environmental design of public space). *Arquitecturas Del Sur* (39):28–43.

Gonzalez, A. and J. Waldron. 2014. How to integrate ergonomics and sustainability in architecture workshops? In M. Soares and F. Rebelo (eds.). *Advances in Ergonomics in Design, Usability & Special Populations*. pp. 412–419. Krakow, Poland: AHFE Conference copyright 2014.

González, A. and M. Trebilcock. 2012. La Sostenibilidad en el Currículo de las Carreras de Arquitectura: implicaciones del concepto de sostenibilidad en el perfil profesional y el plan de estudios de las carreras de arquitectura en América Latina (Sustainability in the curriculum of architec). *Habitat Sustentable* 2 (1):26–35.

Hall, E. T. 1990. *The Hidden Dimension, Anchor Books*. New York: Anchor Books.

Hillier, B. 2007. *Space is the Machine: A Configurational Theory of Architecture*. London, United Kingdom: Space Syntax. http://solo.bodleian.ox.ac.uk/primo_library/libweb/action/dlDisplay.do?vid=OXVU1&docId=oxfaleph012040966.

Hsu, A., J. Emerson, M. Levy, A. de Sherbinin, L. Johnson, O. Malik, J. Schwartz, and M. Jaiteh. 2014. *The 2014 Environmental Performance Index*. New Haven: Yale Center for Environmental & Law Policy.

Janda, K. B. 2011. Buildings don't use energy: People do. *Architectural Science Review* 54 (1):15–22.

Kant, I. 1960. *Observations on the Feeling of the Beautiful and Sublime*. (J. T. Goldthwait, Trans.) (1st ed.). Oakland, California: University of California Press.

Kroemer, K. H. E., H. B. Kroemer, and K. E. Kroemer-Elbert. 1994. *Ergonomics: How to Design for Ease and Efficiency, Prentice-Hall International Series in Industrial and Systems Engineering*. Englewood Cliffs, New Jersey: Prentice-Hall.

Lerner, J. 2003. *Acupuntura Urbana (Urban Acupuncture)* (1st ed.). Rio de Janeiro, Brasil: Grupo Editorial Record Ltda.

Ley 435 de 1998 [Standard 435/1998]. Colombia: Diario Oficial N 43.241 de 19 de febrero de 1998.

Monedero Isorna, J. 2002. *Enseñanza Y Práctica Profesional de La Arquitectura En Europa Y Estados Unidos. Estudio Comparativo Sobre La Situación En El Año 2000* (Education and Architecture Professional Practice in Europe and the United States. Comparative Study about the Situation in 2000). Barcelona, España: Escola Técnica Superior d'Arquitectura de Barcelona.

Muntañola Thornberg, J. 1998. *La Arquitectura Como Lugar (Architecture as a Place)*. 2nd ed. Barcelona, España: Ediciones UPC.

Olgyay, V. 1973. *Design with Climate*, Vol. 41. Princeton, New Jersey: Princeton University Press.

Panero, J. and M. Zelnik. 1979. *Human Dimension & Interior Space: A Source Book of Design Reference Standards*. New York/London: Whitney Library of Design/Architectural Press.

PVG Arquitectos, SAS. 2015. Accessed April 02. http://pvgarquitectos.com/prehome.php.

Ramírez, L. A. 2009. Sostenibilidad o pensamiento ambiental (Sustainability or environmental thinking). *Congreso UPC Sostenible*, (64), 1–8. http://hdl.handle.net/2099/8210.

Real Academia Española. (n.d.). Ciudad (City). Diccionario de La Lengua Española. Edición Del Tricentenario (Dictionary of the Spanish Language. Tercentenary Edition). http://dle.rae.es/?id=9NXUyRH.

Rondón Sandia, L. A. 2009. The environment and sustainable development in Latin American cities. *Investigación y desarrollo* 17 (2):268–287.

Salazar, J., A. García, and A. González. 2006. *Labor Cero. Arquitectura a La Medida (Starting Point. Tailored Architecture)*. 1st ed. Medellín, Colombia: Argos.

Schon, D. A. and V. DeSanctis. 1986. The reflective practitioner: How professionals think in action. *Journal of Continuing Higher Education* 34 (3):29–30.

Soria López, F. J. 2005. Arquitectura y naturaleza a finales del siglo XX 1980–2000. Una aproximación dialógica para el diseño sostenible en arquitectura (Architecture and nature in the late twentieth century 1980–2000. A dialogic approach to sustainable design in architecture). Universitat Politècnica de Catalunya. http://www.tdx.cat/handle/10803/6799.

Torres, S. 2005. Factores Subjetivos de la Construcción Sostenible (Subjective factors of Sustainable Construction). *Ide@Sostenible. Espacio de Reflexión Y Comunicación En Desarrollo Sostenible*, 8, 1–9.

UNDP, United Nations Development Programme. 2010. *The Real Wealth of Nations: Pathways to Human Development. Human Development Report 2010*. Vol. 21. New York: Palgrave Macmillan. doi:10.2307/2137795.

UNDP, United Nations Development Programme. 2011. *Sustainability and Equity: A Better Future for All. Human Development Report 2011*. New York: Palgrave Macmillan. doi:10.2307/2137795.

Vitruvius, P. 1771. *The Architecture of M. Vitruvius* [translated from the original Latin, by W. Newton, Architect]. Pall-Mall, London: J. Dodsley.

9 Virtual Reality Devices Applied to Digital Games
A Literature Review

*Breno Carvalho, Marcelo M. Soares, Andre Neves,
Gabriel Soares, and Anthony Lins*

CONTENTS

9.1 INTRODUCTION

After World War II, scientists and researchers from the U.S. military departments conducted various studies to develop simulators for military training with a view to avoiding accidents to humans, besides reducing costs and pinpointing design flaws. These simulators did not have a system for obtaining visual feedback. All they did was to simulate the ratio of movements in a three-dimensional (3D) spatial perspective when they were being used. With the advances of computer vision technology, the first artifacts were created that enabled reality to be virtualized in a less complex way with regard to rendering graphics more realistic in addition to which they enhanced the user's experience of immersion. The term virtual reality (VR) was coined in the 1980s by the artist and computer scientist, Jaron Lanier, who thereby succeeded in expressing the search for the merger between what is real and what is virtual (Kirner and Torir, 2004). In mid-1982, the movie Tron by Steven Lisberger, spread this concept massively, by presenting the universe of VR as a technology for digital games, aided by quality graphics in visuals that served as the standard for the digital entertainment industry. As an example, there are the games *Halo*, *Crysis*, and *Far Cry 4*, which provide the user with greater interaction with and immersion in the technology.

VR is based on three basic principles (Pinho and Rebelo, 2004): immersion, interaction, and user involvement with the environment and narrative proposed. On using VR, software becomes more interactive, through which the user stars to become part of a virtual space, thus enabling data to be manipulated and exploited in real time by using one's senses in a 3D environment. The objective of this study is to review the state of the art on the concept of VR used in the universe of digital games, whether in the context of research or entertainment, and to do so by setting out from a rereading of scientific articles, books, and online publications on the subject.

In the context of digital games, in the past there was a major difficulty about using equipment such as a helmet with goggles because the ergonomics of these were poor. Currently, we see a new perspective as artifacts with better usability are being developed (Balista, 2013; Dias and Zorzal, 2013) that seek to improve the user's experience of digital games, for example, *Kinect* (Microsoft) and *Oculus Rift* (Oculus VR), together with other interaction mechanisms, such as gloves and the Omni treadmill platform.

This article is divided into four sections. First a rereading of what VR is and its aspects will be presented. Then we present the evolution of VR games without immersion, including the evolution of input artifacts that lead to greater playability and immersivity on comparing the 1980s with the 1990s. The third section describes some serious games applied to several areas that are not entertainment, which demonstrates the extent to which VR games are used. Finally we present some input artifacts for VR with immersion that were developed for the use of researchers, and others that have been developed for the purposes of entertaining the general public.

As to future prospects for using VR in digital games, what is being sought is the development of low cost devices as well as an increase in research studies on the use of nanotechnology applied to virtualizing reality to overcome the limitations of the human body within the virtual world.

9.2 VIRTUAL REALITY: THEORETICAL BASES

In the late 1980s, the artist Jaron Lanier presented the term VR as a technology that sought to merge the real with the virtual. However, research on this technology dates back to the 1960s and 1970s with the development of artifacts in seeking interactive interfaces closer to the human senses. These interfaces were trying to simulate the real world.

Long before the introduction of the term coined by Jaron Lanier, we should highlight the computer scientist Ivan Sutherland who was the forerunner of a series of research and development on artifacts that make the VR applications we know today possible (Kirner and Tori, 2004). It was in the 1960s he developed the VR helmet, *Ultimate Display* (Packer and Jordan, 2002) (see Figure 9.1). In this section we shall discuss the bases of VR presented by several researchers. It should be remembered that in-depth discussion on mixed reality and hyperreality will not be entered into.

According to Rebelo et al. (2011, p. 383), VR is "the use of computer modeling and simulation that allows a person to interact with a 3D visual artificial environment or to have other sensory involvement. In VR situations, the user is located in a computer-generated environment in which

FIGURE 9.1   Sword of Damocles was the first computer aided VR headset *Ultimate Display*. (Mounting the author from images of sites iGyaan.in. Copyright iGyaan.in.)

reality is simulated by using interactive devices. These send and receive information and can be used in the form of goggles, ear-phones, gloves, or body suits. In a typical VR environment, the user who makes use of a helmet with a stereoscopic screen sees animated images of a simulated environment."

According to Hand (1996), VR is the paradigm under which people use a computer to interact with something that is not real, but when they use it, people may consider it to be real. This can be considered as the most advanced digital interface between users and computers, where people can interact with a virtual model in real time (Whyte, 2002), which enables them to visualize and manipulate representations of the real world (Aukstakalnis and Blatne, 1992).

Some authors aimed at the concept of VR as a 3D virtual environment receiving human interactions by means of devices (goggles, gloves, and clothing) (Coates, 1992). These simulated environments are in general visited with the help of a garment with sensors, which feature stereo video goggles and gloves with fingers in optic fiber (Greenbaum, 1992). VR can also be defined as a real or simulated environment in which an observer experiences telepresence (Steuer, 1992).

Through this technology, the user can be given a state of experiencing something that does not exist or is not happening (Piovesan et al., 2011). Other researchers stated that VR is a simulation in which computer graphics are used to create a realistic-looking synthetic world that is not static but which responds to the user's interaction with virtual application (Burdea and Coiffet, 2003). The perspective of VR interaction is related to real-time response, which is an essential feature, and this shows that the computer is able to detect a user's input and modify the virtual world instantaneously (Vilar, 2012).

In 1991, Sega arrived with the first VR headsets for consumers. This device did not arrive on the shelves, because of a tepid response from the press, and fears that it could spoil the eyesight of children. Nintendo developed its version of VR headsets, the Virtual Boy, in 1995 (Nagashetti, 2015) (see Figure 9.2). This device too was a commercial failure, and the users faced discomforts like dizziness, nausea, and headaches after extended use. Other problems were for the lack of a head-tracking feature and color graphics in this device.

Together with interaction, immersion (see Figure 9.3) is another important aspect of VR. This is about the extent to which users' senses are attracted and motivated by the virtual world as if it were real (Duarte et al., 2010; Ragusa and Bochenek, 2001; Witmer and Singer, 1998). Generally, immersion in VR is related to artifacts that go beyond a common monitor and input devices, such as a mouse and keyboard. Thus, there are three levels of VR: immersive, nonimmersive, and semi-immersive (Ramaprabha and Sathik, 2012).

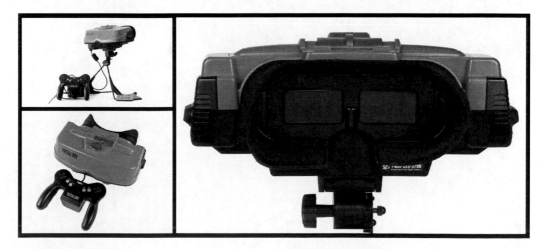

FIGURE 9.2 Virtual Boy by Nintendo released in 1995. (Mounting the author from images of site iGyaan. in. Copyright iGyaan.in.)

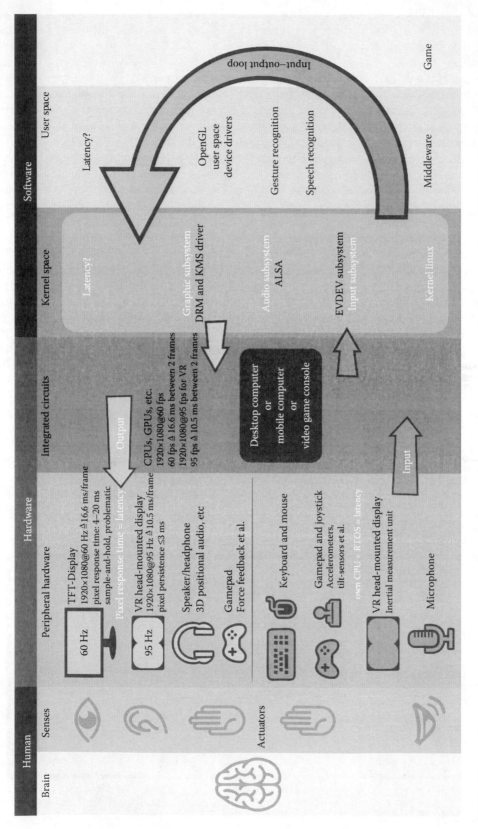

FIGURE 9.3 Paramount for the immersion into VR. (Adapted from Shmuel Csaba Otto Traian. Solve for Interesting, 2016. 24h in VR 0:00–0:23: Electric Monk. Solve for Interesting. http://solveforinteresting.com/24h-in-vr-000-023-electric-monk/ (accessed February 28).)

Immersive VR is when the interaction is mediated by a head-mounted display (HMD) and a position sensor of the tracker. The second classification is the nonimmersive one, where interaction occurs using a common monitor, replacing the HMD (Gorini et al., 2011). This model is the one most experienced by those who play commercial 3D games. The third category, the semi-immersive one, occurs when the HMD is replaced by multiple televisions or large projection screens, like IMAX movie screens (Gutierrez et al., 2008).

Other studies point out a third basic principle on which VR is based: involvement (Pinho and Rebelo, 2004). His perspective proposes that the user could be part of the virtual world. On directly interfering in the outcome of the application, the user could navigate the virtual environment passively or actively during the process of exploring a virtual environment. To involve the user, it is extremely important that there is a relationship between immersion and the narrative context because immersion creates an illusion of space while the context creates an experience that will be controlled by the subject's needs.

In *wayfinding* studies, VR is always a powerful tool because researchers can manipulate variables while considering more aspects than those used in the traditional media, thus allowing the experiments to be more controlled. The same happens with aspects of mental workload during tests of navigation, numbers, sounds, people moving around the building, time, and other variables that could influence *wayfinding* actions (Vilar, 2012).

An important argument about the use of VR is with regard to its flexibility. From this technology, researchers can develop new worlds for the purpose of study in addition to having greater control of the possible variables. Also, with this technology, any changes in the experimental environment can be made at low cost and in less time (Morganti et al., 2007).

We should stress that some studies on VR present the concept of augmented reality (AR) as a technology that prompts research with users based on real-time interactions with virtual objects inserted into the real context (Medeiros et al., 2008). The basic objective of an AR system is to increase the user's perception and interaction with the real world by means of having the real-world interact with virtual 3D objects that appear to coexist in the same space as the real world (Azuma et al., 2001).

Another definition for AR is a variation of the Virtual Environment or VR as it is more commonly called (Azuma, 1997). Besides defining AR, Azuma, Milgram, and Kishino (Milgram and Kishino, 1994) put forward a definition using the Milgram reality–virtuality continuum (see Figure 9.4). AR simply means "augmented reality," and this process is augmented only by adding some additional data so as to perceive the objects around us that are not visible memory. AR is the artificial, seamless, and dynamic integration of a new item of content, or the removal of the existing content of perceptions of the real world (Fahey, 2013).

On the left side of Figure 9.4 there is a reference to the real world. On the far right is the virtual world. A VR environment is a computer-simulated environment that can simulate the real world or an imaginary world in which the environment is virtual, while in AR the surrounding environment is real.

VR allows participants of the study to join the virtual environment from different perspectives and to interact with virtual objects, even those that would not fit into the real situation. The use of

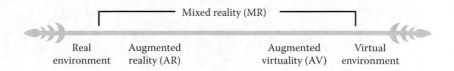

FIGURE 9.4 Milgram reality-virtuality continuum. (Adapted from Milgram, P. and Kishino, F. 1994. A taxonomy of mixed reality visual displays, IEICE (Institute of Electronics, Information and Communication Engineers) Transactions on Information and Systems, *Special issue on Networked Reality*, December.)

this technology allows participants' actions to be precisely recorded ranging from small steps to moments of hesitation which would not be perceived in the real world. By studying the actions of a person in VR, extremely difficult observations can be noted such as the movement of the direction in which the eye is observing (Vilar, 2012). Thus, the kind of observations that would be extremely valuable to make in the real world suddenly become possible in the virtual world (Conroy, 2001).

9.3 DIGITAL GAMES: HOW INTERACTION ENVIRONMENTS EVOLVED

A classic definition of what games are, is presented as a voluntary activity or occupation, performed during certain limits of time and space, with freely consented to and mandatory rules, accompanied by a feeling of tension and delight, as well as an awareness of being different from everyday life (Huizinga, 1993).

Since its emergence as a hobby, as in the classic Pong from Atari, in the 1970s, games has evolved their approaches in terms of mechanics, human–computer interaction (HCI), narratives, and characters. Ever since the first versions in 4 bits of information, with the limitations of two-dimensional (2D) graphics, the learning atmosphere of the games made interactions possible that led some users to get lost in time and space (Carvalho et al., 2013).

The games industry accompanied the technological advances of the past 40 years, by evolving playabilities for greater interaction and graphics that allowed photographs of characters in a 2D environment to be projected, such as the fight game *Mortal Kombat*, released in 1992 by the Nintendo Company (Lewis, 2013). In the 1980s, Ed Rotberg of the Arcades Division of Atari created the first 3D first-person game, *Battlezone*, in order to simulate a challenge from a tank warfare scenario (UOLa, 2014) (see Figure 9.5).

In 1988, Bruce Artwick and his team launched the game *Microsoft Flight Simulator 3.1*, in which the player interacted with 3D graphics and acceleration hardware, mediated by input devices

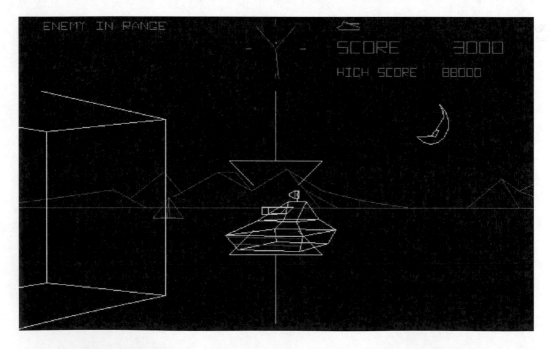

FIGURE 9.5 Game 3D Battlezone by Atari. (From Time. 2015. *Lists All-Time 100 Video Games—1980s Battlezone*. Techland Time. http://techland.time.com/2012/11/15/all-time-100-video-games/slide/battle-zone-1980/ (accessed September 12). Copyright Rebellion.)

that allowed the user to pilot a virtual plane and experience all the control interface of a real aircraft (Microsoft, 2013). This game was a milestone in the gaming industry, as it stimulated the research and development of artifacts for HCI, that were closer to the real world, for example, joysticks, steering controls, and so forth.

Another important milestone was the launch of the title *Doom*, from id Software in 1993 (UOLb, 2014), which popularized the style of first-person games. This game, in 32 bits (an innovation for its time) began the era of VR games without immersion as it allowed users to interact with realistic graphic elements in a 3D environment with the sound of the virtual environment.

Until 2002 there were no significant innovations in the context of HCI, except the improvement of the graphic processing of 3D elements. In 2003, with the launch of the multiplayer title *Second Life*, from Linden Lab, the player could create a virtual copy closer to himself, called an avatar or virtual human (Badler, 1997) and explore virtual environments with aspects of a human being's real and social life. By allowing a transposition of scenarios, emotions, and interpersonal interactions of the real world to the metaverse (Azevedo, 2013), this game sparked several discussions about the concept of the game, the level of the user's immersion without traditional VR devices, and matters related to sociopolitical and cultural rules since the platform allowed a second life to be created.

During the E3 event, the *Electronic Entertainment Expo* 2005, Nintendo revolutionized the way of interacting and playing games with the release of the *Wii* console and wireless controller *Wiimote* (see Figure 9.6), which made the experience much more immersive, richer, and more fun (Oliveira, 2009). Using sensor technology able to detect movement in three dimensions and an accelerometer, the player controlled his/her character by using buttons and carrying out intuitive movements according to the action of the game. Starting with *Wiimote*, other devices were developed through research, usability, and HCI.

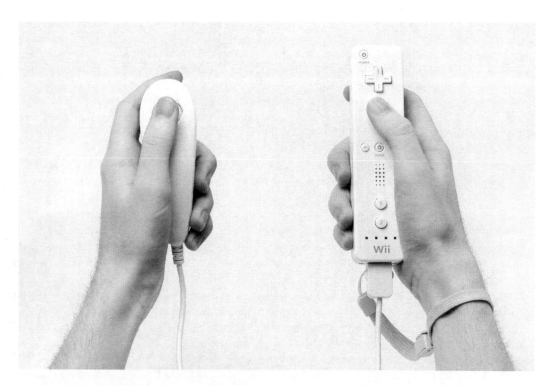

FIGURE 9.6 Wiimote and Nunchuck by Nintendo. (From Amazon. 2015. Wiimote + Nunchuck. Amazon. http://images.amazon.com/images/G/01/videogames/detail-page/B000IMYKQ0-2-lg.jpg accessed September 12.)

Microsoft developed the Christmas project so as to recognize gestures and in 2010 launched a *Kinect* device for the *Xbox-360* console, which allowed the player to control his/her character using body movements without using controls. Natural gestures captured by equipment which is currently cheaper and more accessible to the general public, provide commands and functions that are more oriented to human beings, thus making the experience of interaction with (entertainment and serious) games and interactive applications more immersive. *Kinect* and *PrimeSense* are examples of artifacts for recognizing 3D gestures. Likewise, *Smartboards* such as *Wiimote*, are common to 2D gestures.

These new artifacts of HCI allowed the use of VR games in other areas such as motor rehabilitation, psychological treatments, sports physiology, and business training. Another device that allows HCI through gesture recognition, at low cost is the *Leap Motion Controller*. This is a device made to interpret the gestures of the hands and fingers, whether to shoot in a game, turn the page on a tablet or draw in any kind of software. It is able to recognize movements as small as 0.1 mm. Recently, Google has been developing a project for the creation of 3D models based on the capture of objects (and environments) from the real world. This project is called *Project Tango* (Google, 2014).

A series of reports on the development of new more ergonomic VR glasses began in 2013, which greatly minimized the HCI problems of previous models, with an affordable cost to the consumer of public digital entertainment. One of the first models of these new devices was the *Oculus Rift*, from the Oculus company. To follow this technology, some companies began to adapt and develop their games to interact with these VR glasses, one of these games is the *Minecraft*, the version of the famous 3D game full of objects based on cubes, *Minecraft* for Oculus Rift DK2 (Rudderham, 2015).

9.4   VIRTUAL REALITY WITH SERIOUS GAMES

The term serious games is a class of games that aims to simulate everyday practical situations, with the objective of providing training for professionals, critical business situations, and educating children, teenagers, and adults. The games in this category have a specific purpose, that is, they extrapolate the idea of entertainment and offer other types of experiences, such as those that target learning and training (Machado et al., 2009).

The games known as *serious games* build a graphical interface used in video games to make the software more attractive. In addition, they carry out activities that foster the absorption of concepts and psychomotor skills (Duarte et al., 2012).

The usability of VR without immersion is observed in the use of *serious games* by researchers, scientists, and professionals from various fields such as health and education (Virvou and Katsionis, 2008). Some articles present studies that adapt commercial games; others developers their own digital artifacts.

A survey on the production of articles on *serious games* in Brazil linked to the correlated area of health shows there are studies on psychology, with the topics of learning disorders, attention deficit disorder and hyperactivity, therapy and autism; on speech therapy with topics on therapy of speech disorders, hearing loss and hearing disorder, oral motricity; on dentistry on periodontics; and in physiotherapy in the motor area (Duarte et al., 2012).

Computer games have recently been successfully adapted so as to generate environments and treat specific phobias, for example, trauma caused by traffic accidents (motor vehicle accident—MVA, including those of a nonserious nature but which increase the risk of severe psychiatric morbidity in survivors [Walshe et al., 2003]). According to researchers, the use of VR games in offices provides the psychologist with greater control over exposure stimuli and minimizes the exposure of patients to potential embarrassment (Walshe et al., 2003). Also in the field of psychological disorder treatments, there is also the use of AR technology, in which the patient is subjected to a simulation with spiders and cockroaches (Medeiros et al., 2008).

In the Laboratory of Virtual Reality applied to Design and Ergonomics, in the Department of Design, Federal University of Pernambuco, studies are being undertaken on the application of this technology as an aid in the rehabilitation of patients with Alzheimer disease, work-related musculoskeletal disorders, and measuring the strength limits of elderly users.

A growing field is "cyberpsychology." This discipline can be described in epistemological terms as the study of the connection between psychological processes and virtual action systems (Virole, 2003). When using games in virtual environments, for example, MMORPGs (*massive multiplayer online role playing games*) the unconscious representations of the players can be observed because it is considered that when a player is immersed in a section of the game, how he/she chooses his/her avatar is committed to his/her psyche in a unique way (Donard, 2011). According to this researcher, games like RPG *Divinity 2* have made positive results with her patients possible (Carvalho, 2014).

Soares and Falcão (2013) define games for learning as those that integrate games or the mechanics of a game in educational experiences. Thus, the authors continue, these games involve assigning scores commonly used in digital games, making students become involved so as to do better and to go on to the next level and finally achieve the goals. Thus, they are motivated to learn in a new learning space, which benefits the students' cognitive development and skills such as collaboration, communication, problem solving, critical thinking, and computer literacy (Johnson et al., 2012).

In education, games developed by research teams were found, some of which are initial experiments. One such project is the game *VR-Engage*, similar to the popular game *Doom* mentioned above. The ultimate goal of a player is to navigate through a virtual world and find the book of wisdom (Virvou and Katsionis, 2008). An interesting project is the game *Moon* (*Lua* in Portuguese), which presents a prism of Indigenous Brazilian Culture, with a digital artifact to assist with indigenous education in schools (Tonéis and Corral, 2013).

Another example is a 3D computer game to teach physics. The plot of the game is in the style of an adventure, discovery, and rewards. The character aims to survive the various traps scattered around the scenario and has to solve physics problems. The setting is a castle, where the player goes through different spaces and secret rooms, a basement, etc. (Calegari et al., 2013).

Another application example is the game *Legends of Alkhimia* developed by the Learning Laboratory in Sciences of the National Institute of Education Singapore. The digital game supports the chemistry curriculum for secondary education and students learn the subject through investigation, conducting chemical experiments, while working as chemical apprentices (Jan et al., 2010). The State Government of Pernambuco has very successfully held the Olympics of Digital Games and Education which aims to provide elementary school students and teachers with activities using dialogue and fun provided by the web environment, and emphasizes collaborative and cognitive activities by the participants. Many learning activities are carried out through simulations using virtual and AR environments.

KinectER is an educational game that aims to train nursing, medicine, dentistry, and orthopedics students. This game uses the Kinect device to mediate the interaction between the student and the digital narrative (Lanza et al., 2013). Also developed and commercialized in Brazil is the game *VRUM* of an educational character that enables children and adolescents to absorb the main traffic rules in a fun and intuitive way (Alves, 2012).

From the perspective of business and economics, we also found two games that deliver the study of taking decision factors to purchase products online, namely the *Ultimatum Game* and the *Dictator Game*. The games use an agent of multimodal communication, a virtual human, to interact with and persuade humans (Nouri and Traum, 2013).

Finally, other games for training are the *Celestia* game, geared primarily at astronomy, education, and planetary exploration, and *VT MÄK* of VR systems used military and war training (Cerqueira et al., 2013). It is important to make clear that in the latest National Technology Plan for Education in the United States, the game was named as an ideal method for assessing a pupil's understanding of knowledge (Johnson et al., 2012).

In this perspective, simulators for training naval professionals and officers are also being developed. This is a national passageway simulator, with the goal of simulating Brazilian merchant marine ships, which have important features when compared to other types of vessels (Lage et al., 2012).

9.5   INPUT DEVICES FOR VIRTUAL REALITY WITH IMMERSION

Since the creation of the first helmet for VR by Ivan Sutherland, such artifacts have always been used by private or government companies and research institutes. Due to economic factors this equipment was inaccessible to the general public. There were also ergonomic and usability problems because the helmets were heavy and caused users discomfort or nausea (Cobb et al., 1999). With the advance of research hardware, companies have invested in studies to make VR input devices with immersion more ergonomic and, in some cases, they are mass-produced and accessible to society as devices for entertainment. We present below some examples of products that have recently become available in the market.

9.5.1   OCULUS RIFT

The *Oculus Rift* is a new headset with a visor for VR. The *Rift* makes it possible for players to feel they have entered their favorite games and virtual worlds. The *Rift* uses customized tracking technology, to provide ultralow latency for 360° head movements, thereby allowing the user to look around the virtual world easily, just as one would do in real life. The *Oculus Rift* creates stereoscopic 3D viewing with excellent depth, scale, movement, and moving scenes (see Figure 9.7). The *Oculus Rift* provides a high-end VR experience at an affordable price. The combination of a wide field of vision with head tracking and stereoscopic 3D creates an immersive VR experience.

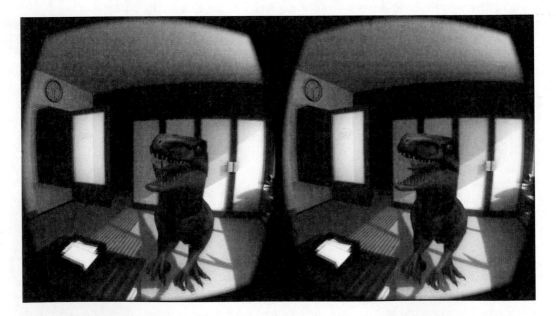

FIGURE 9.7   Velocirator in game Don't Let Go, by Rift developer Yorick van Vliet. (From Rudderham, T. 2015. *Minecrift Released for Oculus Rift DK2*. The Rift Arcade. http://www.theriftarcade.com/minecrift-released-for-oculus-rift-dk2/ (accessed September 12). Copyright Tom Rudderham, The Rift Arcade.)

9.5.2 Google CardBoard

Google also developed its VR glass, presented in its I/O conference in 2014 and named *Google CardBoard*. The device uses smartphones as screens and leverages the accelerometer and gyroscope features of devices with the Android operating system. The structure is of cardboard and can be purchased online or built by the user following the specifications provided by the developer (Google, 2015).

9.5.3 Omni

Omni (see Figure 9.8) is an omni-directional treadmill which can be used in conjunction with other devices for VR, such as *Oculus Rift*. The Omni treadmills promise to take the process of immersivity a step further than that achieved by current motion controllers such as *Wiimote*, *PlayStation Move*, and *Microsoft Kinect* so as to translate movements to an avatar on the screen, such as walking and running in real time. *Omni* is developed by the company *Virtuix* and is a treadmill model designed for home users. Its creators have recently demonstrated its use with *Oculus Rift*, increasing its potential for interactivity so as to provide an immersive VR experience, which allows players to perform real movements.

FIGURE 9.8 Gamer interacting with an artifact. (From Omni. 2015. *Omini*. Virtuix Omini. http://www. virtuix.com (accessed September 12). Copyright CEO, Jan Goetgeluk, Virtuix.)

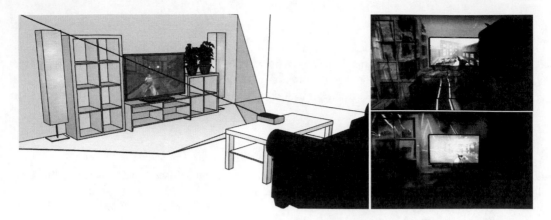

FIGURE 9.9   Device promises to change environment as per the context of the game. (Adapted from Jones, B. R., Benko, H., Ofek, E., and Wilson, A. D. 2013. IllumiRoom: Peripheral projected illusions for interactive experiences, in *Proceedings of the SIGCHI Conference on Human Factors in Computing Systems*. ACM, pp. 869–878. Copyright 2013 Microsoft Research.)

9.5.4   MICROSOFT ILLUMIROOM

Presented by Microsoft in early 2013, the *IllumiRoom* (see Figure 9.9) is a system that is interconnected with *Kinect* which projects details of games out of the TV, filling the entire room with animations and lots of light. However, the company stated to the press that the product is too expensive to be viable soon for the public. The high cost of the system will prevent it becoming popular in the short and medium term.

9.5.5   CAVE AUTOMATIC VIRTUAL ENVIRONMENT

Cave automatic virtual environment (*CAVE*) is a VR system with an advanced visualization solution, the size of a room, which combines high resolution, 3D stereoscopic projection, and computer graphics so as to create a complete sense of presence in a virtual environment. *CAVE* (see Figure 9.10) allows multiple users to have a complete immersive experience, in the same virtual environment, at the same time. This VR system is used to display data or projects, and can be adapted for use with digital games. The name refers to the allegory of the philosopher Plato's cave, which includes concepts on perception, reality, and illusion.

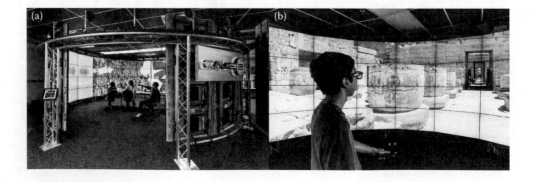

FIGURE 9.10   (a) Current version of the Cave system, the projects molecular dynamics simulation, copyright 2013 University of Illinois at Chicago, Electronic Visualization Laboratory. (b) Luxor (right), copyright 2012 University of Illinois at Chicago, Electronic Visualization Laboratory.

In molecular dynamics simulation of a glass fissure (see Figure 9.10a) participants fly through a close-up of a glass fissure computed in a 5-million atom molecular dynamics nanoscale simulation on the Argonne Leadership Computing Facility (ALCF) at Argonne National Laboratory. Simulation and data are provided by University of Southern California. Visualization is by ALCF. NanoVol software developed by the Electronic Visualization Laboratory (EVL) at the University of Illinois at Chicago is used to display the 3D model in EVL's CAVE2™ System. CAVE2 is a trademark of the University of Illinois Board of Trustees.

In Luxor (see Figure 9.10b), as a demonstration of cultural heritage, participants can view 360° panoramic images of Luxor, in Egypt, in the CAVE2™ Hybrid Reality Environment. Images were taken by researchers at Calit2-Qualcomm Institute (Calit2-QI) at the University of California, San Diego, with the assistance of collaborators from the King Abdullah University for Science and Technology in Saudi Arabia using the CAVEcam camera developed by Dick Ainsworth, Dan Sandin, and Tom DeFanti. Calit2-QI's CalVR software was used to display the panoramic image in the CAVE2 System at the EVL, University of Illinois at Chicago. CAVE2 is a trademark of the University of Illinois Board of Trustees.

9.5.6 PSYCHE

Psyche is the name of the Greek goddess, who represents the human soul, the spirit or the forces that influence thoughts, behaviors, emotions, perceptions, and personality. The *Psyche* project (see Figure 9.11) of the Cyberpsychology group at the Université du Québec en Outaouais (UQO) developed an immersive environment (often called "CAVE," in reference to another virtual environment

FIGURE 9.11 Presentation of the immersive environment Psyché. (Adapted from Uqo. 2014a. *PSYCHÉ (Voûte immersive à 6 faces)*. Université du Québec en Outaouais. http://w3.uqo.ca/cyberpsy/fr/psyche_fr.htm (accessed September 12). Copyright by @ Cyberpsychology Lab of UQO, Gatineau.)

project "Chicago Automatic Virtual Environment"). The system works in the form of a cube, and in which the individual uses 3D glasses in a standing position at the center of the cube, and where the virtual environment will be projected on all four walls, the ceiling, and floor. The 3D goggles enable pieces of information to be integrated, which leads to the individual finding himself/herself fully "immersed" in the virtual scenes.

Given the projects that are emerging both in the academic world and in industry, these initiatives demonstrate that people are trying to develop ways to create their own VR, but at an affordable price. They are doing so by using a game console as a low cost tool and this does not require the level of knowledge required to build a complete virtual environment such as CAVE. Considering that most people do not have the financial resources to build their own totally immersive environment, using a game console is presented as a viable option.

9.6 CONCLUSION

In just 40 years, digital games have gone through several transformations, evolving from 8-bit 2D graphics into 3D realism found in next-generation consoles in the entertainment industry. The 3D environment, interactions, and engagements found in the universe of commercial games fostered a formidable interface for integrating research with immersive VR artifacts for other fields of research, whether for health, education, military, or corporate training.

Research studies to enhance usability problems within the old traditional VR devices (goggles and gloves) have driven the entertainment industry and new developers of hardware and games to make it possible to have immersion in a parallel universe as seen in the science fiction films of the 1980s.

The main objective of this work was to conduct a review of the literature on the use of VR in digital games, in order to guide future research on the subject so as to develop virtual environments, with interaction devices that are less and less invasive for the human body, but which increase the quality of immersivity of these applications at low financial cost.

REFERENCES

Alves, P. 2012. *Lançamento do game VRUM (Launch the game VRUM)*. Games For Change América Latina (Games For Change Latin America). http://gamesforchange.org.br/2012/12/02/festival-tem-oficinas-test-drive-debates-e-feira-de-trocas-e-doacoes/ (accessed September 12).

Amazon. 2015. *Wiimote + Nunchuck*. Amazon. http://images.amazon.com/images/G/01/videogames/detail-page/B000IMYKQ0-2-lg.jpg (accessed September 12).

Aukstakalnis, S., Blatne, D. 1992. *Silicon Mirage: The Art and Science of Virtual Reality*. Berkeley: Peachpit Press.

Azevedo, T. 2013. *10 anos de 'Second Life': relembre a ascensão e a queda do metaverso (10 years of 'Second Life': remind the rise and fall of the metaverse)*. Uol Jogos. http://jogos.uol.com.br/ultimas-noticias/2013/06/28/10-anos-de-second-life-relembre-a-ascensao-e-a-queda-do-metaverso.htm (accessed September 12).

Azuma, R. 1997. *A Survey of Augmented Reality*, Presence: Teleoperators and Virtual Environments. V. 6, n. 4, August, pp. 355–385.

Azuma, R., Baillot, Y., Behringer, R., Feiner, S., Julier, S., and MacIntyre, B. 2001. Recent advances in augmented reality. *Journal of the IEEE Computer Graphics and Applications Archive*, 21(6), 34–47.

Badler, N. 1997. Real-time virtual humans, In *Proceedings IEEE 5th Pacific Conference on Computer Graphics and Applications*. IEEE Computer Society Press.

Balista, V. 2013. PhysioJoy: Sistema de Realidade Virtual para Avaliação e Reabilitação de Déficit Motor (Virtual Reality System for Evaluation and Rehabilitation of Motor Deficit), in *Workshop on Virtual, Augmented Reality and Games at the XII Simpósio Brasileiro de Games e Entretenimento Digital, SBGames*, São Paulo, pp. 17–20.

Burdea, G. C. and Coiffet, P. 2003. *Virtual Reality Technology* (2nd ed.). New York: Wiley-IEEE Press, p. 464.

Calegari, P. F., Quirino, S. S., Frigo, L. B., and Pozzebon, E. 2013. Jogo computacional 3D no ensino de física (3D computer game in physical education), in *Art & Design Track do XII Simpósio Brasileiro de Games e Entretenimento Digital, SBGames*, São Paulo, pp. 558–563.

Carvalho, B. 2014. *Pesquisadora da França visita o Curso de Jogos Digitais da Unicap (Researcher from France visit the Digital Games Course Unicap)*. Jogos Digitais Unicap. http://www.unicap.br/tecnologi-cos/jogos/?p=1799 (accessed September 12).

Carvalho, B. J. A., Soares, M. M., Neves, A. M. M., and Medeiros, R. P. 2013. Interactive doodles: A comparative analysis of the usability and playability of Google Trademark games between 2010 and 2012, in Marcus, A. (Org.). *Design, User Experience, and Usability: Health, Learning, Playing, Cultural, and Cross-Cultural User Experience. Proceedings, Part II.* Heidelberg: Springer, pp. 508–517.

Cerqueira, C. S., Santos, W. A., and Ambrosio, A. M. 2013. Serious game interaction techniques applied to an operational satellite simulator, in *Art & Design Track at the XII Simpósio Brasileiro de Games e Entretenimento Digital, SBGames*, São Paulo.

Coates, G. 1992. Program from invisible site—A virtual show. A multimedia performance work presented by George Coates Performance Works, San Francisco, CA, March.

Cobb, S. V. G., Nichols, S., Ramsey, A., and Wilson, J. R. 1999. Virtual reality-induced symptoms and effects (VRISE). *Presence: Teleoperators and Virtual Environments*, 8(2), 169–186.

Conroy, R. 2001. Spatial navigation in immersive virtual environments. Unpublished doctoral dissetation, University of London, p. 249. http://discovery.ucl.ac.uk/1111/.

Dias, D. A. and Zorzal, E. R. 2013. Desenvolvimento de um Jogo Sério com Realidade Aumentada para Apoiar a Educação Ambiental (Development of a Serious Game with Augmented Reality to Support Environmental Education), in *Workshop on Virtual, Augmented Reality and Games at the XII Simpósio Brasileiro de Games e Entretenimento Digital, SBGames*, São Paulo, pp. 65–68.

Donard, V. 2011, *Enjeux identitaires et relationnels des MMORPG* (Identity and relational issues of MMORPG). Pratiques Psychologiques, Société française de psychologie, Editions Elsevier Masson, 18(1), pp. 23–36.

Duarte, E., Rebelo, F., and Wogalter, M. S. 2010, Virtual reality and its potential for evaluating warning compliance. *Human Factors and Ergonomics in Manufacturing & Service Industries*, 20(6), 526–537.

Duarte, J. M., Vitti, S. R., Prado, C. S., Domenico, E. B. L. De, and Pisa, I. T. 2012, Revisão de serious games na área de saúde (Review of serious games in healthcare), in *XIII Congresso Brasileiro em Informática em Saúde—CBIS*, Curitiba, Brazil.

Fahey, J. 2013. *Augmented vs Virtual Reality: Contrasting Technologies and Tools*. James Fahey website: http://www.jamesfahey.com/2013/05/30/augmented-vs-virtual-reality-contrasting-technologies-and-tools/.

Greenbaum, P. 1992. The lawnmower man. *Film and video*, 9(3), 58–62.

Google. 2014. *The Future is Awesome*. Projeto Tango. http://www.google.com/atap/projecttango/ (accessed September 12).

Google. 2015. Google Cardboard. https://www.google.com/get/cardboard/ (accessed September, 15).

Gorini, A., Capideville, C. S., De Leo, G., Mantovani, F., and Riva, G. 2011. The role of immersion and narrative in mediated. Presence: The virtual hospital experience. *CyberPsychology, Behavior & Social Networking*, 14(3), 99–105.

Gutierrez, M., Vexo, F., and Thalmann, D. 2008. *Stepping into Virtual Reality*. Santa Clara, California: Springer-Verlag Telos.

Hand, C. 1996. Other faces of virtual reality. in P. Brusilovsky, P. Kommers, and N. Streitz (Eds.). *Multimedia, Hypermedia, and Virtual Reality Models, Systems, and Applications*, Berlin, Heidelberg: Springer, Vol. 1077, pp. 107–116.

Huizinga, J. 1993. *Homo Ludens: o jogo como elemento da cultura* (Homo Ludens: the game as cultural element), 4th ed. Tradução João Paulo Monteiro. São Paulo: Perspectiva.

Jan, M. F., Chee, Y. S., and Tan, E. M. 2010. Learning science via a science-in-the-making process: The design of a game-based-learning curriculum, in S. Martin (Ed.), *iVERG 2010 Proceedings—International Conference on Immersive Technologies for Learning: A Multidisciplinary Approach*. Stockton: Iverg Publishing, pp. 13–25.

Johnson, L., Adams, S., and Cummins, M. 2012. *NMC Horizon Report: 2012 K-12 Edition*. Austin, Texas: The New Media Consortium.

Jones, B. R., Benko, H., Ofek, E., and Wilson, A. D. 2013. IllumiRoom: Peripheral projected illusions for interactive experiences, in *Proceedings of the SIGCHI Conference on Human Factors in Computing Systems*. ACM, pp. 869–878.

Kirner, C. and Tori, R. (Eds.) 2004. Realidade virtual: conceito e tendências [Virtual reality: concept and trends], in Livro do Pré-Simpósio SVR 2004, Editora Mania de Livro, São Paulo.

Lage, M., Clua, E., Barboza, D., Taveira, G., Jefferson, W., Ruff, C., Vicente, J. et al. 2012. Simulador de Passadiço (Passadiço simulator), in *Military Simulation Workshop at the XI Simpósio Brasileiro de Games e Entretenimento Digital, SBGames*, Brasília, pp. 5–8.

Lanza, F. F., Lacerda, A. J., and Souza, A. A. 2013. KinectE.R.Desenvolvendo um game educacional com o uso do Kinect (Kinect ER. Developing an educational game using Kinect), in *Art & Design Track at the XII Simpósio Brasileiro de Games e Entretenimento Digital, SBGames*, São Paulo, pp. 541–546.

Lewis, S. 2013. *What it was like to 'go indie' in the 90s*. Gamasutra. http://www.gamasutra.com/view/news/200148/What_it_was_like_to_go_indie_in_the_90s.php (accessed September 12).

Machado, L. S., Moraes, R. M., and Nunes, F. 2009. Serious Games para Saúde e Treinamento Imersivo (Serious Games for Health and Immersive Training), in F. L. S. Nunes, L. S. Machado, M. S. Pinho, C. Kirner (Orgs.). *Abordagens Práticas de Realidade Virtual e Aumentada*. Porto Alegre: SBC, pp. 31–60.

Medeiros, D. C., Silva, W. A., Lamounier, E. A., Ribeiro, M. W., Cardoso, A., and Fortes, N. 2008. Realidade Virtual não-imersiva como tecnologia de apoio no desenvolvimento de protótipos para o auxílio no tratamento de aviofobia por profissionais de psicologia (non-immersive virtual reality as assistive technology in the development of prototypes for the aid in the treatment of aviofobia by psychology professionals), in *8° Workshop de Realidade Virtual e Aumentada, WRVA*. São Paulo, pp. 98–104.

Microsoft. 2013. *Microsoft Flight Simulator*. Microsoft. https://www.microsoft.com/Products/Games/FSInsider/product/Pages/default.aspx (accessed September 12).

Milgram, P. and Kishino, F. 1994. A taxonomy of mixed reality visual displays, IEICE (Institute of Electronics, Information and Communication Engineers) Transactions on Information and Systems, *Special issue on Networked Reality*, December.

Milgram, P. et. al. 1994. Augmented reality: A class of displays on the reality-virtuality continuum. *Telemanipulator and Telepresence Technologies*, SPIE, Vol. 2351.

Morganti, F., Carassa, A., and Geminiani, G. 2007. Planning optimal paths: A simple assessment of survey spatial knowledge in virtual environments. *Computers in Human Behavior*, 23(4), 1982–1996.

Nagashetti, H. 2015. *Here is Everything You Need to Know about VR Headsets*. Igyann. http://www.igyaan.in/93089/vr-headsets-explained/ (accessed September 12).

Nouri, E. and Traum, D. 2013. A cross-cultural study of playing simple economic games online with humans and virtual humans, in A. Marcus (Org.). *Human–Computer Interaction. Applications and Services. Proceedings, Part II*. Heidelberg: Springer, pp. 266–275.

Oliveira, S. 2009. *Top 10: Periféricos e Acessórios em consoles da Nintendo (Top 10: Peripherals and Accessories for Nintendo consoles)*. Nintendo Blast. http://www.nintendoblast.com.br/2009/07/top-10-perifericos-e-acessorios-em.html (accessed September 12).

Omni. 2015. *Omini*. Virtuix Omini. http://www.virtuix.com (accessed September 12).

Packer, R. and Jordan, K. (Eds.), 2002. *Multimedia: From Wagner to Virtual Reality*. New York: W. W. Norton & Company, pp. 396.

Pinho, M. S. and Rebelo, I. B. 2004. Interação em Ambientes Virtuais Imersivos [Interaction in Immersive Virtual Environments]. in: C. Kirner and R. Tori. (Org.). Realidade Virtual—Conceitos e Tendências (Virtual Reality—Concepts and Trends). São Paulo: Editora Mania de Livro, v. 1, pp. 109–132.

Piovesan, S. D., Balestrin, C., Pereira, A. S., Vit, A. R. D., Silva, J., and Franciscatto, R. 2011. Realidade Virtual Aplicada à Educação (Virtual Reality applied to education), in *EATI—II Encontro Anual de Tecnologia da Informação*. UFSM, RS.

Ragusa, J. M. and Bochenek, G. M. 2001. Collaborative virtual design environments: Introduction. *Communications of the ACM*, 44(12), 40–43.

Ramaprabha, T. and Satnik, M. M. 2012. The efficiency enhancement in non immersive virtual reality system by haptic devices. *International Journal of Advanced Research in Computer Science and Software Engineering*, 2(3), 113–117.

Rebelo, F., Duarte, E., Noriega, P., and Soares, M. 2011. Virtual reality in consumer products design: Methods and applications, in W. Karwowski, M. Soares, and N. Stanton (Eds.). *Human Factors and Ergonomics in Consumer Product Design: Methods and Techniques*. Boca Raton: CRC Press, pp. 381–404.

Rudderham, T. 2015. *Minecrift Released for Oculus Rift DK2*. The Rift Arcade. http://www.theriftarcade.com/minecrift-released-for-oculus-rift-dk2/ (accessed September 12).

Solve for Interesting. 2016. 24h in VR 0:00-0:23: Electric Monk—Solve for Interesting. http://solveforinteresting.com/24h-in-vr-000-023-electric-monk/ (accessed February 28).

Soares, M. M. and Falcão, C. 2013. *Design, Ergonomia e Novas Tecnologias na educação (Design, Ergonomics and New Technologies in Education)*, in L. M. Fadel, V. R. Ulbricht, and M. Castro Neto (Orgs.). *Hipermídia e acessibilidade na era da inclusão*. João Pessoa: Editora Ideia.

Steuer, J. 1992. Defining virtual reality: Dimensions determining telepresence, *Journal of Communication*, 42, 73–93.

Time. 2015. *Lists All-Time 100 Video Games—1980s Battlezone.* Techland Time. http://techland.time. com/2012/11/15/all-time-100-video-games/slide/battlezone-1980/ (accessed September 12).

Tonéis, C. N. and Corral, F. C. C. 2013. O game como fonte de diálogo no ambiente escolar. LUA: Uma aventura na mata através dos olhos do indígena (The game as dialog source at school. LUA: An adventure in the woods through indigenous eyes), in *Art & Design Track at the XII Simpósio Brasileiro de Games e Entretenimento Digital, SBGames*, São Paulo, pp. 446–454.

UOLa. 2014. *História do video game (Video game history)*, 1978. Uol Jogos. http://jogos.uol.com.br/ reportagens/historia/1978.jhtm (accessed September 12).

UOLb. 2014. *História do video game (Video game history)*, 1993. Uol Jogos. http://jogos.uol.com.br/ reportagens/historia/1993.jhtm (accessed September 12).

Vilar, E. 2012. Using Virtual Reality to Study the Influence of Environmental Variables to Enhance Wayfinding within Complex Buildings. Unpublished doctoral dissetation, Lisboa, p. 254.

Virole, B. 2003. *Du bon usage des jeux vidéo et autres aventures virtuelles* (Good use of video games and other virtual adventures). Paris: Hachette.

Virvou, M. and Katsionis, G. 2008. On the usability and likeability of virtual reality games for education: The case of VR-ENGAGE. *Computers & Education*, 50(1), 154–178.

Walshe, D. G., Lewis, E. J., Kim, S. I., O'Sullivan, K., and Wiederhold, B. K. 2003. Exploring the use of computer games and virtual reality in exposure therapy for fear of driving following a motor vehicle accident. *Cyberpsychology Behavior*, 6(3), 329–334.

Whyte, J. 2002. Virtual reality and the built environment. London: Architectural Press, p. 150.

Witmer, B. G. and Singer, M. J. 1998. Measuring presence in virtual environments: A presence questionnaire. *Presence: Teleoperators and Virtual Environments*, 7(3), 225–240.

Section II

Human Characteristics in Design

10 An Ergonomics Focus on Built Environments for the Elderly

Vilma Villarouco, Nara Raquel Silva Porto,
Marie Monique Bruère Paiva,
and Thaisa Francis César Sampaio Sarmento

CONTENTS

10.1 INTRODUCTION

The growth of the elderly population has raised awareness of the need for services and products for the elderly, in various branches of activities. Trade and industry see this segment as one that offers major business opportunities, with good profitability. Tourism and the leisure industry have also realized that seniors who travel and seek entertainment form a target group and thus these industries have created various possibilities for seniors.

By being an integral part of the business segment focused on seniors, collective housing called long-stay institutions for the elderly (LSIE), has proliferated in the last 30 or so years and has been well received and attracted large numbers of clients.

Since our line of research is dedicated to the study of these institutions, the project entitled "An Ergonomic look at housing for seniors" has been working in this segment for some years. This has resulted in four master's degree theses (three on design and one on ergonomics) and papers by undergraduate students who are being introduced to research in architecture and town planning. In this survey, a total of six homes for the elderly, located in the city of Recife, Pernambuco, in the northeast of Brazil, have been examined. In this sample, three were private enterprises maintained by monthly fees paid by the residents; two are public institutions maintained with government funds; and one is a charity, which receives funds from various sources.

Given that the architecture and design of homes for the elderly exercise a direct influence on the quality of life and care for their users (Barnes, 2002), LSIEs should be seen to have welcoming environments that resemble residences and foster their autonomy, independence, and privacy. In order to impart such attributes, several other features are required such as security, environmental comfort, accessibility, areas for leisure, which when combined with meeting specific laws and regulations, produce spaces suitable for an elderly person.

To evaluate spaces configured to accommodate seniors in these collective residences, ergonomics applied to the built environment was adopted as the theory and basic methodology for these research studies. Because it is interdisciplinary, ergonomics has become a very important tool for studies and projects targeting the elderly, and has contributed to the safety, comfort, and development of routine activities of everyday life which can be exercised by expending a low amount of physical energy, thereby adding to seniors' autonomy and self-esteem.

This paper describes the research undertaken and the approach taken is divided into two large blocks, for the purposes both of furthering the understanding of the methodology adopted and our findings. First, how one of the LSIEs was evaluated is set out by demonstrating each step in the conduct of the evaluation and each technique used. Then a comparative summary of the results from the six institutions evaluated is given, with comments on the suitability or otherwise that was identified of the measures taken.

Throughout the analysis, the parameters used were the laws and regulations applicable in Brazil to institutions for the elderly as were the ergonomic procedures applied to the built environment. Therefore, this is a research study based on a systemic and global view of the situation, as recommended in the specialist literature on ergonomics.

Although representing a market segment with high profitability and good growth prospects, what is evident in the analysis of the homes is that they are barely suited to their occupants, some of whom pay high amounts to stay in them. Problems are found in all the homes, with regard to the various levels charged to stay in them. Residents do not pay a fee to public or philanthropic LSIEs but those in private LSIEs pay monthly fees which vary greatly. Problems are identified in several areas, both in terms of compliance with legislation and the quality of the spaces for the elderly. These deficiencies are clear from the comparative tables of the results of research carried out to date (see Tables 10.3 through 10.5).

10.2 AGING

The phenomenon of an aging population is visible worldwide. In developing countries, this was only perceived as from 1950, and displayed a much faster pace of growth. This scenario requires economic, social, and political structures that are geared to dealing with the impacts of this demographic transition in order to ensure an adequate quality of life for the growing number of seniors (Tomasini, 2005).

Between 1940 and 1960, there was a significant decline in the mortality rate in Brazil, while fertility remained at very high levels, thus leading to a population that was young and almost stable by age band and with rapid economic growth (Nasri, 2008).

To the extent that a population ages, the proportion of people with ailments increases, and thus a new range of demands to meet the specific needs of this group emerges. Nevertheless, there is a large heterogeneity in the aging patterns, vulnerabilities, and the dependence of the elderly, which are influenced by living standards.

Aging is characterized by several organic alterations such as reductions in maintaining one's balance and mobility, in physiological capabilities. In addition, there are psychological changes which include vulnerability to depression being greater (Nahas, 2006; Maciel, 2010). A reduction is sometimes evident in the full capacity to perform the activities of daily living, thus underscoring the enormous importance that the physical environment plays in the aging process (Paiva et al., 2015a).

According to the Ministry of Health of the Brazilian Government—Law no. 10.741/2003 (Brasil Ministério da Saúde, 2006), seniors are defined as people who are aged 60 or over. As people get older, their quality of life is seen to be determined, to a large extent, by their ability to maintain their autonomy and independence. Most seniors fear old age since it may make them dependent on others because of illness or because they cannot perform their everyday activities (Paiva et al., 2015b). Such an eventuality strengthens the approach of maintaining a healthy life, which means compressing morbidity, thereby preventing incapacities (Freitas et al., 2010).

According to the *Age-friendly City Guide*, issued by WHO in 2008, in general, it is considered important that seniors should live in homes that are built with appropriate materials and are structurally sound; that have level surfaces; that have an elevator if there are stairs to climb; that the bathroom and the kitchen are adapted; that they are large enough to move around in; that there is adequate storage space; that the home has corridors with doors wide enough to allow the passage of a wheelchair; and that it is adequately equipped to meet environmental conditions (Organização Mundial de Saúde, 2008).

10.3   ERGONOMICS OF THE BUILT ENVIRONMENT

The study of the system of the environment and its users is a significant area of interest in the pursuit of improving the quality of people's life. This improvement to the built environment comes about because of ergonomics which uses the technology of the human–environment interface and therefore several methods to assess the user–environment relationship.

In addition to the physical issues, the ergonomics of the built environment seeks to consider variables such as ease of orientation, accessibility, furniture design, graphic optimization, and lighting design, it is being understood that architecture and design can improve the quality of space (Mont'alvão and Villarouco, 2011).

By doing so, the ergonomics of the built environment extends beyond architectural issues, and focuses on the adaptability and compliance of space to the tasks and activities that will be undertaken in it (Villarouco and Santos, 2002).

According to Furtado et al. (2013), inhabiting a building means more than using it physically and functionally. The configuration of the space of a dwelling makes gestures and behaviors possible or precludes them and thereby it reflects forms of sociability and has a role to play in forming identities and integrating collective and individual memories.

The physical user–environment relationship is what determines the success of planning spaces for the elderly, whether or not these are institutional spaces. Therefore, to ensure that people remain active as they get older, it is important to consider the physical environment in which they live. Taking into account the conditions of habitability, accessibility should be present to ensure the integration of people in general, but particularly of the elderly (Fernandes and Botelho, 2007).

By developing an ergonomic approach, in order to understand, evaluate, and modify the environment and continuous interaction with its user, the Ergonomic Methodology of the Built Environment (EMBE) sets out to analyze the physical environment and to identify the user's perception of the physical space in which he/she lives. To this end, it is based on comparing results obtained from the phases of physical and perceptual order, and thereby to generate an ergonomic diagnosis and subsequently recommendations that are necessary to improve the environment and to make it suitable (Paiva and Villarouco, 2012).

10.4 LSIE AND PERTINENT LEGISLATION

There is no doubt how important it is that the elderly live within a family environment and in the community. However, LSIEs arise as a housing option when it is impossible, for various reasons, for elderly people to remain in their homes.

According to the Brazilian Public Health Agency—ANVISA (Brasil, 2005), LSIEs are defined as "governmental or non-governmental institutions, of a residential character, intended for a collective household of people aged 60 or over, with or without family support, and which provide seniors with freedom and dignity and citizenship" (RDC no. 283, 2005, p. 1).

According to the Brazilian Institute of Applied Economic Research (IPEA, 2010), it is estimated that there is a growth in demand for forms of collective residence that may serve both seniors who are independent but lack sufficient income and/or family, as well as those who need special care due to having difficulties in carrying out everyday activities.

The institutional environment, of the LSIEs, needs to be made adequate so as to meet the main functional limitations of seniors, thereby guaranteeing the comfort, safety, and independence needed to carry out their daily activities. The spaces should also avoid and prevent falls, which generate different forms of immobility: fractures, fear, dependence, and a number of limiting consequences for the life of an LSIE resident (Farias et al., 2005; Villarouco et al., 2012).

Currently, issues related to the housing for seniors are regulated by Law 10.741/03, which sets out the terms of the Seniors' Statute (Estatuto do Idoso, in Portuguese); Law 8.842/94 which does likewise for the National Policy for Seniors; and the Resolution of the Board of the Directors (DRC no. 283, 2005) of the Public Health Agency—ANVISA, which specifically defines the minimum criteria for the operation and evaluation of LSIEs.

10.5 METHODOLOGY

Ergonomics of the built environment directs studies toward the human being as the user of the given space and the suitability of this environment for promoting his/her well-being, by using space in the best possible way, it being accessible to anyone and offering quality of life. Thus, this paper is part of a broader investigation that seeks to evaluate and understand the human–environment–activities relationships, where this person is a senior and the environment is a collective dwelling that targets this population.

From this perspective, this study set out to make a comparative analysis between an LSIE of a private character with regard to financial maintenance, which is considered of a high standard in terms of monthly expenditure, and five houses that are collective dwellings. Of these five, two are public, two are private, and one is philanthropic. The latter were previously studied by the Research Group in Ergonomics Applied to the Built Environment | Laboratório ErgoAmbiente (in Portuguese), which is linked to the post-graduate program in design at the Federal University of Pernambuco.

As for the ethical aspects, the study was submitted to and approved by the Ethics and Research Committee of the Federal University of Pernambuco—UFPE under no. 740.908. The seniors who volunteered to take part in the study were asked to read and sign a Term of Free and Informed Consent, as a prerequisite of their participation in the research.

The qualitative study used the EMBE (MEAC, in Portuguese) (Villarouco, 2008, 2009) in a LSIE which has a high standard in terms of monthly expenditure, and thereafter the data were compared to five other similar institutions surveyed previously.

TABLE 10.1
Structure of Applying the MEAC

	MEAC—EMBE	
	Step	**Objective**
Phase I	Global analysis of the environment	To make observations on the organizational structure, the dynamics of the institution, and the work processes
	Identification of the environmental configuration	To identify and measure physical and environmental conditioning factors
		To obtain information of a physical and organizational order, as well as a description of prescribed tasks
	Assessment of the environment in use	To identify the suitability of the environment (the extent to which it facilitates or hinders activities being undertaken) by analyzing the flows used to conduct tasks
Phase II	Environmental perception	To identify users' wishes with regard to the environment being studied by using an Environmental Psychology technique

10.5.1 EMBE (MEAC in Portuguese)

EMBE (MEAC) is a methodology that develops an ergonomic approach with a view to understanding, evaluating, and modifying the environment and continuous interaction with its user. It is developed in two phases—one is of a physical order and the other of a perceptual kind. The data collected generate an ergonomic diagnosis of the environment diagnostics, and this is followed by putting forward propositions for ergonomic adjustments (Table 10.1).

The physical analysis is divided into three stages: Global analysis of the environment; assessment of the environmental configuration; and assessment of the environment in use. These steps are aimed at understanding the environment–man–activity system. This comprises a survey of the physical and environmental conditioning factors, which may adversely affect the suitability of the environment and the performance of tasks that determine how the environment functions.

The analysis of the user's perception seeks to collect the user's vision, from the users themselves, on the space-user interaction, with cultural influences and those arising from the user's memory and personality. For this evaluation, environmental psychology techniques are used such as the *Wish Poem*, the *Constellation of Attributes*, a *Mind or Cognitive Map*, and so forth.

The ergonomic diagnosis presents itself as a stage of collating the problems found from the data gathered in phases 1 and 2. The methodology ends with making ergonomic recommendations that are propositions for design or attitudinal adjustments that aim to contribute to improving the quality of the environment analyzed.

10.6 CASE STUDY: PRIVATE INSTITUTION LSIE 6

The research study titled "An ergonomic look at housing for the elderly," prompted by the growth in the elderly population in Brazil, focuses on the quality of the environment of collective housing for seniors. Therefore, five institutions were studied. All of them are located in Recife, Brazil, and they have different sources of financial maintenance, namely from public, private, and philanthropic funds.

In seeking to verify the existence of a relationship between the monthly charges to elderly residents and the spatial quality of the institutions as to comfort and compliance with existing laws, a private LSIE (LSIE 6) was assessed which is considered of a high standard since its monthly fee is well above the average of institutions previously surveyed. Thus, the collective housing unit (LSIE 6) is presented below, in accordance with the phases of the EMBE (Villarouco, 2008, 2009), and during the description of the findings, there will also be discussion of the five institutions already investigated.

Photographic records were made, and also a cognitive screening test called the Mini-Examination of the Mental State was applied so as to identify the seniors who might be able to take part in the survey on the perception of the environment undertaken by applying the Wish Poem technique.

10.6.1   PHASE I: ANALYSIS OF THE GLOBAL ENVIRONMENT

The initial stage of physical analysis is presented as the first contact with the space to be analyzed, with the initial observations of the researcher who seeks to understand the environment–human–activity system from the perspective of a broad approach. This favors a systemic vision of the environment based on understanding the dynamics of the activities done in the spaces, with observation of possible spatial inadequacies. This also seeks to understand what is done in the environment and how it is done.

LSIE 6 is situated in the north zone of Recife, in an upper middle-class neighborhood, located on a corner, with pedestrian access which faces a quiet street, a sidewalk, and near a school. As it has elderly residents through spontaneous demand or mostly brought by family members, the LSIE is characterized as being private, the monthly charges are above the average paid at other institutions in this city. The monthly cost of board at LSIE 6 is approximately US$ 1500 for those who make use of individual accommodation units and US$ 1100 for twin units.

LSIE 6 consists of two buildings: (a) block A (Figure 10.1a)—an adapted house and (b) block B (Figure 10.1b)—a construction designed and built for the purpose of receiving the elderly. Block B originated from the existing demand to meet a waiting list for rooms in the house. On the ground floor of this block, there are 10 double rooms and a private bathroom. The upper floor provides for the construction of two more double rooms with bathrooms.

This collective housing also offers a partial board service with daily rates, with meals, a physiotherapy service, and recreational activities during the period of stay in the institution. During the research, 24 seniors with full-time board and 10 seniors in the partial board system, who do not use the facilities in the bedrooms, were identified.

The main access of the institution is located in block A, and this is achieved by climbing four broad steps or via a ramp, both of which are equipped with handrails, although these are not anatomical handles, and so do not offer safety to the user. The access does not have a canopy to protect users from bad weather conditions. The area is arid and without much vegetation, and takes people directly to the reception of the home.

There are no walkways or floors with tactile signage, except for a warning surface at the start of the external metal ramp giving access to the upper floor of block A, thus hindering the ambulation of seniors with low vision or who are blind throughout the environments of the home.

The workforce of the institution consists of the director (owner), administrator, supervisor, and 21 outsourced employees in shifts of 12 h of service, of whom nine work during the day and 12

(a)   (b)

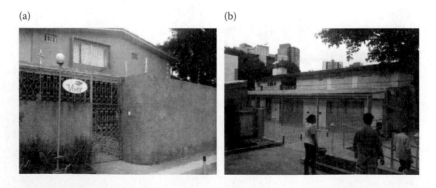

FIGURE 10.1   (a) Main access (block A) of LSIE 6 and (b) block B of LSIE 6.

(a) (b)

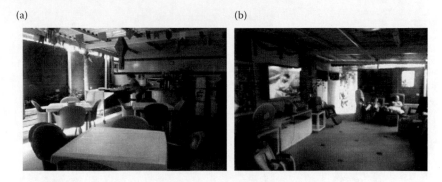

FIGURE 10.2 (a) Refectory and (b) social living area of block A of LSIE 6.

at night. In addition to these employees, if the need arises and by choice, the institution provides families the possibility of funding caregivers for their seniors. The elderly are attended to by a geriatrician who sees them once a fortnight or when requested. The physiotherapy, speech therapy, and occupational therapy services are included in the monthly cost and take place two to three times a week, in accordance with the needs of the elderly.

Next to the refectory is the space in which the seniors socialize. It is called the social life space and has a TV for entertainment (Figure 10.2a). In this environment, the seniors sit on individual armchairs or rocking chairs, and receive visits from their family. It is in this space that recreational activities take place (Figure 10.2b), and also where some seniors have physiotherapy and speech therapy.

10.6.2 PHASE I: IDENTIFICATION OF THE ENVIRONMENTAL CONFIGURATION

LSIE 6 has a built area of 313.70 m^2, distributed in blocks A and B, and on two floors (Figure 10.3). Block A, of the greater flow, has 10 bedrooms, on two floors, a central reception, an administrative area, a kitchen, a breakfast room, a staff restroom, and a social life area and a dining area, in addition to an open space where there are shrubs and trees.

Access to the upper floor is by way of an internal staircase provided with handrails, and of a metal ramp located in the external area. There are five single bedrooms and one double bedroom (Figure 10.4a), all of which have private bathrooms (Figure 10.4b), and the director's office.

The rooms are distributed on the ground floor and on the first floor. Two single accommodation rooms and a double occupancy room are located on the ground floor. For access to the rooms on the upper floor the elderly usually use the external ramp, with protection to sun and rain, with a nonslip floor. This access is also used by elderly people who use wheelchairs.

Block B also consists of two floors, but only the ground floor had been finished and was in operation. The floor plan shows the configuration of the rooms is more uniform than that of those in block A, because it is not a converted house but designed with the specific purpose of a collective residence for seniors. Thus, the extension of LSIE 6 has eight bedrooms, four of which are singles and four doubles, and all of which are side by side. Besides the bedrooms there is a covered social living area (Figure 10.5a), similar to the one in block A, and an open space in front of the bedrooms (Figure 10.5b).

As to the cladding materials, the floors of the entire institution, including the bathrooms and ramps, are clad with nonslip ceramic tiles, which is in compliance with RDC 283, which advocates the use of one type of nonslip ceramic throughout an institution, especially in bathrooms so as to avoid accidents and falls because these are wet areas.

10.6.3 EVALUATION OF LIGHTING COMFORT

The illuminance level was measured for the LSIE 6 environments of the bedrooms, bathrooms, and social areas (blocks A and B), reception, and the administration and services areas, and the results

FIGURE 10.3   Floor plan of the ground floor of block A of LSIE 6.

compared with the levels set out in NBR 8995/13. Of the 16 points measured, 12 of them had luminance below that recommended by the standard, and four—reception, the kitchen, the rest area, and
service area—were in accordance with what the legislation lays down as law, or above this.

It is important to point out the low luminance in environments such as the bedrooms where residents receive medical care and are given injectable or oral medications. In addition, this increases
the likelihood of falls and accidents, since most users have low vision. The nursing station does not
meet the standard either, which increases the possibility of mixing up medical records and medications, as well as contributing to employees having to increase their efforts as they are working in a

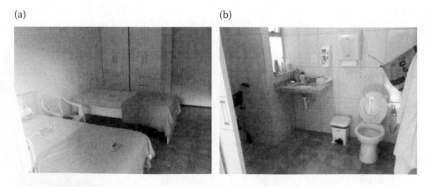

FIGURE 10.4   (a) Double bedroom of LSIE 6 and (b) bathroom of LSIE 6.

(a)   (b)

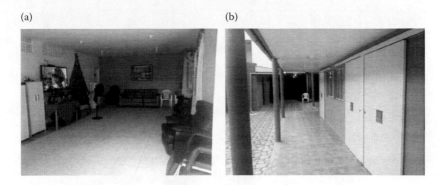

FIGURE 10.5   Block B of LSIE 6. (a) Social living area and (b) bedroom access.

dimly lit place. Also the administrative area was dimly lit as the lighting difference is almost 200 lux below the NBR recommendation.

10.6.4   EVALUATION OF ACOUSTIC COMFORT

In order to verify the adequacy of the LSIE 6 environments with regard to acoustic comfort, in accordance with Brazilian law NBR 10152/87, a comparison was made between the values measured in the institution and what the standard recommends. Of the 18 points measured, only one of them—a bedroom in block A—was in accordance with the limit set down, possibly because it is far from the places of greatest flow of people located on the ground floor in this block.

The following were inadequate in terms of the legislation: the environment of the nurses' station which has a very high level of noise (71 dB) as does the rest area (69 dB). These two noise levels are well above the parameters recommended between 35 and 45 dB.

All the bedrooms of the house positioned on the ground floor had inadequacies in relation to NBR 10152/87. Although the difference is slight, the measurements in the bedrooms of block A (67 dB) were higher than those recorded in block B (61.4 dB) for a 35–45 dB range laid down by the legislation. Also, the social areas were found to be inadequate as the values measured were 77 dB (block A) and 77 dB (block B), when what is laid down is between 40 and 50 dB.

The same difference was found when comparing the social areas, in which 77 dB (40–50 dB standard) was measured in block A, while in the same sector in block B, the measurement was 74 dB, and although this is not an environment aimed at rest, this inadequacy cannot be accepted.

10.6.5   EVALUATION OF THERMAL COMFORT

Thermal balance is considered the first condition for comfort, that is, the amount of heat gained by the body should be equal to the amount of heat transferred to the environment. The comfort zone is set between the effective temperatures of 20°C and 24°C, for heat-adapted organisms, in temperate countries. During the winter, with the body adapted to cold, this comfort zone is between 18°C and 22°C (IIDA, 2005).

Following the parameters published in the literature, when the thermal comfort was being assessed, it was observed that none of the environments analyzed complied with the temperature published by IIDA (2005).

High temperatures can harm the elderly, especially those above 85 years old. Since this is a vulnerable stage of life, overheating can aggravate their state of health or even cause illness (IIDA, 2005; Marto, 2005; Soares et al., 2012).

The highest measurements were concentrated in the kitchen environments (30.7°C) and service area (30.1°C) when the recommendation sets a maximum temperature of 24°C. A temperature of

TABLE 10.2

Sizing of LSIE 6 and Recommendations of RDC 283/2005

RDC 283/2005	Block A	Block B
Single room ≥5.50 m²	5.62 m²	5.57 m²
Twin room—7.50 m²/bed	6.50 m²	7.09 m²
Space between beds ≥0.80 m	0.75 m²	0.45 m²
Space bed—wall ≥0.50 m	None	None
Night-time safety light	Ok	Ok
Alarm bell	Ok	Ok
Window sill ≥1.00 m	0.60 m	0.60 m
Useful width of door opening ≥1.10 m	0.80 m	1.00 m
Bathrooms A ≥3.60 m²	2.97–3.60 m²	3.65 m²
Handrail 2 sides—circulation L ≥1.50 m	–	No
Handrail 1 side—circulation L < 1.50 m	Ok	–

above 27°C was recorded in all bedroom environments. Especially in the afternoon, the areas in which the seniors spend most of their time, the social areas, were very hot and stuffy and the temperatures measured were 27.24°C for block A and 27.35°C for block B.

10.6.6 Sizing

According to RDC Resolution 283/2005, the analysis of LSIE 6 showed some inadequacies as to some sizing items, which are highlighted in gray in Table 10.2.

10.6.7 Accessibility

Inadequacies were identified, relating to NBR 9050/2004 legislation on accessibility, in all bathrooms of ILPI 6 (Figure 10.4b), regarding the items about support bar items and the installation height of the basin, seat, and grab bars for the shower box, height of the wash-hand basin and the mirror. Only the items of floor cladding, height of the discharge, and box complied in full. As to the faucet, the distance from the outside edge was respected but it is not activated automatically nor is it a model with levers.

The handrails installed on the ramp and entrance staircase to the institution as well as the handrail of the staircase that leads to the first floor are of a format that makes it difficult to grasp, especially for elderly people with pathologies that affect their hands, which is in disagreement with NBR 9050/2005.

10.6.8 Phase I—Analysis of the Environment in Use

The administration reported that all the seniors' utensils are listed by color and other details to avoid belongings being mixed up. Space in the bedrooms is very limited, most beds being against a wall, making it difficult to attend to bedridden or dependent seniors because this prevents circulation around the bed.

All rooms have a mobile control that makes an audible signal at the nursing station when triggered, and serves to permit a senior to call a caregiver to the bedroom.

Medications are prepared in the infirmary which serves the two blocks of the house. All medical history records are there, space for which is very little and dimly lit.

Despite there being two social areas, collective activities are concentrated in the area of block A, where the environment is dimly lit and the heat is intense, due to the absence of natural ventilation. This also the area of intense and continuous flow because seniors spend most of the day there.

The refectory is an environment adjacent to the social area, and so has the same physical characteristics. This area has square tables with capacity for four places at each one, and is where the elderly have their meals, as well as the employees that provide services for the institution.

Since everyone uses it, there is a schedule by which the elderly have lunch between 11:30 and 12:30, and the staff have their meal thereafter because the space does not fit everyone together.

10.6.9   Phase II—Perception of the Environment

The instrument used to assess environmental perception was the Wish Poem (Sanoff, 1991) after cognitive screening was applied by using the version of the Mini-Examination of the Mental State (Bertolucci et al., 1994) adapted for Brazil for all seniors of the institution who were able to interact and respond to it.

Of the 20 seniors in the institution, seven were unable to respond to the cognitive screening test, and others were not present when the test was applied. Thus, of the 13 seniors assessed by the Mini Mental Exam, only six were considered eligible to take part in the survey. Among these, it was found that most of them made some reference to family, thereby demonstrating that they do make a link to the concept of home.

The results revealed that three seniors of those surveyed wished some change in the environment; one senior would like the house to have three bedrooms in order to house her family; one senior wished for a larger house, with more furniture; wider (hospital type) beds, and one senior wished a change that would lead to there being more green areas to the house. However, the seniors showed they had accepted the environment around them, as they did not wish great changes, or even no change in the environment.

10.6.10   Ergonomic Diagnosis and Recommendations for LSIE 6

The ergonomic evaluation of the environment ends with the diagnosis of the situation analyzed. Thus, the ergonomic diagnosis compares the data found, based on the researcher's observations, the interactions with the various stakeholders questioned, and the elements of the users' environmental perception. Thus, the result of this comparison of findings provides an overview of faults and problems as well as strengths such that this enables suggestions for improvements and solutions to issues to be made that represent bottlenecks in the performance of the system.

The built environment of LSIE 6 is satisfactory for some users but when this is evaluated in the light of ergonomics and specific legislation for LSIEs, some inadequacies are identified.

The fact that the institution is private, and of a high standard due to the monthly fee paid by the seniors created the expectation that the housing would present a high level of adequacy as to the standards pertinent for buildings of this nature. However, inadequacies were observed regarding the laws RDC 283/2005 and NBR 9050/2004, even in block B (expansion of the collective housing) which had recently been built and designed for this purpose.

Block A (an adapted house) features limited environments and the intense flow of people in confined spaces. It was also noted that activities overlap in the same environment (the refectory, the social and relaxation areas, and those for fun thinking activities, recreational, and celebratory activities and physiotherapy treatment). The existing open area is small and with few shrubs and trees, thus making it likely that the seniors remain indoors.

As for the environmental comfort, a low luminance index in 75% of the evaluated points was identified in ILPI 6, which is in disagreement with law NBR 8995/13, despite all environments having natural lighting.

Not much different are the results for acoustic comfort, which had rates above those set out in the NBR 10152/87 standard in 10 of the 16 environments analyzed. Only the bedrooms of the upper floor of block A had noise levels below the limit laid down.

As for thermal comfort, it was found that heating and ventilation was poor, mainly in the social area of block A, the place in which the seniors spend most of their time.

The sizing of the dormitories in block A of the LSIE is poor, except for those twin rooms that are occupied by only one senior. There was also the absence of standardization of the furniture with improved important features for users (quantity, height, and width of beds, chair coverings, etc.).

With respect to accessibility, irregularities were found regarding the handrails (type of grip, height, and location of installation) as was their absence in block B. Support bars were found in the bathrooms of block A, but these do not comply with the legislation; in block B they do not exist. The doors also have an irregularity with regard to the useful free width recommended by NBR 9050/2004.

10.7   COMPARATIVE ANALYSIS BETWEEN THE LSIEs STUDIED

This topic succinctly presents each of the institutions, it being emphasized that for every one of them, the same survey as in the previous section for ILPI 6 was carried out. The methodology for all assessments was the EMBE. The tables summarize the numbers and comparative data between all of them.

In order to maintain the anonymity of the institutions, they will be identified in this paper by numbers and in all the photographs in which the resident seniors appear, their faces have been blurred in order for them not to be recognized.

Of the LSIEs surveyed, two are of a public nature, one is philanthropic, and three are private, one of which is considered of a high standard. The latter, called LSIE 6, has a monthly cost of around US$ 1500 for expenditure on food, accommodation, caregivers, and the health team. For the institutions, the cost per senior varies greatly according to the services offered and the type of resource received by the institution.

LSIE 1 is a public institution, located in the west of the city, for up to 24 seniors who arrive by spontaneous demand or are brought in from the streets on which they have been abandoned, the acceptance criteria being that they do not have a fixed income.

Seniors can go out and undertake activities outside the LSIE. Therefore, this institution houses only independent seniors. The building is a converted residential house, and divided into a house where there are bedrooms and another that holds the services. Figure 10.6 is a photograph of its exterior (a) and the sectored floor plan of the house (b), with the built area highlighted in light gray.

LSIE 2 is also a public institution and it too is located in the west zone of the city. It mostly houses seniors whose family has abandoned them or seniors who are homeless. It has the capacity for 40 seniors of both sexes, most of whom depend on special care and do not have the independence to perform routine activities. The house is an adaptation of a residence. Figure 10.7 shows the floor plan of LSIE 2.

LSIE 3 is located in the metropolitan region, is the farthest from the center of the city, and houses about 120 seniors, there being 50% of each sex, who are separated into male and female pavilions, and of whom 85 are considered independent. It is a philanthropic institution, and one of the oldest in Pernambuco, which has a large group of buildings and a large green area in the outdoor area, and occupies a very large site. Figure 10.8 shows the spatial configuration of LSIE 3.

LSIE 4 is located in the west zone and is private. Its seniors are either retirees from the town hall or those brought by family members, who pay a monthly fee and provide medications and other features that may be needed. It functions in a space adapted from two neighboring residences. It accommodates 47 residents, most of whom depend on assistance to move around and in the activities of everyday living. Figure 10.9 shows the floor plan of LSIE 4.

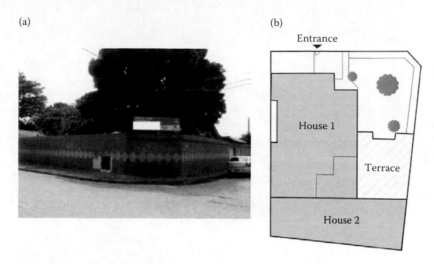

FIGURE 10.6 (a) Façade of LSIE 1 and (b) sectored floor plan of LSIE 1.

FIGURE 10.7 (a) Sectored floor plan and (b) external area of LSIE 2.

(a)   (b)

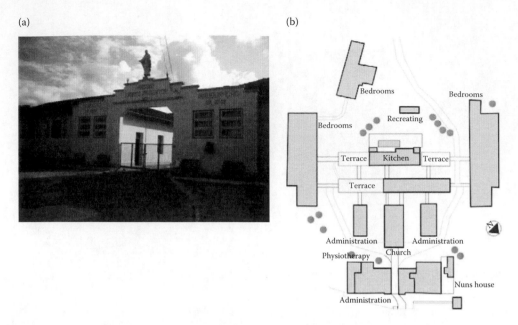

FIGURE 10.8   (a) Main access of LSIE 3 and (b) sectored floor plan of LSIE 3.

LSIE 5 is located in the north of the city, and is formed by the junction of three houses. There are 36 residents, male and female, nearly half of whom are totally dependent. The seniors are brought by family or come by spontaneous demand. The monthly cost is approximately US$ 500 which does not include a caregiver for those who are dependent nor are medications included. Figure 10.10 shows the floor plan of LSIE 5.

(a)   (b)

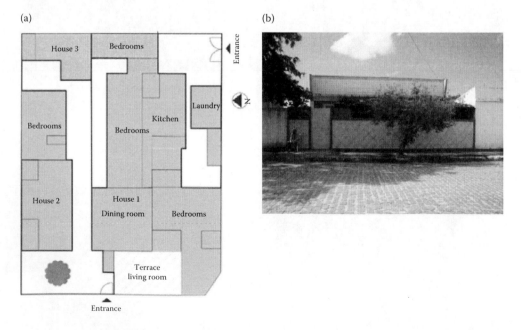

FIGURE 10.9   (a) Sectored floor plan of LSIE 4 and (b) main access of LSIE 4.

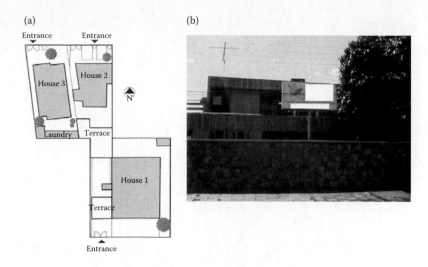

FIGURE 10.10   (a) Sectored floor plan of LSIE 5 and (b) main access of LSIE 5.

10.7.1   COMPARISON OF THE DATA FOUND IN THE LSIEs EVALUATED

We start by showing the data on area constructed for each institution. In the left column they are grouped by color, corresponding to institution type in accordance with the financial cost. The colors are explained in the key.

Of the six institutions surveyed, the site they occupy varies in size as shown in Table 10.3. LSIE 3, a philanthropic institution, has the largest area, a conserved green area and large indoor spaces. The second largest is LSIE 2, a public institution, which has a total area of 1428 m². LSIEs 4 and 5, which are private, have areas of 640 and 817 m² respectively, while LSIE 5 is divided into three houses. LSIE 1 covers 395 m², while the smallest of the institutions surveyed is LSIE 6, a private institution, which is considered to have a high standard. A very wide variation is seen between institutions, yet this has no relationship as to how they are priced.

In order to compare the data found in the six institutions surveyed, tables were compiled by looking at the data from each institution in relation to Brazilian standards, thereby showing the extent to which the data do or do not comply with these. Table 10.4 deals with the standards as to illumination, acoustic comfort, and temperature. The first line cites the regulatory standard; the second line, the environments in which the measurements were made; and the third line, the values

TABLE 10.3
Comparison of the Built Area between the LSIEs Surveyed

Institution	Built Area (m²)
LSIE 1	395.00
LSIE 2	1428.00
LSIE 3	7941.99
LSIE 4	640.00
LSIE 5	817.00
LSIE 6	313.70

Note: Public institutions in light gray; philanthropic institution, no colors; and
private institutions in dark gray.

TABLE 10.4
Comparison between Findings with Regard to Environmental Comfort

LSIE	Ilumination (lux) NBR 8995/13			Noise (dB) NBR 10152/87			Temperature (°C) (IIDA, 2005)		
	BR	LR	DR	BR	LR	DR	BR	LR	DR
	200	300	300	35–45	40–50	45–55	20 a 240	20 a 240	20 a 240
LSIE 1	Fem. 55	1811	203	F. 63.0	73	73.3	30.6°	31.13°	31°
	Masc. 15			M. 61.0			30.8°		
LSIE 2	Fem. 96	20	47	F. 67.4	71.2	75.5	F. 27.1°	27.6°	27.6°
	Masc. 56			M. 72.1			M. 27.2°		
LSIE 3	Block G 80	33	95	Block G 59	73	62	Not measured	Not measured	Not measured
	Block H 36			Block H 71					
LSIE 4	7	164	75	60.8	76.5	71.7	30.01	30.9	30
	222			59.2			29.4		
LSIE 5	395	50	Not measured	55	86	75	32.6	31.9	31.7
LSIE 6	Block A ground—108	20	69	A1-67	A-77	71	A-27.8	A-27.24	27.9
	Block A 1st floor—150			A2-44					
	Block B 62	120		B-61.4	B-74		B-27.68	B-27.35°	

Note: BR—bedroom, LR—living room, and DR—dining room. Colors: In light gray, the public institutions, no colors is the institution of mixed costs, and in dark gray, the private institutions.

recommended by the standards per environment. The columns are highlighted in dark gray for matters that are inadequate as to compliance with the recommendations of the standards and in light gray for those that are adequate.

10.7.2   LIGHTING COMFORT CONDITIONS

After comparing the data obtained with the standard NBR 8995/13, as regards the lighting in relation to the bedrooms, only one of the units of LSIE was in compliance with the standard. Another unit showed very low luminance of 7 lux, which raises the possibility of making this adequate, rather than because of concern about meeting the standard. LSIE 5 also fully met NBR 8995/13, but the other institutions (LSIE 1; LSIE 2; and LSIE 6) were well below the 200 lux recommended. Note that this inadequacy was identified regardless of the legal or financial characteristic. The social areas of five institutions were below 300 lux, which is the index rate recommended for this type of environment. Only LSIE 1 was in compliance but its luminance was much higher than that laid down, around 1811 lux. With regard to the refectory, this piece of data on the environments highlighted in this table was not measured for LSIE 5, and the other five institutions are below or well below the 300 lux recommended.

10.7.3   ACOUSTIC COMFORT CONDITIONS

According to NBR 10152/87, of all the environments identified as relevant—bedroom, social area, and refectory, in the six LSIEs studied, only the bedroom on the first floor of block A of LSIE 6 (a private institution of a high standard) followed the norm, 44 dB was the measurement. This is close to the limit, as the standard recommends between 35 and 45 dB.

On analyzing the levels of inadequacy found, in the item bedrooms, the highest value found was 72.1 dB in the male bedroom in LSIE 2. In the item social area, the standard lays down 40–50 dB. The lowest value obtained was 71.2 dB and the highest was 86 dB (measured in the social area of LSIE 5). In the refectories, the standard lays down 45–55 dB. The highest value (75.5 dB) was in LSIE 2 and the lowest (62 dB) in LSIE 3.

10.7.4   THERMAL COMFORT CONDITIONS

IIDA (2005) recommends that for human comfort, the internal temperature in environments should be between 20°C and 24°C. Except in LSIE 3, in the five other LSIEs, no environment can be considered as being at a suitable temperature. The environments of LSIE 5 were at temperatures higher than 31.5°C in all environments, and reached 32.6°C. In LSIE 1, the temperature ranged from 30.6°C to 31.3°C. In LSIE 2, the temperature did not exceed 27.6°C. In LSIE 4, the range was between 29.4°C and 30.9°C in its environments and temperatures in LSIE 6 were around 27.24°C and 27.9°C. The temperatures ranged from 31.7°C to 32.6°C in LSIE 5.

10.7.5   SIZING CONDITIONS

In Brazil, there is specific legislation regarding buildings intended for collective housing for the elderly, defined by ANVISA (the Brazilian National Agency for Public Health). This is the Regulatory Standard RDC 283 of 2005. Table 10.5 shows the extent to which the six LSIEs are or are not in compliance (the latter highlighted in dark gray) with respect to DRC 283/2005, which deals with internal sizing.

The table identifies that many criteria of the standard are not met in the institutions, regardless of whether they are public or private, or even if the charge is higher or lower. Some of the requirements are not met by any of the institutions. It was not possible to identify whether the owners or managers are unaware of the standards but the lack of attention to the sizing criteria causes very small environments, thus hindering their smooth functioning. On the other hand, there is no effective

TABLE 10.5

Comparison of the LSIEs as to Dimensioning of Physical Areas in Accordance to RDC 283/2005

Environment	Requirements of RDC 283/2005 (Sizing of Physical Areas)	LSIE 1	LSIE 2	LSIE 3	LSIE 4	LSIE 5	LSIE 6
Environment Internal Circulation	≥1.50 m, has a handrail on both sides	No	No	No	No	No	No
	≤1.50 m, has a handrail on one side	No	No	No	No	No	No
	Free breadth of the span ≥1.10 m Simple locking mechanism	No	No	No	No	No	No
Doors Dormitories	Separate ones by gender I maximum for 4 people	Ok	No	Ok	Ok	Ok	Ok
	Has a bathroom	Ok	No	No	No	No	Ok
	For 2 to 4 people (area/bed, including area for clothes wardrobe)—5.50 m²	No	No	Ok	No	No	Ok
	Has a permanent safety lamp lit	No	No	Ok	Ok	No	Ok
	Has an alarm bell	No	No	Ok	Ok	No	Ok
	Distance between 2 beds—d ≥ 0.80 m	No	No	Ok	No	No	No
	Distance between side of the bed and wall (d = 0.50 m)	No	No	No	No	No	No
Activity Areas	Social room has 1.30 m² per person	Ok	No	Ok	Ok	Ok	Ok
	Individual support room—9.00 m²	No	No	No	Ok	Ok	No
	Area 3.6 m²—has a transfer area	Ok	Ok	Ok	Ok	Ok	Ok
Bathrooms	Area per user—(LSEI increased by snack storage area)—1.0 m²	Ok	Ok	Ok	Ok	Ok	Ok
Refectory	Separated by gender	Ok	No	No	No	Ok	Ok
	1 basin I 1 wash-hand basin I 1 shower—for every 10 employees with 3.60 m² and changing area of 0.5 m²	No	No	No	No	No	No

Note: Full compliance—represented by Ok; Is not in compliance with respect to DRC 283/2005—highlighted in dark gray.

inspection system by the regulatory agencies, which leads to many institutions continuing to function even though they do not meet the legal requirements.

10.7.6 ACCESSIBILITY CONDITIONS

Accessibility was also a criterion of analysis used to compare the six institutions. The findings are listed in Table 10.6. To be considered adequate, the institution should meet the recommendations of NBR 9050/2004. This standard (NBR—the Brazilian Standard) has the force of law in Brazil, and is issued by ABNT (the Brazilian Association for Technical Standards) which regulates all conditions to make spaces accessible to people with disabilities or reduced mobility. The revised NBR 9050 was published on September 11, 2015. However, the research studies discussed here were completed prior to the publication of this new version.

No institution complies fully with NBR 9050/2004. Problems are identified with regard to the support bars for toilet bowls, floors, types and heights of toilet bowls, activation of flushing cisterns, areas of the box for bathing, existence of a bath seat, grab bars in shower boxes, and wash-hand basins, both in dimensions and in space for approach. Also the taps do not meet standards in most of the homes, and there are no mirrors.

TABLE 10.6

Comparison of the LSIEs as to Accessibility in Accordance with NBR 9050/2004

Item	NBR 9050/2004	General Recommendations NBR 9050/2004	LSIE 1	LSIE 2	LSIE 3	LSIE 4	LSIE 5	LSIE 6
Support bars of the basin	Diameter; 3 and 4.5 cm; C = 0.80 m; H = 0.75 m	Fixation to the basin, lateral and bottom; articulated or fixed bars are allowed	No	No	No	No	No	No
Flooring		Regular, firm surface, stable, and nonslip	No	Ok	No	No	Ok	Ok
Toilet bowl	H = 0.43 to 0.45 m from the floor based on the edge without seat 0.46 m with seat	Foresees areas of lateral, perpendicular, and diagonal transfer	No	No	No	No	Ok	No
Discharge	H = 1.00 m	Preference, lever or automatic	No	No	No	No	Ok	Ok
Box	0.90 m × 0.95 m	Foresees area of external transfer; door with resistant to impact material	Ok	Ok	Ok	Ok	Ok	Ok
Seat of the box	P = 0.45 m; H = 0.46 m from the floor; C = 0.70 m	Articulated or removable; nonslip and waterproof surface	Ok	Ok	Ok	Ok	Ok	Ok
Support bars of the box	Vertical and in "L": H = 0.75 m from the floor; C = 0.7 m; Distance = 0.85 m from the side wall; Horizontal: C = 0.6 m; H = 0.7 m	Vertical, horizontal, or "L"-shaped bars; located on the wall to which the seat is fixed and on the side wall	No	No	Ok	No	No	No
Wash-hand basin	H = 0.78–0.80 m from that of the edge above the floor; Siphon: 0.25 m from the front face	Foresees area of frontal approximation; the siphon should be protected; the use of columns to the floor or cabinets is not permitted; bars of joint support should be installed	No	No	No	No	No	No
Tap	Max. of 0.50 m from the external front face of the wash-hand basin	Triggered by an automatic lever or sensor	No	No	No	Ok	No	No
Mirror	H = 0.90 from the bottom edge and 1.80 m from the top edge to the floor (per/vertical mirror)	A mirror can be used with a 10% incline from the vertical plain	No	No	No	No	No	No

Note: Full compliance—represented by Ok; does not meet the recommendations of NBR 9050/2004—highlighted in dark gray.

In addition, there are ramps with slopes greater than those permitted, doors narrower than recommended, handrails that are outside the standards and loose because they have been poorly fixed.

10.7.7 ANALYSIS OF THE PERCEPTION OF THE ENVIRONMENT

Although considered an important phase of the ergonomic evaluation of environments, the understanding of the seniors' perception about the spaces in which they live was not successful. In four LSIEs, perception of the environment was assessed by the Constellation of Attributes technique, and did not obtain satisfactory results. The responses obtained from the Constellation of Attributes were as vague as the answers found from using the Wish Poem technique, adopted in LSIE 6 because of the need to innovate perceived by the team.

Most seniors could not answer or responded with texts that did not lead to a secure interpretation of their perception. Some seemed to repeat something that they had heard from someone else, such as expressions of: that place is good, was as the family wanted it to be. It was not possible to assess the real perception of the environment, and also the number of seniors who managed to answer was very small within the sample. The interviews with the staff were limited because most refused to participate, probably because they understood that they could harm themselves if their opinion was reported.

In one of the houses, the philanthropic one, the tool used was that of the structured and semi-structured questionnaire. Ten elderly users were surveyed, and they identify the space for socializing as the place in which they spend most time and most prefer. Most users show that they are satisfied with the physical spaces. In relation to the furniture, thermal comfort, and colors used on the floors and walls of the environment, in general, the seniors in the totality of their answers say they are satisfied.

On analyzing the answers that we managed to obtain, it is found that when the users idealize the environment, this points to a large, organized, and comfortable environment, practically the opposite of what was reported as the real environment in which they live together. The interviewees also pointed out that a perfect LSIE is one that has better facilities, the items most recalled being large spaces, having an outside area, cleanliness, green area, accessibility, good care, adequate furniture, and qualified employees. Also the need for improvements and renovations, such as offering spaces and leisure activities to the elderly, and the lack of medical care and difficulties with the public system of health were cited. Some answers came from caregivers and not the resident seniors.

Of the three tools used to study the perception of the built environment, namely the Constellation of Attributes (Ekambi-Schmidt, 1974), the Wish Poem (Sanoff, 1991) or the traditional questionnaires, none of them gave a real insight into the perception of the seniors of the LSIEs that they live in. The research team is continuing its investigation in order to identify, or even develop tools that can do this efficiently.

10.8 ERGONOMIC DIAGNOSIS

Ergonomics with its focus centered on the user will always give priority to activities and environments that lead to comfort, safety, and well-being. By inserting these concepts into the evaluation of environments for those who are growing old, the research undertaken made it possible to compare the findings from six institutions for collective housing for the elderly, thus enabling the spatial configuration of these homes to be understood. For each of the cases studied, recommendations for improvements were generated based on ergonomic assumptions applied to built environments.

By making comparisons between the institutions, it is seen that Brazilian legislation is not fully complied with. Some of the items of the built environment are the focus of great concern of the institutions, who are trying to comply with the legislation. These items include grab bars, tactile flooring, access ramps, width of the doors and handrails, and even so no institution is perfectly matches up to what is required.

Items such as the environmental comfort of temperature, lighting, and acoustics are put aside. Perhaps because they require greater specialization if they are to be measured accurately and because greater investment is needed to solve this, they go beyond what is necessary. As an example we mention the issue of inappropriate temperatures, where 100% of the environments of the six surveyed institutions are considered to be above the comfort temperature. Regarding acoustic comfort, only one environment of LSIE 6 complied with the standard. The situation is similar when lighting comfort is compared. There is only one environment of LSIE 1, one of LSIE 4, and one of LSIE 5 where the values were greater than the minimum measurement. All the others were seen to show a deficit in this respect.

Most of the bedrooms are impersonal. Some have a picture frame with a photo of the family, but there are few of these, and yet, just one picture-holder does not make a small bedroom a place that is similar to one's own home. In addition, visits from the family become increasingly rare, which increases the sense of abandonment. To the twin bedrooms is added the lack of individuality; the space is limited, almost impassable, which prevents personal furniture being placed in it, yet this would make the bedroom closer to what the senior would like.

Some elderly people have no notion of the place that they are living in because their routine is always the same: they go from their bedroom to the social area or refectory. They have no familiarity with other bedrooms or other areas of the house or even the environment outside the home. One of the seniors interviewed related in her wish poem that the house should have three bedrooms so she could live with her family but it does not. One can question whether the lack of perception is actually a limitation of age, or if the fact that she always has her wheelchair pushed through the same environments has given her the impression that there were no other bedrooms.

Difficulties stand out also in relation to the staff, the large majority of whom refused to take part in the assessment of the perception of the environment. In addition, there is the major limitation of the elderly as to their cognitive capacity to answer the test and thus to measure information about their perception of the environment.

One of the greatest lacks is identified as being the lack of green areas, gardens, well-sized socializing spaces, and the offer of diversified activities. In some houses, the social area is the same location as that for meals, watching TV, playing board games, and for recreational activities. There are no spaces defined based on the activity that it accommodates.

Finally, what is clear from these six case studies, is that there is a great similarity of problems, even in the LSIE that was building a new extension. In the new space, the same old problems were found.

10.9 CONCLUSIONS

As it represents a promising branch of the economy, caring for the elderly has been gaining prominence. Along these lines, providing housing to shelter very elderly people can be a lucrative business, especially when many families do not manage to meet the needs of their elderly relatives.

However, given the data collected, it became evident that the amount paid by the user or the type of costing made in a long-term care facility, does not have a positive relationship with the adjustments to the standards prevailing in Brazil, nor with ergonomic criteria for the environment. Many structures are poorly conducted adaptations, where full compliance is not given priority either in private or public institutions or in philanthropic ones.

The investment made by institutions whose purpose is to serve the elderly, needs to take fully on board a greater concern with meeting the special needs of this population. To do so, the standards need to be fulfilled in their entirety, with a view to targeting comfort, safety, and accessibility.

The research proceeded without major difficulties, although many other institutions contacted refused to open their doors to our work. They seem to be afraid that weaknesses will be exposed and that they will be fined.

In conducting the survey, the authors became aware that the administrator of the most expensive institution was concerned about trying to meet the standards but many inadequacies were found,

similar to those found in simpler institutions. Increased investment is needed to improve complying with standards and laws, to provide comfort, healthy environments, and well-sized spaces. The
evaluation of a high standard ILPI allowed us to compare how many improvements there might be
given the high charges paid by the family. However, despite the value, minimum comfort conditions
are not offered to these residents. When compared to a public or philanthropic LSIE, significant differences in the quality of spaces were not identified. Perhaps in those with less financial input, meals
have little variety and not always to the liking of all seniors and hygiene is deficient, there being a
bad smell in some LSIEs. However, the lack of agreeable, tree-planted, landscaped, and ventilated
spaces is a constant in all institutions.

After the assessment of the built environment of the LSIEs studied, it is clear that the seniors'
housing conditions would be much better if the legislation were complied with. In most cases, adaptations that do not require large investments could bring great benefits, but they do require a rather
greater concern with keeping an eye on the quality of life of the seniors, and not only so as not to
take in a greater number of these people to a life that is inhumane, devitalized, and given no worth.

The people who live in these houses have already had their own space, built dreams, formed their
families, some have achieved economic power and thus can pay for a high standard institution but
the weight of age and the very demands of this phase of life sometimes prevent them from living
within a family. In this situation, what is expected is that relatives, employers, and authorities strive
to make such a life dignified and comfortable. The elderly cannot be seen as a source of abundant
income. The spaces that are set aside for them need to be thought out carefully by a team that has
skills in architecture, engineering, and gerontology, acting jointly. The concepts and principles of
ergonomics can contribute much to this effort, as it places the human being as the center, as an element to which all variables should be adapted. It is by doing so that we may be able to be happier
as we grow older.

REFERENCES

Barnes, S. 2002. The design of caring environments and the quality of life of older people. *Ageing and Society*
    22:775–789.
Bertolucci, P.H.F., Brucki, S.M.D., Campacci, S.R., and Juliano, Y. 1994. O Mini-exame do Estado Mental em
    uma população geral. Impacto da escolaridade (The mini-mental state examination in a general popula-
    tion. Impact of schooling). *Arq. Neuropisiquiatra* 52:1–7.
Brasil. 2005. *Resolução da Diretoria Colegiada – RDC no 283, de 26 de Setembro de 2005.* Estabelece o
    padrão mínimo de funcionamento das Instituições de Longa Permanência para Idosos. Ministério da
    Saúde. Brasília, 2005. Legislação em Vigilância Sanitária. Brasília. Disponível em: http://e-legis.anvisa.
    gov.br/leisref/public/showAct.php?id=18850 (acesso em Abril 2014).
Brasil Ministério da Saúde. 2006. Estatuto do Idoso/Ministério da Saúde Law no 10.741/2003 (Statute of the
    Elderly / Ministry of Health Law No. 10.741 / 2003); 2.ed. rev. Brasília: Editora do Ministério da Saúde
    (Série E. Legislação de Saúde).
Ekambi-Schmidt, J. 1974. *La percepción del hábitat* (The perception of habitat). Barcelona: Editorial Gustavo
    Gili, S.A.
Farias, S.F., Guimarães, A.C.A., and Simas, J.P. 2005. O ambiente asilar e a qualidade de vida do idoso (The
    environment asylum and quality of life of the elderly). *A Terceira Idade* 16(33):55–68.
Fernandes, A.A. and Botelho, M.A. 2007. Envelhecimento activo, envelhecimento saudável: o grande desafio
    (Active aging, healthy aging: the great challenge). *Fórum Sociológico* n17 (II série):11–16.
Freitas, M.C., Queiroz, T.A., and Souza, J.A.V. 2010. O significado da velhice e da experiência de envelhecer
    para os idosos. *Ver. Esc. Enferm. USP* 44(2):407–412.
Furtado, J.P., de Tugny, A., Baltazar, A.P., Kapp, S., Generoso, C.M., Campos, C.B., Guerra, A.M.C., and
    Nakmura, E. 2013. Modos de Morar de pessoas com transtorno mental grave no Brasil: uma avaliação
    interdisciplinar (Living modes of people with severe mental disorders in Brazil: An interdisciplinary
    evaluation). *Ciência Saúde Coletiva* 18(12):3683–3693.
Iida, I. 2005. *Ergonomia: projeto e produção* (Ergonomics: Design and production). 2nd Ed. São Paulo: Ed.
    Edgard Blücher.
IPEA, Instituto de Pesquisa Econômica Aplicada. 2010. http://ipea.gov.br/portal. (acesso em Março).

Maciel, M.G. 2010. Atividade física e funcionalidade de idosos. *Motriz,* Rio Claro, 16 (out/dez). (4): 1024–1032.

Marto, N. 2005. Ondas de calor—Impacto sobre a saúde. *Acta Médica Portuguesa.* http:// www.actamedica-portuguesa.com/pdf/2005-18/6/467- 474.pdf (acesso em Setembro 2011).

Mont'Alvão, C. and Villarouco, V. 2011. *Um novo olhar para o projeto: a ergonomia no ambiente construído* (A new look at the project: Ergonomics in the built environment). Teresópolis, RJ: FAPERJ, 2AB, 16.

Nahas, M.V. 2006. *Atividade física, saúde e qualidade de vida: Conceitos e sugestões para um estilo de vida ativo.* 4ª Ed. Londrina: Editora Mediograf.

Nasri, F. 2008. O envelhecimento populacional no Brasil (Population aging in Brazil). *Einstein* 6(1):4–6.

NBR 8995/13. 2013. *Iluminação de ambiente de trabalho.* Norma Brasileira—Associação Brasileira de Normas e Técnicas—ABNT.

NBR 9050/2004. 2004. *Acessibilidade a edificações, mobiliário, espaços e equipamentos urbanos.* Norma Brasileira—Associação Brasileira de Normas e Técnicas—ABNT.

NBR 10152/87. 1987. *Níveis de Ruído para conforto acústico.* Norma Brasileira—Associação Brasileira de Normas e Técnicas—ABNT.

Organização Mundial de Saúde. 2008. Guia global: cidades amigas dos idosos. Disponível em: http://www. who.int/ageing/GuiaAFCPortuguese.pdf (acesso em: Abril, 4 2014).

Paiva, M., Ferrer, N., and Villarouco, V. 2015b. The process of aging: A case study approach implementing an ergonomics evaluation of the built environment for the elderly in Brazil. *Work* (Reading, Massachusetts), 50(4):595–606.

Paiva, M., Sobral, E.R., and Villarouco, V. 2015a. The elderly and environmental perception in collective housing. In: *Proceedings of the AHFE 2015: 6th International Conference on Applied Human Factors and Ergonomics 2015 and the Affiliated Conferences,* Las Vegas, United States.

Paiva, M.M.B. and Villarouco, V. 2012. Accessibility in collective housing for the elderly: A case study in Portugal. *Work* 41: 2012; Disponivel em: http://www. ncbi.nlm.nih.gov. (acesso em 17 de abril 2012).

Sanoff, H. 1991. *Visual Research Methods in Design.* New York: Van Nostrand Reinhold.

Soares, F.V., Greve, P., Sendín, F.A., Benze, B.G., Castro A.P., and Rebelatto, J.R. 2012. Relação entre alter-ações climáticas e fatores determinantes da mortalidade de idosos no município de São Carlos (SP) em um período de dez anos (The relationship between climate change and determinants of mortality in elderly in São Carlos (SP) in a period of ten years). *Revista Ciência & Saúde Coletiva* 17(1):135–146.

Tomasini, S.L.V. 2005. Envelhecimento e planejamento do ambiente construído: em busca de um enfoque interdisciplinar (Aging and environmental planning built: In search of an interdisciplinary approach). *Revista Brasileira de Ciências do Envelhecimento Humano, Passo Fundo* (jan./jun):76–88.

Villarouco, V. 2008. Construindo uma metodologia de Avaliação Ergonômica do Ambiente—AVEA. In: *Anais XV Congresso Brasileiro de Ergonomia, VI Fórum Brasileiro de Ergonomia,* Porto Seguro, Porto Seguro, ABERGO.

Villarouco, V. 2009. An ergonomic look at the work environment. In: *Proceedings of IEA 09: 17th World Congress on Ergonomics,* Beijing, China.

Villarouco, V., Ferrer, N., Paiva, M., and Oliveira, M. 2012. Ergonomics and the environment for the elderly. In: *Proceedings of the AHFE 2012: 4th International Conference on Applied Human Factors and Ergonomics (AHFE),* San Francisco, California.

Villarouco, V. and Santos, N. 2002. Ergonomia do Ambiente Construído (Ergonomics Built Environment). In: *Anais do II ERGODESIGN—II Congresso Internacional de Ergonomia e Usabilidade de Interfaces Humano—Tecnologia: Produtos, Programas, Informação, Ambiente Construído, Rio de Janeiro, RJ Proceedings of ERGODESIGN II – II International Congress of Ergonomics and Usability Human Interface—Technology: Products, Programs, Information, Built Environment, Rio de Janeiro, RJ.* Rio de Janeiro, RJ: LEUI – Depto. De Artes & Design—PUC-RIO.

11 Biomechanics and Psychophysics of Manual Strength Design on Different Interfaces

Bruno M. Razza, Luis Carlos Paschoarelli, Cristina C. Lucio, José Alfredo C. Ulson, and Danilo Silva

CONTENTS

11.1 INTRODUCTION

Despite increasing automation in the industrial environment, many tasks still demand great magnitude of manual efforts, such as maintenance tasks, manual material handling, patients' transportation in hospitals, etc. (Imrhan and Jenkins, 1999; Kim and Kim, 2000). Tasks and products that require inappropriate application of manual strength are considered risk factors for the development of occupational diseases (Kattel et al., 1996), and are responsible for much of the total injuries in industry (Aghazadeh and Mital, 1987). Specifically, the use of pinch grips in several industrial tasks, and also torque strength, have been associated with high levels of occupational diseases (Armstrong and Chaffin, 1979; Chao et al., 1976; Eksioglu et al., 1996), hence they are considered risk factors in ergonomic evaluation (Keyserling et al., 1993).

In addition to reported accidents and work-related diseases, there are also data associating consumer product accidents to incorrect design of manual forces demand. According to Crawford et al. (2002), in England, in the year 1994, there were 550 accidents related to the opening of glass jars and 610 accidents associated with the opening of plastic bottles due to the use of sharp tools in an attempt to open lids and closures that require high demand of strength. This circumstance is more evident for elderly consumers, who do not have the strength required to open such packages or even to perform common daily activities (Voorbij and Steenbekkers, 2002).

Biomechanical, physiological, and psychophysical approaches have been used to establish recommended capacity thresholds for specified task demands (Fischer and Dickerson, 2014). The ISO 6385 (ISO International Organization for Standardization, 1981)—which deals with ergonomic guidelines for product design, workplace, and tasks—recommends that, along with objective assessments (countable), subjective assessment (psychophysical metric) should be taken as a complementary measure. Therefore, the use of methods to subjectively evaluate tasks and determine acceptable workloads has become a common procedure in ergonomic approaches, combining physical variables to the subjective perception of users.

11.2 BIOMECHANICS OF MANUAL STRENGTH

Despite of increasing automation in industrial environments, many tasks still require a great deal of manual effort. Activities and products that demand inappropriate levels of manual strength are considered risk factors for the occurrence of occupational diseases (Kattel et al., 1996), which can become even more serious when it involves digital grip (Armstrong and Chaffin, 1979; Chao et al., 1976; Eksioglu et al., 1996; Keyserling et al., 1993).

The position, size, and shape of an object can modify the posture of wrist and fingers in the prehension, thus influencing the joints and the length of the muscles required to grasp the object (Shih and Ou, 2005) and consequently the strength (Dempsey and Ayoub, 1996; Richards, 1997).

The body posture can also affect the strength of the upper limbs. Teraoka (1979) and Balogun et al. (1991) shown that the standing posture generates higher strength in comparison to sitting or lying postures. However, Catovic et al. (1989, 1991) and Richards (1997) have not found any differences in strength for these related body postures.

Beyond the body posture, the position of the upper limb has also been thoroughly investigated. Wrist deviation can cause compression of the finger-flexor muscles against the carpal tunnel and other inner structures, which can lead to diseases such as carpal tunnel syndrome for example (Armstrong and Chaffin, 1979; Imrhan, 1991). The neutral posture of the wrist (or slightly extended and with ulnar deviation—considered as the free wrist posture by O'Driscoll et al. [1992]) is the more advantageous for the exertion of strength while the flexed wrist causes the higher detriment to the exerted strength (Dempsey and Ayoub, 1996; Imrhan, 1991; Lamoreaux and Hoffer, 1995; Kraft and Detels, 1972; Shih and Ou, 2005) and there seems to be an inverted relation between the intensity of the deviation and the magnitude of the force (Kattel et al., 1996).

The position of the elbow and shoulder has also been investigated. Higher scores of strength have been obtained with the elbow fully extended in comparison to other degrees of flexion (Balogun et al., 1991; Kim and Kim, 2000; Kuzala and Vargo, 1992; Su et al., 1994), although another study has shown discordant results, with a higher magnitude of manual force with the elbow flexed at 90° (Mathiowetz et al., 1985b). For the shoulder, the neutral position offers the higher results of strength, and positions of abduction, flexion, or extension have a detrimental effect on strength (Imrhan and Jenkins, 1999; Kattel et al., 1996; Su et al., 1994).

The variety of prehension also significantly affects manual strength due to the different configuration of joints, muscles, and bones. Handgrip generates significantly higher forces than pinch grips (Kraft and Detels, 1972; Lamoreaux and Hoffer, 1995; Mathiowetz et al., 1985a, 1986). The type of pinch grip used can significantly affect the strength; taken from the strongest to the weakest, pinch grips can be ordered as follows: lateral pinch (key-pinch); chuck pinch (three jaw pinch—index and middle fingers in opposition to the thumb); pinch 2 (index finger in opposition to the thumb); and pinch-3 (medial finger in opposition to the thumb) (Dempsey and Ayoub, 1996; Imrhan, 1991; Imrhan and Jenkins, 1999; Imrhan and Rahman, 1995; Mathiowetz et al., 1985a).

For torque strength, biomechanical action has a strong influence on the force. The flexion–extension movement of the wrist has great magnitude of torque exerted than the pronation–supination movement of the forearm (Imrhan and Jenkins, 1999; O'Sullivan and Gallwey, 2002).

11.2.1 Task-Related Variables

One important variable in the design of tasks that require strength is the size and shape of the object. The size of the handle will determine the grip spam and affect the configuration of the muscles and joints in the hand and the upper limb as well. For the handgrip strength, many studies investigated the influence of grip spam to the exerted force. In general, intermediate spams generated the higher magnitude of grip strength; for the Jamar Dynamometer it is from 47 to 60 mm (Härkönen et al., 1993; O'Driscoll et al., 1992) and for cylindrical handles it is from 40 to 50 mm approximately (Blackwell et al., 1999; Edgren et al., 2004; Peebles and Norris, 2003). For torque strength in the flexion–extension movement, the 50 mm handle has also presented the best results (Kong and Lowe, 2005a,b; Pheasant and O'Neill, 1975).

For the pinch grips the influence of the grip spam is less clear. Although Imrhan and Rahman (1995) and Dempsey and Ayoub (1996) have presented higher measures of force in intermediate spams (44–50 mm), there is a tendency reported in the literature of higher strength with the increment of the grip spam (Shih and Ou, 2005; Shivers et al., 2002). The pulling strength with pinch grips has also been evaluated and the 40 mm handle has shown the best results of strength (Peebles and Norris, 2003).

The shape of the handle has been investigated for torque strength. The results indicate that the forms that tend to present a higher difference in the width–height proportion, that is, triangular shape in comparison to cylindrical shape, generate a higher momentum and therefore higher levels of strength (Crawford et al., 2002; Fothergill et al., 1992).

On the other hand, the effects of pinch grips associated with pulling strength are not well explored by the literature, in spite of being very common in industrial activities (especially when the object is tiny, the access to the object is restricted, or when the use of tools is prevented) and in daily activities (pulling plastic or long-life package ribbons, removing sealing wax from bottles, tearing plastic bags, breaking vacuum packing, opening drawers, etc.). One substantial example is the study of Peebles and Norris (2003) which investigated three different kinds of pinch grip (pinch 2; chuck pinch, and lateral pinch) in the three handprints thicknesses (40, 20, and 1 mm); it is also important to mention Fothergill et al. (1992) who analyzed the strength to pull with different handprint formats (circular handle, handle, and cylindrical bar); and Imrhan and Sundararajan (1992) who evaluated the pulling strength associated with three pinch grips (pinch 2; index-finger thumb, chuck pinch, and lateral pinch) which was measured with a dynamometer associated with a ribbon cloth.

Manual strength has also been investigated in interfaces that represent or simulate real work tasks or daily activities. Torque strength was evaluated in many tasks such as opening jar lids (Carse et al., 2010; Crawford et al., 2002), opening soft drink packages (Silva and Paschoarelli, 2013), and the twisting force on door handles (Paschoarelli and Santos, 2011; Paschoarelli, 2012b). The usage of hand tools was investigated by Kim and Kim (2000). As a result, in decreasing order of the strength generated, the tools can be ordered as follows: wheel, wrench, cylindrical handle, knob, and screwdriver. Mital (1986) and Mital and Sanghavi (1986) have previously demonstrated that wrenches generate more strength than screwdrivers. Pheasant and O'Neill (1975) have investigated the shape of screwdrivers and Fransson and Winkel (1991) investigated the forces generated by pliers.

11.3 PSYCHOPHYSICS OF MANUAL STRENGTH

Psychophysics is a branch of psychology concerned with the relationship between sensations and their physical stimuli. According to psychophysical theory, the perceived strength of a sensation is directly related to the intensity of its physical stimulus by a power function (Garg et al., 2014). Psychophysical methods are commonly used to establish guidelines and limits for task acceptability or to indicate task demands. Despite their widespread use and practical application, the subjectivity of psychophysical methods can limit their perceived benefit to ergonomists, engineers, and designers (Fischer and Dickerson, 2014). According to Dempsey (1998), it was believed that these approaches yielded different or conflicting information regarding a given task. However, recent research has begun to uncover associations between biomechanical and psychophysical approaches. Growing evidence demonstrates that acceptable psychophysical loads and forces are related to underlying joint loading exposures, particularly at the most biomechanically limiting joint (Fischer and Dickerson, 2014).

According to Garg et al. (2014), some advantages of the psychophysical approach include the ability to simulate industrial work; allowing the study of intermittent or repetitive tasks; psychophysically determined maximum acceptable weights and forces are reproducible and can predict back injuries. Disadvantages of the psychophysical approach include: the subjectiveness of the method based on self-reported data from subjects; the method is time consuming and expensive for collecting data for very infrequent tasks.

In ergonomic assessments, psychophysical data such as the individual perception of interface variables acting on it and its activity or product usage may provide relevant information that complement or assist in understanding the results of physical assessments. Thus, psychophysics offers an opportunity to examine worker perception of tasks involving multiple occupational stressors by allowing the worker to "integrate" this information. In the psychophysical approach, the human serves as an instrument of observation. Like any instrument, the human observer can be biased and lead to inaccurate measures due to subjectivity of the metric (Fernandez and Marley, 2014).

Nevertheless, the ergonomic uses of psychophysical data continue to grow and the major applications are on the determination of subjective acceptability and/or tolerance and for the design of systems or products (Fernandez and Marley, 2014). For example, psychophysical data has been used to develop the NIOSH (National Institute for Occupational Safety and Health) lifting guide as well as the Liberty Mutual models for manual material handling tasks. In both cases, the data were generated using whole body exertion (Fernandez and Marley, 2014). The Method of Adjustment in particular has been widely used in establishing acceptable weights of lift (Ciriello et al., 1993; Snook, 1985a,b).

Accordingly to Tullis and Albert (2008), the most efficient way to record subjects' perception in ergonomic assessments is with some type of rating scale such as the Visual analog scale (VAS) (Bacci, 2004; Björkstén et al., 1999; Collins et al., 1997), Borg's scales (Borg, 1998), Likert scales (Likert, 1932), among others, to the detriment of open questions, which are more difficult to analyze.

The VAS scale is widely used in subjective evaluation of many variables. It consists of a row (horizontal or vertical), with a certain length (often 10 cm) and anchors that represents the maximum and minimum criteria of the variable to be measured (Bacci, 2004). Collins et al. (1997) recommends the use of this type of scale at the expense of categorical scales, claiming to be more precise. Huskisson (1983 apud Björkstén et al., 1999) also recommends the use of this method since VAS scores are highly correlated, simple to measure and interpret, sensitive, easily reproducible, and universal.

One crucial aspect to be considered when using this method is the correct choice of terms that will be used as anchors. In the field of ergonomic design, at least two criteria of perception must be respected, one of which may involve a negative concept—for example, discomfort or greater effort—and another with a positive concept—for example, comfort and less effort.

Ayoub and Dempsey (1999) and Fernandez and Marley (2014) provided an excellent summary of the advantages and disadvantages of the use of psychophysical methods in occupational activities. The advantages of psychophysics metrics include (1) reproducible results; (2) realistic simulation of activities and conditions; (3) results that take into consideration the whole job and integrate biomechanical and physiological factors; and (4) results that appear transferrable to similar tasks as guidelines. Similarly, these psychophysical studies also have disadvantages which include (1) fundamentally, these results are subjective and can be influenced by many personal factors; (2) there is risk that the results may overestimate or underestimate long-term working limits, particularly for high-frequency or very intense task requirements; and (3) several studies were based upon nonindustrial (student) populations.

In this study we aimed to evaluate the individual perception of effort in simulated tasks for torque and pulling strength. The influence of the personal characteristics of gender and laterality were considered. The tasks' elements consisted of four different shapes of the handle in torque strength, three types of pinch grips, and three handle heights for pulling strength. Two samples were recruited for two subsequent trials. The first consisted of the investigation of the influence of gender on perceived strength and the second trial compared perceived reported strength for right and left-handed individuals.

11.4 MATERIAL AND METHODS

The procedures for this study were approved by the Committee of Ethics in Research (CEP-FMB-UNESP n. 373/2005) and the recommendations of the Brazilian National Health Council (Resolution 466/CNS/MS/CONEP, 2012) and the Brazilian Association of Ergonomics (Associação Brasileira de Ergonomia, 2009) for research involving humans were met.

This study consisted of two different samples. In the first approach, the influence of gender on perceived exertion in pulling strength was investigated. This measure was assessed with three different pinch grips and three different handles. In a second approach, the influence of laterality on the perception of effort was investigated. In this second approach, we also investigated the psychophysics of the strength exerted (reported perception) when performing the torque in different handles.

11.4.1 Subjects

11.4.1.1 Sample 1

Sixty right-handed unpaid volunteers participated in the experiment, as follows: 30 women in the mean age of 21.60 years (SD 3.05), ranging from 18 to 30 years, and 30 men in the mean age of 21.83 years (SD 2.46), ranging from 18 to 28 years. None of the subjects reported any history of musculoskeletal disease in the upper limbs in the last year. The subjects' written consent was obtained and all procedures were completely explained to them. The Edinburgh Inventory (Oldfield, 1971) was applied to certify that the whole sample was right handed.

11.4.1.2 Sample 2

Sixty unpaid male adults volunteered in this second research. Thirty of them were right handed and 30 left handed. The excluding criteria was the same as the previous sample and the Edinburgh Inventory (Oldfield, 1971) was also applied to certify the laterality of the individuals. The mean age of the left-handed subjects was 21.7 years (SD 3.05) in a range 18–30 years. The right-handed subjects were the same recruited in sample 1.

11.4.2 Apparatus

The record of strength was performed by a digital dynamometer AFG500 (Mecmesin Ltd., England), with maximum capacity of 500 N, accuracy of 0.1% (full scale), analogical communication interface +4 … 0 … −4 V full scale, digital communication interface RS-232, and maximum sample rate of 5000 Hz. The data were obtained by a personal computer with a Windows XP operating system (Microsoft®, version 2002), and a specific software (SADBIO—System of Acquisition of Biomechanical Data, Labview 7.0, National Instruments®, England) was developed for this study.

Four handles with different formats were used to measure the torque strength (Figure 11.1, in the left). They were prismatic shaped with 50 mm length and based with the geometrical forms of a circle (cylindrical), a triangle (triangular), a square (cubic), and a hexagon (hexagonal).

For the measurement of pulling strength, three handles of three different heights were employed, one corresponding to a height of 40 mm high ($40.0 \times 40.0 \times 40.0$ mm), another one 20 mm high ($20.0 \times 40.0 \times 40.0$ mm), and the last one having an extension in fabric of 1 mm thickness

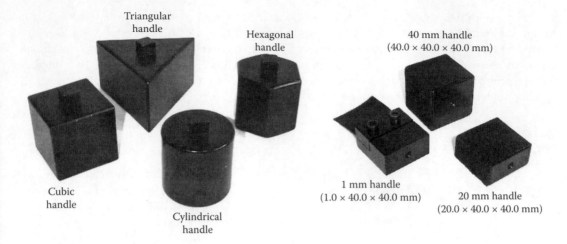

FIGURE 11.1 Handles used for the assessment of perceived torque (left) and pulling strength (right).

$(40.0 \times 40.0 \times 1.0$ mm). In the 20 and 40 mm handles, fabric was applied all over the surface in contact with the hands, for a standardization of texture in the hand–object interface; besides, corners were made round to avoid pressure concentration on the hands of the subjects (Figure 11.1, in the right).

Printed questionnaires were used for the registry of reported data, as follows: personal information of the subjects; written consent to participate voluntarily in the study; assessment of laterality score; and the registry of perceived exertion for each trial.

11.4.3 PROCEDURES

The procedures of this study followed recommendation related in the literature (Caldwell et al., 1974; Chaffin and Andersson, 1990; Mathiowetz et al., 1984; Mital and Kumar, 1998a,b, among others). Subjects in this experiment were asked to exert their maximum pulling strength (maximum isometric voluntary contraction) with each hand, alternately, and hold the strength for a 5 s period. The sequence of measurements was randomized and an interval of at least 1 min was provided among trials to avoid fatigue.

For the evaluation of maximum voluntary isometric contraction for torque and pulling strength with pinch grips, each subject was asked to remain in the standing posture facing the equipment, the elbow of the upper limb flexed at 90°, the forearm in neutral position—horizontally aligned—and the wrist positioned freely, according to the preference of the subject. The equipment was positioned at the height of the subject's elbow. In the measurement of torque strength, the subjects were oriented to grasp the handle freely. For the pulling strength, however, three types of pinch grips were evaluated. The pinch grips used were

- Pinch 2 (thumb opposed to index finger)
- Three jaw chuck pinch (index and middle finger opposed to thumb)
- Lateral pinch (thumb opposed to the lateral side of index finger, also called key-pinch)

The subjects were asked to keep the fingers that were not active in the grip flexed to the palm of the hand. In the pulling strength measurement for pinch 2 and chuck pinch, the wrist remained in extension and slight ulnar deviation. Figure 11.2 exemplifies the types of grips employed in pulling strength.

The recorded strength of these measurements and detailed procedure for this assessment was reported in previous studies (Paschoarelli et al., 2012a; Razza et al., 2012). After the exertion of

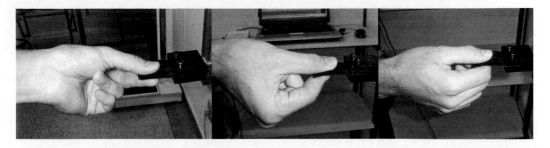

FIGURE 11.2   Pinch grips evaluated in the study: chuck pinch; pinch 2; and lateral pinch, respectively.

strength for each variable, subjects were asked to report the perceived effort in a VAS scale of 100 mm, in which 0 corresponds to no effort and 100 corresponds to the maximum effort. The intention was to report the effort applied with each condition and not to try to estimate the resulting force. Thus, it was expected that the least favorable condition would be more difficult to perform and higher effort scores must have been registered.

We used descriptive statistics analysis across all the data. The Mann–Whitney U test ($P \leq 0.05$) for nonparametric group comparison was used to verify statistically differences between genders and lateralities. Wilcoxon's test for nonparametric sample comparison was applied to analyze the influence of different handle shapes for torque strength, different heights, and different pinch grips for pulling strength.

11.5 RESULTS AND DISCUSSION

Results of the subjective evaluation are presented in Tables 11.1 and 11.2. The mean values presented dimensions measured on the VAS, in which 0 means no effort and 100 means maximum effort.

Table 11.1 presents the results (mean and standard deviation) of reported effort for pulling strength with pinch grips for gender comparison. The handle in which the least effort was necessary to perform the maximum exertion was the 40 mm handle, and the 1 mm handle presented the worst condition. For the pinch grips, we observed that the lateral pinch had better results, and the worst reported effort was obtained for the pinch 2 grip. The last column indicates the results of Mann–Whitney U test. The gender effect was responsible for statistical differences only on the pinch 2 grip ($P \leq 0.05$).

TABLE 11.1
Results of Perceive Effort for Pulling Strength with Pinch Grips: Gender Comparison

	Female (Right-Handed)		Male (Right-Handed)		P Level— Male × Female
	Mean	SD	Mean	SD	Mann–Whitney U test
40 mm handle	63.2	22.8	56.1	23.8	0.234
20 mm handle	54.8	24.3	51.4	20.8	0.564
1 mm handle	46.2	28.2	48.5	30.6	0.679
Pinch 2	78.3	17.3	66.4	22.6	0.033*
Chuck pinch	54.0	16.7	58.8	15.5	0.337
Lateral pinch	31.3	22.5	37.5	27.7	0.496

*$P \leq 0.05$.

TABLE 11.2
Results of Perceived Effort: Laterality Comparison

	Right-Handed (Male)		Left-Handed (Male)		P Level—Right × Left-Handed
	Mean	SD	Mean	SD	Mann–Whitney *U* test
Cylindrical	59.4	26.8	65.4	28.2	0.240
Hexagonal	52.7	25.8	44.3	26.1	0.183
Cubic	51.5	21.3	49.9	21.1	0.790
Triangular	42.5	32.2	40.0	26.9	0.824
40 mm handle	56.1	23.8	56.6	24.7	0.965
20 mm handle	51.5	20.8	42.2	24.0	0.126
1 mm handle	48.5	30.7	45.8	28.7	0.647
Pinch 2	66.4	22.7	71.8	23.5	0.225
Chuck pinch	58.9	15.5	51.0	18.7	0.206
Lateral pinch	37.5	27.7	22.2	20.3	0.008*

*$P \leq 0.05$.

Table 11.2 presents the results (mean and standard deviation) of perceived effort for torque strength and pulling strength with pinch grips for laterality comparison. In reported torque effort, the best results can be seen with the triangular prismatic shape and the worst results were recorded for the cylindrical handle. For pulling strength with pinch grips, the effects of the handle height and type of pinch grip were the same as the results reported in sample 1. The last column indicates the results of the Mann–Whitney U test. Laterality was responsible for differences in reported strength only for the lateral pinch ($P \leq 0.05$).

The effect of handle shape for perceived torque strength was assessed with Wilcoxon's test. The results of this analysis are presented in Table 11.3. The cylindrical handle was perceived to be significantly different from the others in terms of reported effort only for left-handed subjects ($P \leq 0.05$). For right-handed subjects, the shape of the handle was not different in reported effort.

The effect of handle height for perceived pulling strength with pinch grip was assessed with Wilcoxon's test. The results of this analysis are presented in Table 11.4. Statistical differences were found only for the comparison of the 20 mm in opposition to the 40 mm for male left-handed subjects and for the 40 and 1 mm handles for right-handed female subjects.

TABLE 11.3
Handle Comparison for Perceived Torque Strength

	Right-Handed (Male)			Left-Handed (Male)		
	Hexagonal	Cubic	Cylindrical	Hexagonal	Cubic	Cylindrical
Triangular	0.213	0.120	0.069	0.478	0.116	0.007*
Hexagonal	–	0.789	0.329	–	0.484	0.001*
Cubic	–	–	0.221	–	–	0.030*

Wilcoxon test: *$P \leq 0.05$.

TABLE 11.4
Handle Height Comparison for Perceived Pulling Strength

	Right-Handed (Male)		Left-Handed (Male)		Right-Handed (Female)	
	20 mm	40 mm	20 mm	40 mm	20 mm	40 mm
1 mm	0.422	0.267	0.673	0.165	0.150	0.025*
20 mm	–	0.579	–	0.028*	–	0.241

Wilcoxon test: *P ≤ 0.05.

TABLE 11.5
Pulling Strength Perceived According to the Type of Grip

	Right-Handed (Male)		Left-Handed (Male)		Right-Handed (Female)	
	Chuck pinch	Lateral pinch	Chuck pinch	Lateral pinch	Chuck pinch	Lateral pinch
Pinch 2	0.131	0.001*	0.000*	0.000*	0.000*	0.000*
Chuck pinch	–	0.001*	–	0.000*	–	0.000*

Wilcoxon test: *P ≤ 0.05.

Table 11.5 presents the results for the analysis of the effect of grip type for the perceived effort. The results indicated that the grip used significantly influenced the perceived effort in most conditions ($P ≤ 0.05$). The one exception found was between pinch 2 and chuck pinch for right-handed male subjects.

Subjects perceived that a greater effort is exerted with pinch 2 in comparison to the chuck or lateral pinch. This indicates that the volunteers considered the pinch 2 grip to be the most demanding or fatiguing of all. It has been demonstrated in the literature that pinch grips may cause tensions in the tendons of the deep digital flexor muscle (Chao et al., 1976; Eksioglu et al., 1996), and since there is less muscle active in the exertion of strength in pinch 2 than in the chuck or lateral pinch, it may explain the more pronounced perceived effort.

Additionally, the size of the contact surface may also have influenced this perception. The lateral pinch grip offers a larger contact surface with the handle, providing a more stable and better friction grasp. Comparatively, pinch 2 grip has the lowest contact surface, which leads to a less stable hold. Consequently, the lateral pinch demands more space in the handle to be used than pinch 2 or chuck pinch.

In previous publications (Paschoarelli et al., 2012a; Razza et al., 2012), the results of the measured strength were reported for this very sample. The comparison of the results for the measured strength and the reported perceived subjective effort may be useful in verifying the accuracy of using psychophysical metrics to evaluate effort or strength.

In Razza et al. (2012) the handle height effect was weaker than the type of grip in determining the strength. The strongest grip was the lateral, followed by the chuck, and then the pinch 2. The handle height differed significantly only in some cases and was revealed to be subdued by the influence of the handle height. The same results were observed for handle height and type of pinch grip.

For the torque strength, Paschoarelli et al. (2012b) demonstrated that the handle shape influenced the strength measured. In this study, the triangular handle showed the higher torque strength, followed by the cubic, and the hexagonal. The cylindrical handle was responsible for the lowest records of strength. The same results were obtained with the use of psychophysics data.

The small or no difference found between gender and laterality in this study was expected. Since the subjects were asked to exert their maximum strength, independently of the magnitude of the strength measured, the effort reported was considered maximum for each individual. This can be a valid psychophysical metric to assess fatigue or muscular stress in work environments or in daily activities, but it is inadequate to predict the magnitude of strength.

In summary, the results suggest that the type of pinch grip used in pulling strength situations has greater effects. In this case, the best design solution is to provide enough contact area on the surface of the object or handle in order to permit the use of this grip, since it has a greater demand of area to be performed in comparison to other pinch grips. The results of the handle shape in torque strength can be easily understood by the momentum provided by some handles. The triangular handle provides the best momentum and consequently will transmit more torque. Thus, more torque applied can be interpreted as less effort exerted, as it can be seen in the results of this study.

11.6   CONCLUSION

The similarities found between the psychophysical results of this study and the metric results for the same sample (Paschoarelli et al., 2012a; Razza et al., 2012) supports the use of psychophysical measures to evaluate perceived fatigue and effort in work or daily situations. Particularly, the psychophysics measured was useful to identify task-related elements such as handle size, handle shape, and type of grip. However, for interrelated subjects' characteristics, such as laterality or gender, it has proved to be a weak measure.

Psychophysical metric is cheaper and more accessible to assess than traditional metrics that require equipment in laboratories and frequently cannot be adapted easily to the workplace. The perceived effort has shown to be useful to investigate workload and to design tasks and products in situations in which the magnitude of the strength exerted is less important than the context in which the force is applied. The perceived effort can be read as an index of ergonomic cost for physical tasks and can be used as a guideline for the workload in industry.

Psychophysics remains a powerful tool for establishing guidelines to dimensioning tasks and preventing injuries by using information from workers. A better understanding about the information inherent in a psychophysical response and how it is related and affected by subjective individual experiences will help to improve the effectiveness and efficacy of the usage of this metric in product and task design.

ACKNOWLEDGMENT

This research was supported by FAPESP (Proc. FAPESP 05/58600-7 and 05/59941-2).

REFERENCES

Aghazadeh, F. and Mital, A. 1987. Injuries due to hand tools: Results of a questionnaire. *Applied Ergonomics*, 18: 273–278. http://dx.doi.org/10.1016/0003-6870(87)90134-7.
Armstrong, C. A. and Chaffin, D. B. 1979. Carpal tunnel syndrome and selected personal attributes. *Journal of Occupational Medicine*, 21 (7): 481–486.
Associação Brasileira de Ergonomia. 2009. *ERG BR 1002 Norm: Deontology Code of Certified Ergonomists*. Retrieved from http://www.abergo.org.br/arquivos/normas_ergbr/norma_erg_br_1002_deontologia.pdf.
Ayoub, M. M. and Dempsey, P. G. 1999. The psychophysical approach to manual materials handling task design. *Ergonomics*, 42 (1): 17–31. http://dx.doi.org/10.1080/001401399185775.
Bacci, A. V. F. 2004. *Comparação da escala CR10 de Borg com a escala analógica visual (VAS) na avaliação da dor em pacientes com disfunções temporomandibulares*, Master's thesis. Retrieved from BV-CDI Fapesp. (Proc. No. 01/11173–6).

Balogun, J. A., Akomolafe, C. T., and Amusa, L. O. 1991. Grip strength: Effects of testing posture and elbow position. *Archives of Physical Medicine and Rehabilitation*, 72 (5): 280–283. Retrieved from http://www.researchgate.net/profile/Joseph_Balogun/publication/21147053_Grip_strength_effects_of_testing_posture_and_elbow_position/links/54da92440cf2ba88a68d4a11.pdf?origin=publication_detail.

Björkstén, M. G., Boquist, B., Talbäck, M., and Edling, C. 1999. The validity of reported musculoskeletal problems. A study of questionnaire answers in relation to diagnosed disorders and perception of pain. *Applied Ergonomics*, 30 (4): 325–330. http://dx.doi.org/10.1016/S0003-6870(98)00033-7.

Blackwell, J. R., Kornatz, K. W., and Heath, E. M. 1999. Effect of grip span on maximal grip force and fatigue of flexor digitorum superficialis. *Applied Ergonomics*, 30 (5): 401–405. http://dx.doi.org/10.1016/s0003-6870(98)00055-6.

Borg, G. 1998. *Borg's Perceived Exertion and Pain Scales*, 1st ed. Champaign, IL: Human Kinetics Publishers, pp. 120.

Caldwell, L. S., Chaffin, D. B., Dukes-Dobos, F. N., Kroemer, K. H., Laubach, L. L., Snook, S. H., and Wasserman, D. E. 1974. A proposed standard procedure for static muscle strength testing. *American Industrial Hygiene Association Journal*, 35: 201–206.

Carse, B., Thomson, A., and Stansfield, B. 2010. Use of biomechanical data in the inclusive design process: Packaging design and the older adult. *Journal of Engineering Design*, 21 (2–3): 289–303. http://dx.doi.org/10.1080/09544820903303456.

Catovic, A., Kosovel, Z., Catovic, E., and Muftic, O. 1989. A comparative investigation of the influence of certain arm positions on hand pinch grips in the standing and sitting positions of dentists. *Applied Ergonomics*, 20 (2): 109–114. http://dx.doi.org/10.1016/0003-6870(89)90132-4.

Catovic, E., Catovic, A., Kraljevic, K., and Muftic, O. 1991. The influence of arm position on the pinch grip strength of female dentists in standing and sitting positions. *Applied Ergonomics*, 22 (3): 163–166. http://dx.doi.org/10.1016/0003-6870(91)90155-b.

Chaffin, D. B. and Andersson, G. B. J. 1990. *Occupational Biomechanics*, 2nd ed. New York: John Wiley & Sons.

Chao, E. Y., Opgrande, J. D., and Axmear, F. E. 1976. Three dimensional force analysis of finger joints in selected isometric hand functions. *Journal of Biomechanics*, 9: 387–396. http://dx.doi.org/10.1016/0021-9290(76)90116-0.

Ciriello, V. M., Snook, S. H., and Hughes, G. J. 1993. Further studies of psychophysical determined maximum acceptable weights and forces. *Human Factors*, 35 (1): 175–186. Retrieved from http://hfs.sagepub.com/content/35/1/175.full.pdf.

Collins, S. L., Moore, R. A., and Mcquay, H. J. 1997. The visual analogue pain intensity scale: What is moderate pain in millimetres? *Pain*, 72 (01–02): 95–97. http://dx.doi.org/10.1016/s0304-3959(97)00005-5.

Crawford, J. O., Wanibe, E., and Laxman, N. 2002. The interaction between lid diameter, height and shape on wrist torque exertion in younger and older adults. *Ergonomics*, 45 (13): 922–923. http://dx.doi.org/10.1080/00140130210162243.

Dempsey, P. G. 1998. A critical review of biomechanical, epidemiological, physiological and psychophysical criteria for designing manual materials handling tasks. *Ergonomics*, 41: 73–88. http://dx.doi.org/10.1080/001401398187332.

Dempsey, P. G. and Ayoub, M. M. 1996. The influence of gender, grasp type, pinch width and wrist position on sustained pinch strength. *International Journal of Industrial Ergonomics*, 17 (3): 259–273. http://dx.doi.org/10.1016/0169-8141(94)00108-1.

Edgren, C. S., Radwin, R. G., and Irwin, C. B. 2004. Grip force vectors for varying handle diameters and hand sizes. *Human Factors*, 46 (2): 244–251. http://dx.doi.org/10.1518/hfes.46.2.244.37337.

Eksioglu, M., Fernandez, J. E., and Twomey, J. M. 1996. Predicting peak pinch strength: Artificial neural network vs. regression. *International Journal of Industrial Ergonomics*, 18: 431–441. http://dx.doi.org/10.1016/0169-8141(95)00106-9.

Fernandez, J. E. and Marley, R. J. 2014. The development and application of psychophysical methods in upper-extremity work tasks and task elements. *International Journal of Industrial Ergonomics*, 44 (2): 200–206. http://dx.doi.org/10.1016/j.ergon.2012.09.004.

Fischer, S. L. and Dickerson, C. R. 2014. Applying psychophysics to prevent overexposure: On the relationships between acceptable manual force, joint loading, and perception. *International Journal of Industrial Ergonomics*, 44 (2): 266–274. http://dx.doi.org/10.1016/j.ergon.2012.09.006.

Fothergill, D. M., Grieve, D. W., and Pheasant, S. T. 1992. The influence of some handle designs and handle height on the strength of the horizontal pulling action. *Ergonomics*, 35 (2): 203–212. http://dx.doi.org/10.1080/00140139208967807.

Fransson, C. and Winkel, J. 1991. Hand strength: The influence of grip span and grip type. *Ergonomics*, 34 (7): 881–892. http://dx.doi.org/10.1080/00140139108964832.

Garg, A., Waters, T., Kapellusch, J., and Karwowski, W. 2014. Psychophysical basis for maximum pushing and pulling forces: A review and recommendations. *International Journal of Industrial Ergonomics*, 44: 281–291. http://dx.doi.org/10.1016/j.ergon.2012.09.005.

Härkönen, R., Piirtomaa, M., and Alaranta, H. 1993. Grip strength and hand position of the dynamometer in 204 finnish adults. *Journal of Hand Surgery* (British and European volume), 18B (01): 129–132.

Huskisson, E. 1983. Visual analogue scales. In: R. Melzack (Ed.), *Pain Measurement and Assessment*, 1st ed. New York: Raven Press, p. 309.

Imrhan, S. N. 1991. The influence of wrist position on different types of pinch strength. *Applied Ergonomics*, 22 (6): 379–384. http://dx.doi.org/10.1016/0003-6870(91)90079-w.

Imrhan, S. N. and Jenkins, G. D. 1999. Flexion-extension hand torque strengths: Applications in maintenance tasks. *International Journal of Industrial Ergonomics*, 23 (4): 359–371. http://dx.doi.org/10.1016/s0169-8141(98)00052-3.

Imrhan, S. N. and Rahman, R. 1995. The effect of pinch width on pinch strengths of adult males using realistic pinch-handle coupling. *International Journal of Industrial Ergonomics*, 16 (2): 123–134. http://dx.doi.org/10.1016/0169-8141(94)00090-p.

Imrhan, S. N. and Sundararajan, K. 1992. An investigation of finger pull strengths. *Ergonomics*, 35 (3): 289–299. http://dx.doi.org/10.1080/00140139208967814.

ISO International Organization for Standardization. 1981. ISO 6385, Ergonomic principles in the design of work systems—Part 4.2.

Kattel, B. P., Fredericks, T. K., Fernandez, J. E., and Lee, D. C. 1996. The effect of upper-extremity posture on maximum grip strength. *International Journal of Industrial Ergonomics*, 18: 423–429. http://dx.doi.org/10.1016/0169-8141(95)00105-0.

Keyserling, W. M., Stetson, D. S., Silverstein, B. A., and Brouwer, M. L. 1993. A checklist for evaluating ergonomic risk factors associated with upper extremity cumulative trauma disorders. *Ergonomics*, 36 (7): 807–831. http://dx.doi.org/10.1080/00140139308967945.

Kim, C. H. and Kim, T. K. 2000. Maximum torque exertion capabilities of Korean at varying body postures with common hand tools. *Proceedings of the International Ergonomics Association*, 44: 157–159. http://dx.doi.org/10.1177/154193120004401710.

Kong, Y. K. and Lowe, B. D. 2005a. Evaluation of handle diameters and orientations in a maximum torque task. *International Journal of Industrial Ergonomics*, 35 (12): 1073–1084. http://dx.doi.org/10.1016/j.ergon.2005.04.009.

Kong, Y. K. and Lowe, B. D. 2005b. Optimal cylindrical handle diameter for grip force tasks. *International Journal of Industrial Ergonomics*, 35 (6): 495–507. http://dx.doi.org/10.1016/j.ergon.2004.11.003.

Kraft, G. H. and Detels, P. E. 1972. Position of function of the wrist. *Archives of Physical Medicine and Rehabilitation*, 52: 272–275.

Kuzala, E. A. and Vargo, M. C. 1992. The relationship between elbow position and grip strength. *American Journal of Occupational Therapy*, 46 (6): 509–512. http://dx.doi.org/10.5014/ajot.46.6.509.

Lamoreaux, L. and Hoffer, M. M. 1995. The effect of wrist deviation on grip and pinch strength. *Clinical Orthopaedics and Related Research*, 5 (314): 152–155. http://dx.doi.org/10.1097/00003086-199505000-00019.

Likert, R. 1932. A technique for the measurement of attitudes. *Archives of Psychology*, 140: 55.

Mathiowetz, V., Kashman, N., Volland, G., Weber, K., and Dowe, M. 1985a. Grip and pinch strength: Normative data for adults. *Archives of Physical Medicine and Rehabilitation*, 66: 69–74.

Mathiowetz, V., Rennells, C., and Donahoe, L. 1985b. Effect of elbow position on grip and key pinch strength. *The Journal of Hand Surgery*, 10 (5): 694–697. http://dx.doi.org/10.1016/s0363-5023(85)80210-0.

Mathiowetz, V., Weber, K., Volland, G., and Kashman, N. 1984. Reliability and validity of grip and pinch strength evaluations. *Journal of Hand Surgery*, 9 (2): 222–226. http://dx.doi.org/10.1016/s0363-5023(84)80146-x.

Mathiowetz, V., Wiemer, D. M., and Federman, S. M. 1986. Grip and pinch strength: Norms for 6 to 19-year-olds. *American Journal of Occupational Therapy*, 40 (10): 705–711. http://dx.doi.org/10.5014/ajot.40.10.705.

Mital, A. 1986. Effect of body posture and common hand tools on peak torque exertion capabilities. *Applied Ergonomics*, 17 (2): 87–96. http://dx.doi.org/10.1016/0003-6870(86)90245-0.

Mital, A. and Kumar, S. 1998a. Human muscle strength definitions, measurement, and usage: Part I—Guidelines for the practitioner. *International Journal of Industrial Ergonomics*, 22 (1–2): 101–121. http://dx.doi.org/10.1016/s0169-8141(97)00070-x.

Mital, A. and Kumar, S. 1998b. Human muscle strength definitions, measurement, and usage: Part II—The scientific basis (knowledge base) for the guide. *International Journal of Industrial Ergonomics*, 22 (1–2): 123–144. http://dx.doi.org/10.1016/s0169-8141(97)00071-1.

Mital, A. and Sanghavi, N. 1986. Comparison of maximum volitional torque exertion capabilities of males and females using common hand tools. *Human Factors*, 28 (3): 283–294. Retrieved from http://hfs.sagepub.com/content/28/3/283.full.pdf.

O'Driscoll, S. W., Horii, E., Ness, R., Cahalan, T. D., Richards, R. R., and An, K. N. 1992. The relationship between wrist position, grasp size, and grip strength. *Journal of Hand Surgery*, 17 (1): 169–177. http://dx.doi.org/10.1016/0363-5023(92)90136-d.

O'Sullivan, L. W. and Gallwey, T. J. 2002. Upper-limb surface electro-myography at maximum supination and pronation torques: The effect of elbow and forearm angle. *Journal of Electromyography and Kinesiology*, 12 (4): 272–285. http://dx.doi.org/10.1016/s1050-6411(02)00014-7.

Oldfield, R. C. 1971. The assessment and analysis of handedness: The Edinburgh inventory. *Neuropsychologia*, 9 (1): 97–113. http://dx.doi.org/10.1016/0028-3932(71)90067-4.

Paschoarelli, L. C., Razza, B. M., Lucio, C. C., Ulson, J. A. C., and Silva, D. C. 2012a. Laterality and usability: Biomechanical aspects in prehension strength. In: M. M. Soares and F. Rebelo (Eds.), *Advances in Usability Evaluation—Part I*, 1st ed. Boca Raton, FL: CRC Press, pp. 181–190.

Paschoarelli, L. C. and Santos, R. J. H. S. 2011. Usability evaluation of different door handles. In: D. Kaber and G. Boy (Orgs.). *Advances in Cognitive Ergonomics*, 1st ed. Miami: CRC Press, Vol. 1, pp. 291–299.

Paschoarelli, L. C., Santos, R. J. H. S., and Bruno, P. 2012b. Influence of door handles design in effort perception: Accessibility and usability. *Work* 41: 4825–4829.

Peebles, L. and Norris, B. 2003. Filling "gaps" in strength data for design. *Applied Ergonomics*, 34 (1): 73–88. http://dx.doi.org/10.1016/s0003-6870(02)00073-x.

Pheasant, S. and O'Neill, D. 1975. Performance in griping and turning: A study in hand/handle effectiveness. *Applied Ergonomics*, 6 (4): 205–208. http://dx.doi.org/10.1016/0003-6870(75)90111-8.

Razza, B. M., Paschoarelli, L. C., Silva, D. C., Ulson, J. A. C., and Lucio, C. C. 2012. Pulling strength with pinch grips: A variable for product design. In: M. M. Soares and F. Rebelo (Eds.), *Advances in Usability Evaluation—Part I*, 1st ed. Boca Raton, FL: CRC Press, pp. 428–436.

Richards, L. G. 1997. Posture effects on grip strength. *Archives of Physical Medicine and Rehabilitation*, 78: 1154–1156. http://dx.doi.org/10.1016/s0003-9993(97)90143-x.

Shih, Y. C. and Ou, Y. C. 2005. Influences of span and wrist posture on peak chuck pinch strength and time needed to reach peak strength. *International Journal of Industrial Ergonomics*, 35: 527–536.

Shivers, C. L., Mirka, G. A., and Kaber, D. B. 2002. Effect of grip on lateral pinch grip strength. *Human Factors*, 44 (4): 569–577. http://dx.doi.org/10.1518/0018720024496999.

Silva, D. C. and Paschoarelli, L. C. 2013. Usability in the opening of soft drink packagings: Age influence in biomechanical forces. In: M. M. Soares and F. Rebelo (Orgs.). *Advances in Usability Evaluation—Part I*, 1st ed. Boca Raton, FL: CRC Press, Vol. 1, pp. 171–180.

Snook, S. H. 1985a. Psychophysical considerations in permissible loads. *Ergonomics*, 28 (1): 327–330. http://dx.doi.org/10.1080/00140138508963140.

Snook, S. H. 1985b. Psychophysical acceptability as a constraint in manual working capacity. *Ergonomics*, 28 (1): 331–335. http://dx.doi.org/10.1080/00140138508963141.

Su, C., Lin, J., Chien, T., Cheng, K., and Sung, Y. 1994. Grip strength in different positions of elbow and shoulder. *Archives of Physical Medicine and Rehabilitation*, 75 (7): 812–815. Retrieved from http://www.ncbi.nlm.nih.gov/pubmed/8024431.

Teraoka, T. 1979. Studies on the peculiarity of grip strength in relation to body position and aging. *Kobe Journal of Medical Science*, 25 (1): 1–17.

Tullis, T. and Albert, B. 2008. *Measuring the User Experience: Collecting, Analyzing, and Presenting Usability Metrics* 1st ed. San Francisco, CA: Morgan Kaufmann.

Voorbij, A. I. M. and Steenbekkers, L. P. A. 2002. The twisting force of aged consumers when opening a jar. *Applied Ergonomics,* 33 (1): 105–109. http://dx.doi.org/10.1016/s0003-6870(01)00028-x.

12 Digital Hand Model for Grip Comfort Evaluation

Peter Gust and Aydin Ünlü

CONTENTS

12.1 INTRODUCTION

Comfort evaluation for work equipment handles frequently fails in practice because of high temporal complexity and development costs, coupled with the pressure of shorter development times. Handle design also has a strong subjective element rooted in the communication of comfort as sensation. Finally, handle design lacks an objective method to make comfort measurable.

The aim is, therefore, to develop an objective method for comfort evaluation of work equipment, based on a digital hand model, taking into consideration pressure discomfort models in handle design. With this new method, the constructor should already have information on the CAD (computer-aided design) model about handle design parameters for the design of comfortable work equipment grips, thus enabling the number of test subjects and expensive prototypes to be reduced.

This chapter presents a new method for achieving comfort evaluation of handles on a digital hand model. After performing pressure analysis, a so-called discomfort model for comfortable pressure distribution is derived. The chapter then focuses on the creation of a hand model for simulation of pressure distribution on the palm. Finally, handle design parameters such as shape, size, material, and surface can be derived from the hand model in relation to comfortable pressure distribution.

12.2 LITERATURE REVIEW

The human hand is constantly in contact with products. In particular, the handle of work equipment, as a direct interface to the user, greatly influences the perception of fatigue, pain, and comfort (Lindqvist and Skogsberg, 2008). The designer's job during handle development is to design handles that can be felt to be comfortable by the users. Comfortable handles need the right decisions regarding handle design parameters such as shape, size, material, and surface. In addition, the designer must consider many factors such as grip types, coupling types, and target groups (Strasser and Bullinger, 2008).

Ergonomic handle design entails a methodical and systematic consideration of all important factors. The designer must always perform a general and a detailed analysis and question all conditions and factors that may finally have an impact on the design (Bullinger and Solf, 1979). The general analysis is about the investigation of body position and movement possibilities, and motion assignment. The detailed analysis is about the investigation of gripping types (crush grip, pinch grip, support grip), coupling types (form and force closure), and hand posture (neutral or nonneutral posture). Design parameters will finally be determined by considering all these factors, with the help of guidelines about, for example, hand anthropometry (Strasser and Bullinger, 2008).

Ergonomic handle design also generally entails using many subjects for the evaluation of prototypes. This evaluation is purely subjective, and justifiably so, as handle design has a strong subjective element rooted in the communication of comfort as sensation. But this also causes high development costs and lengthy development times.

Handle design lacks an objective method for making comfort measurable. Take, for example, the case of vehicle seat comfort. A seat can be checked for comfortable pressure distribution and can be compared with other seats (Hartung, 2006). Thus the seat design can be objectively improved. It is also possible with FEM (finite element method) simulation to predict the discomfort of sitting and to improve on the CAD model (Mergl, 2006). Research indicates that pressure distribution while gripping has a major impact on discomfort (Kuijt-Evers et al., 2004).

12.2.1 PRESSURE EFFECT

When pressure acts on the skin over a longer period of time, the blood vessels dilate to maximize blood flow. This is shown by the increasing redness of the skin. After a while a second stage begins, where the blood starts to coagulate in the capillaries. If tissue deformation limits are exceeded, inhibition of the blood flow in the tissue occurs. This may in the worst case lead to cell destruction and an associated release of toxic substances that stimulate pain. Pain occurs when the body's trained nociceptors signal that tissue-damaging stimuli noxious to the human body are acting (Murray et al., 2001).

Dents in the range of 0.01 mm already evoke sensations of pressure in the palm. Tissue damage can occur if more significant localized pressure is applied. Thus, pressure from 0.42 to 0.5 N/cm² applied for a duration of 2.5 min can result in a pressure collapse or obstruction of blood flow (Hartung, 2006). However, with interim relief to the skin, higher pressures are tolerable.

Bennet et al. (1979) found that with high shear forces only half the pressure is required to occlude blood flow. The results relate to an area on the back of the hand, but appear to be transferable in principle to the palm. Goosens (1994) observations on the sacrum in young healthy subjects record hypoperfusion following 1.16 N/cm² of pure compression. Moreover, a shearing load of 0.31 N/cm² is already equivalent to a pressure load of 0.87 N/cm² and is sufficient to cause critical constriction of the blood flow.

In addition, adaptation behavior comes into play in relation to pressure stimulus level and duration. A distinction is made here between very fast, medium-fast, and slowly adapting receptors. Schmidt (1977) applied three pressure stimulus levels of 995, 525, and 155 g on the skin and measured the receptor discharge (Imp/s) over a time of just 40 s. He found adaptation of the pressure receptors to pressure stimulus. For example, receptor discharge converged from 100 to 10 Imp/s.

12.2.2 PRESSURE DISCOMFORT MODEL

A pressure discomfort model is derived from subjective ratings and objective pressure values (Hartung, 2006, p. 22). Pressure values are referred to as pressure discomfort thresholds (PDTs)

(Aldiena et al., 2005). The term discomfort stands in relation to biomechanical and fatigue factors (Zhang et al., 1996). Another key parameter is pressure pain thresholds (PPTs). PPTs describe the level of pressure that is perceived by the subject as painful (Rodday et al., 2011).

Based on this pressure pain research, Hall and Kilbom (1993) examined the PPT (pressure pain threshold) of 64 skin areas and showed that the heel of the hand and the fold of skin between the thumb and index finger are sensitive to pressure. Further pressure pain investigations were carried out by Stevens and Mack (1959) and Brennum et al. (1989). Women generally had significantly lower PPTs than men. The PPTs were mostly performed with a pressure algometer. But PPTs vary considerably in different publications depending on the methodology, the anatomical region of the pressure surface, and the pressure increase. In addition a dependency of pressure sensation in relation to the form, size, and material of the pressure stamp has been shown (Hall and Kilbom, 1993).

The designer basically has to fall back on the PDTs of Hall and Kilbom (1993) and Johannson et al. (1999). In the work of Johannson et al. (1999) the PDTs for three points of the palm were determined by the pressure stamp. The results show that the index finger, the center of the hand area, and the thumb have different sensations, and the thumb as opposed to the index finger and the center of the hand area is more sensitive to pressure. The PDT for the middle finger is 188 kPa, for the center of the hand area is 200 kPa, and for the thumb is 100 kPa. Using color subtractive printing films (Prescale) Hall also identified a PDT of 104 kPa for the entire palm (Hall and Kilbom, 1993). Overall, Lindqvist and Skogsberg (2008) recommended a uniformly distributed pressure distribution of 200 kPa on the palm as a rule for work equipment.

Work equipment handles produce different pressure sensitivity on the palm depending on time. Thus the existing PDT, which lies between 100 and 200 kPa, cannot be used to rate handle comfort. For this purpose, the study of Gust and Ünlü (2014) has shown different PDTs exemplarily on an iron bender. The PDTs during the investigation were below the reference value of 104 kPa, but still felt painful. Established PDTs are also irrelevant for handles that produce shear forces or vibration on the palm surface, because these factors can raise the pressure sensation (Bennet et al., 1979).

12.2.3 DIGITAL HAND MODEL

Digital hand models are generated by computer representations of the hand and can be simulated using either the multibody systems (MBS) method or the FEM. It is now also possible to couple the methods of FEM and MBS. In contrast to MBS models, FEM models are deformable and can calculate mechanical stresses such as pressure in certain parts of the body. MBS models consist of rigid nondeformable bodies connected to one another by kinematic joints. Using the MBS method, it is only possible to determine the kinematics of the body and the contact forces. These data are used, for example, as input data for FEM simulation (Merten, 2008, p. 45ff).

Numerous digital human models that reflect the human hand are already available as commercial software packages. MBS models are primarily used in RAMSIS and ANYBODY to support ergonomic design optimization processes. An example of a coupled FEM–MBS human model is MADYMO, which was developed primarily for simulating crash tests. The data from MADYMO come from the human model HUMOS (Keppler, 2003, p. 10f). Material properties relevant for hand models, such as density, Young's modulus, and Poisson's ratio for bone, skin, muscle, and tendons, are also known.

Other digital hand models focus on realistic simulation. For example, Wu et al. (2012) developed a combined FEM and MBS finger model of three finger segments with realistic bones, nails, and soft tissue. In the course of validation, pressure distribution was simulated in relation to handle stiffness and changes in geometry (Wu and Dong, 2005). Han et al. (2008) developed an FEM fingertip and simulated the pressure distribution on the fingertip when opening can tabs. The results revealed that over a large contact area pressure distribution was reduced to the fingertip. Xie et al. (2013) presented a hand model for the simulation of nonlinear contact deformations, in particular in relation to realistic overlapping of the skin. The results are pressure distribution on the grip surface. In sum, there is no simulation program that has improved handle design parameters with respect to the pressure distribution of the palm.

12.3 EXPERIMENT

The literature review cites many studies relating to PPTs. Among these, Hall and Kilbom (1993) found differing sensations of pressure for most hand areas (64 in all), which is considered close to experimental procedure.

To establish a discomfort model for pressure loads of the palm, the PPTs of Hall and Kilbom (1993) were first classified according to the BodyMap of Corlett and Bishop (1976). This contains hand areas A through W. Here, attention is limited to the areas of thumb folds (Q), palmar (P), and hypothenar (O), as these are the ones that are essentially loaded when pressing a handle. Figure 12.1 shows the PPTs for the three hand areas in three different colors.

To establish a generally applicable discomfort model, a so-called percentage distribution of pressure was derived from the consideration of PPTs. This percentage pressure distribution is known in the discomfort model for seating comfort as percentage load distribution (PLD). It indicates the percentage of the total load of the contact surface that should impact a specific body area (Mergl, 2006). The PLD for the selected hand areas is divided into global PLD (see table in Figure 12.1). This PLD can be calculated from the ratio of the individual PPTs to the sum of all PPTs. From the average values of the global PLD it can be seen that all three regions have different sensitivity thresholds. Hand area P is the least sensitive compared to O and Q. Hand area Q is highly sensitive. The local PLD gives the ideal pressure distribution inside the hand area.

Global PLD is checked experimentally against the general statement about handle discomfort. For this purpose, a pressure analysis of the palm is performed. This involves the generation of various sensations of pressure, as well as evaluating handle discomfort and measuring pressure

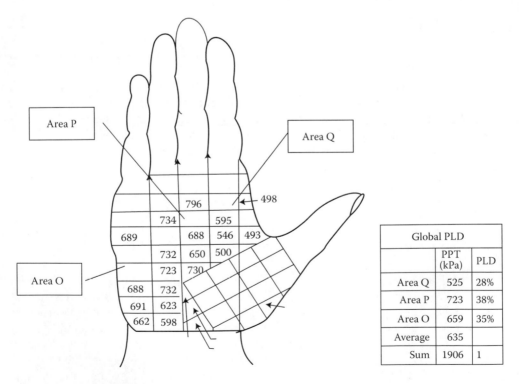

Global PLD		
	PPT (kPa)	PLD
Area Q	525	28%
Area P	723	38%
Area O	659	35%
Average	635	
Sum	1906	1

FIGURE 12.1 Determination of PLD.

distribution between the contact surface of the handle and the side of the hand. The results are then compared with the determined PLD values.

12.3.1 PRESSURE ANALYSIS

A pressure analysis was performed by means of a drill stand. Three different handles were attached to the handle of the drill stand. These handles were printed, and varied in relation to the mean radius (minimum R = 12 mm, medium R = 16 mm, maximum R = 18 mm). All handles had the same surface roughness and stiffness. The force transmissions of the drill stand were measured with a scale on the drill bit. According to the force balance, a force of 80 kg was required on the scale to transmit a force of about 200 N onto the palm.

Pressure distribution on the hand area was measured with three FSRs (force-sensing resistors). FSRs consist of a conductive polymer that changes resistance following application of force to its surface. The hand areas treated were selected and limited zones that were under pressure during the process. The FSR were bonded to an elastic band and programmed with an Arduino (Nano) microcontroller. Their output voltages were recorded with PLX-DAQ (Parallax Data Acquisition tool). Pressure values were calculated from the output voltages by means of one-exponential approximation function.

To measure pressure distribution, the handles were attached to the handle of the drill stand and pushed with the predetermined factors such as body posture, grip type, and hand posture in a vertical direction. The working height of the drill stand was set to the body size of the subjects. The force transmission of 80 kg and the pressure force of about 200 N on the handle were checked with the measuring scale on the balance. The handles were pressed approximately with 200 N and held for 1 min, during which time three pressure values for each of the three grips were recorded with the PLX-DAQ. Each of the three handles was measured three times.

Pressure was investigated with two male subjects between 27 and 30 years of age. The subjects had roughly the same hand proportions. The discomfort sensation was then evaluated for the three hand areas. The evaluation was conducted with a scale of 1–3 (3 = high, 2 = medium, and 1 = low discomfort sensation). It was also permitted to evaluate discomfort with 0.5, 1.5, and 2.5. Subjects were previously instructed about the relevant factors and were obliged to comply with them. The influence of variables such as body posture, grip type, hand posture, and force transmission was also controlled (Figure 12.2).

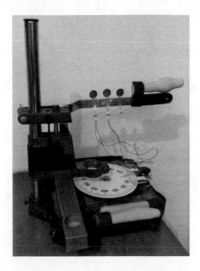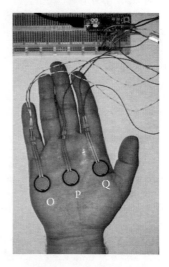

FIGURE 12.2 Measuring device for generating pressure and discomfort on the thumb folds (Q), palmar (P), and hypothenar (O).

TABLE 12.1

Average of Discomfort Ratings and Pressure Measurements at 200 N Pressing Forces

Pressure	O	P	Q	Sum
Sharp (kPa)	18.55	249.49	14.92	282.96
Oval (kPa)	4.71	199.33	59.36	263.40
Flat (kPa)	16.62	89.47	58.41	164.50
Discomfort	**O**	**P**	**Q**	**Average**
Sharp	0.25	3	2	1.75
Oval	1	2.25	1.5	1.58
Flat	1.5	2	2.25	1.92
Pressure Distribution	**O (%)**	**P (%)**	**Q (%)**	**Sum (%)**
PLD sharp	7	88	5	100
PLD oval	2	76	23	100
PLD flat	10	54	36	100
PLD comfort	35	38	28	100

12.3.2 RESULTS

The three pressure values and discomfort sensations for each handle yielded different pressure distributions. Table 12.1 shows the measurement results for pressure distribution and discomfort evaluation for all three types of handle. The pressure distributions were converted into PLD. The sum of the PLD is 100%. Overall high discomfort can be seen in relation to the hand areas when pressure distribution lies above the PLD. High discomfort was created for hand area P, as shown by the maximum pressure measurement of 249.49 kPa. This may be due to the fact that the tendon plate in the middle of the hand, unlike the fat padding on the hypothenar and thenar has a higher stiffness and thus produces a higher reaction force. Moreover, field P has the greatest form variation, so the greatest pressure is generated there.

If the pressure distribution was below a PLD, the affected hand area was evaluated as experiencing little discomfort. It is also recognizable that at hand area O the pressure distribution lay within the PLD values for all handles. At pressures that are under the PLD, the affected hand area was evaluated as experiencing moderate discomfort. For example, a comfortable distribution of pressure was felt at the oval handle. Here the pressure distribution, except for hand area P, is under the PLD. The pressure distribution on hand area Q is no greater either. In Table 12.1, average discomfort ratings are placed in the right-hand column. Overall, it can be seen that discomfort and pressure rise together with the total load. The oval handle is more comfortable than the sharp or flat handle. This is confirmed by the approximation to comfortable pressure distribution.

12.4 SIMULATION

The derived discomfort model will now be used with a digital hand model to evaluate the handle. This is a research innovation: earlier research did not have access to a digital hand model to simulate pressure distribution on the human palm, or to evaluate grip design in relation to a discomfort model, or to the percentage of load distribution. There have been hand models to calculate pressure distribution, and studies on PPTs, but there has until now been no digital hand model with which to observe, measure, and evaluate the sensation of pressure between hand and handle.

The conceptual transfer from the discomfort model to the digital hand model takes pressure sensitivities (or the ideal pressure distribution) of the palm to make a statement about grip comfort.

First, a digital hand model is created for this purpose. This goes beyond the creation of geometry and the determination of the mechanical properties of the numerical model. Section 12.5 deals with the application of this model for grip comfort evaluation. The application of the model exemplifies the way in which the computational model can be used.

12.4.1 MODELING

For the creation of the numerical model the overall behavior of human tissue (with muscles, tendons, connective tissue, etc.) is represented as simply as possible. The processes and interactions in tissues or between muscles are disregarded. Thus, in this model, only the bones and the soft tissues are modeled as two separate bodies, without mapping internal structures. If you were to try to simulate the displacement of muscle strands, tendons, and tissues against each other, the result would be a very complex model with many possible sources of error. Such complex models have high computation times on conventional computers. Since short computation times are desirable, the modeling of internal structures is confined to soft tissue.

12.4.1.1 Geometric Properties of the Model

The skeleton model of the hand was derived from the biomechanical human model "OpenSim" and adjusted to soft tissues. The soft-tissue data is based on computed tomography (CT) images. The CT images of soft tissues were read with the software Invalius. From the individual images a realistic contour of the hand was generated. Starting from point clouds, the CAD software CATIA (Dassault Systèmes, IBM) was used to produce solid models of the soft tissues and bones. The positioning of the bones within the soft tissues was based on the CT images. Finally, the skeleton model was subtracted from the soft-tissue geometry.

Furthermore, meshing of the hand model was carried out. For the meshing of the CAD model, CATIA Advanced Meshing Tools (AMT) workbench was used. For the connection of the internal structure of the soft tissue to the outer structure of the bone, the soft-tissue model was meshed in 3D tetrahedral elements. For the selected hand areas Q, P, and O a total of 35 faces were selected and defined for accurate results, with an element size of 5 mm. The rest of the body of the soft-tissue model was meshed with an element size of 15 mm. Before meshing, the model was excised down to the palm of the hand, in order to save fewer nodes and hence computation time. The meshed soft-tissue model with 3031 nodes and 11,236 elements was imported from AMT to Recurdyn as a flexible body. The bone model was not used because the internal nodes were fixed in space (Figure 12.3).

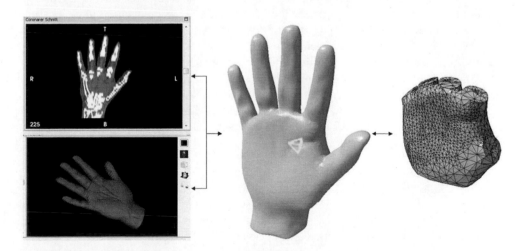

FIGURE 12.3 Generation of a numerical model of the soft tissue from CT images.

12.4.1.2 Physical Properties of the Model

The first question about the model is what material behavior can be attributed to it. According to current literature the behavior of human tissue is anisotropic. Anisotropic means depending on direction. A muscle is very stiff in the longitudinal direction, but in the transverse direction it shows very little resistance. When impacted by an attacking force, the muscle will dodge to the side. Pressure is only exercised in the longitudinal direction of the hand muscles. Therefore, isotropic material behavior is selected as a simplification.

Next, there is the question of the linearity of material behavior with respect to an applied force. In the literature it is known that the behavior of human tissue is nonlinear. However, the compression tests show from Brinckmann et al. (2012) that the initial path and pressure force are linearly related. For planar pressure, as in the handle, the initial path is important. Therefore, a linear material model is chosen for modeling. As tissue after exercise regularly returns to its original shape, the material behavior is modeled as elastic.

Human tissue is therefore simplified for the model to be isotropic, linear, and elastic. In order to describe this material behavior in the simulation, the definitions of Young's modulus, Poisson's ratio, and density, as well as damping property are required. Poisson's ratio is determined on the basis of Kayser and Jarewski (1996) to the value $\mu = 0.49$. This represents a nearly incompressible behavior. Since human tissue consists largely of water, this is understandable. Density is also determined on the basis of Kayser and Jarewski (1996), based on a value of $\rho = 1100$ kg/m³. As skin behaves like rubber, a damping constant of 0.05 is chosen for the attenuation of the model. This corresponds to a default value for rubbery materials.

This leaves Young's modulus as the last variable. This is determined on the basis of compression tests, also known as stamp attempts. In a first step, stiffness distribution along the palm is determined, using compression tests. In a second step, the stiffness of the numerical model is scaled to the measured stiffness, so that the measured pressure distribution is in agreement with the calculated pressure distribution. In order to eliminate influences on pressure distribution by different hand shapes, these tests are carried out with the same person from whom the geometry of the model was derived.

12.4.1.2.1 Compression Test

Compression testing, known from the Shore hardness test, is often carried out to determine the elastic properties of the skin. A stamp is pressed onto the skin area to be examined, and the indentation depth measured with respect to the indentation force. For the compression tests a load cell (induct) was used in conjunction with a stamp (diameter 12 mm). Measurement was extended to the marked test points in 1 mm increments to a maximum of 10 mm, and the indentation force measured by the load cell and the indentation depth on the measurement scale. For each indentation depth five replicates were performed and subsequently analyzed using Microsoft Excel.

As a result, it can be stated that the force–displacement curves in the initial region as described by Brinckmann et al. (2012) are approximately linear and rise exponentially on closure. The force–displacement curves show the different counterforces against the rising pressure on the hand areas. A possible reason for this is the low subcutaneous fat levels in region P, and thus the stiffening in the tendons of the palm. Total stiffness is shown by the initial increase of the force–displacement curves (Figure 12.4).

12.4.1.2.2 Compression Simulation

The behavior of the human palm as determined by the compression test will now be transferred to the numerical model. For this purpose, the stamp was simulated in Recurdyn. First, the numerical model was divided into three areas, each with its own Young's modulus. The areas Q, P, and O correspond to the Bodymap of Corlett and Bishop (1976). The stamps were also modeled in the multibody simulation software as an undeformable body. For the translational movement of the stamp, translation joints were defined in the focus of the stamp. Contact conditions were introduced

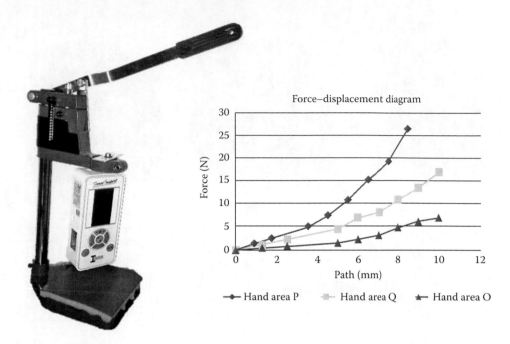

FIGURE 12.4   Stiffness measurements: measuring device and force–displacement curve.

between the end-face of the stamp and the palm. The plan area was so oriented that there were no pressure peaks. The increment of the translational motion was recorded/represented with a step function for a total duration of 5 s.

The comparison between experiment and simulation is as follows: the stamp is measured in increments of 1 mm against the palm. The resulting force–displacement curve of the stamp determines the indentation force. The relevant Young's modulus value is reached after six iterations of the compression test. The following Young's modulus values are the results for hand areas Q, P, and O after compression simulation:

- Hand area Q: $E_Q = 4.79$ kPa
- Hand area P: $E_P = 11.47$ kPa
- Hand area O: $E_o = 1.41$ kPa

Simulated stress is then compared with measured pressures. The measured pressures are for $P_Q = 9.8$ kPa, $P_P = 10.83$ kPa, and $P_O = 3.25$ kPa. It can be seen that the simulated pressure stresses correspond to the pressure measurements. The pressure measurements are obtained by the ratio of the measured force and the surface of the stamp.

Young's modulus values as measured are thus transferred onto the numerical model. The values determined by this test design are well within the field specified in the literature. Thus, Yamada (1973) reports a Young's modulus of 20 kPa and Daly (1982) a general value of 5 kPa.

12.4.2   MODEL APPLICATION

For the model application, the samples from the experiment are simulated. These handles are modeled as rigid bodies. The surfaces lie on the hand regions O, P, and Q. The same force transmission was modeled: squeezing the handles with a torque on the palm. In order to minimize computation time, a pressure of 10 N was chosen for the simulation. Figure 12.5 illustrates the simulation model with relevant conditions.

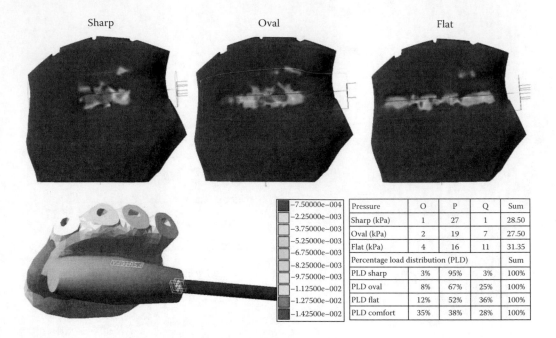

Sharp   Oval   Flat

Pressure	O	P	Q	Sum
Sharp (kPa)	1	27	1	28.50
Oval (kPa)	2	19	7	27.50
Flat (kPa)	4	16	11	31.35
Percentage load distribution (PLD)				Sum
PLD sharp	3%	95%	3%	100%
PLD oval	8%	67%	25%	100%
PLD flat	12%	52%	36%	100%
PLD comfort	35%	38%	28%	100%

−7.50000e−004
−2.25000e−003
−3.75000e−003
−5.25000e−003
−6.75000e−003
−8.25000e−003
−9.75000e−003
−1.12500e−002
−1.27500e−002
−1.42500e−002

FIGURE 12.5 Simulation model for the application and pressure simulation at 10 N pressing forces.

For the determination of the percentage distribution of pressure, three maximum pressures from the simulation results of the palm are formed to a total load of the sum of the three pressure levels. The ratio of maximum pressure to the total load of the three pressure levels is the PLD. The PLD for each simulation is calculated via Excel and compared with the comfortable pressure distribution (PLD).

It is clear that a sharp grip, as in the investigation, causes the highest pressure on the palm P and exceeds the PLD. The two remaining hand regions are lightly loaded. The flat handle exceeds the hand area on P and Q, just as in the pressure analysis. The most comfortable handle is oval-shaped. Here the pressure distribution, except for hand area P, is below the PLD. In addition, the pressure distribution is no longer on hand area Q. Nevertheless, it is not sufficiently comfortable by PLD. The pressure on hand region P is high and should be reduced, and on hand region O is low and should be increased.

12.5   DISCUSSION

Static grip comfort evaluation determined in this way with an FEM–MBS hand model and ideal pressure distribution is thus an effective instrument to evaluate grip shape. Initially the PPTs of Hall and Kilbom (1993), based on the three hand areas of Corlett and Bishop (1976), are converted to a percentage pressure distribution. The PLD boundaries are then examined to determine whether they allow handle comfort to be concluded. The results show that high discomfort arises when the PLD is exceeded, whereas values below the PLD indicate lower discomfort for the subject. Hence approximation to the PLD makes for handle comfort.

The second part of the research involves the creation of a realistic FEM–MBS hand model. For this purpose, the geometry of the soft-tissue model is first generated from CT images and then the geometry is meshed. The material properties are determined with the compression test and compression simulation. The model validation shows the compliance of the pressure distribution with the measured pressure values. Then the handles from the experiment are simulated with the FEM–MBS hand model. The percentage distribution of pressure then shows the results of the pressure analyses.

In future, the remaining hand areas of Corlett and Bishop (1976) will be investigated with respect to the PLD. Grip comfort optimization will be performed in accordance with the local PLD. Splines for the palm will be calculated stepwise. If simulated pressure distribution is below comfortable pressure distribution, the pressure exerted by the handle shape or by the handle material should be increased. If simulated pressure distribution is greater than comfortable pressure distribution, the pressure exerted by the handle should be decreased.

The remaining handle design parameters—dimensions, material, and surface—are still to be investigated in relation to the PLD and the impact on discomfort and pressure distribution of varying those parameters remains to be clarified. It may turn out that the main determinant of pressure distribution, and thus discomfort, is variation of shape. In this case a PLD model will depend in practice on shape variation.

In addition, the shear forces that occur during the use of work equipment should be included in the PLD. For this work, shear forces could be neglected, as pressure distribution mostly arose in the palm. But screwdrivers, for example, inevitably generate shear forces on the palm, which may affect the pressure sensation to a high degree (Bennet et al., 1979).

Further studies must be performed with respect to age-related factors among elderly people. Currently, there are no PPTs of subjects over the age of 60. It has been found that, because of the lowering of sensitivity associated with aging, older people perceive pressure difference less acutely (Zenk, 2008). Other age-related factors include the reduction of the friction coefficient of the hand surface and the reduction of muscle strength and fat. Further research is needed into the role played by these changes for the PLD model, by investigating, for example, whether older people have a higher or lower PPT and PLD.

For the development of the simulation model, the principle was to reproduce the overall behavior of human tissue (including muscles, tendons, connective tissue, etc.) and material as easily and accurately as possible. The occurrence of processes and interactions in tissues or between muscles was disregarded: only the bones and surrounding tissue were modeled, as two separate entities, without reproducing internal structures. This procedure reduced possible sources of error, as well as high computation time. In future, the simulation model should approximate reality more closely, in order to calculate pressure distributions more accurately. In addition, a hand model is desirable in order to include gripping force and frictional forces in the pressure simulation.

The overall aim of the research is to program an ergonomics software using a readily available hand model. The user should, after importing the handle and making a choice of influence factors, be able to derive proposals for handle design parameters: for example, spline proposals, which can then be imported into the CAD system. Such a program requires a database of different variables such as grip types and gripping forces, as well as hand-type and hand-position-dependent material properties. The program should offer a choice of three hand types with low, medium, and high levels of subcutaneous fat. In addition, it should scale the hand models independently by percentile and gender. The hand model should automatically grip the tool handle by means of a "snapping" function.

REFERENCES

Aldiena, Y. et al. 2005. Contact pressure distribution at hand–handle interface: Role of hand forces and handle size. s.l.: *International Journal of Industrial Ergonomics* 35, 267–286.

Bennet, L., Kaviner, D., Lee, B. K., and Trainfor, F. 1979. Shear vs pressure as causative factors in skin blood flow occlusion. s.l.: *Archives of Physical Medicine and Rehabilitation* 60, 309–314.

Brennum, J. et al. 1989. *Measurements of Human Pressure-Pain Thresholds on Fingers and Toes.* Copenhagen, Denmark: Elsevier.

Brinckmann, P., Frobin, W., Leivseth, G., and Drerup, B. 2012. *Orthopädische Biomechanik.* Münster: Wissenschaftliche Schriften der WWU Münster.

Bullinger, H.-J. and Solf, J. 1979. *Ergonomische Arbeitsmittelgestaltung—2.* Handgeführte Werkzeuge: Fallstudien. Frankfurt: Bundesanstalt für Arbeitssicherheit und Unfallforschung.

Corlett, E. N. and Bishop, R. P. 1976. A technique fpr measuring postural discomfort. *Ergonomics*, 19(2), 175–182.

Daly, C. H. 1982. Biomechanical properties of the dermis. s.l.: *Journal of Investigative Dermatology* 79(Suppl 1), 17–20.

Goosens, R. H. M. 1994. *Biomechanics of Body Support: A Study of Load Distribution, Shear, Decubitus Risk and Form of the Spine*. Rotterdam: Erasmus Universität Rotterdam, 1994.

Gust, P. and Ünlü. A. 2014. Developing a comfort evaluation method for work equipment handles. *Design 2014—13th International Design Conference*, Cavtat, Dubrovnik, Croatia, May 19–22, pp. 2043–2050.

Hall, C. F. and Kilbom, A. E. 1993. Sensitivity of the hand to surface pressure. s.l.: *Applied Ergonomics* 24(3), 181–189.

Han, J. Nishiyama, S., Yamazaki, K., and Itoh, R. 2008. Ergonomic design of beverage can lift tabs based on numerical evaluations of fingertip discomfort. Hrsg. Japan: *Applied Ergonomics* 39(2), 150–157.

Hartung, J. 2006. Objektivierung des statischen Sitzkomforts auf Fahrzeugsitzen durch die Kontaktkräfte zwischen Mensch und Sitz (Objectification of static seating comfort on vehicle seats by the contact forces between man and seat). Dissertation, München.

Johannson, L., Kjellberg, A., Kilbom, A., and Hagg, G. 1999. Perception of surface pressure applied to the hand. *Ergonomics*, 42, 1274–1282. London: Taylor & Francis.

Kayser, A. and Jarewski, J. 1996. *Entwicklung von Erstazmodellen zur Beschreibung der dynamischen Eigenschaften von menschlichen Körpersegmenten der unteren Extremitäten (Development of alternative models to describe the dynamic Characteristics of human body segments of the lower extremities)*. Wuppertal: DFG.

Keppler, V. 2003. *Biomechanische Modellbildung zur Simulation zweier Mensch-Maschinen-Schnittstellen (Biomechanical modeling for the simulation of two man-machine interfaces)*. Tübingen: Eberhard-Karls-Universität zu Tübingen.

Kuijt-Evers, L., Groenesteijn, L., de Loozea, M., and Vink, P. 2004. Identifying factors of comfort in using hand tools, Netherlands: *Applied Ergonomics* 35, 453–458.

Lindqvist, B. and Skogsberg, L. 2008. Ergonomie bei Handwerkzeugen (Ergonomics for hand tools). s.l.: Atlas Copco.

Mergl, C. 2006. *Entwicklung eines Verfahrens zur Optimierung des Sitzkomforts auf Automobilsitzen (Development of a method for optimizing the seating comfort of automotive seats)*. München: Lehrstuhl für Ergonomie.

Merten, v. K. 2008. *Verringrung des Risikos von thoraxverletzungen beim PKW-Sitenaufprall—FEM Simulationen mit dem Mensch-Modell HUMOS (Reducing the risk of thoracic injuries in car side-impact FEM Simulations with the human model HUMOS)*. Dissertation, University of München, München.

Murray, L., Magazinovic, N., and Stacey, M. 2001. Clinical practice guidelines for the prediction and prevention of pressure ulcers. s.l.: *The Australian Journal of Wound Management* 9(3), 88–97.

Rodday, V., Geißler, B., and Ottersbach, H. J. 2011. *Druckschmerzschwellen bei Druckreizen* (Pressure pain thresholds at pressure stimuli). Mainz: GfA.

Schmidt, R. 1977. *Grundriss der Sinnesphysiologie (Ground plan of the physiology of the senses)*, Auflage. s.l.: Springer 3.

Stevens, J. C. and Mack, J. D. 1959. Scales of apparent force. In: s.l.: *Journal of Experimental Psychology*, 58, 405–413.

Strasser, H. and Bullinger, H. 2008. Handgeführte Werkzeuge und handbetätigte Stellteile. s.l.: *Handbuch der Arbeitsmedizin* (Hand-held tools and manually controlled control devices. s.l.: *Handbook of Occupational Medicine*). ecomed Medizin, ISBN 978 3 609 10570 3.

Wu, J., Wimme, B. M., Welcome, E., and Dong, R. 2012. An analysis of contact stiffness between a finger and an object when wearing an air-cushioned glove. *Medical Engineering and Physics* 34(3), 386–393.

Wu, J. Z. and Dong, R. G. 2005. Analysis of the contact interactions between fingertips and objects with different surface curvatures. *Proceedings of the Institution of Mechanical Engineers. Part H, Journal of Engineering in Medicine* 219, 89.

Xie, Y., Kanai, S., and Date, H. 2013. Simulation of contact deformation property of digital hand skin and its experimental verifications. , Hokkaido University, Sapporo, Japan.

Yamada, H. 1973. *Strength of Biological Materials*. Baltimore, Maryland: Williams & Wilkins.

Zenk, R. 2008. *Objectification of Seating Comfort and His Automatic Adjustment*. Münschen: Herbert Utz Verlag.

Zhang, L., Helander, M. G., and Drury, C. G. 1996. Identifying factors of comfort and discomfort in sitting, s.l.: *Human Factors: The Journal of the Human Factors and Ergonomics Society* 38(3), 377–389.

13 Classroom Ergonomics
A Portuguese University Study

Luís Sousa, Maria Eugénia Pinho, Pedro Arezes, and Joaquim Góis

CONTENTS

13.1 INTRODUCTION

The main objective of ergonomics is to provide a safe and efficient environment through the design of work and environment (Mokdad and Al-Ansari 2009). It has a broad scope, being used to improve tools, environment, and electronic devices, thus contributing to dramatically change the modern world.

No matter how strong the advances in technology and ergonomics are, often they are not implemented or, at least, not in the correct manner. So, worldwide researchers have been paying increasing attention to the classroom environment, since the quality of the learning process also depends
on it.

The multifactorial nature of the classrooms environment (e.g., thermal environment, acoustics,
lighting, daylight, indoor air quality, classroom layout, and furniture) increases the complexity
of its ergonomic assessment (Yang et al. 2013). All components are essential to the classroom
environment, however there is a focus on furniture as it has a key importance, considering that
students are spending more time interacting in classes. These interactions or activities, such as
sitting, writing, observing, drawing, reading, and resting, require that students remain seated for
a considerable amount of time (Castellucci et al. 2010, 2015, Dianat et al. 2013, Mokdad and
Al-Ansari 2009, Parcells et al. 1999). The nature of these activities leads to negative impacts in
the students' musculoskeletal system, due to the adoption of static postures, sitting and forward
bending positions for long periods of time, straining muscles, ligaments, and discs, which are also
well-known risk factors for low back pain (LBP) (Jung 2005, Oyewole et al. 2010, Panagiotopoulou
et al. 2004).

Considering that it is of utmost importance to provide our students (children, adolescents, and
adults) with the safest and best learning environment, such activities should be carried out with
furniture designed according with the anthropometry and biomechanics of the student population,
in order to reduce the risk of accidents and overuse disorders (Dianat et al. 2013, Gouvali and
Boudolos 2006, Panagiotopoulou et al. 2004).

One way of reducing the negative impact from the sitting posture is to change the design of
chairs (Chung et al. 2003). In this matter there are plentiful variations, such as seat pan design, seat
fabrics and cushioning, backrest design, angle between the seat pan and the backrest, seat height
(SH), etc.

Even governments have introduced regulations to ensure the standardization of the furniture in
order to provide uniform conditions to students.

To name a few, standards have been published in the European Union (CEN 2006), United
Kingdom (BSI 2006), Japan (JIS 2011), Chile (INN 2002), and Colombia (ICONTEC 1999).

13.1.1 How Sitting Works

The sitting posture is pointed out as being strongly associated with musculoskeletal disorders. These
multifactorial pathologies often arise from spinal loading, duration of sitting, posture, psychosocial
characteristics, obesity, level of physical activity, and backpack use (Gilkey et al. 2010).

The normal sitting posture of a human being is featured by a hip joint flexion of approximately
60°, and a back convexity (kyphosis) in the lumbar region (Murphy et al. 2004). When people sit
the ischial tuberosities support about 75% of the entire body weight. However, if the mass center is
not aligned with the ischial tuberosities some stress may affect other body regions and tissues, as
the weight is no longer being supported by the ischial tuberosities. When the feet firmly rest on the
floor, the legs help to distribute the upper body weight, therefore reducing the loads on thighs and
buttock (Oyewole et al. 2010, Parcells et al. 1999).

The use of seats without cushioning (hard surface) is related with stress to the soft tissues of the
lower body. Due to the hardness of the seat pan, a nonhomogenous tissue deformation and internal
tissue stress and strain of the buttocks occurs, since they are compressed between the ischial tuberosities and the seat pan (Cascioli et al. 2011). The buttocks (muscles and fat) form a cushioning layer
underneath the ischial tuberosities as they begin to compress due to withstanding of the weight of
the upper body. As a result of this compression, that layer is reduced to a thin layer of cushioning
and support (Vos et al. 2006).

When people remain seated for long periods of time, either in a cushioned chair or on a hard flat
seat, discomfort arises. The discomfort, and sometimes pain, arises from the restriction of blood

through the tissues due to the compression of the capillaries (Vos et al. 2006). Cascioli et al. (2011) referred to an interval of 45–180 min to generate in-chair movements. These movements allow the person to redistribute the weight, and release the stress from the tissues.

Despite some authors recommending a seat-to-backrest angle greater than 90° in chairs, in order to reduce the spinal disc pressure (Parcells et al. 1999, Vos et al. 2006), working at a desk requires different postures and different furniture design.

The seat should have a tilt forward to maintain a trunk–thigh angle of 90°, and the desk should be tilted to the user, which would reduce the lumbar and neck flexion (Parcells et al. 1999).

13.1.2 CLASSROOM FURNITURE AND ITS RELATION WITH STUDENTS

The reduction of illiteracy and the supply of broader academic resources have led people to study longer, carrying on their academic careers for more than 15 years.

As school for children starts young (5–7 years), the use of improperly designed furniture may lead to anatomical-functional changes, while poor sitting postures, the use of heavy backpacks, and no exercise may result in intervertebral disk degeneration (Castellucci et al. 2015, Dianat et al. 2013, Jung 2005). Additionally, these musculoskeletal dysfunctions may increase in puberty (Castellucci et al. 2010).

Furniture has an important role in the maintenance of good sitting posture (Agha and Alnahhal 2012, Jung 2005, Panagiotopoulou et al. 2004), which is believed to prevent musculoskeletal symptoms (Agha and Alnahhal 2012, Dianat et al. 2013, Panagiotopoulou et al. 2004).

It is usual to see the one-size-fits-all furniture in classrooms, as it is less expensive, even if it does not fit all the student population. This may result in mismatches between the furniture dimensions and the students' anthropometric dimensions.

All over the world researchers have found situations of these mismatches in primary and high schools (Agha and Alnahhal 2012, Castellucci et al. 2014, Dianat et al. 2013, Van Niekerk et al. 2013). Van Niekerk et al. (2013) also found in their study that chairs were unable to match the basic ergonomic recommendations of furniture design to fit the user.

The use of adjustable furniture should be adopted by schools as it is considered better suited to the learning environment, being capable of fitting students with different dimensions (Straker et al. 2006, Thariq et al. 2010, Yang et al. 2013).

Even if schools are financially unable to acquire adjustable furniture, the people responsible for acquiring furniture, as well as manufacturers, should design furniture according to the anthropometric dimensions of the student population, assuring that the furniture will avoid negative consequences to students (Mokdad and Al-Ansari 2009, Oyewole et al. 2010, Thariq et al. 2010). The problems related to the school environment are resulting in an increase of musculoskeletal symptoms in children and adolescents (Dianat et al. 2013).

A musculoskeletal symptom is the feeling of discomfort and it is hardly prevented from recurring due only to chair design (Thariq et al. 2010). Nonetheless, it is noticeable that alterations in comfort when redesigning a chair may have big impact in interface pressure (Vos et al. 2006), that softness and support are linked with comfort sitting (Cascioli et al. 2011), and adequate design reduces fatigue and discomfort while sitting (Agha and Alnahhal 2012, Dianat et al. 2013).

The discomfort may be benign at the musculoskeletal level but it can hamper students' concentration (Agha and Alnahhal 2012, Castellucci et al. 2015, Dianat et al. 2013, Jacquet and Petitdant 2014, Jung 2005, Yang et al. 2013) even in the most interesting lessons (Castellucci et al. 2010, Thariq et al. 2010). A study conducted by Castellucci et al. (2014) concluded that children sitting in appropriate sized furniture obtained higher scores when compared with others using bigger sized school furniture.

After a while, sitting discomfort may evolve into pain, which appears to be felt differently between males and females (Castellucci et al. 2014, Dunk and Callaghan 2005, Van Niekerk et al. 2013). This pain may be experienced as upper quadrant musculoskeletal pain (UQMP), which Brink et al. (2009) found to have a high prevalence in adolescents due to extreme postural angles in sitting, or as LBP.

The reason for the occurrence of LBP is multifactorial and one factor that is highly associated is the adoption of unnatural spinal postures (Castellucci et al. 2010, Dianat et al. 2013, O'Sullivan et al. 2012, Thariq et al. 2010). Other factors are related with chair design and the maintenance of the arched curvature of the back (Callaghan and Dunk 2002), as well as the high level of mismatch between students and school furniture (Castellucci et al. 2014).

A very disturbing fact evidenced by some epidemiological studies consists of a high prevalence rate of back pain found, including among schoolchildren. Murphy et al. (2004), in their review, observed that several studies of back pain among children indicated a high prevalence of LBP in five countries: 20% in Finland, 26% in England, 33% in Canada, 36% in the United States, and 51% in Switzerland. Also Parcells et al. (1999) found a great number of school children and adolescents frequently experiencing back, neck, and headache pain episodes. Even though these studies refer to back pain in schoolchildren and adolescents, a study conducted by Straker et al. (2008) found that recent research showed that adults who complain of neck/shoulder pain had their first episode before adulthood. In the same line of thought, a study conducted by Gilkey et al. (2010), in their review, found that back pain among populations in the United States, the United Kingdom, and the Netherlands have reported annual prevalence rates for back pain ranging from 15% to 65%, with lifetime prevalence as high as 70%–84%.

13.1.3 ANTHROPOMETRY

According to available publications at an international level, only a few anthropometric studies regarding a wide range of anthropometric dimensions were carried out in Portugal. Some of them rely on previously collected data for biographic means (Cardoso 2008, Padez 2003, 2002, Sobral 1990) while others accomplished an extensive anthropometric data collection (Arezes et al. 2006, Barroso et al. 2005).

The older study from Sobral (1990) aimed to find secular changes in stature in southern Portugal using data from records between 1930 and 1980. In his study the author found a secular shift in height greater in the urban areas than in the rural, and associated with infant mortality and total death rate. Moreover, this author found that, despite the height growth, this was smaller than in other European populations, and that the Portuguese people recorded the shortest heights in Europe.

The study of Padez (2002) used a set of gathered data obtained by military physicians that measured the stature of the 18-year-old young men who went to compulsory military service between 1985 and 1998, as well as the data from Sobral (1990) and the data from a 1904 dissertation. This study found statistical significant differences in height between the Portuguese districts, an average increase of 89.3 mm in height and estimated a 9.9 mm increase per decade.

The same author found stature increasing trend in women in another study (Padez 2003) carried out among a Portuguese female university students population (n = 3366), born between 1972 and 1983 (range 18–23 years).

Cardoso (2008) found a mean increase in height of 15.4 mm per decade, in 10-17-year-old boys, based on data from the archive and medical files of a Portuguese military boarding school (Colégio Militar) for the periods of 1899–1906, 1929–1936, 1961–1966, and 1999–2006.

As for the study from Barroso et al. (2005), an extensive anthropometric data was collected (24 static anthropometric measures plus body weight) from 891 individuals (399 female and 492 male), with ages ranging from 17 to 65 years. In this study, a procedure for data acquisition was developed and the first Portuguese adult workers' anthropometric database was created.

Internationally, several anthropometric studies have been made over the years for different populations, such as the following: 20–30-year-old students of Tehran University (Mououdi 1997), southern Thai population (Klamklay et al. 2008), 19–25-year-old Malaysian University Students (Chong and Leong 2011), and the Bangladeshi male population (Khadem and Islam 2014).

The secular trend, changes in anthropometric characteristics of a population over a period of time (Pheasant and Haslegrave 2005), has been studied and confirmed in Portugal (Arezes et al. 2006, Barroso et al. 2005, Cardoso 2008, Padez 2002, Pheasant and Haslegrave 2005, Sobral 1990) as well in other countries (Pheasant and Haslegrave 2005). This trend is, so far, increasing in Portugal at the rate of 9.9 mm per decade (Padez 2002), although in another countries it may be stabilizing (Ioana et al. 2014) or decreasing, such as in third world countries (Pheasant and Haslegrave 2005). This trend, which may be used as a nation's (or region's) socioeconomic development rate, occurs due to changes in nutrition, socioeconomic status, health care, social practices, climate, and geographical location (Barroso et al. 2005, Chuan et al. 2010, Kaya et al. 2003). Because of this, some authors recommend that anthropometric databases are regularly updated (Chuan et al. 2010, Mokdad and Al-Ansari 2009), while Kaya et al. (2003) propose that they are updated every 5 years.

The majority of the studies, and in particular the Portuguese ones, are more than 5 years old and so these databases may be out of date.

13.1.4 University Studies

Although the majority of studies regarding furniture mismatch and back pain in school are made with primary school children and adolescents, several studies have also been conducted in universities. These university studies have focused mainly on back pain, furniture mismatch, and design of new university classroom furniture.

The bad sitting habits of young students may be irreversible as the sitting habits are acquired in school as a child and are hard to change in adults (Parcells et al. 1999). However, the importance of university studies is related to several implications for the future workforce, including current university students, who may already integrate into the work market suffering from shoulder, neck, and back symptoms or disorders (Murphy et al. 2004).

One symptom of dysfunction between the anthropometry of students and furniture is the existence of reported back pain. The high prevalence of back pain among high-school students is known to increase with age (Van Niekerk et al. 2013) and develop into chronic musculoskeletal pain in adulthood (Brink et al. 2009, Handrakis et al. 2012, Jacquet and Petitdant 2014).

If students continue with inadequate furniture, the back pain may get worse, as a great percentage of adults who have back pain had their first episodes of back pain in their teenage years or in their 20s (Murphy et al. 2004).

The ergonomic design of learning environments, beyond classrooms physical and environmental characteristics, must also take into account students' perception on the issue (Yang et al. 2013). Accordingly, students' perception, which is an important factor to consider in higher education, has not been neglected. Several studies on the assessment of students' perception were found.

For instance, Gilkey et al. (2010) studied the relationship between psychosocial stressors and back pain, in college students from Colorado state, and found high prevalence rates when compared with the general population.

Another study, conducted in France, shows that the participant students felt more discomfort than comfort in the studied classroom, an amphitheater (Jacquet and Petitdant 2014). Authors concluded that this was a result of under-evaluation of the main task performed in that classroom. When authors analyzed the 13 criteria used to evaluate the conditions, they concluded that 9 of them did not match. An Indian study on the evaluation of fixed-type university furniture concluded that students preferred the furniture height to be adjustable (Khanam et al. 2006b). A recent study from Tirloni et al. (2014) surveyed the perceptions of university students about the seated body posture and their preference between two models of chairs.

They verified that more than 60% of students admitted to stay in a forward bending position for a long period (more than half of the class time).

There are also several studies regarding the design of university furniture, since it is believed that the designing of good ergonomic furniture for university students has been neglected (Thariq et al.

2010). Examples are of a study conducted in Sri Lanka to design chairs with mounted desktop for university students from Thariq et al. (2010), two Indian studies, one of anthropometric consideration for designing students desks in engineering colleges (Qutubuddin et al. 2013) and another of designing student's seating furniture for a classroom environment (Khanam et al. 2006a).

The lack of an anthropometric database of the Portuguese student population highlights the relevance of the current study. A threefold objective was set for this study: (1) to build an anthropometric database for the student population of the Faculty of Engineering of the University of Porto (FEUP), (2) to assess the level of mismatch between student's anthropometric dimensions of the Portuguese students of FEUP and the furniture dimensions, by comparing their anthropometric dimensions with the dimensions of the available furniture, and (3) to find out if there is any statistically significant association between students' perceived discomfort and the use of the classroom furniture, as well as to assess the mismatch between students' anthropometric measurements and the school furniture dimensions.

13.2 MATERIAL AND METHODS

13.2.1 Subjects

One hundred and thirty one men and 75 women (n = 206) from a universe of 7295 FEUP students were measured. The sample was mostly composed by students from the integrated master programs, but also includes students from undergraduate and master programs. Their ages range between 18 and 35 years and they were selected when passing through the hall of the building where most of the classes are taken.

13.2.2 Equipment Used for the Anthropometric Data Collection

The equipment used for the purpose of the current study composed of

a. A stationary anthropometer combined with a bench (Figure 13.1)—built from wood panels which were arranged as a corner and covered with graph paper, for the specific purpose of this study.

FIGURE 13.1 Stationary anthropometer with 1200 × 1500 × 2100 (depth, width, height) (mm).

To guarantee the calibration of the anthropometer a self-retracting tape measure was used, ensuring the match of the paper lines with the tape marks. To aid visually the measurement process, some gridlines were drawn every 100 mm, marking the corresponding value.

b. A portable anthropometer (Holtain's Harpender anthropometer)—as the quality of the measurement in the static anthropometer of some anthropometric dimensions could be compromised due to the adoptions of certain postures, the portable anthropometer was used to guarantee that this possible influence would not occur.

13.2.3 Data Collection Procedure

A Total of 14 anthropometric dimensions, were measured for each individual. Six out of the 14 dimensions were measured with the individual standing, while the remaining were obtained while the individual was seated (Table 13.1). The anthropometric measures were taken with the subject in a relaxed and erect posture. Each student was measured in thin clothes (T-shirt, shirt, or thin sweatshirt), jeans, skirts, or dresses. The standing dimensions were taken with the student standing erect to the anthropometer without shoes. The sitting dimensions were taken with the student seated erect onto the anthropometer, with knees bent 90°, and feet (without shoes) flat on the floor. The body dimensions were measured as described in ISO Standardization (2008).

13.2.4 Procedure for Data Treatment

First, a Kolmogorov–Smirnov test was applied to study the normality of the anthropometric data's distribution. Then, mean and standard deviation were calculated for all the measured dimensions. Later, a Student's t-test was used to test if the anthropometric dimensions of female and male populations were statistically different.

In order to measure data dispersion, the coefficient of variation (CV) was calculated as shown in

$$CV = \frac{s}{m} \times 100\% \qquad (13.1)$$

where CV = coefficient of variation; s = standard deviation; and m = mean.

TABLE 13.1
Dimensions Measured in Standing and Sitting Positions

Dimensions	
Standing	**Sitting**
Abdominal depth (AbD)	Buttock–knee length (BKL)
Elbow–knuckle length[a] (Ekl)	Buttock–popliteal length (BPL)
Eye height (EH)	Popliteal height (PH)
Forward grip reach (FGR)	Sitting elbow distance (SED)
Hip breadth[a] (HB)	Sitting eye height (SHE)
Shoulder breadth (bi-deltoid)[a] (ShB)	Sitting height (SH)
Height (H)	Thigh thickness[a] (TT)

[a] Measured with the portable anthropometer.

13.3 FURNITURE DATA GATHERING

13.3.1 TYPE OF FURNITURE

FEUP's school furniture is diverse, therefore five different types of chairs were measured: (C1) old design chair with flat perpendicular surfaces (seat and back support); (C2) new design chair with rounded front edge and concavities in the seat and back support, and an angle between the seat and the back higher than 90°; (C3) new design chair with a more eccentric design, a rounded front edge and concavities in the seat and back, and an angle between the seat and the back slightly larger than 90°; (C4) chair with adjustable height, with wheels and footrest (similar design as in C2 chair); (C5) side-mounted desktop (on the right side of the chair) chair with a folding seat.

For the tables, three different types were measured, as follows: (T1) old design table with metal frame and wood table top; (T2) new design table with metal frame and wood table top; (T3) adjustable table with new design metal frame and wood table top; (T4) table from the chair type 5 with armrest and a small space to lay notebooks.

In FEUP classrooms, the following furniture combinations are used: (1) C1 with T1 (Comb1); (2) C2 with T2 (Comb2); (3) C3 with T2 (Comb3); (4) C4 with T3 (Comb4); (5) C5 with T4 (Comb5).

13.3.2 DIMENSIONS OF CHAIRS AND DESKS

The dimensions of the classroom furniture were taken with a metal tape measure. The adopted criteria, in Figure 13.2, used for the measurement of each dimension are defined as follows:

- Seat height (SH)—vertical distance from the floor to the middle point of the front edge of the seat
- Seat depth (SD)—distance from the back to the front of the seat
- Seat width (SW)—horizontal distance between the lateral edges of the seat
- Seat to desk clearance (SDC)—distance from the top of the front edge of the seat to the lowest structure point below the desk
- Seat to desk height (SDH)—vertical distance from the top of the middle of the seat to the top of front edge of the desk

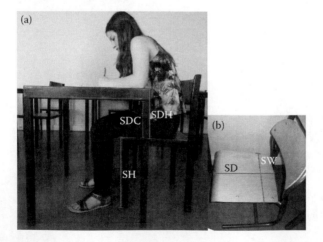

FIGURE 13.2 Representation of the classroom furniture measures: (a) lateral view and (b) top view.

13.4 MISMATCH ASSESSMENT

13.4.1 Considered Anthropometric Dimensions

Only 6 out of the 14 anthropometric dimensions measured were taken into account for the mismatch criteria, as the equations only considered these dimensions: hip breadth (HB), buttock–knee length (BKL), buttock–popliteal length (BPL), popliteal height (PH), sitting elbow height (SEH), and thigh thickness (TT).

13.4.2 Match and Mismatch Criteria

Applied anthropometry and ergonomic principles were considered to evaluate the classroom furniture. Equations enabling the determination of each furniture dimension were used. These equations are based in two types of criteria: "one way" or "two way." "One way" means that only a minimum or a maximum value is required. Two levels ("match" and "mismatch") were defined. In its turn, "two way" criteria require the establishment of two limits, and three levels: a minimum limit (high mismatch) and a maximum limit (low mismatch), and, in between the limits, the dimension is adequate (match).

13.4.2.1 PH and SH Mismatch

Based on published literature (Castellucci et al. 2010, Dianat et al. 2013), the mismatch between PH and SH can be defined by

$$(PH + 30)\cos 30^\circ \le SH \le (PH + 30)\cos 5^\circ \tag{13.2}$$

13.4.2.2 BPL and SD Mismatch

Based on existing studies (Castellucci et al. 2010, Dianat et al. 2013, Parcells et al. 1999), the mismatch between BPL and SD can be defined by

$$0.80 \times BPL \le SD \le 0.95 \times BPL \tag{13.3}$$

13.4.2.3 Hip Width and SW Mismatch

Based on existing studies (Castellucci et al. 2010), the mismatch between hip width (HW) and SW can be defined by

$$HW < SW \tag{13.4}$$

13.4.2.4 TT and SDC

Based on published studies (Castellucci et al. 2010, Parcells et al. 1999), the mismatch between TT and SDC can be defined by

$$TT + 20 < SDC \tag{13.5}$$

13.4.2.5 SEH and SDH Mismatch

Based on recent studies (Castellucci et al. 2010, Dianat et al. 2013), the mismatch between SHE and SDH can be defined by

$$SEH \le SDH \le SEH + 50 \tag{13.6}$$

13.5 DISCOMFORT QUESTIONNAIRE

13.5.1 The Questionnaire

During the measurements the students were queried about the feeling of discomfort during the classes, using the following question: "Do you feel any discomfort using the furniture available in classrooms and/or computer rooms?"

13.5.2 Data Treatment

The multiple correspondence analysis (MCA) is a method for the factor analysis of data, of which the objective consists in the description of the relational structures subjacent to the starting data.

MCA allows us to visualize, through bi-dimensional charts (factor plans) obtained by reducing the special dimensionality of the starting data, not only the internal relation system of each and every one of the variable sets (matrix rows) or the individuals (matrix lines) but also the existing relation system sets between variables and individuals. Even though one of the advantages in the use of MCA over other factor methods lies in the possibility of simultaneous visualization of the structures between variables and individuals, the use of MCA will be limited to the description of the structural relation between variables.

Once our initial data matrix that contained variables of different nature (e.g., the variable stature is quantitative and the variable discomfort is qualitative) was required to ensure the variable homogeneity through a previous encoding of the initial data, this encoding, with the transformation of some measurable ordinal variables subdivided into several classes (variable categories), conformed to previously defined criteria in order to form classes that were important to observe in detail.

Q being the total number of variables and rj the number of categories in which the variable I is subdivided, the total number of data matrix rows (total of Q variable categories) is given by

$$p = \sum_{j=1}^{Q} r_j \tag{13.7}$$

If X is the matrix with n lines (n individuals) by p rows (p categories) filled in terms of presence/absence by the binary encoding present in

$$x_{ij} = \begin{cases} 1 - \text{if individual i has occurrence in category j} \\ 0 - \text{otherwise} \end{cases}, \quad \forall x_{ij} \in X \tag{13.8}$$

It becomes possible to build the logic description board (encoded matrix) presented in Table 13.2.

This coding system ensures that whatever the nature of the variables, the sum in line of the values that come up in the table is constant and equal to the number of variables Q, bringing a statistic homogeneity that is required for subsequent processing.

For this case, our table is a matrix with 48 rows of which sum in line is always equal to 16 (number of variables) and which sum in row gives us the absolute frequency of each variable's category. For each variable, the sum of the absolute frequencies of its categories is always equal to the number of individuals, n, therefore the total in line and in row reproduces nQ. This is an important property as the table of data may be taken as a juxtaposition of contingency tables. This method is based on the research study of Burt (1950), and developed by Benzécri (1973) and Lebart (1975), with the objective of questionnaire treatment.

TABLE 13.2
Logic Description Board (Presence/Absence), Matrix with Q Variables (Q = 16), and p Categories (p = 48)

Variable → (16)	Gender		Age (Years)		...	Elbow–Knuckle Length (mm)			Discomfort	
Categories → (48) Cases (206) ↓	M (Male)	F (Female)	Age1 (18–23)	Age2 (≥23)	...	Ekl1 (≤334)	Ekl2 (335–357)	Ekl3 (>357)	Y (Yes)	N (No)
1	1	0	1	0	...	0	1	0	0	1
2	1	0	1	0	...	1	0	0	1	0
3	1	0	0	1	...	0	0	1	1	0
...	...	...	...	...	...	...	...	...	...	...
204	0	1	0	1	...	0	1	0	1	0
205	0	1	1	0	...	1	0	0	0	1
206	1	0	1	0	...	0	1	0	0	1

All the anthropometric dimensions were encoded in three categories (e.g., Ekl1, Ekl2, Ekl3): (1) smaller (~30%), (2) average (~40%), (3) higher (~30%). As for the other variables (gender, age, discomfort) they were divided in two categories.

13.6   RESULTS

13.6.1   Anthropometric Database

A great majority (89%) of the participants is younger than 25 years old. The sample used includes subjects from several courses of FEUP, and several curricular years. That is reflected on the range of ages of the sample, 18–35 years old. However, there is a predominance of younger students, 18–24 years old. According to statistics from FEUP, the sample characteristics (age and female percentage) ensure the intended representativeness of the student population (see Table 13.3).

The CV (Table 13.4) of this study was compared with the CV obtained in the previous study of Barroso et al. (2005) and with the characteristic value range of Pheasant and Haslegrave (2005).

Comparison between the results obtained and the previous study from Barroso et al. (2005) indicates a significant difference for the CV values in almost 80% of the anthropometric dimensions. It can be explained by the uneven sample size, which could be corrected by increasing the sample size. When compared with the characteristic ranges defined in Pheasant and Haslegrave (2005), a lower CV was found in one anthropometric dimension (male and female stature). Nevertheless, about 70% of the dimensions are between the recommended ranges. Exceptions also occur with the values obtained for forward grip reach, HB, PH, TT, and abdominal depth whose values are higher than the reference values by Pheasant and Haslegrave (2005).

TABLE 13.3
Distribution of the Participants by Age Group and Gender

Age Group (Years Old)	Male		Female	
	N	%	N	%
Less than 20	47	35.9	18	24.0
20–24	71	54.2	48	64.0
25–29	8	6.1	9	8.0
30–35	5	3.8	3	4.0

TABLE 13.4

Coefficient of Variation: Results of the Current Study and the Characteristic Value Range According to Literature

	Coefficient of Variation (%)				
	Current Study		Barroso et al. (2005)		Pheasant and
Dimensions	Male	Female	Male	Female	Haslegrave (2006)
Stature	3.7	3.4	4.5	4.2	4–11
Abdominal depth	11.8	14.9	12.1	13.7	5–9
Buttock–knee length	5.1	6.4	5.6	5.6	4–11
Buttock–popliteal length	5.7	6.2	6.7	6.3	4–11
Elbow–knuckle length	5.3	4.9	5.1	5.4	4–11
Eye height	4.0	3.5	4.7	4.5	3–5
Forward grip reach	5.2	4.6	8.6	5.0	3–5
Hip breadth	7.4	9.4	6.4	6.8	5–9
Popliteal height	6.6	5.1	6.3	6.3	3–5
Sitting elbow height	9.9	10.6	11.8	11.8	4–11
Sitting eye height	4.0	3.7	4.2	4.6	4–11
Sitting stature	3.5	3.5	4.1	4.1	3–5
Shoulder breadth (bi-deltoid)	7.0	7.6	6.4	7.0	5–9
Thigh thickness	12.0	16.4	9.9	9.2	5–9

For the last two, a possible explanation can be the fact that these dimensions are associated with soft body tissue, particularly fat and muscle. Other possible explanation for the differences between them can be the fitness habits of different people.

With regard to the other dimensions, explanation for the CV values found is possibly associated with the need for a larger sample.

Table 13.5 displays mean and standard deviation values for the 14 anthropometric dimensions measured in both genders, as well as the p-value obtained by Student's t-test. The results indicate that males have greater anthropometric dimensions, except for the hip breath. In this case, there is no significant difference between both genders. This can be explained by the participant's young age, at which bodies are still developing.

The p-value for SEH was very close to the value of 0.05 but it was still below the significance level established. In all the other dimensions, males are bigger than females.

Tables 13.6 and 13.7 compare the mean values, for the male population, obtained in this study with the Portuguese population anthropometric database from Barroso et al. (2005), with British adults aged 19–25 years (Pheasant and Haslegrave 2005), with Malaysian adults aged 19–25 years (Chong and Leong 2011), and with Iranian adults aged 20–30 years (Mououdi 1997).

The comparison between the results of the current study and the Portuguese anthropometric data by Barroso et al. (2005) shows a few differences, some of which are significant. For example, the stature has an increment of 61 mm in male and 55 mm in female. This difference may be related to three probable causes: (1) the age range of the sample once the spinal curvature increases above 40 years old (Pheasant and Haslegrave 2005); (2) the secular trend, which is known to occur in other studied populations (Pheasant and Haslegrave 2005), and was also observed in Portugal by Padez (2002); and (3) the provenience of the sample (university students who, due to different financial conditions, may adopt distinct dietary habits).

As for the other populations, it might be observed that some differences exist, some of which are statistically significant.

TABLE 13.5
Mean (M), Standard Deviation (SD, in mm) of the Male Population (n = 131) and the Female Population (n = 75), and the Comparison of the Means

Dimensions	Male		Female		
	M	SD	M	SD	p-Value
Stature	1751	65	1625	56	<0.00001
Abdominal depth	264	31	249	37	0.00460
Buttock–knee length	615	31	580	37	<0 0001
Buttock–popliteal length	490	28	466	29	<0.00001
Elbow–knuckle length	357	19	328	16	<0.00001
Eye height	1635	65	1515	53	<0.00001
Forward grip reach	733	38	676	31	<0.00001
Hip breadth	347	26	346	33	0,76219
Popliteal height	420	28	392	20	<0.00001
Sitting elbow height	245	24	238	25	0.04678
Sitting eye height	808	32	758	28	<0.00001
Sitting stature	925	32	868	30	<0.00001
Shoulder breadth (bi-deltoid)	482	34	424	32	<0.00001
Thigh thickness	190	23	171	28	<0.00001

TABLE 13.6
Mean Values (mm) for the Anthropometric Dimensions from the Current Study and Different Young Adult Male Populations

Dimensions	Current Study		Barroso et al. (2005)		Pheasant and Haslegrave (2005)		Chong and Leong (2011)		Mououdi (1997)	
	M	SD	M	SD	M	SD	M	SD	M	SD
Stature	1751	65	1690	76	1760	73	1713	48	1725	58
Abdominal depth	264	31	265	32	240	26	227	27	–	–
Buttock–knee length	615	31	590	33	595	32	561	21	579	27
Buttock–popliteal length	490	28	485	32	500	34	450	23	–	–
Elbow–knuckle length	357	19	350	18	–	–	–	–	–	–
Eye height	1635	65	1585	74	1650	72	1589	52	1616	56
Forward grip reach	733	38	730	62	790	36	764	57	–	–
Hip breadth	347	26	380	24	350	31	354	29	–	–
Popliteal height	420	28	400	26	445	30	431	16	431	21
Sitting elbow height	245	24	255	30	245	32	229	28	260	26
Sitting eye height	808	32	810	34	795	36	786	37	–	–
Sitting stature	925	32	920	37	915	37	874	112	912	26
Shoulder breadth (bi-deltoid)	482	34	475	30	465	29	435	31	455	24
Thigh thickness	190	23	175	17	160	16	–	–	–	–

TABLE 13.7

Mean Values (mm) for the Anthropometric Dimensions from the Current Study and Different Young Adult Female Populations

Dimensions	Current Study		Barroso et al. (2005)		Pheasant and Haslegrave (2005)		Chong and Leong (2011)		Mououdi (1997)	
	M	SD	M	SD	M	SD	M	SD	M	SD
Stature	1625	56	1565	66	1620	61	1560	61	1597	101
Abdominal depth	249	37	260	36	220	22	195	21	–	–
Buttock–knee length	580	37	570	32	565	29	529	26	563	29
Buttock–popliteal length	466	29	470	30	475	29	434	28	–	–
Elbow–knuckle length	328	16	320	40	–	–	–	–	–	–
Eye height	1515	53	1465	66	1515	60	–	–	1486	58
Forward grip reach	677	31	675	33	705	31	701	46	–	–
Hip breadth	346	33	400	27	350	29	351	22	–	–
Popliteal height	392	20	365	23	400	27	402	31	365	22
Sitting elbow height	238	25	250	28	230	28	217	30	259	37
Sitting eye height	758	28	760	35	745	33	721	40	–	–
Sitting stature	868	30	865	35	855	35	828	32	861	36
Shoulder breadth (bi-deltoid)	424	32	445	31	395	24	383	19	392	24
Thigh thickness	171	28	165	15	150	16	–	–	–	–

For instance, for stature, the highest differences found are those registered between the Portuguese and the Malaysian population, 38 and 65 mm for female and male populations, respectively. The contrast found in the anthropometric dimensions of the different populations highlights the usefulness of this study and of the presented results.

13.6.2 FURNITURE DIMENSIONS

Tables and seats measures are presented in Table 13.8. The dimensions consider the height, depth, and the width of the seat of each chair, as well the space between the seat pan and the bottom top, and the height between the seat pan and the bottom top.

TABLE 13.8

Furniture Dimensions (mm)

Type		SH (mm)	SD (mm)	SW (mm)	SDC (mm)	SDH (mm)
Chairs	C1	444	409	382	–	–
	C2	473	390	400	–	–
	C3	480	420	396	–	–
	C4	396–522[a]	390	400	–	–
	C5	418	428	410	–	–
Chair + Table	Comb1	–	–	–	229	279
	Comb2	–	–	–	243	294
	Comb3	–	–	–	238	289
	Comb4	–	–	–	<590[a]	290–402[a]
	Comb5	–	–	–	224	202

[a] Values vary due to the adjustability of the furniture.

13.6.3   MISMATCH BETWEEN FEUP's FURNITURE AND STUDENTS' ANTHROPOMETRIC DIMENSIONS

For the SH (see Table 13.9), a "high mismatch" was found for most male students (92% in C3, 72% in C2) while only 1% of the students are included in the "low mismatch" group. In the case of the "high mismatch" most of the students were not able to support their feet on the floor, generating increased pressure on the soft tissues of the thighs (Gouvali and Boudolos 2006). Even though, chair C1 has a lower "high mismatch" (47%) it still has a low "match" with only 51% of the students. Chairs C4 (98%) and C5 (88%) have high levels of match with the male population.

For the female population, the case worsens with larger "high mismatches" in chairs C1 (91%), C2 (95%), and C3 (99%). Chair C4 remains with a good "match" for 95% of the female population while the C5 "match" lowers to 51%.

For the SD (see Table 13.10) chair C4 has the lowest "match" with only 46% of the male population.

All the other chairs have "match" levels of 70% of the population (C2) or more (C1, C3, and C5). With the female population only chairs C1 (84%), C3 (82%), and C5 (72%) have a better "match" with the HB. The C2 chair has a "high mismatch" with 69% and C4 with 61% of the population. The "high mismatch" may cause compression on the thighs and block the blood circulation causing discomfort (Gouvali and Boudolos 2006) and also restrain the use of the back rest inducing kyphotic postures (Castellucci et al. 2010).

Concerning the SW (see Table 13.11), all the chairs have high levels of "match," since all of them are larger than 90%. For the female population only C1 chair has a lower "match" with 88% of the population, unlike all the other chairs that have a "match" with more than 90%.

The SDC (see Table 13.12) is compatible with 86% of the male population for Comb1, 71% for Comb5, and with more than 90% for Comb2 to Comb4. All combinations are compatible with more than 90% of the female population. This mismatch situation produces mobility restriction due to the contact of the thighs with the desk (Parcells et al. 1999).

As for the SDH (see Table 13.13), Comb2 has a "high mismatch" with 50% of the male population and Comb3 with 38%. This requires them to work with shoulder flexion, causing muscle work

TABLE 13.9
Mismatch between PH and SH

	Male Population					Female Population				
	C1 (%)	C2 (%)	C3 (%)	C4 (%)	C5 (%)	C1 (%)	C2 (%)	C3 (%)	C4 (%)	C5 (%)
High mismatch	47	72	92	1	5	91	95	99	5	48
Match	51	27	7	98	88	9	5	1	95	51
Low mismatch	2	1	1	1	7	0	0	0	0	1

TABLE 13.10
Mismatch between BPL and SD

	Male Population					Female Population				
	C1 (%)	C2 (%)	C3 (%)	C4 (%)	C5 (%)	C1 (%)	C2 (%)	C3 (%)	C4 (%)	C5 (%)
High mismatch	4	30	5	1	7	11	69	16	61	27
Match	79	70	85	46	87	84	31	83	39	72
Low mismatch	17	0	10	53	6	5	0	1	0	1

TABLE 13.11
Mismatch between HB and SW

	Male Population					Female Population				
	C1 (%)	C2 (%)	C3 (%)	C4 (%)	C5 (%)	C1 (%)	C2 (%)	C3 (%)	C4 (%)	C5 (%)
Match	92	00	96	100	98	88	100	93	100	99
Mismatch	8	0	4	0	2	12	0	7	0	1

TABLE 13.12
Mismatch between TT and SDC

	Male Population					Female Population				
	Comb1 (%)	Comb2 (%)	Comb3 (%)	Comb4 (%)	Comb5 (%)	Comb1 (%)	Comb2 (%)	Comb3 (%)	Comb4 (%)	Comb5 (%)
Match	86	94	91	100	71	93	97	95	100	93
Mismatch	14	6	9	0	29	7	3	5	0	7

TABLE 13.13
Mismatch between SEH and SDH

	Male Population					Female Population				
	Comb1 (%)	Comb2 (%)	Comb3 (%)	Comb4 (%)	Comb5 (%)	Comb1 (%)	Comb2 (%)	Comb3 (%)	Comb4 (%)	Comb5 (%)
High mismatch	5	50	38	0	0	29	65	53	0	0
Match	66	48	57	100	4	67	34	45	100	47
Low mismatch	29	2	5	0	96	4	1	2	0	53

load, discomfort, and pain in the shoulder region (Parcells et al. 1999). There is a "low mismatch" for Comb5 with 96% of the population and with 29% for Comb1. Only Comb4 has a perfect "match" with 100% of the students. In the female population the larger "high mismatch" belongs to Comb2 (65%), Comb3 (53%), and Comb1 (29%). Comb5 has a "low mismatch" with 53% of the female population and, as for the male population, Comb4 has a 100% "match."

13.6.4 Reported Discomfort for Classroom and Computer Room Furniture

The questionnaire results showed that the majority of the students (69.9%) admitted to have felt discomfort while using the classrooms' and the computer rooms' furniture (tables and chairs). The perceived discomfort is higher in females (74.7%) than in the male participants (67.2%).

The analysis of the obtained graphic outputs (through the application of CA) obeys some interpretation rules, which passes through the selection of the number of inertial axis to retain.

In order to develop the simplicity of interpretation, position, and absolute contribution of the variables to the construction of the factorial axis (the relative contribution intervenes on the analysis of categories or individuals), the shape of the projection cloud of which spatial distribution may indicate the subjacent structure to the starting scenario, etc.

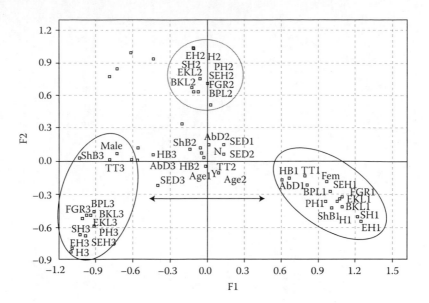

FIGURE 13.3 MCA. Factor plan F1, F2.

From the enormous profusion of outputs obtained, we chose to retain and interpret only those considered relevant to our work. The F1 axis (see Figure 13.3) represents about 25% of the total variability of the considered data. Analyzing the projections of the different categories, along the F1 axis and considering that each category is only considered if its absolute contribution, for that axis, is over the value of 100/p (where p is the number of categories, in our case p = 48), we can establish the following associations.

The categories projected along the F1 negative semi-axis are positively correlated: Male, ShB3, TT3, BPL3, FGR3, BKL3, EKL3, SH3, PH3, EII3, SEH3, and H3. The categories projected on the positive semi-axis F1 allow us to identify a strong association between the categories Fem (Female), HB1, TT1, AbD1, BPL1, SEH1, PH1, FGR1, EKL1, BKL1, ShB1, H1, SH1, and EH1. These two groups are negatively correlated.

In Figure 13.4, the projections through axis F1, the projection that separates men, and their respective anthropometric characteristics (negative semi-axis) and women, and their respective anthropometric characteristics (positive semi-axis) is clearly perceptible. This result allows us to assess the consistency of the data collected. On the positive semi-axis F2 the association between the average size categories is obvious.

As for the factor plan composed by the F1–F3 axis (see Figure 13.4), that comprises 7.9% of the sample, we will only interpret the categories projected on the F3 axis, as the F1 was already interpreted. There is a positive association between the categories AbD3, TT3, HB3, ShB3, on the positive semi-axis. On the negative semi-axis a positive association between AbD1, HB1, and TT1 exists. These two groups are negatively correlated. Beyond the obvious findings it demonstrates the important association between the abdominal depth, the TT, and the HB, whether in smaller or bigger students. The latter group shows also a correlation with the bigger shoulder breadth.

Even though the F7 axis is responsible for a reduced percentage (3.8% of the sample) in the original total data variability explanation, in part due to the large considered categories, we were able to establish the following correlations. In Figure 13.5 it is noticeable, on the negative semi-axis F7, that the projected categories N (no feeling of discomfort), Male, AbD2, Age1, and FGR1 are positively associated. On the other hand, on the positive semi-axis F7 a strong association exists between the categories Y (feel discomfort), Fem, PH3, BPL, Age2, FGR2. The two groups are negatively correlated.

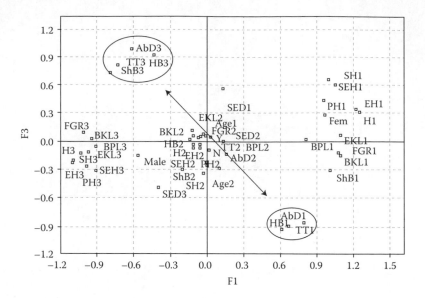

FIGURE 13.4   MCA. Factor plan F1, F3.

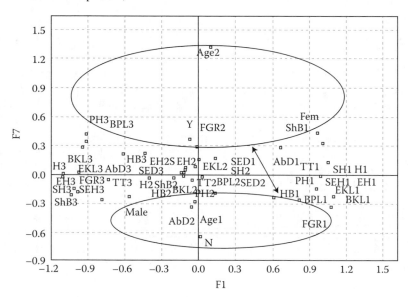

FIGURE 13.5   MCA. Factor plan F1, F7.

13.7   CONCLUSIONS

Anthropometric data were collected for 14 body dimensions on a sample of Portuguese university students and it is summarized in this chapter. These data enabled the anthropometric characterization of the Portuguese student population of the studied Portuguese faculty. This can be used for the design of more ergonomic classroom and auditoria furniture.

The contrast between the data of this study and the data of the Portuguese adult anthropometric database emphasizes the need to extend the database, as well as its stratification for different age groups for more accurate use of this resource.

One of the main limitations of this study is the sample size that, due to the lack of time, is smaller than recommended. In the authors' opinion, this is also the reason for the obtained CV values.

This study aimed to analyze the relationship between some anthropometric dimensions of 206 students from a Portuguese faculty and the dimensions of the classroom furniture. The results show that none of the chairs are adequate for the user population. However, the C4 chair, with its adjustability, has characteristics that allow it to better adjust to the students' needs. With a seat with a lesser depth it would be adequate to most of the students.

SH and SDH were the furniture dimensions with a higher level of mismatch, which may result in discomfort and pain on the posterior surface of the knee and shoulder.

The questionnaire results revealed that the majority of students feel discomfort while using the classroom furniture, validating the obtained results for the mismatch.

The MCA showed that, for 25% of the sample, the male students are the ones with bigger dimensions, in general, and the female students have the smallest dimensions.

Also for 7.9% of the sample, it was possible to conclude that there is an association between the abdominal depth, TT, and HB, both in shorter and taller students. This may occur due to the fact that these are soft tissues which depth is associated with body fat.

For a smaller percentage of the sample (3.8%), it was concluded that there is a correlation between the absence of discomfort and male younger students with average abdominal depth and smaller forward grip reach. As for the presence of the discomfort, there is a correlation with the older female students with smaller shoulder breadth, average forward grip reach, and higher PH and BPL.

The results of this study underline the fact that most of the time classroom furniture is usually acquired and selected without any previous concern for ergonomics and without considering the specific user population, which ultimately can result in its inadequacy.

REFERENCES

Agha, S. R. and M. J. Alnahhal. 2012. Neural network and multiple linear regression to predict school children dimensions for ergonomic school furniture design. *Applied Ergonomics* 43 (6):979–984.

Arezes, P. M., M. P. Barroso, P. Cordeiro, L. G. Costa, and A. S. Miguel. 2006. *Estudo Antropométrico da População Portuguesa, Lisboa: Série Estudos em Segurança e Saúde no Trabalho (Anthropometry Study of the Portuguese Population)*. Lisboa: Instituto para a Segurança, Higiene e Saúde no Trabalho (ISHST).

Barroso, M. P., P. M. Arezes, L. G. da Costa, and A. Sérgio Miguel. 2005. Anthropometric study of Portuguese workers. *International Journal of Industrial Ergonomics* 35 (5):401–410.

Benzécri, J. P. 1973. *L'analyse des données: L'analyse des correspondances* (Data analysis: Correspondence analysis). Vols 1 and 2. Paris: Dunod.

Brink, Y., L. C. Crous, Q. A. Louw, K. Grimmer-Somers, and K. Schreve. 2009. The association between postural alignment and psychosocial factors to upper quadrant pain in high school students: A prospective study. *Manual Therapy* 14 (6):647–653.

BSI. 2006. BS EN 1729-1: 2006 Furniture—Chairs and tables for educational institutions—Part 1: Functional dimensions. *BSI*, UK: BSI (British Standard Institution).

Burt, C. 1950. The factorial analysis of qualitative data. *British Journal of Statistical Psychology* 3 (3):166–185. Doi: 10.1111/j.2044-8317.1950.tb00296.x.

Callaghan, J. P. and N. M. Dunk. 2002. Examination of the flexion relaxation phenomenon in erector spinae muscles during short duration slumped sitting. *Clinical Biomechanics* 17 (5):353–360.

Cardoso, H. F. V. 2008. Secular changes in body height and weight of Portuguese boys over one century. *American Journal of Human Biology* 20 (3):270–277. Doi: 10.1002/ajhb.20710.

Cascioli, V., Z. Liu, A. I. Heusch, and P. W. McCarthy. 2011. Settling down time following initial sitting and its relationship with comfort and discomfort. *Journal of Tissue Viability* 20 (4):121–129.

Castellucci, H. I., P. M. Arezes, and C. A. Viviani. 2010. Mismatch between classroom furniture and anthropometric measures in Chilean schools. *Applied Ergonomics* 41:563–568.

Castellucci, H. I., P. M. Arezes, and J. F. M. Molenbroek. 2014. Applying different equations to evaluate the level of mismatch between students and school furniture. *Applied Ergonomics* 45 (4):1123–1132.

Castellucci, H. I., P. M. Arezes, and J. F. M. Molenbroek. 2015. Analysis of the most relevant anthropometric dimensions for school furniture selection based on a study with students from one Chilean region. *Applied Ergonomics* 46 (Part A [0]):201–211.

CEN. 2012. prEN 1729-1:2006: Furniture—Chairs and tables for educational institutions—Part 1: Functional dimensions. *CEN*, CEN (European Committee for Standardization).

Chong, Y. Z. and X. J. Leong. 2011. Preliminary findings on anthropometric data of 19–25 year old Malaysian university students. In *5th Kuala Lumpur International Conference on Biomedical Engineering 2011*, edited by N. A. A. Osman, W. A. B. W. Abas, A. K. A. Wahab, and H.–N. Ting, pp. 193–196. Berlin: Springer.

Chuan, T. K., M. Hartono, and N. Kumar. 2010. Anthropometry of the Singaporean and Indonesian populations. *International Journal of Industrial Ergonomics* 40 (6):757–766.

Chung, M. K., I. Lee, and D. Kee. 2003. Assessment of postural load for lower limb postures based on perceived discomfort. *International Journal of Industrial Ergonomics* 31 (1):17–32.

Dianat, I., M. A. Karimi, A. A. Hashemi, and S. Bahrampour. 2013. Classroom furniture and anthropometric characteristics of Iranian high school students: Proposed dimensions based on anthropometric data. *Applied Ergonomics* 44 (1):101–108.

Dunk, N. M. and J. P. Callaghan. 2005. Gender-based differences in postural responses to seated exposures. *Clinical Biomechanics* 20 (10):1101–1110.

Gilkey, D. P., T. J. Keefe, J. L. Peel, O. M. Kassab, and C. A. Kennedy. 2010. Risk factors associated with back pain: A cross-sectional study of 963 college students. *Journal of Manipulative and Physiological Therapeutics* 33 (2):88–95.

Gouvali, M. K. and K. Boudolos. 2006. Match between school furniture dimensions and children's anthropometry. *Applied Ergonomics* 37 (6):765–773.

Handrakis, J. P., K. Friel, F. Hoeffner, O. Akinkunle, V. Genova, E. Isakov, J. Mathew, and F. Vitulli. 2012. Key characteristics of low back pain and disability in college-aged adults: A pilot study. *Archives of Physical Medicine and Rehabilitation* 93 (7):1217–1224.

ICONTEC. 1999. Norma Técnica Colombiana 4641. Muebles Escolares. Pupitre con Silla para Aulas de Clase (Colombian Technical Standard 4641. School furniture. Desk with chair Classrooms). *ICONTEC*. Bogotá, Colombia: ICONTEC (Instituto Colombiano de Normas Técnicas y Certificación).

INN. 2002. Norma Chilena 2566. *Mobiliario Escolar e Sillas Y Mesas Escolares e Requisitos dimensionales (Chilean Standard 2566. School Furniture and Chairs And Tables School and dimensional requirements). INN*, Santiago de Chile: INN (Instituto Nacional de Normalización Chile).

Ioana, O., G. C. Liliana, and M. Cozeta. 2014. Secular trend of growth in height, weight and body mass index in young Romanians aged 18–24 years. *Procedia—Social and Behavioral Sciences* 117:622–626.

Jacquet, M. and B. Petitdant. 2014. Amphithéâtres: des étudiants en mauvaise posture (*Amphitheatres: the students in trouble*). *Kinésithérapie, la Revue* 14 (150):22–33.

JIS. 2011. JIS S 1021. School furniture—Desks and chairs for general learning space. *JIS*, Tokyo, Japan: JIS (Japanese Industrial Standards).

Jung, H. S. 2005. A prototype of an adjustable table and an adjustable chair for schools. *International Journal of Industrial Ergonomics* 35 (10):955–969.

Kaya, M. D., A. S. Hasiloglu, M. Bayramoglu, H. Yesilyurt, and A. F. Ozok. 2003. A new approach to estimate anthropometric measurements by adaptive neuro-fuzzy inference system. *International Journal of Industrial Ergonomics* 32 (2):105–114.

Khadem, M. M. and Md A. Islam. 2014. Development of anthropometric data for Bangladeshi male population. *International Journal of Industrial Ergonomics* 44 (3):407–412.

Khanam, C. N., M. V. Reddy, and A. Mrunalini. 2006a. Designing student's seating furniture for classroom environment. *Journal of Human Ecology* 20:241–248.

Khanam, C. N., M. V. Reddy, and A. Mrunalini. 2006b. Opinion of students on seating furniture used in classroom. *Journal of Human Ecology* 20:15–20.

Klamklay, J., A. Sungkhapong, N. Yodpijit, and P. E. Patterson. 2008. Anthropometry of the southern Thai population. *International Journal of Industrial Ergonomics* 38 (1):111–118.

Lebart, L. and J. P. Fénelon. 1975. Statistique et informatique appliqées (Applied Statistics and Informatics). Paris: Dunod.

Mokdad, M. and M. Al-Ansari. 2009. Anthropometrics for the design of Bahraini school furniture. *International Journal of Industrial Ergonomics* 39 (5):728–735.

Mououdi, M. A. 1997. Static anthropometric characteristics of Tehran University students age 20–30. *Applied Ergonomics* 28 (2):149–150.

Murphy, S., P. Buckle, and D. Stubbs. 2004. Classroom posture and self-reported back and neck pain in schoolchildren. *Applied Ergonomics* 35 (2):113–120.

O'Sullivan, K., R. McCarthy, A. White, L. O'Sullivan, and W. Dankaerts. 2012. Can we reduce the effort of maintaining a neutral sitting posture? A pilot study. *Manual Therapy* 17 (6):566–571.

Oyewole, S. A., J. M. Haight, and A. Freivalds. 2010. The ergonomic design of classroom furniture/computer work station for first graders in the elementary school. *International Journal of Industrial Ergonomics* 40 (4):437–447.

Padez, C. 2002. Stature and stature distribution in Portuguese male adults 1904–1998: The role of environmental factors. *American Journal of Human Biology* 14 (1):39–49. Doi: 10.1002/ajhb.10017.

Padez, C. 2003. Social background and age at menarche in Portuguese university students: A note on the secular changes in Portugal. *American Journal of Human Biology* 15 (3):415–427. Doi: 10.1002/ajhb.10159.

Panagiotopoulou, G., K. Christoulas, A. Papanckolaou, and K. Mandroukas. 2004. Classroom furniture dimensions and anthropometric measures in primary school. *Applied Ergonomics* 35 (2):121–128.

Parcells, C., M. Stommel, and R. P. Hubbard. 1999. Mismatch of classroom furniture and student body dimensions: Empirical findings and health implications. *Journal of Adolescent Health* 24 (4):265–273.

Pheasant, S. and C. M. Haslegrave. 2005. *Bodyspace: Anthropometry, Ergonomics and the Design of Work*, Third edition. Boca Raton, FL: Taylor & Francis.

Qutubuddin, S. M., S. S. Hebbala, and A. C. S. Kumarb. 2013. Anthropometric consideration for designing students desks in engineering colleges. *International Journal of Current Engineering and Technology* 3:1179–1185.

Sobral, F. 1990. Secular changes in stature in Southern Portugal Between 1930 and 1980 According to conscript data. *Human Biology* 62 (4):491–504.

Standardization, International Organization for (ISO). 2008. *ISO 7250-1:2008 Basic Human Body Measurements for Technological Design—Part 1: Body Measurement Definitions and Landmarks*. Geneva, Switzerland: International Organization for Standardization.

Straker, L., C. Pollock, and R. Burgess-Limerick. 2006. Excerpts from CybErg 2005 discussion on preliminary guidelines for wise use of computers by children. *International Journal of Industrial Ergonomics* 36 (12):1089–1095.

Straker, L. M., P. B. O'Sullivan, A. J. Smith, M. C. Perry, and J. Coleman. 2008. Sitting spinal posture in adolescents differs between genders, but is not clearly related to neck/shoulder pain: An observational study. *Australian Journal of Physiotherapy* 54 (2):127–133.

Thariq, M. G. M., H. P. Munasinghe, and J. D. Abeysekara. 2010. Designing chairs with mounted desktop for university students: Ergonomics and comfort. *International Journal of Industrial Ergonomics* 40 (1):8–18.

Tirloni, A. S., D. C. Reis, M. Soares, and A. R. P. Moro. 2014. Influence of the school furniture design on the body posture of college students. In *Advances in Ergonomics in Design, Usability & Special Populations: Part II*, edited by F. Rebelo and M. Soares, AHFE Conference, Krakow, Poland, Copyright 2014, pp. 364–370. USA: AHFE.

Van Niekerk, S.-M., Q. A. Louw, K. Grimmer-Somers, J. Harvey, and K. J. Hendry. 2013. The anthropometric match between high school learners of the Cape Metropole area, Western Cape, South Africa and their computer workstation at school. *Applied Ergonomics* 44 (3):366–371.

Vos, G. A., J. J. Congleton, J. S. Moore, A. A. Amendola, and L. Ringer. 2006. Postural versus chair design impacts upon interface pressure. *Applied Ergonomics* 37 (5):619–628.

Yang, Z., B. Becerik-Gerber, and L. Mino. 2013. A study on student perceptions of higher education classrooms: Impact of classroom attributes on student satisfaction and performance. *Building and Environment* 70:171–188.

Section III

Methodological Issues

14 Designing User-Oriented Future Ship Bridges
An Approach for Radical Concept Design

Mikael Wahlström, Hannu Karvonen, Eija Kaasinen, and Petri Mannonen

CONTENTS

14.1 INTRODUCTION

In design for industrial work, it is typically vital that the design ideas generated support the existing aims of industrial workers. This is because the aim of the activity in designing for work activities, as opposed to designing for consumers, is usually fixed. Whereas consumers may engage in activities that are wholly new to them, such as playing new games or embarking on a new dietary regimen, industry workers respond to certain basic needs and goals, among them habitation, energy, security, health care, production, and logistics. When considering the shipping industry, for example, one may assume that operations will continue to serve their current main mission—the circulation of goods via the oceans of the world. All this implies the usefulness of studying the existing work activities when one is designing for industry professionals: the design solutions should correspond with the existing aims of the activity. In other words, user-centered design (Gould and Lewis, 1985), in which the needs, desires, and capabilities of the users are taken into account, appears clearly justified.

It is noteworthy, however, that user-centeredness in design has been criticized for not providing new types of design solutions with the potential to surprise the people involved and offer them new possibilities (Norman and Verganti, 2014). Indeed, it typically offers users what they already knew they wanted, since the design solutions respond to the issues identified by the users (Keinonen, 2009). That is, because the solutions correspond too closely with the existing models of activity or "user paradigms," studying users' activities and needs does not seem to provide new kinds of solutions or radical, revolutionary innovations. Assumedly, when work-related innovations are under development, the users are not aware of all the forthcoming technical opportunities or trends. In addition, they may be too wedded to current practices (perhaps even heavily invested in them) to ideate radical changes. It is thought, therefore, that user studies predominantly provide incremental or evolutionary design solutions, that is, mere modifications to existing designs. In contrast, new kinds of design solutions should offer greater potential to provide business advantage by significantly enhancing or modifying existing work activities. Accordingly, one may identify a topical problem to be solved: how to generate design solutions that both (1) support the existing activities of professional workers and (2) surprise the users with innovativeness by offering new possibilities. One can conclude that applying a design approach that solves this problem creates potential for the positive renewal of industries.

The design approach presented here is aimed at addressing that problem. It is based on the following procedures:

1. *Reformulation of user-study findings* in such a manner that sufficient "distance" is achieved for the study's findings, by not "directly dictating" the design solutions, thereby allowing for radical design
2. *Foresight of technology trends and future developments*, enabling future-oriented design solutions
3. *Codesign and coevaluation with actual users and experts in the relevant field of application after creation of the initial design ideas*, thereby allowing higher quality and better specification of design ideas

In the following discussion, we illustrate each of the procedures employed through examination of a ship-bridge concept design case. The bridge is the place from which the ship is commanded, navigated, and maneuvered. The ship's surroundings are observed from the bridge, with the watch-keeper looking out the bridge windows to note any potential dangers, such as rocks or other vessels. In addition, the vessel's steering and communication devices are found on the bridge. Modern ships' bridge equipment includes, for example, electronic chart displays (in ECDIS format, used for navigation), radar displays, and dynamic positioning systems.

In the design assignment reported here, the aim stated by the industry partner with whom the concepts were generated, Rolls-Royce, was for the solutions to stimulate the field of maritime operations in general by providing future-oriented ship-bridge alternatives. An additional aim was for the concepts to be user-oriented in the sense that, while providing highly novel types of solutions, they would still have to be accepted and appreciated by mariners who possess practical knowledge of hands-on maritime activity. It was, however, agreed that the project need not take maritime legislation into consideration: the objective was to supply alternatives—options for possible futures—instead of strictly accommodating existing realities. The goal was that the design concepts would represent the ship bridges of the year 2025. Three ship types were considered, these being tugboats, platform supply vessels (PSVs), and cargo ships. The results of the design project are described online (Rolls-Royce, 2014, 2015).

Overall, the design approach resembles those termed "contextual design" (Beyer and Holtzblatt, 1998), "experience design" (Hassenzahl, 2010), and "codesign" (Sanders and Stappers, 2008): it is based on (1) studying of the work context, (2) user-experience-related goal setting, and (3) collaborative design involving the users, respectively. The deviation from these approaches is found in the purposeful aim of future orientation and radical design. In addition, we employed a specific analysis method called core-task analysis (CTA) (Norros, 2004, 2013) to make sense of the relevant professional activity. We contrast our design approach against that of others in Section 14.3.

14.2 OUR APPROACH FOR RADICAL YET USER-ORIENTED DESIGN SOLUTIONS

Our design approach combines thorough understanding of the users and their situations, visions of the future from several fields, and cocreation to produce radical design solutions. The process included a group interview, 12 one-on-one interviews and four field observations. Interviewed were maritime experts from diverse fields, with varied professions, among them designers, researchers, shipping company directors, trainers, officers, seamen, and sea captains. The interviews were conducted both before and after expression of the initial concept ideas. The field studies, which took place in Finland, Norway, and Estonia, included observation of actual tugboat operation and observation of PSV (platform supply vessel) operations in a simulator context. Additionally, a future-studies workshop was held, wherein researchers from various disciplines came together to discuss emergent trends related to maritime transportation, maritime technology, interaction technologies, and general societal developments.

14.2.1 Reformulation of User-Study Findings

As has been discussed elsewhere (Wahlström et al., forthcoming), creating a "reasoned departure" from user-study findings can serve as a useful means of avoiding the phenomenon of designers being "trapped within the current paradigms" (Norman and Verganti, 2014). In other words, we propose that in the preliminary phases of design, the design indications drawn from the users' explications should be meaningful but also purposefully broad. This broadness allows room for the users' ideas not to dictate the creative process of design too directly and specifically. This is important for its facilitation of creation of ideas that are new to the users. In practice, the idea is for user activity to be modeled and understood rather than for users' ideas to be directly applied as design indications.

Another means of creating radical design ideas in work proceeding from study of users is to focus on user experience (UX) rather than on product features. This approach draws from experience design (Hassenzahl, 2010), whose output serves the purpose of users' ideas not directly translating into product ideas. The key prerequisite for experience-driven design is determination of what experience to design for. UX goals concretize the intended experience (Kaasinen et al., 2015). An experience goal describes the momentary emotion that is intended to be experienced during use of the product or service or, alternatively, a person's emotional relationship to the designed product or service (Lu and Roto, 2014). In the design of industrial systems, there are several stakeholders involved, and they should hold shared design goals. The UX goal setting serves the objective of bringing together the divergent viewpoints of the stakeholders and gaining their commitment to the said goal setting as a strategic design decision (Kaasinen et al., 2015). UX goals reflect intended user feelings; in so doing, they do not inhibit radical design by directly dictating the design idea creation.

Product use can be conceptually divided into the inherent instrumental and noninstrumental qualities (Mahlke, 2005). The former involve utilitarian elements such as usefulness and ease of use, and noninstrumental elements have to do with the emotional aspects of product use. Similarly, in design, the aim could be a certain feeling, such as a sense of comfort, or it might be a certain practical task, such as efficient communication between individuals. We assume CTA (Norros, 2004, 2013) to be a useful method for pinpointing instrumental task- and activity-related design goals for certain work domains and at a systemic level. This assumption appears justified because in studies of risk-intensive work it has been used to identify interconnected elements influencing the way in which the aims of a specific work activity have been reached, where these identified issues include elements related to the work activities, the tools used, and the work environment in general. In other words, the CTA method is useful for identifying pertinent systems usability issues (Savioja, 2014). As has been discussed by Norros et al. (2015) analyzing systems usability is beneficial in varied phases of design, including ideation and concept design. Under the CTA model, challenging and safety-critical work activity entails generic control demands related to (1) dynamism (i.e., temporal demands, such as the need to make quick decisions), (2) complexity (i.e., multiple, reciprocally connected influencing elements, such as weather, technology, and human behavior), and (3) uncertainty (i.e., unexpectedness of events, which implies that decisions must be made in the absence of sufficient information). In addition, the CTA model assumes three basic features of work activity to be the means (i.e., resources) through which these control demands are managed: (1) skill, (2) knowledge, and (3) collaboration. Work activity can be analyzed through examination of how these control demands and resources connect one with another; the connections found are referred to as core-task demands of the relevant work domain. The core-task demand findings represent both enacted ways in which the control demands are addressed (i.e., as expressed in the interviews or observed by the researchers) and potential ways (i.e., as inferred or suggested by the researchers or interviewees). In this way, an "analytical grid" is formed of these interrelations (see Figure 14.1); the interrelations can be used as indications of the instrumental UX and systems usability goals. Indeed, this was done in the ship-bridge design case. The model can be visualized in both graphical (see Figure 14.1) and tabular (see Table 14.1) form. Figure 14.1 and Table 14.1 present the core-task demands identified when we studied PSV operations for the purpose of concept design ideation.

We propose that the systems usability issues, which influence UX on a functional level and can be distinguished via the CTA method, can be directly applied as useful concept design indications. In addition, we believe that feeling-related (noninstrumental) UX goals may benefit from another kind of reformulation of data. Reflecting on design that applies UX personas—that is, fictional characters representing certain target demographics (Cooper, 1999)—or stories (Carroll, 2000), one may assume that the goal of designing for a certain emotion is not, in itself, sufficiently inspirational for designers. The stories and UX personas aid the designers in grasping the abstract emotion-related ideas by giving these a human face (Pruitt and Adlin, 2006).

Somewhat similarly, feeling-related findings were reformulated into an inspirational theme in our ship-bridge design case. A general finding was that feelings of togetherness and unity are important for mariners at embodied, cognitive, and social levels. At the embodied level, the mariners operate the vessel with an intuitive feel in their bodies of how it interacts with the environment as the vessel rocks on the waves and of how it reacts when it is being maneuvered in various conditions. At the cognitive level, the mariners have profound understanding of the features of the environment and

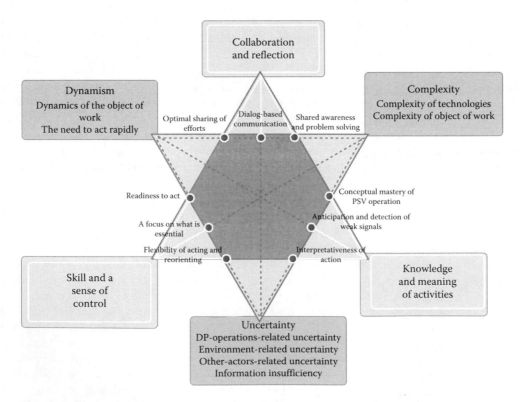

FIGURE 14.1 Core-task demands of PSV operation. The model includes control demands, means of managing them, and the associated core-task demands; see Table 14.1 for more detailed descriptions of the core-task demands in PSV operation.

TABLE 14.1

Description of the Core-Task Demands of PSV Operation

Resources ⟹

Demands ⇓	Collaboration	Skill	Knowledge
Dynamism	Optimal sharing of efforts • Sharing responsibility with others • Trusting in one's own competence in resolving problematic operation situations • Switch-offs in who is in charge of DP operations, to counter boredom/fatigue • Good relationships among the people onboard and trust in their expertise	Readiness to act • Development of new skills and situational sensitivity through practical experience from both simulator and real-world conditions • Taking situational attributes into account • Confidence and calmness in PSV operation at all times • Good depth perception and spatial awareness	Anticipation and detection of weak signals • Understanding of chains of events (e.g., those related to GPS satellites lining up nonoptimally) • Checklists that prepare both the crew and the vessel for the upcoming DP operations
Complexity	Shared awareness and problem-solving • Radio communication (e.g., with the deck crew and the rig crane operator) in order to maintain situation awareness • Clear communication to confirm safety of operation • Combination of competencies in demanding situations • Appreciation for others' work contribution • Sharing of experiences with colleagues	A focus on what is essential • Rapid comprehension of the essential information provided by the displays • Ability to "shut down" surrounding distractions on the bridge, to enable focusing solely on the task at hand • Utilization of tacit knowledge • Utilization of professional experience from diverse situations	Conceptual mastery of PSV operations • Knowledge of the technical solutions implemented and their limitations • Understanding of the environmental demands and the physical forces relevant in the operations • Knowledge of the surrounding operation environment
Uncertainty	Dialogue-based communication • Professional appreciation for others and oneself • Experience of unity and good interaction as a motivational factor • Radio communication that enables following the input of all parties involved in DP operations	Flexibility of action and reorientation • Training that creates preparedness to act in various situations • Maintaining a standard bridge environment, to enable effective application of routines and skills • Confirmations and checks to ensure safe operation • Ability to adapt to the situation and perseverance for completion of the tasks at hand	Interpretiveness of action • Readiness to draw knowledge from training • Sufficient training and operation experience in DP operations and exceptional situations • Management of routines (e.g., attending to safety checklists)

Note: Control demands run down the leftmost column (in **boldface**) and means of managing the control demands are along the top (also in **boldface**), with the core-task demands <u>underscored</u> in the description cells.

FIGURE 14.2   Augmented crane operations concept. (Copyright 2013 Rolls-Royce plc.)

the vessel. Finally, at the social level, the crew members feel strong social unity; they are together on these vessels 24 h a day. Because "a feeling of togetherness" is a rather abstract idea to apply in design, an inspirational theme reflecting this noninstrumental UX goal was generated in our case study. The theme, referred to as "being one with the ship and the sea," served as a reference to how the mariners' should feel with the aid of the design solutions. In practice, the theme reflects both the instrumental CTA-derived findings and the intuitive interpretations of the noninstrumental emotion-related elements in the nautical work; in the practical design work, the instrumental and noninstrumental elements are inferred simultaneously and cannot be distinguished one from another.

The augmented crane operations concept (see Figure 14.2) exemplifies this design theme and use of the CTA model in design: it reflects an enhanced sense of unity between the ship operator and the environment, and it addresses some of the systems usability issues identified. The aim with the concept was to generate a solution that could support the creation of collaborative understanding during container-lifting operations involving rigs. The rigs have a container crane that is used when a PSV is positioned in the correct place. Currently, the collaboration between the crane and PSV operator is conducted mostly via radio communication. The concept idea was for the PSV operator to see exactly the same view (the bottom-left display in Figure 14.2) that the rig crane operator sees from the crane cabin and vice versa. Furthermore, the PSV operator should see where the container is supposed to land on the aft deck—this presentation would take place via augmented reality (AR) lines on a heads-up display (HUD) (see the middle portion of Figure 14.2). In the concept, the rig crane operator has the same view available in the cabin. Assumedly, through all these supportive systems, the feeling of continuity between self and environment could be increased: the operators would be more aware of one another's view of the situation. This concept design solution draws from the "shared awareness and problem-solving" and "dialogue-based communication" core-task demands (see Table 14.1), identified in the CTA of PSV operators' work.

14.2.2   FUTURE STUDIES

In parallel with user studies, we carried out studies of technological, maritime, and general societal trends. Visions of new user interaction tools were of special interest, since it was apparent that these might change work processes by providing new possibilities. Trends in the maritime industry shed light on what kinds of vessels the future might hold, where they may operate, and what purposes they could serve. All these issues affect the work on ship bridges. Our aim was to create a shared understanding within the project group of relevant trends. Project participants studied relevant future studies from their own perspective. The results were used in combination with the user-study findings to inspire ideation in a cocreation workshop.

Commissioned from a consulting company, a study of emerging interaction technologies and techniques was performed. The user interaction technologies explored included personal projection, large display areas, deformable devices, wearable devices, gestures, tangible user interfaces, hover sensing, tactile feedback, touch input, brain–computer interaction, AR, HUDs, motion simulators, and gaze tracking, along with speech, ambient sound, biosignals, and implanted user interfaces. For each technology, pros and cons were assessed, as was technology readiness.

Maritime transportation and technology trends were analyzed by a researcher focusing specifically on this area and by the company participant (Rolls-Royce). The most important trends identified were these: (1) globalization of markets (greater competition), (2) environmental concerns (entailing a need to reduce emissions and risks), (3) exhaustion of natural resources (bringing about a quest for new sources and a need to reduce consumption), and (4) navigation in Arctic conditions (since new shipping lanes in the Arctic region are gradually opening).

The general societal trends found had to do with future users in particular. Trends were identified on the basis of forecasts by Frost and Sullivan (2010), Gartner Research (2009), JWT Intelligence (2012), and Frog (2012). These general trends encompassed, for example, silence and minimalism as counterforces to information, media, and technology overflow; ubiquitous and embedded computing; and the values of today's young adults.

After all the trend analysis work had been carried out and the results were shared within the group, we organized a workshop in which both researchers and company representatives participated. The participants were assigned a task for completion before the workshop: they were asked to choose their "favorite" trends (1–2 user trends, 1–2 collaboration trends, 1–2 interaction trends, and 1–2 other trends) and prepare to indicate why these are important. At the workshop, the technology and general societal trends were presented each on one sheet of paper on the wall. The workshop participants were asked to mark their favorite trends, after which each participant could distribute five votes among the trends so marked. Even if the ranking of the trends was of interest, a more important result was characterization of the process that engaged the participants in considering the trends and discussing them.

After the voting, maritime trends, trends in user interaction tools, and the findings from the field studies were presented and discussed. The workshop continued with the participants divided into two groups, each of which was asked to identify themes (challenges/possibilities/solutions/concepts) to process further. The groups were also to choose related trends and discuss how these could affect the theme. Though their task was the same, the two groups chose slightly different approaches. The first group proposed radically new, challenging "what–if" concepts and solutions rooted in domain trends and technical enablers. In contrast, the second group identified new ways of carrying out today's tasks, resolving the core-task demands identified, and defining UX targets. By combining the results produced by the two groups, we identified the following ideas for further work:

- What if there were no joysticks? Could the operations be carried out with other kinds of tangible objects?
- What if there were no separate screens? Instead of screens, the necessary information could be presented on a HUD integrated into the bridge's windows or on AR glasses worn by the operator.
- Visibility could be enhanced via placement of navigation/steering workstations in a "glass bubble" at the front of the bridge. The vertical view downward could be enhanced with a glass floor.
- Data overflow is an issue on the bridge. How could this be reduced?

FIGURE 14.3   Sea-ice analyzer concept. (Copyright 2013 Rolls-Royce plc.)

- The bridge environment could be a less "machine-like" and "working environment" type of place, that is, it could be more cozy, homelike, and personalized.

Several design concepts reflect the work in the trend workshop. One example is the Sea-ice analyzer concept (see Figure 14.3). The capability of operating in Arctic conditions was one of the main maritime domain trends identified; therefore, we aimed to enhance the associated operations through novel technologies. Indeed, in icy conditions, it can be difficult to know whether a ship is able to break the ice in front of it, especially in darkness or fog. The intent with the sea-ice analyzer concept is to assist in this estimation: the thickness and strength of the ice around the vessel are calculated, and a computer estimates whether it is possible for the ship to proceed and shows the best route. This information is presented on a large HUD in the front window. The AR HUD data are organized such that the display shows the data needed for decision-making overlaid on the actual view of the outside.

14.2.3   Codesign and Coevaluation

By intuition, one might think that design in collaboration with users is not beneficial from the standpoint of generating radical design solutions. This is because, by definition, radical design ideas are those that provide users with wholly new kinds of activities—one cannot assume that these new kinds of possibilities can be easily imagined by users submerged in the existing modes of work in day-to-day operations (Norman and Verganti, 2014). During the creation of future ship-bridge design solutions, however, it was found that applying codesign and expert users' evaluations in the final part of the concept design process aids in promoting radical design. This is because, first, the awareness that the preliminary design solutions is going to be evaluated by expert end users frees the designers to imagine even potentially "bad" design solutions; it does not matter if some of the design solutions do not yield potential, since the experts will cull these nonfunctional or uninteresting solutions from the pool of design ideas. By affording diminished self–censorship, this knowledge promotes generation of more ideas more rapidly—and, eventually, since many ideas will arise, there will also be solutions appreciated by the actual users. The future ship-bridge design case produced several design ideas that were rejected by the end users and therefore not refined further.

The second way in which expert users provide beneficial design input is via the refinement of design ideas. A good example of this is the intelligent towing concept. The initial idea was that a HUD would present the unit of tug plus towed ship (as visible in the map box at the left in Figure 14.4). We had discovered that when a tug pushes or tows another vessel, where that other vessel starts to rotate and head is not always self-evident. To assist in estimating this, the bridge could indicate the forces influencing the pushed or towed ship. Upon discussion with real-world tugboat operators, however, the concept idea was developed further. It was explicated that actually the tugboats and the escorted cargo ships often share towing-relevant information via radiophone. The tugboat operator requests relevant information from the cargo ship's crew, such as rate of turn (ROT), speed, and course. An immediate design implication of these accounts was that the intelligent towing concept should also incorporate direct presentation of these verbal exchanges. In other words, a direct data link between the vessels would provide the tugboat with indications of the ROT, speed, rudder direction, and course of the cargo ship (see the box and the semicircle at the center of Figure 14.4).

In the future ship-bridge design case, the initial concept solutions were presented to the expert users by means of pictures and user scenarios. The creation of scenarios was itself an iterative and collaborative process. We first imagined certain kinds of scenarios and then discussed these with certain users. If the scenarios seemed plausible, they were applied when the concepts were discussed with a larger sample of users. For instance, the intelligent towing concept was explained to the users via a scenario in which the rudder of the towed vessel becomes jammed, thereby steering it in an undesired direction; in such a situation, communication between the two vessels is essential.

FIGURE 14.4 Intelligent towing concept. (Copyright 2013 Rolls-Royce plc.)

14.3   DISCUSSION

Figure 14.5 presents the overall workflow of the design approach suggested in this chapter. The approach is based on the following premises, which are presented in the figure. First, we assume that it is useful to distinguish the challenges and strengths involved in the activity that the design solution is to serve; this can be accomplished by the study of users (Step 1 in Figure 14.5) and with core-task analysis (Step 2). We believe that it is beneficial for the design solutions to draw on and support these existing strengths that reflect professionalism and human capabilities. Assumedly, current industry professionals are likely to take a positive view of a design solution if that solution allows the user to apply his or her existing potential and/or provide support for resolving real-world challenges in the work—that is, if the so-called systems usability design goals are addressed (Step 3a). These goals can then be arranged visually—for instance, in tabular form as shown in Table 14.1 (Step 4a). Furthermore, we find it beneficial if the design solutions reflect the most important emotion-specific elements related to the work domain (that is, addressing issues such as work-related identities and emotion-laden ideals). We do not suggest a particular method for this. Instead, we encourage intuitive consideration of interview- and observation-based findings; empathy is needed for derivation of emotional UX goals from the various accounts given by professionals (Step 3b). Reformulation of data into broad function-oriented goals and into emotion-related stories, themes, and/or user personas (Step 4b) allows that measure of "distance" from user data that is necessary for radical design. Furthermore, the design solutions should not merely address the existing domain–specific challenges; future challenges and needs too are important (Step a). This further assures that the design solutions are future-oriented, which again leaves room for more radical design alternatives. The future studies should include technology foresight (Step b1), industry-domain-related foresight (Step b2), and examination of general trends and likely future values within society (Step b3); the future studies may take place in parallel with user studies. Applied together, these procedures lay a foundation for initial design ideas (Step 5), which can be visualized or prototyped for purposes of user evaluations. It might be beneficial if the visualizations are embedded in user scenarios (i.e., stories) so that users may more easily imagine themselves applying the concept solutions (Step 6). By these means, cocreation and coevaluation take place; considering the solutions alongside the users enables the better ideas to be distinguished far more readily from the worse.

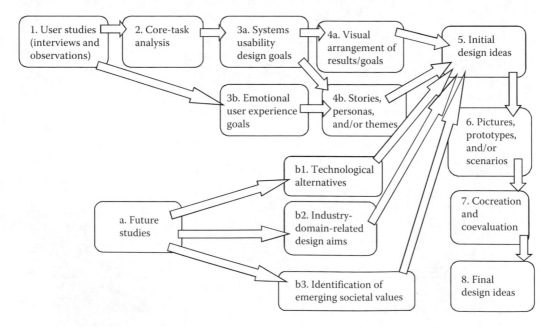

FIGURE 14.5   Overall workflow for user-oriented radical concept design.

Those deemed worthy can be further developed (Step 7). This ultimately leads to the final concept design solutions (Step 8).

The approach presented in Figure 14.5 reflects some existing design approaches, contextual design (Beyer and Holtzblatt, 1998) chief among them. Contextual design assumes that good design solutions are created through profound understanding of the use (or work) context. The idea is that "contextual inquiry" precedes the actual design process. This includes studying the work or use context that the new design is to serve. Among the methods are interviews and observations: thereby, shared understanding of the content of the work is developed with the users. Notess (2005) summarizes contextual design in general by suggesting that it applies four core principles. First is the assumption that data on work activity are largely contextual and, therefore, it is the actual work context that is to be studied. The second principle is that designers should work in partnership with users who act as experts. Third, contextual design applies visualizations; that is, the findings from contextual inquiry are presented via diagrams for the purpose of aiding the design process. The fourth principle, that of iteration, implies that the design process is not entirely linear. Instead, it involves paper prototypes that may lead to further refinement of the product. The approach presented in Figure 14.5 is largely in line with these principles of contextual design. Some noteworthy differences from the contextual design approach do, however, exist, these serving the purpose of generating radical design ideas.

The first key difference is that the design approach we present applies the CTA method as a means of analyzing the contextual inquiry data. Typically, CTA has been applied for the purpose of studying various work contexts, such as nuclear power plant (Norros, 2004) and metro train (Karvonen et al., 2011) operation, without the explicit purpose of providing the relevant domains with new concept design solutions. Nonetheless, there are elements of the method that render it useful for provision of indications for radical design. This is because the method can be used to generate broad design goals. The core-task demands pinpointed—that is, the ways in which the control demands and resources uncovered connect one with another—can be seen as design goals. These general design goals can be considered to be systems usability goals, ideas as to how the overall work system, including its users, the environment, and the technologies employed, could and should function together. Furthermore, the method provides visualizations of the findings, thereby responding to the common conclusion in design studies (Schön, 1992; Findeli, 2001) that visual representations are especially beneficial in design activity. In our experience, the visual models aid in conceiving of and communicating the model and the findings in the concrete design work.

Second, the approach we suggest here emphasizes the importance of considering UX in design. The latter reflects the experience design approach. It suggests that, rather than products, it is the UX that should be at the focus in design work (Hassenzahl, 2010). "User experience goal driven design" as described by Karvonen et al. (2012) and Koskinen et al. (2013), in turn, is more specific in suggesting that "user experience goals" should be defined at the very beginning of the design process. In other words, the designers should, in principle, begin by determining what kind of activity or emotion should be supported by the design; only after this should product-related design ideas be generated. This, indeed, might allow "thinking outside the box," since the design process is not bound to the existing product when one considers the future design. Both CTA–derived systems usability goals and the feeling-related noninstrumental UX goals draw the design focus from the preexisting product to workers' potential future activity. As has been suggested elsewhere (Wahlström et al., forthcoming), user studies may provide a basis for radical innovations if the resultant findings are reformulated as models, goals, or themes in such a manner that sufficient distance from the findings themselves is achieved. This implies the findings not directly dictating the design solutions so much as informing them by providing broad but justified design indications. These assumptions are in line with the thinking of Hekkert et al. (2003), who propose that innovative product design can be achieved when one first abandons presuppositions about the product and then develops the product by formulating three "visions": an initial vision of the user and the context of use is elaborated into an interaction vision, which describes how the user interacts with the product, then a product vision. The latter approach arguably forces designers to free themselves of apparent restrictions or requirements and, instead, encourages seeking desirable possibilities. The approach of Hekkert et al. also entails the designer empathizing with the future user; however, the user is not directly involved in the design process. They state that in this way that undesirable constraints resulting from users' fixation on familiar solution directions are avoided.

Indeed, design through empathetic understanding of users has been a common theme in the design literature. Leonard and Rayport (1997) introduced empathic design as an approach complementary to marketing research, one that contributes to the flow of ideas that still need further testing. When a company representative explores the customers' worlds through the eyes of a fresh observer, the company can redirect existing organizational capabilities to new markets. Wright et al. (2008) remind that good experience-centered design requires designers to engage with the users and their culture in rich ways in order to understand how the users make sense of technology in their lives. Kouprie and Sleeswijk Visser (2009) propose a framework for empathy in design, formulated as "stepping into and out of the user's life." Proceeding from the psychological literature, they distinguish two components of empathy: the affective and the cognitive. The affective component includes the emotional response, feeling and identifying with the user: "becoming the user." The cognitive component, on the other hand, includes understanding, adopting a perspective, and imagining the other: "staying beside the user." Sleeswijk Visser (2009) emphasizes that knowing the users' world is important for designer motivation, and stories are tools that contribute well to this understanding. Successfully communicated user information provides empathy and inspiration for product ideas.

A third way in which our design approach differs from typical contextual design is in our emphasis on the inclusion of future studies in derivation of design indications. Foresight is addressed in the innovation management literature and as a separate field, but it seldom features explicitly in user-centered design literature or in the design literature generally. Design literature usually relies on brainstorming and reframing methods and approaches for creating novel ideas and breaking free from current constraints (e.g., Krippendorff, 2006). However, an understanding of technology and societal trends can be crucial for a product or company's success. For example, Christensen (1997) has identified multiple cases of existing market leaders failing to understand the effect of emerging technologies. Overall, our approach reflects the technology research of Nieminen and Mannonen (2005) and more general trend analysis carried out by Salovaara and Mannonen (2005). Nieminen and Mannonen suggest that a separate element of technology research should be part of user-centered concept development projects, for tying the design ideas to a meaningful level of technology. Salovaara and Mannonen, in turn, attempt to achieve balance between the future–orientation and user-centeredness requirements of concept development by dividing the design-supporting information into information on upcoming changes (i.e., trends in society and working life) and stable features of the context. Our foresight involved a wider view of the future than a pure technology review in that it also covered general trends analysis. However, as has been discussed elsewhere (Wahlström et al., 2014a,b), very central to our approach is the combination of user-study findings with future studies (of interaction technologies in particular); it can be assumed that novel interaction technologies are those with the most potential to enhance work activities.

Fourth, in a similarity to contextual design, our approach accepts the utility of working alongside the users in design. The terms "participatory design," "cocreation," and "codesign" have been used in the design literature. The associated concepts are employed with varying meanings, but we follow the interpretation by Sanders and Stappers (2008), in which cocreation refers to any creative act involving more than one individual while codesign refers to cocreation in a design process specifically. Participatory design, in turn, can be seen as a Northern-Europe-based design movement that views the user as a partner in the design process. We embrace the underlying message implied by these concepts but suggest that users should be involved only after the generation of the initial design ideas. However, the design ideas do have to be based on thorough and empathic understanding of the users' work and work environment. This is vital for the purpose of creating radical design ideas, that is, it is important because the aim is to create solutions not yet imagined by the users. Here, the users are involved in two phases of the concept design: in the beginning as informants on the work domain and in the final stage as codesigners evaluating and enhancing the initial design concepts. In our experience, visually illustrated design ideas have served well as "boundary objects" that deliver understanding of a joint object across boundaries between disciplines (Star and Griesemer, 1989). Star and Griesemer emphasize the communicative nature of boundary objects in enabling collaborating parties to represent, transform, and share knowledge. In our codesign activities, wherein the stakeholders had very different backgrounds, the visualized future concepts facilitated and encouraged communication. The future concepts served well in the three roles of boundary objects proposed by Star (2010): offering personalized value to each party, facilitating understanding of the task at hand in the same manner, and offering an understanding of all relevant options related to the task.

Our contribution with the design approach presented in this chapter has potential to yield concept designs that are radical yet remain grounded in the actual activity and work-domain-specific needs at hand. As is arguably visible in the ship-bridge design solutions presented here, the approach has worked for us. Future studies, however, would be needed to confirm the approach's generalizability, and its broader utility.

ACKNOWLEDGMENTS

This study is part of the User Experience and Usability in Complex Systems (UXUS) research and development program, which is one of the research programs of the Finnish Metals and Engineering Competence Cluster, FIMECC.

This study was preliminarily reported in a conference paper (Wahlström et al., 2014b). This chapter further elaborates the argument and includes new information on the ship-bridge design project.

The authors would like to thank all those who have provided input to the design of future ship bridges and to this article, among these being Leena Norros (VTT), Iiro Lindborg (Rolls-Royce), Jussi Lehtiniemi (Troll VFX), Zhang Daoxiang (Aalto University), Hanna Askola (VTT), Göran Granholm (VTT), Frøy Birte Bjørneseth (Rolls-Royce), and Karno Tenovuo (Rolls-Royce).

REFERENCES

Beyer, H. and K. Holtzblatt. 1998. *Contextual Design: Defining Customer-Centered Systems.* San Francisco: Morgan Kaufmann.

Carroll, J. 2000. *Making Use: Scenario-Based Design of Human–Computer Interactions.* Cambridge, Massachusetts: MIT Press.

Christensen, C. M. 1997. *The Innovator's Dilemma: When New Technologies Cause Great Firms to Fail.* Boston, Massachusetts: Harvard Business School Press.

Cooper, A. 1999. *The Inmates are Running the Asylum.* Indianapolis: Morgan Kaufmann.

Findeli, A. 2001. Rethinking design education for the 21st century: Theoretical, methodological and ethical discussion. *Design Issues* 17:5–17.

Frog. 2012. 12 tech trends that will define 2012, selected by Frog's design minds. Fast Company Website: http://www.fastcodesign.com/1668912/12-tech-trends-that-will-define-2012-selected-by-frogs-design-minds.

Frost & Sullivan. 2010. World's top global mega trends to 2020 and implications to business, society and cultures. Frost & Sullivan Website: http://www.frost.com/sublib/display-report.do?id=M65B-01-00-00-00.

Gartner Research. 2009. A day of your life, 2028. Gartner Website: https://www.gartner.com/doc/933312/day-life-.

Gould, J. and C. Lewis. 1985. Designing for usability: Key principles and what designers think. *Communications of the ACM* 28:300–311.

Hassenzahl, M. 2010. *Experience Design: Technology for All the Right Reasons.* San Rafael, California: Morgan & Claypool.

Hekkert, P., M. Mostert, and G. Stompff. 2003. Dancing with a machine: A case of experience-driven design, in: *Proceedings of the 2003 International Conference on Designing Pleasurable Products and Interfaces (DPPI '03).* New York: ACM.

JWT Intelligence. 2012. 100 things to watch in 2012. JWT Intelligence Website: http://www.jwtintelligence. com/2011/12/100–things-to-watch-in-2012/.

Kaasinen, E., V. Roto, J. Hakulinen, T. Heimonen, J. Jokinen, H. Karvonen, H. Koskinen et al. 2015. Defining user experience goals to guide the design of industrial systems. *Behaviour & Information Technology* 34:976–991.

Karvonen, H., I. Aaltonen, M. Wahlström, L. Salo, P. Savioja, and L. Norros. 2011. Hidden roles of the train driver: A challenge for metro automation. *Interacting with Computers* 23:289–298.

Karvonen, H., H. Koskinen, and J. Haggrén. 2012. Defining user experience goals for future concepts. A case study, in: *Proceedings of 7th Nordic Conference on Human–Computer Interaction (NordiCHI2012), UX Goals 2012 Workshop—How to Utilize User Experience Goals in Design?* Tampere, Finland: TUT Publication series.

Keinonen, T. 2009. Immediate and remote design of complex environments. *Design Issues* 25:62–74.

Koskinen, H., H. Karvonen, and H. Tokkonen. 2013. User experience targets as design drivers: A case study on the development of a remote crane operation station, in: *Proceedings of the 31st European Conference on Cognitive Ergonomics (ECCE2013)*. Toulouse, France: ECCE.

Kouprie, M. and F. S. Sleeswijk Visser. 2009. A framework for empathy in design: Stepping into and out of the user's life. *Journal of Engineering Design* 20:437–448.

Krippendorff, K. 2006. *The Semantic Turn: A New Foundation for Design*. New York: Taylor & Francis.

Leonard, D. and J. F. Rayport. 1997. Spark innovation through empathic design. *Harvard Business Review* 75:102–113.

Lu, Y. and V. Roto. 2014. Towards meaning change: Experience goals driving design space expansion, in: *Proceedings of the 8th Nordic Conference on Human–Computer Interaction*, pp. 717–726. New York: ACM.

Mahlke, S. 2005. Understanding users' experience of interaction, in: *Proceedings of the 2005 Annual Conference of the European Association of Cognitive Ergonomics (EACE '05)*. Athens, Greece: University of Athens.

Nieminen, M. P. and P. Mannonen. 2005. User-centered product concept development. in: *International Encyclopedia of Ergonomics and Human Factors*, 2nd edition, ed. W. Karwowski, pp. 1728–1732. Boca Raton: CRC Press/Taylor & Francis.

Norman, D. A. and R. Verganti. 2014. Incremental and radical innovation: Design research versus technology and meaning change. *Design Studies* 30:78–96.

Norros, L. 2004. *Acting under Uncertainty: The Core-Task Analysis in Ecological Study of Work*. Espoo, Finland: VTT Publications.

Norros, L. 2013. Developing human factors/ergonomics as a design discipline. *Applied Ergonomics* 45:61–71.

Norros, L., P. Savioja, and H. Koskinen. 2015. *Core-Task Design—A Practic–Theory Approach to Human Factors*. Milton Keynes, UK: Morgan & Claypool.

Notess, M. 2005. Using contextual design for digital library field studies, in: *Proceedings of ACM/IEEE–CS Joint Conference on Digital Libraries Workshop Studying Digital Library Users in the Wild: Theories, Methods, and Analytical Approaches*. Denver, Colorado: ACM.

Pruitt, J. and T. Adlin. 2006. *The Persona Lifecycle: Keeping People in Mind Throughout Product Design*. San Francisco, California: Elsevier.

Rolls-Royce. 2014. Rolls-Royce presents the future of tug bridge controls. Rolls-Royce YouTube Channel. https://www.youtube.com/watch?v=27uCL90s20o.

Rolls-Royce. 2015. Ship intelligence for cargo vessels. Rolls-Royce YouTube Channel. https://www.youtube. com/watch?v=_nApv-C7qSg.

Salovaara, A. and P. Mannonen. 2005. Use of future-oriented information in user-centered product concept ideation, in: *IFIP TC13 International Conference on Human–Computer Interaction (Interact 2005)*, September 12–16, pp 727–740, Rome, Italy.

Sanders, E. and P. Stappers. 2008. Co-creation and the new landscapes of design. *CoDesign: International Journal of CoCreation in Design and the Arts* 4:5–17.

Savioja, P. 2014. Evaluating systems usability in complex work. Doctoral dissertation, available via VTT Website: http://www2.vtt.fi/inf/pdf/science/2014/S57.pdf.

Schön, D. 1992. Designing as reflective conversation with the materials of a design situation. *Research in Engineering Design* 3:131–147.

Sleeswijk Visser, F. S. 2009. *Bringing the Everyday Life of People into Design*. Doctoral dissertation, available via TU Delft Website: http://www.narcis.nl/publication/RecordID/oai:tudelft.nl:uuid: 3360bfaa-dc94-496b-b6f0-6c87b333246c.

Star, S. L. 2010. This is not a boundary object: Reflections on the origin of a concept. *Science, Technology, & Human Values* 35:601–617.

Star, S. L. and J. R. Griesemer. 1989. Institutional ecology, "translations" and boundary objects: Amateurs and professionals in Berkeley's Museum of Vertebrate Zoology, 1907–39. *Social Studies of Science* 19:387–420.

Wahlström, M., H. Karvonen, and E. Kaasinen. 2014a. InnoLeap—Creating radical concept designs for industrial work activity. *NordiCHI 2014 Workshop WS4: The Fuzzy Front End of Experience Design*, October 26, Helsinki. A workshop position paper available via VTT Website: http://www.vtt.fi/inf/pdf/technology/2015/T209.pdf.

Wahlström, M., H. Karvonen, E. Kaasinen, and P. Mannonen. 2014b. Designing for future professional activity—Examples from ship-bridge concept design, in: *Proceedings of the 5th International Conference on Applied Human Factors and Ergonomics AHFE 2014*, Kraków, Poland, July 19–23.

Wahlström, M., H. Karvonen, L. Norros, J. Jokinen, and H., Koskinen. (Forthcoming). *Radical Innovation by Theoretical Abstraction—A Challenge for the User-Centred Designer*, under peer review.

Wright, P., J. Wallace, and J. McCarthy. 2008. Aesthetics and experience-centered design. *ACM Transactions on Computer–Human Interaction* 15:1–21.

15 New Color Planning Methodology for Urban Furniture as an Ergonomic Factor

Margarida Gamito and Fernando Moreira da Silva

CONTENTS

15.1 INTRODUCTION

This chapter approaches color application to urban furniture as an ergonomic factor, bearing in mind that pertinent color application to urban furniture can optimize its use. For this purpose a new methodology for color planning is presented, which seeks to apply color to urban furniture in a way that originates a system that will function simultaneously as an identification factor for different city quarters and as an orientation factor for its inhabitants and visitors. In parallel, color application to urban furniture will also become an inclusivity factor, by incrementing the visibility of these elements. This new methodology is intended to be applied solely to urban furniture, which will become an ergonomic and inclusive factor, and not interfering with other elements of city signage, which are encoded by road legislation.

Usually, city orientation systems are not appropriate to fulfill the needs of the entire population. They tend to be generalized, following the directives prescribed by the AIGA (American Institute of Graphics) for U.S. airports, which are considered as a basis for pictograms in several countries. This internationalization, reinforced by the standardization of pictograms prone by ISO (International Standardization Organization) facilitates the recognition of signage symbols by an increasingly itinerant population that finds the same signs all over the world. So, we can accept that an internationalization of signage may be a right decision, although Baines and Dixon (2003:14) expresses some doubts: "(…) Gratifyingly, even within countries which follow the protocol, there is a considerable variation in the drawing and implementation of these signs which allows a little of the characteristics (or aspirations) of individual countries to shine through."

When referring to signage chromatism, which must be "an integration factor between signage and environment" (Costa 1987:182) it is usually restricted to the shape and background contrast, following the options described on the roads code, or using the black and white achromatic contrast.

15.2 SUBSTANTIATION

Urban furniture is not just a simple ensemble of decorative elements to embellish the city, it must accomplish an amount of functional requirements in order to assure its functionality and fulfill the needs of the population, facilitating their lives, and contributing to their comfort. So, when urban furniture accomplishes its functions, it contributes to protect the health and well-being of the city inhabitants; facilitates the accessibility and use to people with visual or motor difficulties; reinforces the local identity, representing a formal *family*, that is, coherent, and values the surroundings (Águas 2003:10–11). However, while recognizing its necessity, the functional possibilities of urban furniture have not been used to their fullest extent, and the choice of its color or form only rarely conforms to a logic that has been thought out.

The connotations of color with the understanding of the environment have already been considered in applications to architecture. However, these concerns are rarely extended to urban furniture plans, despite multiple warnings about their lack of visibility made by various authors.

15.2.1 INCLUSIVE DESIGN–ERGONOMIC DESIGN

Inclusive Design is a way of designing products and environments so that they are usable and appealing to everyone regardless of age, ability or circumstance by working with users to remove barriers in the social, technical, political and economic processes underpinning building and design (DPTAC 2003).

Concerning urban furniture, this definition may be applied to Ergonomic Design, as it implies the adaptation of all its elements to the city's population needs, erasing, as much as possible, the differences between disabled and able-bodied people, and contributing to the improvement of everyone's quality of life. The objective of an ergonomic design must be considered as an "interaction between the individual and the environment," and "can be described in terms of personal control that can be exerted by the individual over the environment" (Brown et al. 1989 apud Brown 1998:34).

We must consider that it may be impossible to contemplate all the needs of people with a high level of disability. However, adaptive environments should be designed in order to ensure that a higher percentage of the population can enjoy all the environmental facilities, and the widest percentage of human beings must benefit from the improvement of visibility in urban furniture. Whenever we design for disabled people, we are improving the quality of life for the entire population. Everyone, disabled or not, will feel more comfortable if the bus stop, the bench, or the waste bin, they are looking for, stands out from the environment without the need of a detailed search. As Brown (1998:75) writes, "Integration does not imply that every conceivable option open to all unimpaired people can be made equally available to every impaired person. It does demand, however, that there should be a sufficient range of options open to any impaired individual to enable him or her to function as a mature person and pursue a personal lifestyle as satisfying in its own way as his/her neighbour's" (Figure 15.1).

The estimates of Portuguese population, published by the *Instituto Nacional de Estatística* (Statistics National Institute) in June 2014 (Serviço de comunicação e imagem—INE 2014), shows that Portuguese people live longer and, consequently, a group of the population tends to age. So their requirements must be taken in account because, despite their limitations, the elderly, and visually disabled people must be able to get out and about locally in order to age well and live independently. The desire to get out does not diminish with old age and older people can continue

FIGURE 15.17 Examples of recorded samples from buildings, pavements, vegetation, and occasional elements.

FIGURE 15.18 Examples of sky color percentage in different urban areas.

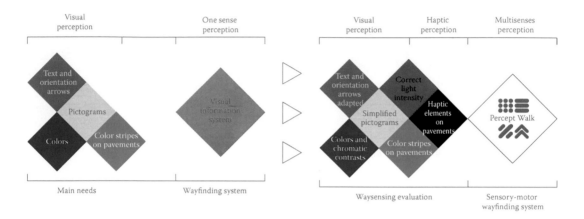

FIGURE 19.3 Wayfinding system and Percept Walk wayfinding system. (By author Aboim Borges.)

FIGURE 19.4 Percept Walk diagram. (By author Aboim Borges.)

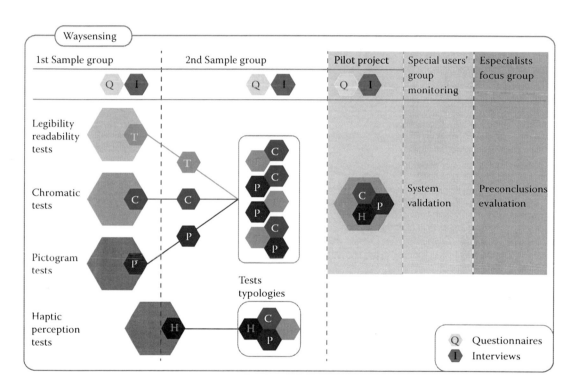

FIGURE 19.6 Waysensing: Steps in evaluation process. (By author Aboim Borges.)

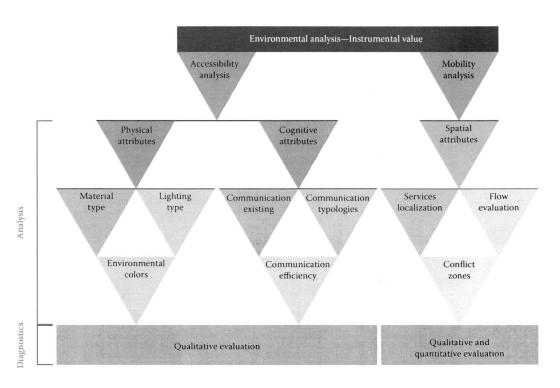

FIGURE 19.7 Environmental analysis and diagnostics. (By author Aboim Borges.)

```
2013
    Existing population—10,427,301              | Women—5,469,281
                                                | Men—4,958,020

    Births decrease—7.9%
    Deaths decrease—1.0%

    Life expectance(2001/2003)—76.98 years        | Women—80.12 years
                                                   | Men—73.55 years

    Life expectance(2011/2013)—80.00 years           | Women—82.79 years
                                                      | Men—76.91 years

2001/2003 – 2011/2013

    People older than 65 years—increased
    People younger than 15 years—decreased
    Aging index: Older people / Younger people = 106–136/100

Previsions for 2060

    Aging index: Older people/Younger people = 307/100
```

FIGURE 15.1   2014 Instituto Nacional de Estatística (Statistics National Institute) estimates for the Portuguese population. (From Serviço de comunicação e imagem—INE. 2014. *Destaque, Informação à comunicação social.* www.ine.pt Accessed June 16.)

practicing a large variety of outdoor activities whenever the environment allows it. On contrary, when it is not easy or enjoyable to get outdoors, their quality of life as well as their physical health will diminish. The difficulty to get around is often due to the poor design of the environment. Older and visually disabled people move about more frequently on foot and this presents big difficulties in poor design environments. Accessible open spaces, with good paths, safe crossings, plentiful seats, and visible signage will improve older people's perception of supportiveness and safety (PDGo, 2012).

Inclusive and ergonomic designs, in their recent development, are primarily focused on people with motor limitations and tend to forget visually disabled people. However, we must consider that the urban population is constituted of an extensive variety of people, with different visual acuities and limitations and, also, by a high percentage of older people. Insofar as people grow older, their ability to see small details decreases, and the eyes have a crescent difficulty of adaptation to sudden changes of light or a quick change in focus. Bearing in mind the visually limited population, only a small percentage is unable to see any color and the main part is able to distinguish luminosity differences (Lindemann et al. 2004). Therefore, to have better visibility conditions, under an inclusive design perspective, urban furniture must present a good chromatic and luminosity contrast. Considering this, Mollerup (2005:161) states that "color can be seen from longer distances than other graphic elements" and that "in signage differentiation is the first and foremost role of color" (Figure 15.2).

FIGURE 15.2   Examples of urban furniture with lack of visibility.

15.2.2 Urban Furniture and Ergonomics

When populations increased and towns became organized, there was a need to implant systems of elements that could offer support and orientation to the cities' occupants. This systems that were institutionalized from the mid-twentieth century are denominated—*urban furniture*—and comprehend every element, placed on the public space, constituting a wide range that includes, among others, benches, litter bins, street lamps, bus stops, kiosks, cabinets, telephone booths, drinking fountains, bollards, and signs. The urban furniture elements do not have a mere decorative function, as Màrius Quintana Creus (apud Serra 2000:6) claims, "These elements are objects which are used and which are integrated in the urban landscape, and they must be comprehensible to citizens. Use, integration and comprehension are thus basic concepts when it comes to assessing any set of objects we might come across in a city's public spaces." In fact, in order to accomplish its functions, it *needs to be seen* and an appropriate color application improves its visibility considerably. Also, when urban furniture chromatism is the same for a city area, they may be converted into effective signage and identification elements that will contribute to a better orientation within the city.

Considering the whole population and, within it, the percentage of people with visual impairments, there are some constraints that must be respected. In accordance with the prescriptions of the *Royal National Institute for the Blind* (RNIB), UK, pedestrian paths must be easily identifiable and differentiate themselves from the adjacent walls. Likewise, all present objects must detach themselves from the background, in order to be recognized as obstructions. Every urban furniture element—fences, bollards, lampposts, litter bins, benches, etc.—must present a strong color and tone contrast with the environment, because they are the most effective means of improving visibility, with tone contrast usually the more effective. A conventionally pleasing coordinated color scheme can usually be significantly enhanced with good tonal contrast (Barker et al. 1995:7–51), in order to stand out and be more easily recognized, among others, by visual disabled people.

So, the application of a chromatic planning to urban furniture, may originate a system which will function simultaneously as an identification factor for the different city quarters and as an orientation factor for its inhabitants and visitors. In parallel, color application to urban furniture will also become an inclusivity factor, by incrementing the visibility and use of these elements.

15.2.3 Ergonomic Color and Visibility

Color in the environment is not a simple element for definition and unification; it becomes a visual characteristic that stands out from the chaos and complexity of the visual field. It is, also, the easiest way to achieve the identification of different city zones, and to promote the orientation of the population, permanent or temporary, because color is the characteristic object which the eye first perceives, even before form or texture—that is why color and ergonomics come together. Color utilization as a means to show the way, has been employed successfully in interior and exterior spaces, therefore we can assume that a sensible and general application to urban furniture may be a way to the successful resolution of the orientation problem within the city.

Despite a recent growing concern about color psychophysiologic connotations and its application to the environment, color urban plans scarcely refer to color application in urban furniture and signage. In parallel, a bad use of color in urban furniture and signage systems contributes to a lack of visibility that is an impeachment to the fulfillment of their functions, as well as it is a factor of social exclusion for people with deficient and aging vision.

As it was stated, in order to be used, *urban furniture must be seen* and, therefore, it must stand out from the environment. However, regardless of color being the easiest and more appropriate tool to this end, it is rarely used with that intention. Usually the urban furniture suppliers prone the uniformity or the color elimination in these elements, maybe as a reaction to the excessive color multiplicity present in the city, but this solution is an impeachment to the satisfactory accomplishment of its functions.

Signage also manifests a reserve on color application, presenting a dominant concern about environment integration, and, as a result, becoming less visible and unable to accomplish its function completely, and creating barriers between inhabitants with and without vision disabilities. Also, signage chromatism frequently restrains itself to the form and ground contrast, the black and white achromatic contrast, or even the chromatic road standards. However, within the city, these chromatic options may lose visibility and confound themselves with build environment colors, becoming illegible for a pedestrian population that has different degrees of visual acuity.

So, the application of a chromatic planning to urban furniture, may give rise to a system which will function simultaneously as an identification factor for the different city quarters and as an orientation factor for its inhabitants and visitors. In parallel, color application to urban furniture will also become an inclusivity factor, by incrementing the visibility and use of these elements.

15.3   COLOR AND ORIENTATION

Color, inseparable from light, is an integral part of our total sensory and perceptual experience. It not only conveys information about our surroundings, but also has great impact on our psychological reactions and physiological well-being (Durão 2002:162).

In nature, color often appears as a means for protection and species conservation, constituting a defense camouflage system, or a warning, as it happens with poisonous animals and plants, whose showy color warns potential predators. Color may also have an attraction function that allows the reproduction of plants or invites the mating of animals.

In their evolution, human beings have inherited psychophysiological reactions which, even if they cannot be controlled or objectively explained, make color act as a necessary mean of information, communication, and comprehension of the environment, as was stated by Lancaster (1996:8): "(…) The functions of color are to attract attention, to impart information, to aid deception and to stimulate the emotions."

The orientation within the city, the problem of finding a location and (or) the way to a destination, is not always easy to solve, regardless of the individual mode of locomotion. Several authors seek to show the way for the solution of this problem. Machnow and Reuss (1976:21) put the question of a resolution through color, Color—architecture—cities—colorful cities—color on the urban scene— how does it all fit together? Is a colorful urban backdrop enough, will more color really change our living and working environment? This is the most basic question of all. To put it differently: Is it possible to raise a city's visual accessibility, the quality of experience and orientation, without merely underlining its character as a huge, conglomerate consumer object?"

Modern cities usually display a color amalgam that is due not only to the diversity of its buildings, but also the profusion of multicolored advertising. This chromatic confusion tends to make their inhabitants less sensitive to particular colors. On this subject, Tosca (1994:155) asserts that "publicity and traffic are chiefly responsible for the depersonalization of the urban place."

Craig Berger (2005:121) stresses that cities adapted themselves to being tourism and convention centers, originating the necessity for urban orientation systems. These systems that gather tourism centers, maps, symbols, and graphic indications are intended to make *navigation* within the city easier to their visitors and inhabitants.

Till the middle of twentieth century, the boundaries between architecture and design were well defined, however influences from other fields, like industrial design and urban planning, cause these boundaries to attenuate and their merging came to be known as *environmental graphic design*. With the development of cities complex traffic and transportation webs appear which caused orientation difficulties, and created the necessity to design comprehensive directional systems that could help in guiding people and, simultaneously, renew their identification with the city. The usual signage and directional systems, however, are designed for the distance and fast vision that one has when driving a car on roads and highways. Within the cities, these options besides losing visibility and merging with the building colors, are not appropriate to be seen by pedestrians because, as Minah (2005:401) states, "As a pedestrian, colours are experienced in a continually changing visual field. Planners have succeeded in achieving visual order in cities by implementing repetitive architectural typologies, zoning to form hierarchy in patterns of blocks and public spaces, and similar building heights."

Gallen Minah (2005:401) explains that most cities which managed to control the visual order are compact historical cities where a hierarchy defined among their spatial elements already existed. Modern cities are generally more dispersed and their architectonic elements are fragmented and autonomous. Because their hierarchy is less clear, order and harmony become difficult experiences from the pedestrian point of view. The development of the contemporary city originates great diversities and complexities in their architecture which compete for visibility. Color exceeds its function of a definition and unification element and becomes a visual characteristic in the midst of the visual field complexity and chaos. The same author also states, "Colour is one of the repetitive visual elements that define the formal, spatial, and material phenomena in the city. One experiences colour in a city through its combination with, and definition of, architectural elements in the visual field."

Color application as a means to show a way has already been successfully used, although occasionally, in interior and outdoor spaces. Its judicious and widespread application appears to be a way to solve the orientation problem.

Concerning the orientation within the city, and the identification of its different zones, we may consider city maps that differentiate them through the use of different colors. However, on the urban physical space these colors do not show up, and there is no concern in establishing the correspondence to a real use on this space. The ideal would be to identify the city different areas by specific colors which may differentiate them and, as well, accentuate the different urban furniture elements. Despite a recent growing concern about color psychophysiologic connotations and its application to the environment, color urban plans scarcely refer to color application in urban furniture and signage. In parallel, a bad use of color in urban furniture and signage systems contributes to a lack of visibility that is, an impeachment to the fulfillment of their functions, as well as being a factor of social exclusion for people with deficient and aging vision.

A coherent and structured color application to urban furniture can act as an orientation system that will contribute to the orientation within the city and become a factor of inclusivity, especially for visually disabled people, which constitute a large percentage of the urban population.

15.4   WAYFINDING/WAYSHOWING

Cities are, generally, a complex mass of roads and buildings and the orientation within them—the wayfinding and signage problems—are like its identification of different zones, not always easy to solve. As Juanita Dugdale (apud Berger 2005) stated, "Visitors and occupants were having difficulty navigating spaces on their own; they needed visual prompts to find their way around."

The growth of cities, the transformation of medium settlements into big cities, diminished the direction sense considerably, especially in cities in which the architecture is more or less similar in the different quarters, and where there is a lack of obvious reference points. About this subject, Charles Higenhurst (1971, apud Berger 2005) wrote, "today we are the strangers in our towns. We do not know and cannot see how things work. Our support systems… are remote. The information supplied in the environment is largely irrelevant to our immediate purposes or to an understanding of the world in which we live."

These conditions aroused the requirement for the installation of a wide urban furniture ensemble, as well as signage systems that could give support and orientation to the city users, showing directions and identifying places. However, these elements are not always enough to completely achieve this function.

Other contemporary cities, as Minah (2005:401) explains, originate a great diversity and complexity in architecture that fights against visibility. Color bypasses its function as an element for definition and unification, and becomes a visual characteristic within the chaos and complexity of the visual field.

Concerning the orientation within the city, and the identification of its different zones, we may consider city maps that differentiate them through the use of different colors. However, on the urban space those colors do not show up, and there is no concern in establishing the correspondence to a real use on this space. The ideal would be to identify the city quarters by specific colors which may differentiate them and, as well, accentuate the different urban furniture elements, such as dustbins, benches, telephone boxes, bus stops, street lamps, bollards, etc.

Golledge (1999:1–7) focus on the process of drawing cognitive maps by the manipulation of selective information which, despite the existence of generalized information (maps, written descriptions, etc.), allows each person to choose the references that will help in the marking out of their route. He also explains that the wayfinding process depends on legibility, the ease with which a path becomes known through a pattern of nominations. Moreover according to the same author, "For successful travel, it is necessary to be able to identify origin and destination, to determine turn angles, to identify segment lengths and directions of movement, to recognize on route and distant landmarks, and to embed the route to be taken in some larger reference frame."

In his study of three American cities, Lynch (1960) seeks to point out directions through the network of streets that in new towns may be elements that stand out, but which are not in most European cities. He also proposes orientation through specific buildings (the corner street shop or a striking building) which in big cities, where architecture tends to become uniform, does not provide great guidance data. On the other hand, these data could only be used by the quarter's inhabitants, or regular visitors, as they will not be recognized by those who use the route for the first time.

Mollerup (2005:17) refers to buildings, in several cities, which are so characteristic that become real signaling milestones. However, not every city has buildings such as the Eiffel Tower or the Centre Pompidou in Paris, the Empire State Building in New York, Sydney's Opera House, or even the Guggenheim Museums, so they require other systems in order to facilitate the orientation of their visitors.

15.5 CITY PLANNING

"Cities have been considered in many different ways, in terms of town planning, architectural form, as commercial and social structures, as human organisms and circulation systems. Rarely have they been considered as color compositions. But if we compare one with another—or different parts of one—this aspect becomes obvious" (Taverne and Wagenaar 1992:12). However, there are several urban plans that are concerned with the application of color in the city, and some of them even consider the use of color to indicate a path, or as a mean to solve the orientation problem. Although almost all are focused solely on architecture, not encompassing urban furniture (Figure 15.3).

The first systematized chromatic plan was designed for the Italian city of Turin, between 1800 and 1850, and published by *Consigli degli Edili*. It was constituted by a color palette with 20 colors, which were considered the most frequent in the city, with corresponding code numbers, and the main streets and squares, which had a uniform architecture, were colored in order to show the way to the city center—*Piazza Castelo*—with a chromatic sequence of eight different colors to the back streets. Although it was applied to the built environment, this chromatic plan demonstrated a concern with the orientation within the city and constituted an inspired source to the establishment of chromatic plans in other cities (Figure 15.4).

At the beginning of the twentieth century, Bruno Taut was a precursor in establishing chromatic plans for new urban areas. He believed that every architecture should be chromatic, and that color should be employed to emphasize the city zones spatial dimension, avoiding monotony and, by the color luminosity and different colors effects, allowing the expansion or the understanding of some city areas and contributing to the happiness of its inhabitants. He studied the relations between colors and forms, and between colors and the incident light. For him the main color combinations commandment was the correspondence of color and architectonic elements, in order to conjugate

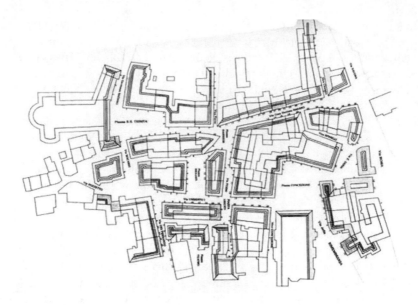

FIGURE 15.3 Chromatic plan for the city of Turin. (Adapted from Noury, L., *La couleur dans la ville*, Éditions Le Moniteur, France, 2008; Linton, H., *Color in Architecture. Design Methods for Buildings, Interiors and Urban Spaces*, McGraw-Hill, New York, 1999.)

FIGURE 15.4 Bruno Taut colors for Argentinish Allee and Uncle Tom's cabin. (Modified from Taverne, E. and Wagenaar, C. (Ed.), *The Colour of the City*, Laren V + K Publishing, The Netherlands, 1992.)

FIGURE 15.5 Barcelona colors. (Modified from Els Colors de L'Eixample 1993.)

space and harmony. However, despite the numerous applications of his concepts, Bruno Taut ideas have not taken root.

Barcelona, by the end of twentieth century, experienced an environment reorganization where color took an organized role. In 1988, Barcelona municipality developed the Barcelona Color Plan Project, in exchange with educational institutions and industry, integrating color schemes on a large scale (Taverne and Wagenaar 1992:92–95) (Figure 15.5).

Moscow is a city that over time often changed its chromatic characteristics, and during the communist era there was no concern with chromatism and, therefore, the city became gray. Since the

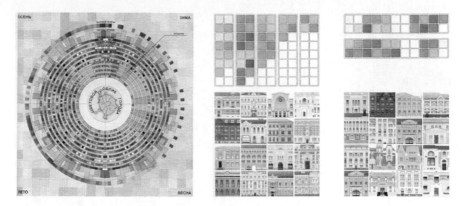

FIGURE 15.6 Chromatic plan for Moscow. (Modified from Noury, L., *La couleur dans la ville*, Éditions Le Moniteur, France, 2008.)

1980s, a new chromatic plan was scientifically projected and applied. The new Moscow color plan is based on a chromatic analysis of its neighborhoods and sets of streets, taking into consideration the city's history and urban structure. The city was divided in three concentric zones: the central historical zone, the intermediate zone, a "shield" area, and the new residential, suburban zone. For each zone a list of recommendations was established with a choice of colors corresponding to the building style palette, which may not correspond with its former color (Noury 2008:98–103) (Figure 15.6).

When referring to city planning, we must study the work of Professor Jean-Philippe Lenclos, who created his own methodology for urban chromatic plans. It was in Japan, in 1961, that Lenclos started to shape his *Color Geography* when he compared the Japanese and French chromatic palettes. This comparison led him to the finding of specific local colors, and it is on the determination of these colors that the methodology developed since 1965 is based, where he selects samples with 25 buildings from the chosen zone, making a systematic collection of all the existent colors and material from the sample (Figure 15.7).

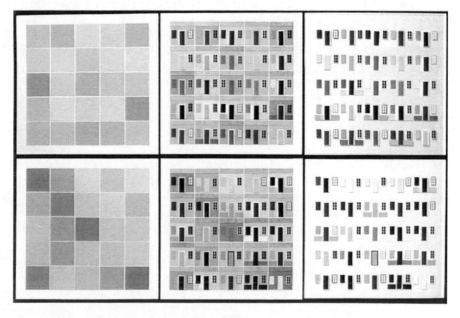

FIGURE 15.7 Codification of recorded samples. (Modified from Porter, T. and Mikellides, B. (Eds.). 2009, *Color for Architecture Today*, Taylor & Francis, Oxon.)

This methodology constitutes an inspiration source to other authors of urban chromatic plans, and Michael Lancaster, Tom Porter, Shingo Yoshida, and Grete Smedal, among others, have applied methodologies similar to the Lenclos' Color Geography when creating their own chromatic plans projects.

In her project for Longyearbyen, Greta Smedal intended to prove the importance of color as an identification factor for the environment, and the chromatic plan was established from the observation of the colors of the natural environment and the characteristic regional light. The found colors were then tuned because, as Greta Smedal alleged, "the language of nature colors can't be directly applied to environmental projects." So the applied colors are a synthesis of the colors found in nature (Noury 2008:84).

Michael Lancaster created chromatic plans for different zones of England, like Ilfracombe in North Devon, the city of Norwich, and the River Thames waterfront. The chromatic plan for Ilfracombe was based on a survey of the environmental colors during the different seasons, and a study of the existing colors in the different parts of town.

On the River Thames project, Lancaster sought to reveal the importance of a policy for color coordination in harmonic compositions in order to prevent disturbances on color applications. With this purpose, the extent and the differences of the various areas of London that the river crosses were taken into consideration, with indications for color application on natural and artificial materials being included. This chromatic plan considers a palette of background colors corresponding to the natural environment, to which were added the building materials colors of the existent industrial complexes and another palette that created a contrast with the background colors (Linton 1999:151–156).

The city of Norwich was chosen as a model for a case study of chromatic plans for European cities by reason of its size, its historical richness, and its location in one of the regions with the highest tradition of using color in the British Isles. In this project Michael Lancaster used cooperation between all stakeholders, including population queries, in order to establish a strategy based on the Lenclos *Color Geography* and on the regional chromatic history, with the aim of achieving a visual coherence for the city and its surroundings.

Tom Porter, when creating chromatic plans, considered color as an adaptable language for a defined context. In his work of reconstituting Oslo's palette, he recorded systematically the existent colors and consulted Harald Sohlberg the painter's work with the intention of evaluating the traditional colors. This palette, with 225 tones, constituted a color guide for Oslo architects.

Porter also created the chromatic plan for the city of Newhall in Essex (1995–2000), where he recorded the existing colors of the land and flora, as well as the specific colors of the nearby villages and farms. This palette was reviewed in 2004, becoming more focused on the colors of building materials and on the mosses that coated the walls (Noury 2008:114–116).

In Tokyo, in 1994, Shingo Yoshida researched the built environment colors in order to understand color distribution and create a chromatic plan, that is the result of a harmonic and subdued blend of this colors. After this first chromatic plan, others were designed to different city zones with the aim of assuring a visual harmony (Linton 1999:146–150).

All these referred urban chromatic plans consider color applied exclusively to architecture, whilst color in urban furniture and signage is only contemplated very specifically. Nevertheless, urban environmental chromatism consists not only of its buildings, as José Aguiar (2002:316) states, "A chromatic study cannot be restricted to the facades of buildings, although these are the most visible elements. There are many other factors that will condition the urban image and, together, create a site specific color."

There are orientation systems, independent from architecture, which are applied in a specific way. Friedman and Thompson (apud Porter and 1976:34) point out an experiment in Boston, where the route around several monuments were marked by red lines—*The Boston Freedom Trail*—converting the confusion of city streets into a pleasant experience. The same authors also mention an intervention from the painter Gene Davies, who decorated the path to the *Philadelphia Museum of Art* with his characteristic colored stripes that acted as a focal point for the museum facade.

FIGURE 15.8 Examples from the *Sentieri Urbani* project. (Modified from Google images.)

A more structured project, planned to optimize the orientation within cities, was the *Sentieri Urbani* designed by Gabriele Adriani. In this project pedestrian walkways were redesigned and transformed with ornamental drawings, different for each city zone, seeking to inform users about zone identity, tradition, history, culture, and commercial activity, without neglecting the technical aspects of ergonomics and road security. This project was presented in 2008 to the city of Turin and was applied in several cities, such as Rome, London, New York, and Tokyo. Despite being innovative, it did not contemplate the use of color that could be a differentiation element of urban areas (Figure 15.8).

In the Japanese city *Shiogama*, the revitalization program combined signage with a chromatic identification system by means of two fundamental elements: color and local characteristic forms. As the city was associated with salt refinement, the chosen form was the salt crystal while color was divided in two components: a background component harmonized with the built environment, and another component linked to primary colors and remembering the city nautical connections. This second component was applied mainly in signage. This system, that was designed by *Maahiko Kimura, gk Graphics*, was first implanted on the port terminal buildings, and then extended to other city zones because it was flexible enough to be adaptable to great variety of architectonic styles (Wildbur and Burke 2001:26) (Figure 15.9).

In Philadelphia a signage system for pedestrians was developed that includes public illumination, pavements, and urban furniture. In this project, from *Joel Katz—Katz Wheeler Design*, the area was divided into five districts, inspired from the William Penn division where the main streets were boundaries. The signage, which is inscribed in big circles with maps, combines symbols with the codified color of each district (Wildbur and Burke 2001:28) (Figure 15.10).

A good example of signage, where color contributes to the identification of city zones, is the project designed by *Rudi Baur* for the city of Lyon (France). In this project a three-color palette

FIGURE 15.9 Signage system for Shiogama. (Modified from Wildbur, P. and Burke, M. 1998. *Information Graphics, Innovate Solutions in Contemporary Design*, London: Thames & Hudson.)

FIGURE 15.10 Signage system for Philadelphia. (Modified from Wildbur, P. and Burke, M. 1998. *Information Graphics, Innovate Solutions in Contemporary Design*, London: Thames & Hudson.)

FIGURE 15.11 Signage system for Lyon. (Adapted from Mollerup, P., *Wayshowing*, Lars Müller Publishers, Baden, 2005.)

shows the river direction with a blue color, while green points to the park, and gray identifies other directions. This signage is completed with another one that shows the points of interest to visit in the different neighborhoods, but without color as an identity value (Mollerup 2005:305) (Figure 15.11).

Considering the several chromatic plans, whether architectural or for signage and urban furniture, only very few apply color as a mean to increase visibility, orientation, and identification, despite the fact that color is a most appropriate tool to achieve these purposes.

15.6 CHROMATIC PLANNING EXISTENT METHODOLOGIES

The concern to establish a coherent urban image through color studies and chromatic plans is relatively recent, despite some pioneer cases, and led to the conception of chromatic planning methodologies, gathering the necessary steps for the selection of a color palette that would constitute the urban image. Urban plans that are concerned with color application to cities, generally employ

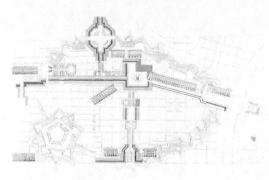

FIGURE 15.12   The successive Turin chromatic plans. (Modified from Taverne, E. and Wagenaar, C. (Ed.), *The Colour of the City*, Laren V + K Publishing, The Netherlands, 1992; Linton, H., *Color in Architecture. Design Methods for Buildings, Interiors and Urban Spaces*, McGraw-Hill, New York, 1999.)

methodologies directly related with the cities different characteristics and are mainly focused in architecture.

In the recovery or restoration of historic cities, the chromatic plans are usually based on file collections that tell the architectonic and chromatic history of the zone or city researched, and on samples of the building coatings extracts, which allow the determination of each building color during their life time. Examples of this methodology application are the well-known chromatic plans of *Turin* and *Barcelon*a (Figure 15.12).

Concerning *Turin*, we may consider three chronologic chromatic plans. The first one was a nineteenth century pioneering plan where the colors along the streets and places were chosen in order to show the ways to the city center—*Piazza Castelo*. This plan was reviewed and developed, between 1978 and 1983, by *Giovanni Brino* who created a data base defining the main chromatic typologies that should be applied to the Turin neoclassic zone. The limitations of the Brino plan led to the necessity of a third chromatic plan—*Progetto-Colore de Torino*—developed by *Germano Tagliaschi and Ricardo Zanetta* at the end of the 1980s. These chromatic plans were very important because, despite considering only architecture, they demonstrated the establishment of urban chromatic plans that were also concerned with the orientation within the city. The chromatic plan for Barcelona, which was conceived and applied by 1992, is important because it is a model for the recuperation of historic cities.

A precursor methodology, was the *Coloroid System* created in Budapest by Antal Nemcsics in 1962. This system's theoretical foundation was based on psychophysiological connotations and historical research of the preferential relations between human beings and color, and the system color parameters were *hue*, *saturation*, and *brightness*, numerically indexed in a tridimensional atlas of 1647 color samples. This system was an innovation in color research in architecture, because it is a subjective evaluation of chromatic plans, by means of determination of objective factors (Figure 15.13).

The chromatic planning methodologies meant to be applied to new cities or zones, which are yet to be built, cannot use historic file collections except for the mention of the region's traditional colors. Also, they only use the colors of the natural environment, and not surveys of existing building colors.

Color Geography is a modern and very complete color methodology designed and developed, since 1965, by Jean-Philippe Lenclos expressly to establish urban chromatic plans. It is focused on the search and definition of specific local colors—the *environmental color*—that may include the survey of the chromatic palette both of the existent materials and local vegetation, in order to create harmonic, or similar, sets which will allow the preparation of chromatic plans, considering color as an adaptive language to a defined context (Figure 15.14).

As Lenclos (1995:86) says, "The color analysis of a site may involve various types of architectural ensembles, on the scale of a country, a region, a city, a city neighborhood, a village, or

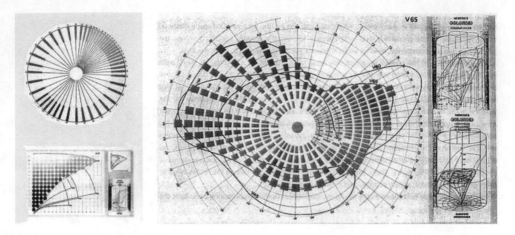

FIGURE 15.13   Coloroid system of Antal Nemcsics. (Modified from Taverne, E. and Wagenaar, C. (Ed.), *The Colour of the City*, Laren V + K Publishing, The Netherlands, 1992.)

a dispersed settlement," and "these studies can be undertaken following orders when they are intended for application on construction or rehabilitation programs, or when creating swatches for new industrial products for the buildings. Otherwise, they are the result of our own research on the 'Color Geography'...."

The application of the methodology starts by the selection of a street, or a collection of representative buildings, that will allow one or more chromatic records of 20 study cases. Whenever it is possible this analysis obeys the buildings numeric order, reproducing the landscape faithfully, and is developed in two phases: in the first one, on the ground, the analysis of the site is made; and in the second phase we make the visual synthesis of ground chromatic records.

Lenclos applies his methodology in two different ways. Whenever the methodology is intended to be applied to architectonic groups that are yet to be built, the research is based on local environment color, making a survey of the chromatic palette of flora and existing materials to create a set that harmonizes or integrates its environment. In contrary, in chromatic plans for industrial sites, Lenclos uses colors that stand out from the surrounding environment, creating real architectural sculptures.

As it was stated before, the methodology developed by Jean-Philippe Lenclos was an inspiration source to other colorists. One of them, Michael Lancaster calls his chromatic plans

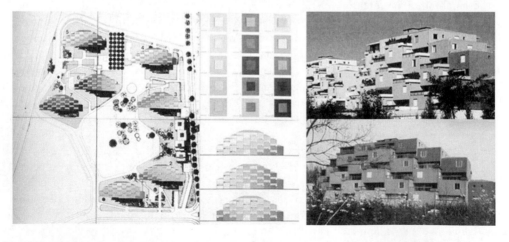

FIGURE 15.14   Lenclos chromatic plan for Château Clos urbanization. (Modified from Linton, H., *Color in Architecture. Design Methods for Buildings, Interiors and Urban Spaces*, McGraw-Hill, New York, 1999.)

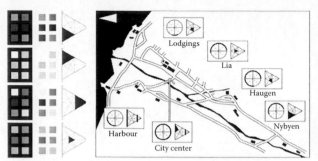

FIGURE 15.15   Greta Smedal chromatic plan for Longyearbyen. (Modified from Porter, T. and Mikellides, B. (Eds.). 2009, *Color for Architecture Today*, Taylor & Francis, Oxon.)

methodology—*Strategy*—because, as he says, the proceeding of chromatic planning evolves a great number of people and it must be, simultaneously, sufficiently prescriptive to attain the objectives, and sufficiently flexible to adapt to changes of use, occupation, structure of the buildings, fashion, and taste. As he says, "A colour strategy implies not only a plan but also the establishment of procedures—working methods that are both practical and economically viable. Above all it depends upon the prediction and communication of an acceptable vision of the future" (Lancaster 1996:88).

The methodology applied by Greta Smedal to her chromatic plan for Longyearbyen is similar to the one applied by Michael Lancaster and contemplates the following steps: the record of existing colors, analysis of possibilities and ambits, development of a global chromatic concept, presentation to the client or public, detailed description of the color plan, implementation. In this project, the existing colors were observed under light variations, throughout the day and year, and the environmental colors were recorded along the different climatic seasons. All the colors were recorded using the *NCS—Natural Color System* (Figure 15.15).

15.7   NEW CHROMATIC METHODOLOGY FOR URBAN PLANNING

This methodology arises from a PhD research which defined the need to create a new methodology for urban furniture color planning, which will make it possible to create color plans for urban environments, allowing urban furniture to stand out from its background, contributing to their better legibility, and transforming them into identification elements that will improve orientation within cities. The development and implementation of this new methodology will allow the determination, with a higher scientific approach and rigor, of the color planning to be applied to urban furniture in each district or urban area, of a city. The present research project is focused on Portuguese cities, with different characteristics, applying the methodology in development, and establishing as a result color plans that can be applied whenever there is a need to design urban furniture chromatic plans.

The new methodology applies an extensive direct observation to the case studies, with the use of mechanical devices, including photographic mapping of both urban furniture and signage, in order to evaluate their visibility and legibility, as well as their color applications (Figure 15.16).

For each urban area, and to facilitate the study, a sample area is to be defined, including the main streets and places and, also, some secondary ones, with the intention of encompassing the most representative zones, those with specific characteristics. Along the chosen area, an exhaustive record of all the environmental colors is made, including material samples not only from buildings, but also from pavements, vegetation, and any additional elements present with a relative permanence in the urban environment—*the nonpermanent colors*—that must be taken into account for the spatial chromatic readings, which are then classified using the NCS, that was chosen because it allows the easy identification of every color, even when they are located out of reach, and without needing additional equipment. It must be underlined that the recorded colors are *perceived colors*, not always coincident with the *inherent colors* (the real colors belonging to pigments and materials)

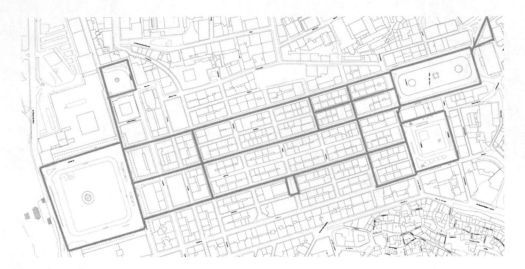

FIGURE 15.16   Example of a quarter sample area.

and that the perceived colors may, also, be a partitive synthesis, particularly in the case of vegetation and tile-coated walls (Figure 15.17).

Among the environmental colors we must take into account the percentage of which the sky color will interfere with the urban area color and, also perceptive factors related with color interactions, as well as geographic and atmospheric conditions and the chromatic variations along the different climatic seasons. With this purpose, the palette is tested along the seasons' changes to judge the chromatic alterations that arise from the different colors of the vegetation as well as day light variations, and sky colors according to weather changes to evaluate the pertinence of the chromatic plan (Figure 15.18).

These collections are completed by photographs of environmental elements and panoramic views from different blocks, using urban plans, architectural elevations, and sections of the selected paths as well, which act as elements of the environmental color components. All these records are methodically indexed in forms and maps, previously designed and tested, which allow the creation of a data base guided by scientific rigor, in order to determine a chromatic palette for each quarter, or urban area and, consequently, to establish a coherent chromatic plan that may be applied to urban furniture.

In order to guarantee the scientific rigor of the determination of each quarter's chromatic plan, we consider the dominant colors, proportionally represented, choosing colors for the urban furniture which may establish an adequate chromatic and luminosity contrast with the dominant colors and, also, respect the traditions, culture, identity, and history of the quarter. These contrasts must be

FIGURE 15.17   (See color insert.)   Examples of recorded samples from buildings, pavements, vegetation, and occasional elements.

FIGURE 15.18   (See color insert.)   Examples of sky color percentage in different urban areas.

observed under the possible variations of local illumination, in order to be sure that they accomplish their functions efficiently.

The urban furniture chromatic plan, which will be different for every quarter, must stand out from the environment, contributing to a better legibility and identification of these elements and, in the same way, will become a city's area identification element which may be used in different supporting areas and, in this way, facilitate the orientation and wayfinding within the city.

15.8   CONCLUSIONS

With this project we aim to define and underline the importance of color application to urban furniture, taking into consideration that a pertinent chromatic plan can contribute to better visualization and, consequently, turn urban furniture into an ergonomic factor, contributing to a better utilization of its elements and, simultaneously, improving the orientation within the city and identifying its different zones. We expect that this methodology, which establishes the importance of a pertinent and structured color application to urban furniture, will contribute to the enlargement of the perspective of urban chromatic plans, allowing them to become more holistic and comprehensive.

This project's empirical phase will focus on some case studies, where we want to implement the use of color plans to urban furniture as a strategy to achieve a better and inclusive design project, optimizing the visibility and use of this equipment, contributing to the identification of city quarters and the orientation of users. The case studies include three municipalities around Lisbon—Barreiro, Loures, and Oeiras—and, in each one of these were chosen three settlements with different specifications. For each settlement sample areas will be defined which will encompass their most representative zones where the new methodology will be applied to all furniture elements, in order to validate the results and the new color planning.

In addition to the inclusion of all the environmental colors, being they from architecture, vegetation, the sky, and all other elements that constitute urban spaces, this methodology takes into account perceptive factors related with color interactions, as well as geographic and atmospheric conditions. In consequence, urban chromatic plans will gain a higher scientific approach and rigor.

We also aim to establish a color plan which may contribute to differentiate each city quarter, respecting the local history and symbolism, and achieving a good contrast with the environment.

Being a research project there is a need to evaluate established color plans. As evaluation methodology we will constitute focus groups composed of city inhabitants of different ages and gender, experts on color application, municipality technicians, such as architects, urban planners, equipment designers, landscape architects, managers, engineers, etc.

The focus groups will interact with an indoor and an outdoor presentation. The first evaluation will consist of an indoor presentation of the guidelines for the elaboration of the chromatic plans, and the focus groups will discuss and evaluate the pertinence of the color choice for urban furniture. In the outdoor presentation the focus groups will be confronted with these color applications in predetermined city areas. Their feedback, their contribution, will be incorporated in the color plans model. We expect that they will recognize the importance of taking into account the chromatic programs of urban furniture in urban color plans, as a way of inclusive design implementation and city zones differentiation.

REFERENCES

Águas, S. 2003. Urban furniture design: A multidisciplinary approach to design sustainable urban furniture, MSc dissertation, University of Salford, Salford.

Aguiar, J. 2002. *Cor e cidade histórica, Estudos cromáticos e conservação do património*, Porto: FAUP publicações.

Baines, P. and Dixon, C. 2003. *Signs, Lettering in the Environment*, London: Laurence King Publishing Ltd.

Barker, P., Barrick, J., and Wilson, R. 1995. *Building Sight: A Handbook of Building and Interior Design Solutions to Include the Needs of Visually Impaired People*, London: Royal National Institute for the Blind (RNIB).

Berger, C. 2005. *Wayfind: Designing and Implementing Graphic Navigational Systems*, Switzerland: Rotovision SA.

Brown, R. (Ed.). 1998. *Quality of Life for People with Disabilities*, UK: Stanley Thornes (Publishers) Ltd.

Costa, J. 1987. *Señalética: de la señalización al diseño de programas*, Barcelona: Ediciones CEAC, SA.

Disabled Persons Transport Advisory Committee (DPTAC). 2003. *Inclusive Projects: A Guide to Best Practice on Preparing and Delivering Project Briefs to Secure Access*. London. HMSO. http://www.dptac.gov.uk/inclusive/guide/index.htm (accessed May 10, 2013).

Durão, M. J. 2002. Colour in the built environment. *Fabrikart: Arte, Tecnologia, Industria, Sociedad*, 2: 162–169.

Golledge, R. 1999. *Wayfinding Behavior—Cognitive Mapping and Other Spatial Processe*, Baltimore: The John Hopkins University Press.

I'DGo—Inclusive Design for Getting Outdoors. 2012. http://www.idgo.ac.uk/pdf/Intro-leaflet-2012-FINAL-MC.pdf (accessed March 06, 2013).

Lancaster, M. 1996. *Colourscape*, London: Academy Editions.

Lenclos, J. 1995. *Couleurs de l'Europe. Geographie de la couleur*, Paris: Publications du Moniteur.

Lindemann, G. et al. (Eds.). 2004. *Regulated Agent-Based Social Systems*, Germany: Springer.

Linton, H. 1999. *Color in Architecture. Design Methods for Buildings, Interiors and Urban Spaces*, New York: McGraw-Hill.

Lynch, K. 1960. *A Imagem da Cidade*, Lisboa: Edições 70.

Machnow, H. and Reuss, W. 1976. *Farbe im Stadtbild*, Berlin: Abakon Verlagsgesellschaft mbH.

Minah, G. 2005. Memory constellations: Urban colour and place legibility from a pedestrian view, in *AIC Colour 05—10th Congress of the International Colour Association*, pp. 401–404. Granada, Spain.

Mollerup, P. 2005. *Wayshowing*, Baden: Lars Müller Publishers.

Noury, L. 2008. *La couleur dans la ville*, France: Éditions Le Moniteur.

Porter, T. and Mikellides, B. (Eds.). 1976. *Colour for Architecture*, London: Studio Vista.

Porter, T. and Mikellides, B. (Eds.). 2009. *Color for Architecture Today*, Oxon: Taylor & Francis.

Sentieri Urbani. http://www.google.com/images?source=ig&hl=en&rlz=1G1TSEFCENUS341&q=sentieri+u rbni&btnG=Google+Search&oq=sentieri+urbani&aq=f&aqi=&aql=&gs_sm=s&gs_upl=319311831010 121133115115101510101297l180911.5.4110&oi=image_result_group&sa=X (accessed June 20, 2011).

Serra, J. M. 2000. *Elementos urbanos, mobiliário y microarquitectura*. Barcelona: Editorial Gustavo Gili.

Serviço de comunicação e imagem—INE. 2014. *Destaque, Informação à comunicação social*. www.ine.pt (accessed June 16).

Taverne, E. and Wagenaar, C. (Ed.). 1992. *The Colour of the City*, The Netherlands: Laren V + K Publishing.

Tosca, T. 1994. Dreams of light for the city. *Color Research and Application*, 19: 155–170.

Wildbur, P. and Burke, M. 1998. *Information Graphics, Innovate Solutions in Contemporary Design*, London: Thames & Hudson.

16 Design Requirements for a Spectacle-Type Device in Rapid Visual Referencing

Daigoro Yokoyama, Takahiro Uchiyama, Yusuke Fukuda, Miyuki Yagi, and Miwa Nakanishi

CONTENTS

16.1 INTRODUCTION

With the recent expansion of the mobile device market, increasing demand for spectacle-type wearable displays (SWDs) is expected. Users of SWDs operate chiefly with their normal vision and can readily retrieve information when required. Supporting information, such as work procedures, should be continuously available to industrial workers. Therefore, in this study, we have designed a reference form that enables the user to shift his viewpoint from the main eyesight to the ancillary information presented by a monocular see-through head-mounted display (HMD) within a short time. In this system, the user needs to only "glance" at the necessary ancillary information (hereafter, we refer to our form as "short reference at any time").

Conventionally, material can be referenced at any time from paper or stationary information terminals. However, several studies have suggested that replacing these terminals with HMDs will improve work efficiency (Caudell and Mizell 1992; Nakanishi et al. 2007; Tanuma et al. 2012), chiefly because the ancillary information is consistently presented in the same field of view of the work object. Therefore, viewpoint movement while referencing the material can be saved. On the other hand, ancillary information may be cluttered by irrelevant information captured in the field of view, which may be problematic (Nakano et al. 2006).

We can consider that saccade eye movement occurs during short-time HMD referencing at any time. The time required for saccade motion, when changes in the dynamic characteristics occur according to various conditions (Van Gisbergen et al. 1981), generally depends on the distance and direction (Westheimer 1954; Ebisawa and Ono 1997). From this knowledge, we can consider that when the reference object is positioned at or near the center of the visual field, the efficiency of short-time referencing will be enhanced. Given that retinal ganglion cells are most densely packed at the fovea (Curcio et al. 1990), users acquire large quantities of visual information at the center of their visual field (Watabe and Sakata 1975). Therefore, although the ancillary information is rendered more obvious when centralized in the visual field, the advantages are offset by the visual complexity if the presented objects do not require a reference. Therefore, we consider that a trade-off exists between the efficiency of short-time referencing and complexity of the visual field. This trade-off implies an optimal distance that is offset from the center of the visual field. In fact, some commercially available HMDs are designed to avoid the central visual field when the user views an image, although the designs vary among manufacturers. To date, the best location for video presentation has not been investigated.

Therefore, in this study, assuming that a HMD is adopted for short-time referencing at any time during work, we hypothesize that the optimal position for presenting auxiliary information is the equilibrium point where the reduction in the referencing efficiency is exactly offset by enhanced simplicity of the visual field. We experimentally verify this hypothesis and propose a design.

16.2   METHOD

16.2.1   EXPERIMENTAL TASK

During the experiment, it was proposed that users should glance at supporting information displayed on the HMD only while operating on real targets in a work space. First, the subjects were seated in front of a 23-inch display (Diamond Crysta RDT 23IWM, Mitsubishi). Using their normal vision, they were requested to chase an object moving across the display with a three-dimensional (3D) input device (Phantom Omni, 3D Incorporated). Among six objects randomly moving across the display, the subjects chased objects of specified colors and shapes using 3D directions. Figure 16.1 illustrates a typical display view during the task. The elements that the subjects controlled by operating the 3D input device are displayed. The subjects were required to retain the tip of the operating element at the center of the tracked object. The six objects were presented in different colors (green, yellow, white, and red) and shapes (sphere, cube, and cone). The chased object was made to disappear once the subjects had referenced the HMD (AirScouter, Brother) display for a specified time. The colors and shapes were always enumerated as text in the HMD. After reading

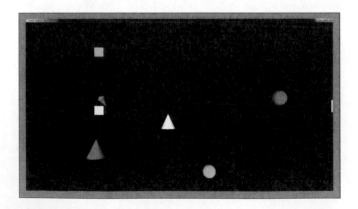

FIGURE 16.1   Image of an actual field of view.

FIGURE 16.2   Text content displayed on the SWD.

these highlighted word sets, the subjects chased the specified object. Figure 16.2 is an example of the strings that were permanently displayed on the HMD and presented in the field of view of the subject. The font size, decided from the results of previous studies (Caudell and Mizell, 1992), was 0°25′46″, which was easily read by the subjects. The strings comprised 48 symbols describing color and shape combinations, one of which is highlighted in the red frame in Figure 16.2. The sequence and highlighted symbol set was switched every 15 s so that the subjects could not remember them during tracking tasks. During reference to the HMD, the target object was wiped from the screen; for example, if the highlighted symbol set was "red Δ," the subjects recognized the red cone as the next target object. To ascertain that subjects had correctly read the symbol set, they were requested to verbally state the recognized target object. If the target object became lost during a task, the 3D input device was automatically locked. Thus, we could clearly distinguish whether the subject had faced the tracking task or had stopped tracking and consulted the HMD. In follow-up tasks, the subject pressed the unlocking button at the time of reading the symbol set and resumed operation. The time of a single task was 1 min 40 s, with five disappearances of the object. The subjects were requested to repeatedly interrupt the tracking task, read the information presented by the HMD, and resume tracking. The flow of a single task is shown in Figure 16.3.

16.2.2   Experimental Environment

Figure 16.4 shows the arrangement of the experimental apparatus and subjects. The viewing distance from the subjects to the 23-inch display (simulating the actual field of view) was 100 cm. We attached an eye mark recorder (EMR-9 NAC) to the subjects' head to measure their eye movement. In addition, to ensure that the center of the visual field of the subject matched the center of the display, we adjusted the height of the chair. The viewing angle of the information presentation area of the display was 28°34′5″ in the horizontal direction (unilateral 14°30′34″) and 16°17′55″ in the vertical direction (unilateral 8°11′26″), as shown in Figure 16.5. The average interior luminance during the experiment was 3.22 lx.

16.2.3   Participants

Twenty-four adults (average age: 21.3 ± 1.29 [SD] years; range: 19–24 years) with no vision problems participated in the study. Because all subjects were right-eye dominant (evaluated by the hole-in-card test), the HMD was mounted at the left eye side, as reported in previous studies (Collewijn et al. 1984).

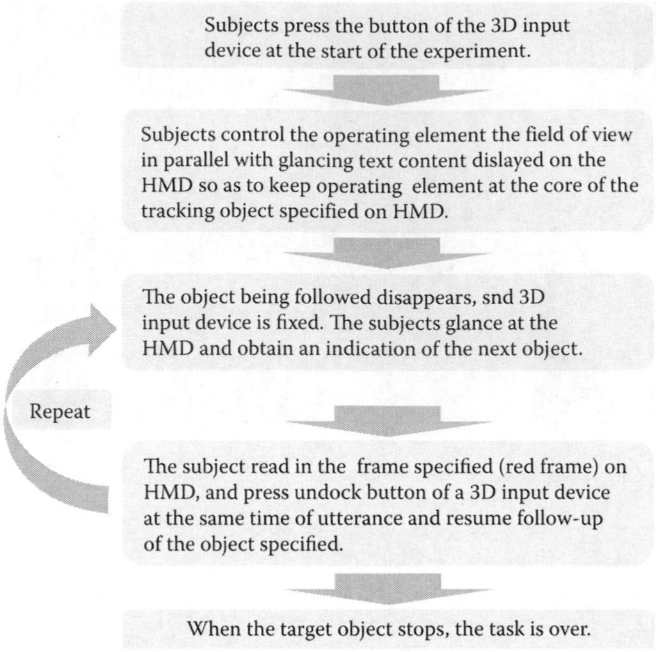

Subjects press the button of the 3D input device at the start of the experiment.

Subjects control the operating element the field of view in parallel with glancing text content dislayed on the HMD so as to keep operating element at the core of the tracking object specified on HMD.

The object being followed disappears, snd 3D input device is fixed. The subjects glance at the HMD and obtain an indication of the next object.

Repeat

The subject read in the frame specified (red frame) on HMD, and press undock button of a 3D input device at the same time of utterance and resume follow-up of the object specified.

When the target object stops, the task is over.

FIGURE 16.3 Flow of a single task.

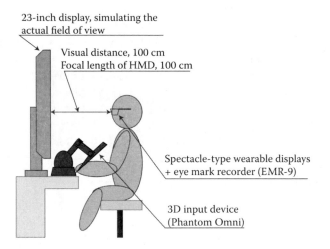

23-inch display, simulating the actual field of view

Visual distance, 100 cm
Focal length of HMD, 100 cm

Spectacle-type wearable displays + eye mark recorder (EMR-9)

3D input device (Phantom Omni)

FIGURE 16.4 Experimental environment.

16.2.4 EXPERIMENTAL CONDITIONS

Positional information by the HMD was presented as eight patterns mimicking the information-receiving characteristics of human vision. This information comprised four patterns ([15°, 0°], [8°, 0°], [−10°, 0°], [−40°, 0°]) in the vertical direction (Figure 16.6) and three patterns ([0°, 15°], [0°, 30°], [0°, 50°]) in the horizontal direction (Figure 16.7), where the center of the visual field is (0°, 0°). Viewing directions were changed by repositioning the HMD adjuster in the vertical direction and the HMD frame in the horizontal direction. In each condition, the viewing angle of the information presented on the HMD was fixed (at ~14°24′ × 10°48′).

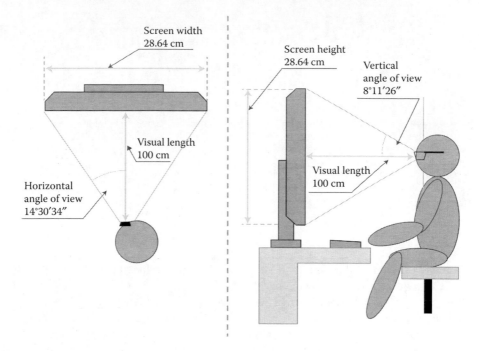

FIGURE 16.5 Viewing angle to the display.

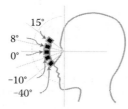

FIGURE 16.6 Vertical positions of the presented image.

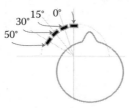

FIGURE 16.7 Horizontal positions of the presented image.

To properly evaluate the efficiency of the experimental performance and to reduce the effect of the viewing sequence, the experiment proceeded through the following steps (Figure 16.8). Prior to an experimental run, the subjects repeatedly practiced the tracking task. We confirmed that increasing proficiency did not alter performance accuracy. The information was then presented to the subjects at different vertical positions. Each subject performed the task three times for each of the five vertical viewing patterns (the standard condition [0°, 0°] and the four vertical patterns described above) in a random order. To offset the order effect, the random order of the five viewing patterns was varied in each of the three trials. The above procedure was then repeated for the four horizontal

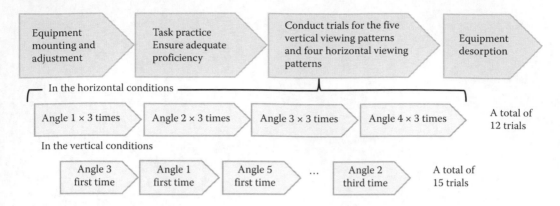

FIGURE 16.8   Experimental procedure.

viewing patterns (the standard condition [0°, 0°] and the three horizontal patterns described above). To offset the order effect, we took a counter balance between the 24 subjects and specified a trial order. In this procedure, the effect of ordering was corrected in the horizontal and vertical directions, although an order effect may have been introduced by viewing from left to right or vice versa. Therefore, we incorporated the standard condition (0°, 0°), which alters the position of information presented in both vertical and horizontal directions, providing a reference during analysis.

16.2.5   MEASUREMENTS

First, to evaluate the efficiency of referencing by the HMD, we measured the time between disappearance of a previously tracked object and the tracking of the next object. Specifically, we recorded the time between automatic locking of the 3D input device (as the $n - 1$th object disappeared) and the unlocking of the device to begin tracking the nth object. Although not all of this time was expended in reading the HMD information; time not spent consulting the HMD was assumed equal under all experimental conditions. Secondarily, the accuracy of the tracking task was considered to indicate the complexity of the visual field when redundant information was presented on the HMD. Specifically, we recorded the distance between the tip of the operator and the center of the tracked object (deviation). Finally, we recorded the saccade distance of the subjects in this task. This is because we needed to verify if the time between disappearance of a previously tracked object and the start of tracking the next object was correlated to the saccade distance.

16.2.6   ETHICS

All participants provided informed consent. Data were encrypted to prevent identification.

16.3   RESULT

16.3.1   REFERENCING EFFICIENCY

The time of referencing the information presented on the HMD was compared among the viewing patterns. Figures 16.9 and 16.10 show the time required for referencing the HMD at different horizontal and vertical positions, respectively. The subjects required significantly more time to reference information at (0°, 50°) than at closer horizontal angles. In the vertical direction, the referencing time was statistically identical at (−10°, 0°), (0°, 0°), and (8°, 0°), but was significantly extended at (−40°, 0°) and (15°, 0°). These results support our hypothesis that the referencing efficiency declines

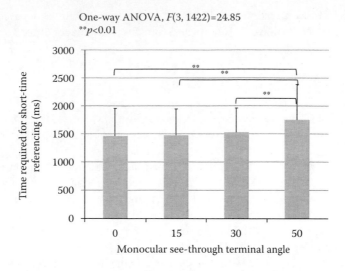

FIGURE 16.9 Time required for short-time referencing of HMD information presented at different horizontal angles.

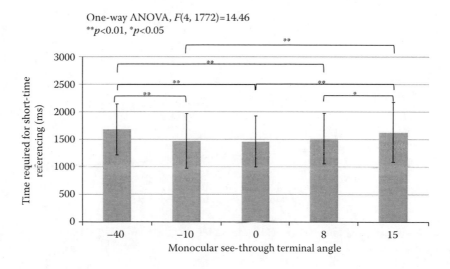

FIGURE 16.10 Time required for short-time referencing of HMD information presented at different vertical angles.

as the information presented by the HMD deviates from the center of the visual field. Furthermore, based on the viewing angle to the display (see Figure 16.5), the viewpoint of the subject during the tracking task shifted by up to 15° in the horizontal direction (unilaterally) and up to 8° in the vertical direction (unilaterally). This result suggests that the referencing efficiency is generally preserved when the information on a HMD is presented within the range of movement of users' actual fields of view.

16.3.2 Complexity of the Visual Field

To evaluate the accuracy of the tracking task, we integrated the distance between the center of the follow-up object and the tip of the operator recorded for each subject at a given information presentation position. Because individual differences were observed in the accuracy of the tracking task, the data were first normalized as follows and compared among conditions:

$$Z = \frac{x - \mu}{\sigma},$$

where x, μ, and σ are the integrated values, average integrated value, and standard deviation of the integrated values, respectively, at each position of the information presented for each subject and Z is the normalized integrated value.

Figures 16.11 and 16.12 show the normalized integrated tracking deviation at each presented position in the horizontal and vertical directions, respectively. Smaller deviations imply higher tracking performance. Figure 16.11 indicates that tracking is considerably more accurate at (0°, 0°) than at (0°, 15°). In the vertical direction, although the differences were not statistically significant, tracking was least accurate at (−10°, 0°) and relatively high at (0°, 0°) and (8°, 0°). These results suggest that the complexity of the visual field relaxes when the information presented by a HMD is viewed at 15° from the center of the visual field (0°, 0°). As the information is presented further from the center of the visual field, no further improvement in the complexity of the visual field occurs. Comparing these results with those of the previous section, we can infer that task performance was

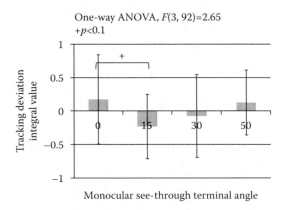

FIGURE 16.11   Normalized integrated tracking deviation for HMD information presented from different horizontal angles.

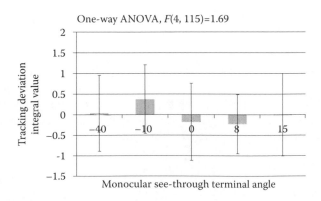

FIGURE 16.12   Normalized integrated tracking deviation for HMD information presented from different vertical angles.

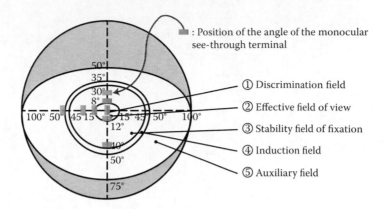

FIGURE 16.13 Information-receiving characteristics of the human eye.

degraded by the decreased efficiency of short-time referencing at wider angles. Supporting this inference, the complexity of the visual field was unaltered in the vertical direction, that is, at (−40°, 0°) and (15°, 0°). Moreover, the complexity of the visual field became relaxed when information was presented at 8° above the center of the visual field, but worsened at the same position below the center of the visual field (at −10°). We suggest that the human eye is designed to receive more information from the depression angles than from the elevation angles, relative to the center of the visual field (see Figure 16.13). Therefore, we consider that the mid-range vertical direction (−10°, 0°) is related to tracking performance. As mentioned in the previous section, during tracking, the subject ranges his real field of view through approximately 15° in the horizontal direction (unilaterally) and approximately 8° in the vertical direction. Therefore, HMD information presented at (−10°, 0°) is unlikely to inhibit the tracking of the visual target. However, when referencing the HMD from the mid-range horizontal and vertical directions, users may have experienced discomfort in their field of view, with consequent reduction in performance. In fact, after completing the experiment, many subjects reported "obstructive" and "in the middle" as their experiences of referencing the HMD from (−10°, 0°).

16.3.3 EYE MOVEMENT

To evaluate the eye movement of the tracking task, we compared the average saccade distance that occurred when the specified object was replaced by a new one. In this study, depending on the definition of the viewpoint that is "the eyes' speed is less than 5°/s," the definition of saccade is "it is that subjects look the same position of less than 1 frame" considering frame rate of the analyzer (62.5 fps). However, we do not describe the result (−40°, 0°) because we could not get the data in this position overlapping with the camera position of the eye mark recorder and the lens of the HMD. The data were first normalized as follows and compared among the conditions because individual differences were observed in the saccade distance:

$$Z = \frac{x - \mu}{\sigma},$$

where x, μ, and σ are the integrated values, average integrated value, and standard deviation of the integrated values, respectively, at each position of the information presented for each subject and Z is the normalized integrated value.

In Figure 16.14, we showed normalized values of the average saccade distance in each position with information presented in the horizontal direction. In Figure 16.15, we showed this in

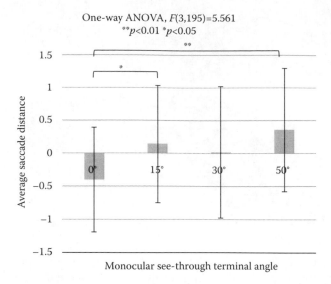

FIGURE 16.14 Normalized average saccade distance for HMD information presented from different horizontal angles.

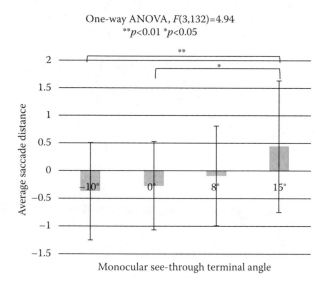

FIGURE 16.15 Normalized average saccade distance for HMD information presented from different vertical angles.

the vertical direction. It should be noted, the smaller values imply shorter average saccade distances. From the results, (0°, 15°) and (0°, 50°) the average saccade distance was significantly increased compared with (0°, 0°) in the horizontal direction. As stated in the section "Referencing Efficiency," these results roughly explained that the time of referencing the information is extended as the information presented by the HMD deviates from the center of the visual field. But according to the results in the section "Referencing Efficiency" (see Figure 16.16), unlike the results in Figure 16.14. (0°, 50°) is stated that the time of referencing the information is significantly extended when compared with an angle other than (0°, 0°). Regarding the function of the eye mark recorder, it was impossible to capture the eyes with a camera attached to the eye mark

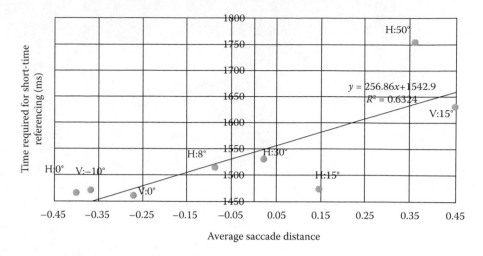

FIGURE 16.16 Relationship of the time required for short-time referencing and average saccade distance.

recorder when the eyes moved to the horizontal angle 50°. So, the cause was considered that it could not obtain data of the saccade distance and there was a difference between the measured value and the actual saccade distance. Also (0°, 0°) and (−10°, 0°) average saccade distance was significantly decreased compared with (15°, 0°) in the vertical direction. The reason was the same as the horizontal direction noted above. Also, unlike the results in section "Referencing Efficiency," there is no significant difference in (15°, 0°) and (8°, 0°). The reason was the same as the horizontal direction that was caused by the function of the eye mark recorder. To show this relationship clearly, we took the average saccade distance on the vertical axis and the time required for short-time referencing on the horizontal axis and summarized it in Figure 16.16 as a scatter plot. The time required for short-time referencing was linearly related to the average saccade distance as shown by the value of R^2 which is 0.63. It is a general tendency that the longer the saccade distance is, the longer time is needed. Also from this analysis, the reasons for the results shown in Figures 16.14 and 16.15 were clarified.

16.4 STUDY OF THE OPTIMAL POSITION OF THE INFORMATION PRESENTED IN SHORT REFERENCE AT ANY TIME

The previous section analyzed the complexity of the visual field and the referencing efficiency of users glancing at a HMD. A trade-off between referencing efficiency and viewing complexity was not confirmed but was suggested. Therefore, we incorporated both factors into a total evaluation index and attempted a comprehensive determination of the optimal position for HMD information presentation.

First, as mentioned in Section 16.2.4, our experimental procedure could not exclude an order effect between left–right viewing and up–down viewing of the presented information. Therefore, the reference condition (0°, 0°) was incorporated in both horizontal and vertical viewing patterns. Here, we examine whether the order effect exists and (if present) to what extent it influences the outcome. To this end, we assume that differences under the same conditions are wholly ascribed to the order effect, and subtract the difference from the results obtained at each horizontal position of information presentation. The corrected values are given by the following equation:

$$V_{all} + (h_0 - V_{-10}) = new V_n,$$

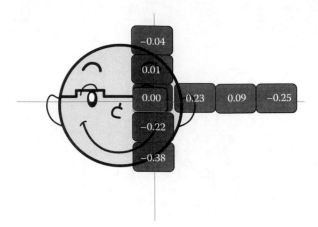

FIGURE 16.17   Evaluation index computed for each position of information presentation.

where V_{all} and $newV_n$ denote the complete and corrected data, respectively, in the vertical direction. V_n and h_n are the average values of n data in the vertical and horizontal directions, respectively and S_{all} is the standard deviation of all data (corrected by the vertical data).

The corrected data were then normalized as follows. The field-of-view complexity index and the referencing efficiency were scaled using the standard condition (0°, 0°).

$$\frac{(h_0 - x_n)}{S_{all} = newx_n},$$

where x_n and $newx_n$ are the data and corrected data, respectively.

Following the above procedures, we obtained the weighted sum of 1:1 for the field-of-view complexity index and the referencing efficiency. We defined the inverse of this value as the total evaluation index. The better the position is for information presentation, the value of this index is larger.

Figure 16.17 shows the total evaluation index at each position of the presented information. The highest evaluation was obtained at (0°, 15°). On the other hand, at (0°, 50°) and (−40°, 0°), which are significantly far from the center of the visual field, and at (−10°, 0°), which is relatively close to the center of the visual field (0°, 0°), evaluation indexes were low. From these results, we infer that the complexity of view is mitigated without compromising the referencing efficiency when the HMD is horizontally positioned at 15° from the center of the visual field (0°, 0°). This result suggests that an optimal position exists for presenting information on a HMD that can be rapidly accessed by workers.

16.5   CONCLUSIONS

Assuming that a HMD is available for short-time referencing at any time during working hours, we have focused on the complexity of the visual field and the referencing efficiency. We expect that a trade-off exists between these two factors. Therefore, we experimentally determined the optimal position for presenting the HMD information that mitigates the complexity of the visual field while preserving performance accuracy. The optimal position was found to be intermediate between the periphery and center of the visual field. Specifically, a horizontal shift of 15° from the center of the visual field yielded the highest evaluation score. As mentioned in the introduction, rapid referencing of continuously accessible information is a distinct advantage of monocular see-through terminals in industrial applications. The proposal of this study could be adopted in guidelines for terminal designs. However, our approach requires further development. When constructing a comprehensive evaluation index of the field-of-view complexity and referencing efficiency, we weighed both factors equally. This weighing may change with the perceived importance of the visual target and the frequency of viewing. The weighing of ancillary information may also depend on the actual field of view. The appropriateness of defining an optimal position for information presentation will form part of our developmental research.

Wearable terminals have been newly introduced to the market (Fujiwara, 2014), and a wide range of terminals and applications are expected. We propose that ergonomics will play an important future role in the use and design of comfortable and versatile wearable terminals.

REFERENCES

Caudell, T. P., Mizell, D. W. 1992, Augmented reality: An application of heads-up display technology to manual manufacturing processes, *Proceedings of the Twenty-Fifth Hawaii International Conference on System Sciences*, Hawaii, USA, Vol. 2, pp. 659–669.

Collewijn, H. et al. 1984, Human smooth and saccadic eye movements during voluntary pursuit of different target motions on different backgrounds, *The Journal of Physiology*, 351, 217–250.

Curcio, C. A., Sloan, K. R., Kalina, R. E., Hendricson, A. E. 1990, Human photoreceptor topography, *Journal of Comparative Neurology*, 292, 497–523.

Ebisawa, Y., Ono, R. 1997, Variability in saccadic dynamics under visual stimulus and instruction conditions, *The Institute of Image Information and Television Engineers*, 51(7), 1106–1113.

Fujiwara, N. 2014, The first year of wearable devices, *The Sankei Shimbun* (Japanese newspaper), February 17, p. 27.

Nakanishi, M., Ozeki, M., Akasaka, T., Okada, Y. 2007, Human factor requirements for applying augmented reality to manuals in actual work situations, *The 2007 IEEE International Conference on SMC (Systems, Man, and Cybernetics)*, Montreal, Canada, on CD-ROM.

Nakano, M., Odagiri, S., Mori, H., Isono, H. 2006, Comparison of PC assembly work using monocular see-through HMD and instruction manual, *Japan Ergonomics Society*, 42, 366–367.

Tanuma, K., Nomura, M., Nakanishi, M. 2012, Effect of the angle of view of a monocular see-through HMD on ease of getting information, *Proceedings of the Annual Meeting of Japan Ergonomics Society*, Fukuoka, Japan, Vol. 48spl, pp. 412–413.

Van Gisbergen, J. A., Robinson, D. A., Gielen, S. 1981, A quantitative analysis of generation of saccadic eye movements by burst neurons, *Journal of Neurophysiology*, 45(3), 417–442.

Watabe, E., Sakata, H. 1975, *Science of Vision* (book in Japanese), Shashin-Kogyo Shuppan, pp. 30–32.

Westheimer, G. 1954, Mechanism of saccadic eye movements, *Archives of Ophthalmology*, 52, 710–724.

17 Ergonomic Design Thinking
A Project Management Model for Workplace Design

Marcello Silva e Santos and Marcelo M. Soares

CONTENTS

17.1 INTRODUCTION

Design thinking may be seen by some as a new fad, a buzzword used as a marketing strategy to achieve some sort of competitive advantage. In fact, the term defines a concept that has ties to both theoretical and practical multidisciplinary themes, such as participatory project, simultaneous engineering, and others. But its originality is not exactly related to its project management adherence, but to the ideation mindset especially present in product design approaches. The idea is to apply the kind of thinking designers employ when solving problems. Instead of Cartesian and straightforward—in terms of sequential actions—ways of sorting out alternatives and merging them into general solutions, designers usually tend to reason in an iterative, holistic manner, not only in innovation projects but also in a broader range of actions. Thus, according to design thinking theory, we can apply innovative schemes and creative actions in the course of carrying out any kind of project, or even business ventures.

Design thinking presupposes a multiphase and nonlinear process known as fuzzy front end, allowing for constant interaction and learning, associated with another nonconventional decision-making process called abductive thinking. Abductive thinking has to do with the way designers formulate inquiries through the apprehension or comprehension of a given phenomenon, when questions and answers derive from information gathered from observation of a real context surrounding the problem. An article in the *Harvard Business Review* by Tom Brown (2008) outlines Design thinking applications. According to the author, what is envisioned is not particularly an invention or innovation process, but an entire system evolving around it. According to him, this would really make a difference because first we set up a marketplace, then a setting for the product to become useful and thrive.

Human factors and ergonomics processes in their turn, also base their actions and subsequent diagnostics of work situations in a thought-through real context, which must incorporate both formal and informal aspects of the workplace. In order to offer comfort, safety, and occupational health to users of a given work environment, it must also incorporate the organizational culture in which the system is. Thus, in order to expedite and potentialize its results as a productivity improvement technique, HFE actions must take into account corporate guidelines, organizational culture and climate, and tune in with the so-called sustainability tripod, which automatically infers special concern to social responsibility.

Ergonomic design must be understood as the type of ergonomic action that anticipates real word inadequacies and produces devices, equipment, workspaces, and work systems, positively embedded with ergonomic principles. Although it is the most cost-effective way of applying human factors and ergonomics in the workplace, it is not widely used by organizations. Instead, what is normally seen is that ergonomists or human factors professionals are called in to consult and help with workplace inadequacies, often when it is not possible to apply HFE principles anymore. Either the work system is fully structured or the cost of engineering is too high, which deems most effective changes unfeasible. The combination of HFE and design thinking might shift decision makers' ideas toward ergonomics, sometimes seen as a minor contributor to occupational health instead of a powerful organizational management tool. In fact, it is quite logical to assume that since ergonomics aims to adapt work means to worker needs, it might as well be powerful enough to address work systems as a whole.

17.2   BACKGROUND

In order to offer full understanding of this proposal, it is necessary to list some concepts from which the ergonomic design thinking approach is formed. It must be emphasized that there is no scientific revolution going on here as to the thoughts of Kuhn (1970) and others. The proposed context does not intend to disenfranchise other theories or bury previous ideas on how to proceed in order to acquire good HFE project results. Its unpretentious, quite humble mindset is due to what real scientists have learned for ages: there is no better way, only the way that serves one present purpose.

17.2.1   DESIGN THINKING

Owing to being a relatively new concept and to its multidisciplinary characteristics, there is no precise definition of Design thinking, even though they all lead to the designers' way of dealing with innovation matters to project management aspects. However, we can use the definition by Brown (2008, p. 86), "Design thinking can be described as a discipline that uses the designer's sensibility and methods to match people's needs with what is technologically feasible and what a viable business strategy can convert into customer value and market opportunity."

Thus, in essence it is a fully participatory design method, employing multidisciplinary actions not only in the developmental stages of the product design process. It goes further to postoccupational evaluation by professionals and end users themselves. After all, when it comes to designing workspaces, the workers will be at the same time the most affected by product results and the ones most likely to contribute with important insights about whatever is being designed for them. Design thinking is considered a natural evolution of design as an applied social science. Traditionally, designers focused their attention on improving looks and functionality of products. In recent years they have broadened their approach, creating entire systems around products and services they work at.

Design thinking incorporates constituent or consumer insights in depth and rapid prototyping, all aimed at getting beyond the assumptions that block optimal solutions. Thus, design thinking is inherently optimistic, constructive, and experiential for users and designers alike. It addresses the needs of the people who will consume a product or service and the infrastructure that enables it. This strategy of involving users in design projects and solving other problems in work systems have been employed by HFE professionals for a long time. It is called ergonomic design.

17.2.2 ERGONOMIC DESIGN

Ergonomic design is understood as the appropriation of ergonomic principles into systems design, which can be any work device, a piece of equipment, or even an entire work environment. Therefore, it is directly related to innovation and project management. Designers usually do not relate ergonomics to work activity adequacy. In fact, from furniture to cars, from appliances to their packages, ergonomics can be incorporated to everything people see around them.

Unfortunately, among all types of ergonomic actions, ergonomic design is the least used by organizations in general. It is a cultural, and not very clever, behavior most people develop by lack of a clear understanding of what HFE really is. In fact, most key people in companies do not think ergonomics should be incorporated in the early phases of a product development process. As a naval engineer once stated: "we usually consider Ergonomics as a design project terminator" (Shipyard Manager, personal communication). This remark represents a quite accurate view of how many people view ergonomics exactly because—contrary to common sense—it is presented to them in the last stages of product development, when attempts to correct bad decisions are usually difficult and costly. Even though we rarely think of ergonomics contributions outside product development processes, it can be applied to service implementations and even work systems. In addition, that adherence to work systems and workstation adequacy leads to the concept of job design.

17.2.3 JOB DESIGN

The process of job design has been adequately defined as, "…specification of the contents, methods, and relationships of jobs in order to satisfy technological and organizational requirements as well as the social and personal requirements of the job holder" (Buchanan, 1979, p. 144). Thus, we can infer that ergonomics and human factors has a very important role to play when it comes to convey the right environment for the right job, which in turn sends us back to ergonomic design. In addition, sorting out alternatives and relationships in the design of work systems is a complex task and it cannot be accomplished without employing a nonlinear design management model, such as Design Thinking.

Everything a worker needs, that is, workstations, work environments, work devices; they all should perform better when created in a planning environment that incorporates an iterative, creatively chaotic, yet productively systemic mindset. That kind of thinking that is usually present when designers translate those "qualities without a name" (Alexander et al., 1977) into a product and other powerful tools needed to handle a multitude of problems, included those unavoidable "design roadblocks" that arise now and then.

17.3 METHODOLOGICAL FRAMEWORK

The idea in design thinking is that there should not be "one" formal methodology to carry out any kind of project. Instead, methods should be considered general guidelines and function as a roadmap to project success, or at least work by averting miscarriage of an original plan. Thus, if one wants to assure a way of successfully carrying out a plan, general guidelines must combine on one hand sequential actions and creative thinking on the other. Creative thinking implies mind abstraction in order to leave doors open not to only good ideas but also to those apparently bad ones too. In terms of HFE actions, naturally we are talking of participatory actions, which means that ideas will come from different actors in the design process. A set of common HFE and design methodologies are presented in the sequence so as to clarify the forthcoming ergonomic design thinking model.

17.3.1 Design Methodologies

Design methodologies tend to attract some controversy, but nothing is as damaging as people known as "methodology worshippers," meaning they place the means (methods) above the ends (objectives). In other words, the problem with any methodology, not only product design ones, is when the methodology becomes more important than what it is supposed to deliver. It creates an achievement contradiction of sorts: instead of facilitating the result, the main objective of the methodology application, rigidity when employing it will most likely bring setbacks and an adverse outcome.

Designers always rely to some sort of sequential method to carry out their design projects. Some even write about them in a more theoretical manner, which seems like a design paradox. Other design theorists prefer to point out different ways to rethink design. Alexander et al. (1977) suggests users to not only participate in their design needs but learn to do it themselves, a path previously, in a philosophical sense, set by Robert Sommer in his book *Design Awareness*, that warned designers, especially architects, of the implications of their mishandling of living spaces. The same author recognizes the importance of user design and sees those initiatives as revolutionary because "user design invites, incites and implies participation of outside observers" (Sommer, 1971).

As an exercise and disciplinary reflection, Amorim et al. (2013) studied and combined different design methodologies (Ambrose and Harris, 2008; Amorim et al., 2013; Munari (2008, apud Amorim et al., 2013); Rittel, 1984) into a first draft of an ergonomic design model. Table 17.1 describes this tentative process, starting from general design project processes. The last column

TABLE 17.1
Combination of Design Processes

Usual Design Process	Horst Rittel	Moacyr Amorin	Bruno Munari	Design Thinking[a]
Ideation	Problematic phase	Problem structuring phase	Problematization phase	Problematization phase
				Priority check point
	Research phase	Focusing and prioritizing	Analysis phase	Research phase
		Synthesis and planning		Priority check point
Development	Execution phase	Design action	Creative phase	Priority check point
		Specifications		Creative phase
		Modeling		Productive phase
Delivery	Communication phase	Presentation	Technical development	Detailing and implementation
		Acceptance	Modeling, evaluation, and implementation	
Evaluation		User evaluation		Feedback
		Tuning and adjustments		

Source: Amorim, M. et al. 2013, Project methodology for design, ergonomic methodology and design thinking: Convergence in developing solutions in design, in *Cadernos Unifoa Especial Design*, Vol. 1, December, Pages 49–66. Adapted by Santos, M. S., Soares, M., Ergonomic design thinking—Approaching ergonomics through a new way for performing innovation in the workplace. In: Soares, Marcelo; Rebelo, Francisco. (Orgs.). *Advances in Ergonomics In Design, Usability & Special Populations*. 1st ed. Krakow: AHFE Conference, 2014, v. II, p. 560–572.

[a] Adapted from Ambrose and Harris (2008).

derives from this line of thought, setting up minimum steps or stages that were shaped by trying to establish an equivalence relationship among all different methodologies.

As it can be noted, all basic stages in every model follow a similar logical sequence. However, only two models take into consideration feedback mechanisms, which adhere to the notion that real user participation is generally small.

17.3.2 SYSTEMS ENGINEERING MODELS

System modeling is an industrial engineering approach of dealing with operational processes. In thesis, every work activity is as transformation process, where inputs (wills) arrive, are then transformed into goods or services and at last delivered (wishes) to society or inside flow production logics (Figure 17.1). However, system modeling is somewhat linked to HFE processes not only by semantics but also because of its sociotechnical equivalency. The system modeling approach is triggered by a quest for answers, just like HFE processes, and presupposes an interrelationship among environmental constraints and job design, as pointed out by some authors (Smith and Carayon, 2000). A model of systems engineering reasoning is showed in the sequence that follows.

Concurrent engineering is another concept that falls into a similar standard in terms of project management. Its systemic roots, however, are not embedded into the model itself but linked to the manner in which a project is carried out. In opposition to usual finish-to-start steps in regular project management approaches, concurrent engineering follows a nonlinear sequence, allowing for multifaceted collaboration in a project (Figure 17.2). In other words, the process is not only participative, but somewhat iterative in the quest for effective decision making.

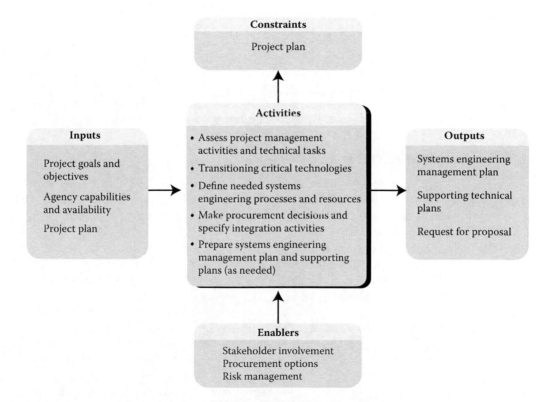

FIGURE 17.1 Overview of system engineering approach on project (or operations) management tasks. (From US Department of Transportation, 2013 (Apud Santos and Soares, 2014). Available in https://www. fhwa.dot.gov/cadiv/segb/views/document/sections/section3/3_4_2.cfm Page last modified on April 17, 2013.)

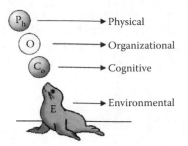

FIGURE 17.2   The Phocoe model for workspace study and design. (From Santos, M. et al. 2010, *4th International Conference on Applied Human Factors and Ergonomics—AHFE 2010 Conference Proceedings*, Miami.)

17.3.3   PHOCOE MODEL

Santos et al. (2010) bring us an ergonomic methodology, or a set of methods and tools, called the Phocoe model, which is actually a more appropriate name for what is meant to be. The model serves as a framework to converge multiple HFE operational aspects and is derived from Carayon and Smith (2000) and their balance theory, which relates work-related musculoskeletal disorders (WMD) with lack of organizational maturity. Starting from the sociotechnical notion that an organization is a system in itself but also a part of a larger whole, the model addresses the need of organizational and environmental balance in order to fulfill the requisites for proper ergonomic and human factors adequacy. A seal, representing environment, trying to keep all other known ergonomic domains in equilibrium (Figure 17.2), illustrates the model.

17.3.4   BALANCE THEORY

As they were trying to understand and relate the effects of work organization to workspace constraints, Smith and Carayon (2000) have conceived their balance theory. In this job design model, they relate WMD to organizational arrangements and other aspects. The model establishes a link between organizational culture and a conditional predisposition to poor job design and poor overall work conditions. Thus, it lends a sense of systemic interdependency between ergonomic actions and organizational maturity.

17.3.5   ERGONOMIC DESIGN-THINKING MODEL

Traditionally, designers have focused their attention on improving the look and functionality of products. Classic examples of this type of design work are Apple Computer's iPod and Herman Miller's Aeron chair. In recent years design professionals have broadened their approach, creating entire systems to deliver products and services.

Design thinking incorporates constituent or consumer insights in depth and rapid prototyping, all aimed at getting beyond the assumptions that block effective solutions. Design thinking—inherently optimistic, constructive, and experiential—addresses the needs of the people who will consume a product or service and the infrastructure that enables it.

Businesses are embracing design thinking because it helps them be more innovative, better differentiate their brands, and bring their products and services to market faster. Nonprofits are beginning to use design thinking as well to develop better solutions to social problems. Design thinking crosses the traditional boundaries between public, for-profit, and nonprofit sectors. By working closely with the clients and consumers, design thinking allows high-impact solutions to bubble up from below rather than being imposed from the top. Ergonomic design thinking must be seen as a model and not a formal methodology. It does combine and employ a series of specific and nonspecific tools and methods. Thus, if one feels like it, it may also be called a methodology of sorts. As mentioned before, a general model is what a formal methodology should always have

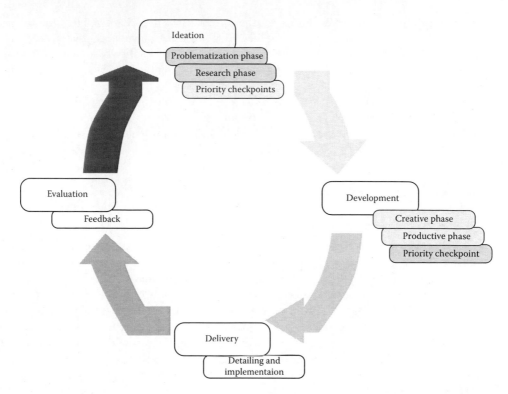

FIGURE 17.3   Ergonomic design thinking model. (From Santos, M., Grecco, C. 2015. *The use of Fuzzy Set Theory for Optimization of Ergonomics Prioritizing Tools: The SIC-Fuzzy Case Study*, Cadernos Unifoa, ISSN:1982-1816, Number 29, December, 2015. Available at: http://web.unifoa.edu.br/cadernos/.)

been: the logics of a way. Therefore, during implementation of ergonomic design thinking models one may undertake a series of actions using a variety of tools and techniques. However, it will not always be possible to utilize all of them and neither there is a rigid order to their use. Because of its implicit characteristics—being a roadmap—the ergonomic design thinking (EDT) model does not have a general framework, but a best practices flowchart. In more simple ergonomic actions, those guidelines convert into a banner, the ergonomics best practices banner, that is suggested to be hung in a well-visited area. It is usually used in occupational training sessions as well.

For the purpose of providing a general model framework, a scheme containing the basic elements and phases in the proposed model is presented below. The steps in Figure 17.3 follow overall convergence of methodologies as shown in Table 17.1.

There are some slight differences when compared to the original design thinking method. In the dsign thinking basic three-stage model, there is one stage prior to ideation and its subsequent stage called implementation. Inspiration is the drive stage, the starting point and reason for carrying out any project.

The main difference is that the EDT model identifies the importance of having three prioritizing phases instead of only one. In fact, the notion here is that of a decision-making demarche, as every milestone represents. Called "priority check points," those milestones are intended for rapid calibration and adjustment of the process. An instrument was specifically designed for helping with priority evaluations. This tool is called "fuzzy-sum of indices of criticality (SIC) prioritizing tool" or F-SIC for short.

17.3.6   Fuzzy-SIC Prioritizing Tool

Presented by Santos and Grecco (2015) the F-SIC tool derived from the SIC Priority Matrix (Santos, 2012) developed with the intent of providing decision-making aids during HFE processes.

The difference from the previous version is that the former uses fuzzy logic applied to the HFE design process to achieve more reliable results. Since HFE processes presuppose user participation, some kind of semiquantitative approach delivers credibility to the quality-based data gathering. It was originated from the idea that HFE actions have to rely on a multipurpose team—HFE professionals, users, and other stakeholders' interests.

With a more credible diagnostics of a work situation, HFE professionals may produce "design opportunities" to improve the work environment. In order to contextualize this, a series of cases in which ergonomic design thinking concepts were employed will be presented next. They range from early stage applications to wide-range examples that include the use of the SIC tool.

17.4 PRELIMINARY EXPERIENCES

17.4.1 Major Energy Company

In this project, the objective was to create conditions for the organization to move up to an upper level of ergonomic maturity (Santos et al., 2009). In order to achieve that, it was agreed the ergonomics team should perform a full EWA (ergonomic work analysis) of the entire company located in a major oil producing area in Brazil. As a subproduct of those evaluations, it was proposed to raise opportunities in terms of ergonomic design. The main obstacle immediately found was the corporate maturity level in the business. Complex, large organizations mean complex structures, which pose direct threat to ergonomics. Since ergonomics usually deals with improving from the inside out, changes must be quickly implemented or fade way in the sea of bureaucracy those enterprises normally share. The passage below illustrates part of what is called an opportunity identification ritual. It happened during the construction of a brand new facility, a six-story building supposed to house the entire administrative workforce spread out in several smaller buildings around the area.

- (EWT) Can we visit the new building with a group of future users and see if there could be any suggestion regarding the work environment, workspace adequacy, or any other user aspect?
- (HSE Manager) I don't think they would allow us...
- (EWT) Couldn't we discuss this ...and who is "they"?
- (HSE Manager) Well, the Floor plans belong to the Architecture, the Building is Engineering's responsibility, but not construction itself, which belongs to Out-Shared Services Division...plus this contract (Ergonomics) is tied to HSE Division, so no chance there...
- (EWT) Couldn't we try to discuss this, set up a meeting with them all... we really think it would be...
- (HSE Manager) (cutting off dialogue) ...I said forget about it!

Eventually construction was over and just past the final days of the HFE contract deadline. It was then realized that it would have to be adapted to serve the work force, since it had grown by more than 10% of what was predicted during the architecture planning phase. Electrical installations and layout inadequacies were spotted even before the allocation of administrative personnel. One manager, who by sheer luck decided to visit the building just prior to its completion, noticed that her office was placed right in front of the main elevator door, from where most of the people come out to that floor. It does not take much intelligence to predict what would have happened if she had not requested the costly change. The ergonomics work team (EWT) watched the building being raised right before their eyes, without being able to prevent what was bound to happen, and it eventually did.

However, this initial experience was not a total failure. We were able to participate in several opportunities in which ergonomics principles and user participation were combined to develop user-friendly work systems. In one of them, it first met a villain of user-friendly positive outcomes. Alexander et al. (1977), an architect and polemical figure, yet admired for his theories, suggests that everything in life has an essence, a shared perception of an optimum solution for a given problem. He calls it "quality without a name," which are translated into pattern (design) languages. In addition, the same author points out those patterns should not be seen as standards.

Setting aside legal and technical conformity to specifications, a standard is truly an antipattern. They only prevail because of some necessary continuity or by sheer desire to keep up with the status quo. There are several examples of antipatterns widely used today, such as glazing, those environmentally unsound glass curtains that cover commercial buildings, and even some residential ones, worldwide. They did not become prevalent because of architects; after all architects hardly visit their creations after they are ready for the public. In fact, bad solutions usually prevail because customers never complain about them, or as an old saying goes, those who shun out, agree upon.

17.4.2 Oil Rig

In this second experience, the focus was more specific. The EWT needed to address certain work inadequacies in the critical processes of an offshore oil rig operation in the same region of the first example. Initially we performed a major walkthrough, in order to understand the nature of the problems and why inadequacies happen in such a controlled production process, an overly regulated activity, in the first place. Similar intercurrences to those in the first example were found. Excessive paper work, segmented job design, and multi-decisional levels all add up to factors that create chances for errors. The excerpt below illustrates one of such problems. In this particular case, the inadequacy brought up both HFE and economic constraints.

- (EWT) What is this big package sitting there?
- (Safety Technician Supervisor) This is the crane's windshield.
- (EWT) Oh, yes. We noticed a broken windshield when we visited up there. But why it has not been replaced yet...It looks it has been sitting there for a while?
- (ST Supervisor) It has indeed...It was ordered about six months ago, but the routing is not that simple...First a request has to be put in here and, once approved internally, it goes to our land based office. Then the office approves it and submits to Purchasing Division. Purchasing has to quote, deal with taxation and other particularities (Note: Equipment is imported with no distributor in Brazil) and eventually place a final purchasing order. When part arrives, it has to go through customs, federal income office, and other regulatory agencies before is allowed to be shipped to us.
- (EWT) It sounds tiresome, but why is it still sitting in storage?
- (ST Supervisor) Well, the windshield has a mounting rubber seal that wraps around the crane frame where the windshield sits in tightly... When they hauled off the broken one, they threw away the fitting along with other debris... Now we've got to repeat the entire process for that part alone!

On an oil rig, time is definitely money, so we could not afford to employ our model to its full extent, but we did resource to creative thinking instead. In one event, a team was set up to analyze a particular problematic work situation. The driller station may be considered the most important job in an oil rig. Aside from reasonable proportions, its role is similar to that of a formula one pilot.

The scudery as a whole wins or loses, but without a good racer, there is no chance for success. Therefore, his workstation is designed as completely appropriate to him, following anthropometric guidance. In fact, the seat alone follows a rare rule: it is designed for the individual's exact body measurements. However, the driller seat in the studied oil rig was far from being adequate in terms of comfort, safety, or functionality. It did not even have an adjustment control. The track where the seat ran was rusted and fixed to a position that did not cater to one single user individually, let alone multiple individual susceptibilities. The EWT investigated utilization history for that workstation and found out it had been changed over time. Figure 17.4 shows the evolution for that particular situation all the way to the team's proposed solution. The photo to the left shows the original problem, when the operator had to improvise a stick to reach the buttons on a switchboard. To the right, is an initial solution—not jointly designed with users—in which a metal switchboard was placed in the way of operators when standing. The red circle in center of that picture shows the point where people constantly hit their heads on the panel.

As pictures show, workstation controls and seat adjustment controls never worked properly. As the seat was originally set in one fixed position, not only did it result in being especially difficult to handle overhead commands, but it also created great discomfort for every operator. Two simple devices were then envisioned: one for facilitating in and out access and another to allow easier seat positioning along the seat's lower track. The first device was basically a movable arm that would move away the console as the operator got up from his seat and the second, a simple adjustable track for moving the seat forward and back. The following excerpt shows part of a conversation that took place during implementation of the design solutions that had been brought up by the EWT actions.

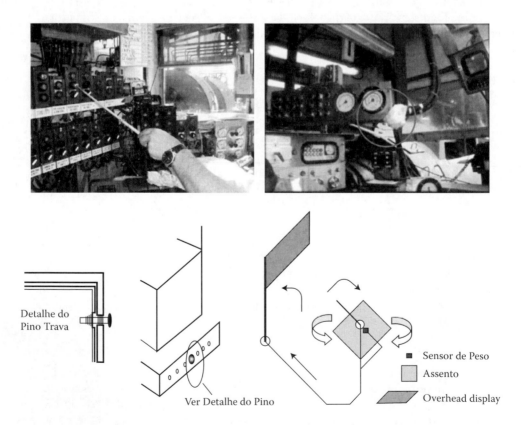

FIGURE 17.4   Resulting participatory-driven evolved solution to job constraint.

- (Drilling Supervisor) I think those proposals are very doable... but explain to me what is the advantage of using those retaining pins for the lower track?
- (EWT) Well, actually the welding shop workers came up with the idea. They said it will do no good keeping repairing the seat tracks since harsh atmosphere most likely will keep on oxidizing everything.
- (Drilling Supervisor) Those guys are full of it...they are there only for eventual repairs and emergencies so they keep having those "creative" ideas...don't they know that those makeshift pins will rust just as much?
- (EWT) Yes they do...and that's the whole point: a cheap, easy to find solution for solving a major discomfort and potential health hazard...it will rust and eventually break, but there will be a whole lot of spares right below deck and not all the way across the ocean*.

* The seat and accessories were manufactured in Sweden.

17.4.3 AUTOMOTIVE MANUFACTURER COMPANY

The purpose here is not a conventional ergonomic design-related project. In this major multinational auto manufacturer in Sao Paulo, Brazil, the contractor asked for a tool for assessing fatigue in various work activities. In a sense, it would still be considered an innovation project, since it involved developing an idea related to technological modernization processes. Many factors contribute to the workload experienced by individuals at work. Factors include the nature of the work, somatic aspects, acquired training, motivation, and environmental influence. They all affect energy throughput through the physiological service function of supplying power and oxygen for muscle metabolism. If work conditions improve, the workload tends to be reduced, even though it may not happen immediately for the work pace takes time to adjust. However, after a set of technological improvements has been implemented for some time, it makes no sense to keep using the same cycle times and planning methods that were designed under other reality without calibrating the data.

In the course of this ergonomic contract, a method intended to prevent distortions when setting up fatigue recovery time was developed. Operational activities were then targeted for the development of the Rfad method. The contractor, a major automotive company in Brazil, decided to review its time planning and wanted more accurate numbers, so that the amount of time allocated to fatigue recovery could become a more reliable variable to be computed into their current time sheets. Initially, a multidisciplinary EWT was established. A physician, an architect, two engineers, two designers, a physiologist—all with a human factors and ergonomics background—and four other certified ergonomists, formed the team.

The team was divided in two groups: one for fieldwork activities and another responsible for data analysis, project management, and coordination of activities along with the customer's project team. As a way of characterizing the situation at hand, it was decided to carry out a series of ergonomic analyses in selected workstations. Those 45 workstations were considered the most critical in a set of production centers. The EWT did not participate in the selection process. The following excerpt briefly illustrates how diverse workload perceptions can be depending on management's role and proximity to shop floor reality.

- (Department Manager) We decided to start analyzing 45 workstations deemed critical in terms of ergonomic impacts. If the pilot program succeeds, then we might extend the project.
- (EWT Project Manager) Well, can you tell us what was the criteria to sort out those workstations?

- (Department Manager) To be honest with you, all I know is that some supervisors and managers chose stations they were more familiar one in 4 or 5 different production centers.
- (EWT Project Manager) But was there any factors considered, like epidemiologic data, accident and work-related medical occurrences, etc.?
- (Department Manager) I really don't know… I would guess this was negotiated with floor personnel based on perceived muscular effort involved in each task…I am sure Medical Department did not get involved… they actually disagreed with this project…
- (EWT Project Manager) Isn't it a little odd? I mean, not choosing workstations REALLY critical will return a wrong diagnostics… the fact Medical Dept. didn't establish work-related root cause for problems and production didn't employ technical considerations, such as energy expenditure measurements, we may not have hard data on this…
- (Department Manager) I understand… but consensus and rationality is kind of difficult to reach in here… let's start something and see what we get.

17.4.4 CONSTRUCTION SITE HFE ACTIONS

Owing to the growth of the construction industry in recent years, this demand is a recurrent one in Brazil for quite some time. However, the question posed is: how could HFE contribute to a type of operational activity so full of particularities and peculiarities? There is no production line in the construction industry and even though we can use multiple project plans—in building construction—every single job is different due to external circumstances. To add up, people move a lot in this business, either for relocation to other companies, or because of change of professional status. One day a person is working in the colder south region as an electrician's assistant, the next day he (or she) is in the warmer northeast as a plumber helping build a dam. Although there is not much recollection of how initial conversations went, it was probably like the following block.

- (HFE Manager) The scope asks for a full evaluation of 400–500 workers in a building site, plus office personnel, but we will have only 3 weeks to finish everything.
- (EWT Member 1) We do a functional analysis of each critical work situation, as pointed out by the workers themselves, then we keep on evaluating all the rest by inference and equivalency. We will gather all the data in one week and we will still have the other two to tweak it out here and there…
- (EWT Member 2) For clerical personnel will be easy. Tasks in nature involve the same kind of physical and cognitive impacts, so we prepare a "10 Ergonomic Errors Cartoon" in where we "hide" work inadequacies in a big banner and post in the coffee room. We've done it before …
- (HFE Manager) And what about the Action Needed Plan? We all know that an Ergonomics & HF Assessment is not complete without a AN Plan.
- (EWT Member 1) We will have one. We will call it Best Practices Guidelines, derived from the Best Practices Flowchart with a set of representative work activities or tasks in terms of each construction phase. Then, we expose the common problems associated to each one of them and suggest alternatives in terms of postures or attitudes to minimize any eventual risks. This is especially important for operational personnel, therefore it should be placed in the cafeteria.
- (HFE Manager) Sounds like a plan… let's go for it! (GENTE/COPPE Lab Reports, 2012.)

17.5   FINAL CONSIDERATIONS

Ergonomic design thinking should be treated as a general model for participatory actions in the workplace. Naturally, ergonomic design, product development and job design are possible terrain for its dissemination. All those actions demand sequential steps loose in nature, in order to allow creative thinking to flourish. In this text, we presented only an initial overview of the model followed by a handful of preliminary experiences. But people involved in those experiences have been "spreading the disease" quite well. Several monographs resulted from the model in the past 5 years included three doctoral theses and one postdoctoral fellowship.

Some of those professionals also have their successful HFE practices thriving because of the model, so it appears the model is at least on the right path so far. The main advantage of ergonomic design thinking is definitely its methodological flexibility and broad outreach for the various contexts one may face in a work environment. In fact, it should be used "outside the box," since everyone is aware that ergonomics and human factors is definitely not a property of one single organizational domain. People that have been consulting in human factors and ergonomics for many years have seen good ideas fade way in the course of an imposed "perfect" methodology for their actions. However, sometimes people get so tied up to the rigidity of a methodology that the ends become a mere detail of the means. In other words, a methodology is a roadmap in which people rely on to reach from point A to point B without losing track.

As to the design end, conscious professionals have learned for years and years about the power of their actions, for good or bad. "If I'd asked my customers what they wanted, they'd have said 'a faster horse,'" Henry Ford once said. On the other hand, Frank Lloyd Wright (1867–1959), a believer in the so-called organic architecture, solemnly preconizing harmony among men and nature, once said that he could kill any happy marriage with a bad floor plan if he wanted to. A good design is only as good as others, not their creators, perceive them to be. In addition, our lives and our health can be affected by poor design choices. Thus, even HFE decisions not based on users' behavior and nature can lead to disastrous outcomes. Regardless of all constraints that may arise in workplace construction, designers ultimately share responsibility with engineers for most of the work environment inadequacies—due to negligence or omission—we still see out there. As people say, hell is full of well-intended folks.

At last, it is fair to say that companies intuitively use design thinking to an extent, but most stop short of embracing the approach as a way to move beyond today's conventional problem solving. Organizations resist in taking a human-centered approach because they cannot grasp the perspective of trying something entirely new, born in the midst of a balance of users' needs, technology, and organizational constraints. As Brown and Wyatt (2007, p. 35) said, "One of the biggest impediments to adopting design thinking is simply fear of failure." The notion that there is nothing wrong with experimentation or failure as a source of learning can be difficult to accept. However, a design-thinking culture will grow and encourage plain, sometimes even quick makeshift prototyping as part of the creative process, not just as a way of validating finished ideas. Continuous employment of our model should bring its steady improvement, consolidating it as a technique and allowing for its consolidation. Further collaboration is also expected from other researchers, since open source and multidisciplinarity are implicit characteristics of this model. Its continuous and broader use should enlighten darker spots in the methodology and incorporate collective value to the entire process.

REFERENCES

Alexander, C. et al. 1977, *A Pattern Language: Towns, Buildings, Construction*, New York: Oxford University Press.

Ambrose, G., Harris, P. 2008, *Design Thinking*, Chicago: University of Chicago Press.

Amorim, M. et al. 2013, Project methodology for design, ergonomic methodology and design thinking: Convergence in developing solutions in design, in *Cadernos Unifoa Especial Design*, Vol. 1, December, pp. 49–66. Available at: http://web.unifoa.edu.br/cadernos/especiais/design/cadernos-especial-design-2014.pdf

Brown, T., Wyatt, J. 2010. *Design Thinking for Social Innovation, Stanford Social Innovation Review,* Winter 2010. Available on https://www.ideo.com/images/uploads/thoughts/2010_SSIR_DesignThinking.pdf.

Brown, T. 2008, Design thinking, *Harvard Business Review*, June, 84–92.

Buchanan, D. 1979. *The Development of Job Design Theories and Techniques.* New York: Greenwood Publishing Group.

Carayon, P., Smith, M.J. 2000. Work organization and ergonomics. *Applied Ergonomics*, 31 (6), 649–662.

GENTE/COPPE Laboratory. 2012, *Collection of Project Reports (2008–2012)*, Rio de Janeiro: GENTE Ergonomics Lab.

Kuhn, T. 1970. *The Structure of Scientific Revolutions*, 2nd edition, London: The University of Chicago Press.

Rittel, H. 1984, Second-generation design methods. In N. Cross (Ed.), *Developments in Design Methodology*, New York: John Wiley & Sons, pp. 317–327.

Santos, M. et al. 2009, Can we really opt in terms of ergonomic methodologies and/or approaches? *Proceedings of the 17th Triennial Congress of the International Ergonomics Association,* 17th IEA, Beijing, China, Vol. 1. pp.1–12, (CD).

Santos, M., Vidal, M., Carvalho, P. 2010. Ergonomic pattern mapping—A new method for participatory design processes in the workplace. In: Salvendy, G; Karkowski, W. (Org.), *Advances in Cognitive Ergonomics.* 1st ed. Boca Raton: CRC Press, vol. 1, pp. 271–281.

Santos, M. 2012, The PhOCoe model—Ergonomic pattern mapping in participatory design processes, *Work*, 41, 2643–2650.

Santos, M., Grecco, C. 2015. *The use of Fuzzy Set Theory for Optimization of Ergonomics Prioritizing Tools: The SIC-Fuzzy Case Study*, Cadernos Unifoa , ISSN:1982–1816, Number 29, December, 2015. Available at: http://web.unifoa.edu.br/cadernos/.

Santos, M. S., Soares, M. 2014. Ergonomic design thinking—Approaching ergonomics through a new way for performing innovation in the workplace. In: Soares, M. and Rebelo, F. (Orgs.). *Advances in Ergonomics In Design, Usability & Special Populations.* 1st ed. Krakow: AHFE Conference, 2014, v. II, pp. 560–572.

Smith, M., Carayon, P. 2000, A balance theory of job design and for stress reduction, *International Journal of Industrial Ergonomics*, 4, 67–79.

Sommer, R. 1971, *Design Awareness*, New York: Rinehart Press.

18 Older Workers and Virtual Environments

Usability Evaluation of a Prototype for Safety Sign Research

Lara Reis, Emília Duarte, and Francisco Rebelo

CONTENTS

18.1 INTRODUCTION

This chapter presents the definition and results of a pilot study which was carried out to evaluate usability issues regarding an experimental virtual environment (VE) prototype that was specifically designed for conducting ergonomic and design studies with older populations/users (i.e., preretirement middle-aged adults, between 50 and 70 years old) and workplace safety signs. Topics under the VE interaction domain, such as VE navigation, the viewing/reading of visual VE information, and VE simulator sickness, are addressed.

Such a study was developed within the scope of a larger research project which proposes to (1) highlight the use and effectiveness of technology-based safety signs and warning systems as inclusive solutions for compensating and/or assisting age-related deficits that may characterize older, yet still active, populations/users; as well as (2) promote VEs as feasible research tools for conducting research in the fields of human factors and ergonomics, as well as safety communications and warning signs.

The nature of both this pilot study and its sample, as well as the larger research project, is justified by the fact that, as workers grow older, their ability to notice, encode, comprehend, interact, and comply with safety signs (i.e., relevant cues/signals/communications placed in the physical environment), and technology (e.g., computers and digital/virtual objects, products, scenarios, and devices), in everyday work and life in general, is adversely affected by several perceptual and/or cognitive deteriorations. These age-related changes (which characterize old age and thus turn older workers more prone to accidents, as well as injury) include declines in the visual, auditory, and cognitive capacities (e.g., Czaja and Lee, 2007; Mayhorn and Podany, 2006; McLaughlin and Mayhorn, 2014; Rousseau et al., 1998). Consequently, safety for preretirement workers is of great concern, since inadequately designed working environments, as well as safety signs and warning systems, conjointly with such deficits, may not only jeopardize their ability to work, but also degrade their quality of life.

Based on such facts, the present pilot study sought to validate the proposed VE prototype for conducting safety sign research with middle-aged adults. In order to undergo such an evaluation, such a study comprised of three key moments, namely to (1) examine to what extent and/or quality such preretirement workers could perform certain interactions inside the VE, by determining their ability/dexterity in using/manipulating/controlling the prototype's system devices/equipment; (2) analyze whether they could view/read the VE's visual cues/stimuli, by assessing the observation/ viewing distances for each of the VE's graphical information (i.e., different safety signs); and (3) analyze their overall user experience with the prototype.

When compared to conventional and/or traditional evaluation methods (i.e., laboratory- and/or field-based research), this VE prototype encompasses several advantages for research in the fields of human factors and ergonomics, as well as safety communications and warning signs. With an interactive, more engaging and life-like scenery/setting, it provides the means to assess the older users' behavioral and subjective experiences (i.e., level of engagement and state of mind/well-being), while controlling the system's technicalities (i.e., different types of interaction tasks, techniques, devices, and levels of immersion). In other words, such technology provides, as claimed by Rizzo and Kim (2005): (1) an enhanced ecological validity; (2) control and consistency of the experimental conditions, which supports repetitive and hierarchical delivery; (3) real-time performance feedback; (4) self-guided and independent exploration; (5) interface modification contingent on users' impairments; (6) safe testing and training, as well as an "error-free learning" environment; plus (7) gaming factors to enhance motivation.

18.1.1 VEs as Feasible Tools for Safety Sign Research

With the advanced development of various complex technologies, different types of VE systems have become commercially available, thereby giving rise to several types of human performance studies. Together with the expansion of such systems and assessments, several usability principles and evaluation methods have emerged to ensure the optimal creation, effectiveness, and satisfaction of VEs (e.g., Bowman et al., 2002; Hix and Gabbard, 2002; Wilson, 1999).

However, although the applications of VEs in various scientific domains are considerable, in the field of safety communications and warning signs, VE usability knowledge regarding older populations/users is scarce and very much in its infancy. The current body of VE safety sign research (e.g., Duarte et al., 2010) raises some concerns when generalizing and applying their usability principles to real-world problems and users, since the majority of the performed studies used: (a) younger adult populations, that is, mainly university students (e.g., Duarte et al., 2014; Machado et al., 2012); (b) specific target groups as research subjects, that is, participants with various degrees of cognitive and/or health disabilities/disorders (e.g., Cobb and Sharkey, 2007); (c) and/or much older age groups (e.g., Liu, 2009; Pacheco et al., 2010). Furthermore, the few existent studies which used older age groups and VEs highlight important performance differences when compared to younger adults (e.g., Moffat et al., 2001; Nichols and Patel, 2002).

Therefore, in order to assess the feasibility of using a VE for conducting safety sign research with preretirement workers, the current study structured and founded its evaluation under one of the main VE usability taxonomies, which research has identified as the *Behavioral/User Domain* (e.g., Gabbard et al., 1999). In other words, the way the users view/visualize, feel, communicate, behave, and interact with the VE's system interface (i.e., all icons, texts, graphics, devices, locomotion, etc.). Consequently, based on the existing literature (e.g., Bowman et al., 1999; Gabbard, 1997; Stanney et al., 2003), the present pilot study's usability evaluation was two-folded: (1) on the one hand, a *System Analysis* was performed to ensure that the middle-aged workers could interact with the VE prototype's physical, technological, and/or constructional components and (2) on the other hand, a *User Experience Analysis* was conducted to certify that their capacities (i.e., perceptual, cognitive, and motor skills), well-being/satisfaction, and safety were duly accounted for.

By conducting such analyses, this study seeks to provide some design guidelines and usability measures for research on safety signs using VEs. This chapter is divided, therefore, into three main sections: in the first section, the study's experiment and methodology is described; in the second section, its results are presented; and in the last section, the primary conclusions are addressed.

18.2 EXPERIMENTAL APPROACH

18.2.1 Participants

The study used a sample of six preretirement workers (i.e., middle-aged adults), aged between 50 and 67 years old (mean age = 58.2, SD = 4.34). Of these, three were men (mean age = 56.7, SD = 1.89) and three were women (mean age = 59.7, SD = 1.89). Upon arrival to the ERGOLAB (FMH/ULisboa), and before beginning the experiment, the participants were asked to fill in a consent form which specified: (1) that they agreed to participate in the study as volunteers; (2) the study's overall procedure; (3) the benefits and risks of participating in such a study; (4) their freedom to interrupt/end the experiment at any time; (5) that they allowed the experimental sessions to be audio recorded; (6) that the data collected was anonymous and confidential; and finally, (7) if they had any problems regarding their health (i.e., mental and/or physical state) which could prevent them from participating in the experiment. After signing the consent form, the participants were screened for color vision deficiencies, using the Ishihara Test (Ishihara, 1988). In conclusion, none of the participants reported mental/physical conditions nor color limitations which could prevent them from participating in the study.

In addition to this form, the participants filled in a demographic questionnaire which served to collect data regarding their age, gender, use of corrective lenses, experience with videogames, among others. The most significant data gathered from this questionnaire was (1) all participants reported that they used corrective lenses (i.e., they had a corrected 20/20 vision) to watch TV/movies, work on the computer and/or read books/magazines; (2) their professional occupations were completely diverse, ranging from procurement, to mechanical engineering, law, among others; (3) they had had some experience with basic computer games (i.e., they played sporadically), but no experience with videogames with first and/or third person players/controllers; and (4) that none of them had ever used the experiment's interaction equipment/devices before.

18.2.2 VE Prototype

18.2.2.1 Virtual Reality System Set-Up

A semi-immersive virtual reality (VR) system set-up was used to interact with the proposed VE prototype, as well as to automatically collect data regarding the participants' interaction. Inside the laboratory's room (i.e., a small and narrow dark room), the VE was projected onto a large screen, without stereoscopy, using a Lightspeed DepthQ® three-dimensional (3D) video projector and a

Microsoft Windows® graphics workstation, equipped with a ©NVIDIA QuadroFX5800 graphics card. A wireless gamepad, model T-Wireless Black from Thrustmaster®, was used as an interaction device. The VE prototype's scenery/setting was designed in 3D, using Sketchup Pro (owned by ©Trimble Navigation Ltd), and then exported to Unity3D (owned by ©Unity Technologies), where the simulation was defined.

The display's projected image size was 1.72 m wide and 0.95 m tall, with an aspect ratio of 16:9, and a 1280×720 resolution. The participants sat (for comfort reasons) on a chair (in order to be aligned with the display's center point, approximately 1.53 m above the ground), in front of the screen at a viewing distance of 1.50 m, which resulted in a 59.7° horizontal field-of-view (FOV), and a 35.2° vertical FOV. The participants' viewpoint of the VE was egocentric (i.e., the first person/character controller's main camera was set using the display's aspect ratio and vertical FOV) and set at eye-height, which was assumed to be 1.53 m above the ground. This average eye-height was calculated using the data gathered by Barroso et al. (2005), which states that the men's mean eye-height is 1.59 m and the women's mean eye-height is 1.47 m.

The speed at which the participants moved/"walked" from one place to another, inside the VE, was controlled in order to simulate a more natural and life-like gait/movement pace. This speed was set at a maximum of 1.25 m/s, which was calculated using the data defined by Bohannon and Andrews (2011). The simulation's image frame rate, that is, number of rendered frames per second (FPS), was also regulated to avoid flicker or lag effects. Throughout the simulation, the FPSs were above 120 per s, in accordance with the projector's frequency of 120 Hz. The gamepad's control sensitivity, regarding the virtual body's head movements, which was used to direct the participants' viewpoints, was left with the standard default settings, as defined by the actual interaction device and the software used to design the simulation.

18.2.2.2 VE Scenery/Setting

The VE's 3D model was designed using a modular layout, which consisted of a series of square, rectangular, and L-shaped sections (see Figure 18.1). Such a plan was defined with the intention of providing the participants with an intricate environment which required them to perform and train specific VE interactions. Therefore, the model was designed to represent a simple (i.e., one level) and minimalist (i.e., monochromatic colors, as well as plain textures were used for the model's architectural elements) open-space public building, which was free of any contextual or scenario-based concept (see Figures 18.1 and 18.2).

Its space was divided into six main areas. Such areas had the following dimensions: (1) compartments in area 1 varied between 3.4–6 m in width and 3.4–12.6 m in length; (2) areas 2–5 were 4.5 m wide and 9 m long; (3) compartments in area 6 varied between 3–5.7 m in width and 9 m to 12.3 m to 39.3 m in length; (4) circulation corridors/paths in areas 1–5 were 1.5–2 m wide, whereas in area 6 they were 3 m; and (5) the walls were 6 m in height.

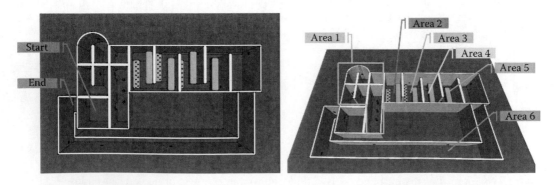

FIGURE 18.1 On the left, the VE prototype's layout/floor plan. On the right, the VE's main areas.

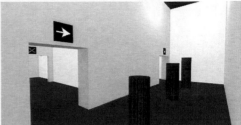

FIGURE 18.2 Screen shots of area 1 of the VE. On the left, the starting point. On the right, the column section.

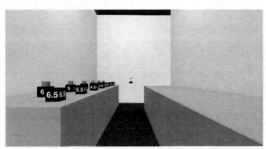

FIGURE 18.3 Screen shots depicting one of the safety sign's placement. On the left, participants' viewpoint, aligned at the center of the corridor, between the two large platforms which leant on opposite walls of the VE, and facing the warning. On the right, a participants' viewpoint when exiting area 4 and entering area 5.

Inside these areas, different types of architectural elements could be found, for example: area 1 had three columns, which differed in height (1.2, 1.5, and 1.8 m), aligned and placed at the center (see Figure 18.2); whereas, areas 2–5 had each two large rectangular-shaped platforms, which leant on opposite walls (i.e., on the left and on the right, respectively) of the VE, and were 1.25 m tall. In each area, one of the platforms had black numbered boxes (20 cm × 20 cm), which served as visual references that were meant to assist the participants' spatial orientation when performing the experiment's tasks (see Figure 18.3).

The VE's graphical appearance/characteristics/attributes were set with the software's maximum quality settings, and the level of lighting (i.e., only a small number of white lights, soft shadows, and lightmaps were used) was equally distributed and controlled (i.e., to avoid affecting the system's overall rendering/processing speed) in order to simulate real-life favorable viewing/reading conditions, as universally/generally defined by most safety standards. For methodological reasons (see topic "Evaluation framework"), the VE had no sound.

18.2.2.3 VE Visual Cues/Stimuli

Four static safety signs (i.e., a danger sign "PERIGO"; a warning sign "AVISO"; a caution sign "CUIDADO"; and a notice sign "ATENÇÃO") (see Figure 18.4), were placed as visual cues/stimuli on the VE's walls, in four separate and consecutive areas, namely areas 2–5 (see Figure 18.1). All four signs were in Portuguese and mounted at eye-height, that is, their center points were set at 1.53 m above the VE's ground (see Figure 18.3).

Given that the areas (i.e., areas 2–5) in which these four visual cues stimuli were placed, were 9 m in length (see Figure 18.3), such signs were designed to have an overall layout that was 35.56 cm × 25.40 cm in size, in accordance with the ANSI Z535.2-2007 (American National

FIGURE 18.4 Four ANSI/ISO type safety signs placed on the VE's walls, and in this specific order, from left to right. Translation: First sign (from left): DANGER. Falling objects. Being hit by falling objects may result in severe injuries or death. Do not pass below or stand in this area; second sign: WARNING. Electrical hazard. Do not touch. Keep away from this area; third sign: CAUTION. Slippery surface. Falling may result in severe injuries such as broken bones or paralysis. Keep away from this area; and forth sign: NOTICE. Authorized personnel only. This elevator leads to restricted areas only. Do not use this elevator.

FIGURE 18.5 Examples of safety signs' overall layout and visual components.

Standards Institute, 2007), ISO 3864-1:2011 (International Organization for Standardization, 2011), and ISO 16069:2004 (International Organization for Standardization, 2004) safety standards.

All four safety signs comprised of three key text components, the: (1) *Signal Word Panel* (*SW*); (2) *Hazard Identification Panel* (*HI*); and (3) *Message Panel* (*M*) (see Figure 18.5). Each of these panels consisted of different graphical information, with distinct typographic fonts and sizes (designed according to the ANSI Z535.2-2007 standard), and which were to be read at specific viewing distances (see Table 18.1). Such observation distances were calculated using the safety standard's ratio of 7.62 m per 2.54 cm of printed text, for favorable reading conditions, in real-life settings, for users with 20/20 (or corrected 20/20) vision.

TABLE 18.1

Safety Signs' Typographic Characteristics and Safe Observation Distances

Text Components	Font	X-Height (cm)	Maximum Viewing Distance (m)
1. Signal word (SW)	Frutiger LT Std—75 black	3.0	9.0
2. Hazard identification panel (HI)	Helvetica LT Std—bold	1.7	5.1
3. Message panel (M)	Helvetica LT Std—roman	0.7	2.1

TABLE 18.2
Safety Signs' Symbol Characteristics and Safe Observation/Viewing Distances

Safety Sign	Symbol Height (cm)	Maximum Viewing Distance (m)
Danger sign—red	12.4	7.44
Warning sign—orange	13.8	8.28
Caution sign—yellow	11.4	6.84
Notice sign—blue	9.3	5.58

As for the *Safety Symbol Panels* (*SSs*) (see Figure 18.5), the four safety pictorials/symbols varied in height size, in light of the quantity of information/text that needed to be included in the safety sign, and their safe observation/viewing distances were calculated using the ISO 3864-1:2011 standard's *Factor of Distance* formula, using the generic value of 60 for Z (see Table 18.2).

This particular pilot study sought to confirm whether the above safe observation/viewing distance references for each of the safety signs' individual components, which were calculated using the safety standards' ratios/formulas, for favorable reading conditions in real-life settings and for users with 20/20 (or corrected 20/20) vision, could be applied/generalized to VEs. In other words, one of the study's main objectives was to confirm whether the safety signs': (1) signal words, which were 3 cm in height, could be read inside the VE at a maximum viewing distance of 9 m; (2) hazard descriptions, which were 1.7 cm in height, could be read inside the VE at a maximum viewing distance of 5.1 m; (3) safety pictorials/symbols, which varied between 9.3 and 12.4 cm in height, could be perceived inside the VE at a maximum viewing distance between 5.58 and 8.28 m; and (4) risk/consequence and/or action statements, which were 0.7 cm in height, could be read inside the VE at a maximum viewing distance of 2.1 m.

In order to make sure that these four visual cues/stimuli were being displayed/projected correctly, preliminary tests with the VR system's set-up and paper-based versions of the safety signs were performed. This is (1) the signs were printed on paper, in their actual size, and placed/stuck on the laboratory's large screen; and then (2) the digital images of the safety signs (i.e., their projection) were aligned/matched with the paper-based versions (see Figure 18.6). Consequently, the projected images of the VE's visual cues/stimuli were being displayed in their correct size and proportion. However, it is to be noted that the VE's visual quality had certain resolution limitations regarding its image projection/display (e.g., the safety signs' text details were pixelated), which was to be expected due to its VR system set-up specifications.

FIGURE 18.6 Examples of safety signs' printed/paper-based versions placed on the laboratory's large screen for alignment.

18.2.3 EVALUATION FRAMEWORK

18.2.3.1 System Analysis: User Performance Metrics

The present pilot study composed of two system usability (i.e., ease of use, learnability, and accuracy of user task performance) test beds, using the same experimental VE prototype, which served to evaluate a set of specific tasks which in turn measured the preretirement workers' performance according to different VE interaction techniques, as described below. Such quantitative data were collected using a log system which automatically recorded the participants' interaction in real-time, by constantly registering their coordinates (i.e., their position in relation to the x-, y-, and z-axis).

The first test bed's main objective was to examine to what extent and/or quality the participants could perform certain interactions inside the VE, in other words, it was designed to determine the participants' skill/dexterity in using/manipulating/controlling the VE prototype's VR system devices/equipment (i.e., the gamepad). In order to evaluate this ability, participants were required to learn how to (a) travel/move/navigate from one location to another, inside the VE, while simultaneously changing their viewpoint accordingly; (b) avoid/contour the VE's architectural elements (e.g., walls, columns, and platforms) and pass through narrow corridors, without colliding with them; (c) circulate through the VE's course with temporal pressure, that is, as quickly and accurately as possible.

These VE interaction performances were measured by (1) *Number of Collisions (NC)*, that is, the number of times the participants collided with the architectural elements present in the VE; (2) *Area Duration (AD)*, that is, the amount of time the participants spent per area, to complete the different tasks defined for the experiment; and (3) *Experimental Session Duration (ESD)*, that is, the total amount of time the participants spent to complete each of the experiment's sessions. It was assumed that the less number of times the participants collided with the VE's architectural elements, the less amount of time they spent in carrying out the tasks, as well as in finishing the experiment's sessions, the better their performance/interaction. In other words, such measurements were interpreted as a clear indication that the participants had acquired the necessary skills/dexterity to simulate a natural and fluid movement inside the VE, as well as to navigate efficiently throughout its different areas.

The second test bed's main goal was to analyze whether the participants could view/read the VE's visual cues/stimuli. This is, it was designed to assess the participants' observation/viewing distances for each of the VE's safety signs' graphical components, which had distinct typographic fonts and sizes. In order to undergo such an analysis, participants were required to perform a series of target identification tasks, namely (a) to detect and discern at what distances they could view/read each of the safety signs' individual visual components (see Figure 18.5) and (b) to define the distance at which they could comfortably view/read all of the safety signs' information, as a whole.

Therefore, five observation distances were measured: (1) *Signal Word Viewing Distance (SWVD)*, that is, the distance at which participants could view/read all four signal words; (2) *Hazard Identification Viewing Distance (HIVD)*, that is, the distance at which participants could view/read all four hazard descriptions; (3) *Safety Symbol Viewing Distance (SSVD)*, the distance at which participants could view all four of the safety pictorials/symbols, that is, correctly perceive their details; (4) *Message Panel Viewing Distance (MVD)*, the distance at which participants could view/read all four risk/consequence and/or action statements; and lastly, (5) *Preferred Viewing Distance (PVD)*, the distance at which participants could comfortably view/read all four of the safety signs' information, as a whole, without image/pixel distortion. Subsequently, with this second test bed, the study sought to assess the differences between these five observation/viewing distances and as those defined in Tables 18.1 and 18.2.

18.2.3.2 User Experience Analysis: User Satisfaction Metrics

In order to evaluate the participants' overall experience with the VE, the present pilot study also gathered qualitative/subjective data regarding the participants' perceptions of the experiment.

Such data served to assess the VE prototype's ability to provide the participants with an engaging, pleasurable, and comfortable experience. Therefore, to undergo such an evaluation, the following methods were applied:

1. *Observation, audio-recording, and self-reporting protocol.* For technical and methodological reasons, the first author was present inside the laboratory's room, and close to the participant, during the entire experiment's procedure. Such a methodological set-up provided the researcher/evaluator with the means to (1) observe and accompany the participants' experiences (i.e., the evaluator could see the users' actions, as well as the graphical environment, at the same time); (2) gather real-time subjective reactions; and (3) instruct the participants, in real-time, on what tasks and techniques to perform inside the VE. Throughout the duration of the experiment, participants were encouraged to talk about their actions, goals, and thoughts regarding the study's different performance test beds. Consequently, they were told to verbalize/articulate their opinions and insights on how they viewed/visualized, felt, behaved, and interacted with the VE prototype, as well as report any difficulties they encountered.

2. *Simulator sickness questionnaire (SSQ).* In order to certify that the participants were comfortable, as well as free of any pain or sickness, during and after the experimental sessions, the *SSQ*, as defined by Kennedy et al. (1993), was applied twice in the course of the experiment's procedure. Such a questionnaire intended to evaluate the occurrence of any *virtual reality induced symptoms and effects (VRISE)* both during (referred to as side-effects) and post (i.e., after-effects) the VE simulation/exposure. In this questionnaire, participants were asked to score 16 symptoms on a four-point scale, with the following numerical and text anchors: (0) none, (1) slightly, (2) moderate, and (3) severe. Such symptoms fell under three general categories: ocular/visual disturbances, disorientation, and nausea.

3. *Demographic questionnaire.* As previously described, the participants filled in a demographic questionnaire which served to collect data regarding their individual characteristics, namely their age, gender, use of corrective lenses, professional occupation, experience with videogames, among others.

4. *Quality of VE interaction questionnaire.* To assess the VE prototype's usability (i.e., ease of use and learnability), this questionnaire, adapted from Witmer and Singer's (1998) *presence questionnaire (PQ)*, was applied to collect the participants' subjective perceptions regarding their overall interaction/behavioral performance inside the VE. In this questionnaire, participants were asked to score the quality of their VE experience according to the following categories: (1) *ease of navigation:* "How easily could you travel/move/navigate inside the VE (e.g., how easy was it for you to get to a certain point in the VE)?"; (2) *navigation control:* "To what extent could you control your movement inside the VE (e.g., how accurately could you position and/or stop yourself at a desired place)?"; (3) *natural expression of visual behavior:* "How natural was the system's visual behavior (e.g., 'To what extent could you look to/at different architectural elements inside the VE, in a natural and/or fluid manner')?"; (4) *viewpoint control:* "To what extent could you control your viewpoint (e.g., how easy was it for you to accurately direct your head to a certain point inside the VE)?"; and (5) *VE performance:* "How would you classify your overall performance inside the VE?". Such questions were ranked using a Likert-type seven-point scale, with the following numerical and text anchors: (1) very difficult/poor; (2) difficult/poor; (3) slightly difficult/poor; (4) average/moderate; (5) fairly easy/good; (6) easy/good; and (7) very easy/good.

5. *VE's visual quality questionnaire.* To evaluate the overall graphical quality of the VE prototype's visual cues/stimuli, this questionnaire, adapted from Witmer and Singer's (1998) *presence questionnaire (PQ)*, was applied to collect the participants' subjective perceptions

regarding their ability/capacity to view/read the different types of visual information. In this questionnaire, participants were asked to score the VE's visual quality according to three categories: (1) *object identification:* "To what degree could you identify the different architectural elements present in the VE (e.g., columns, pathways, walls, and platforms)?"; (2) *sign visibility:* "To what extent could you locate and follow the signs present in the VE (e.g., the way-finding arrows)?"; and (3) *safety signs' observation :* "To what extent could you view/read safety signs' visual components?". Such questions were ranked using the same Likert-type seven-point scale referred to above.

18.2.4 PROCEDURE

The study's experiment was divided into five major phases: (1) *introduction to the study*; (2) *first experimental session*; (3) *first posthoc questionnaire*; (4) *second experimental session*; and (5) *second posthoc questionnaire*. The whole procedure lasted approximately 45 min in total.

1. *Introduction to the study phase.* After signing the consent form and completing the color deficiency detection test, the participants were given a brief explanation about the study and its different phases (i.e., they were told that the study's main objective was to validate a new VR software which was being developed at the laboratory; thus, they were unaware of the study's real objective) and were introduced to its VR system set-up.

2. *First experimental session.* This session was divided into two key moments: (1) a training period, in which the participants practiced using the system's equipment/device (i.e., gamepad), as well as learnt how to interact inside the VE and (2) a second moment in which the participants were asked to view/read the VE's visual cues/stimuli (i.e., four different safety signs).

   The experiment began in a square-shaped area (see Figures 18.1 and 18.2), preceding area 1, where the participants were given instructions on how to use the gamepad and practiced how to navigate from one place to another (i.e., use the device's thumbstick on the left-hand side to travel/move inside the VE; and the thumbstick on the right to direct their viewpoint). They were asked to explore this area freely, with no time restrictions, until they felt that they were able to control their movement. Once the participants verbally stated that they felt at ease to continue with the rest of the exercises, they were instructed to enter the subsequent section, that is, area 1 (see Figure 18.2). In this part of the VE, participants were asked to perform a chicane task around the columns, that is, they were asked to circulate, in an s-shaped manner, around the three columns. They were told to perform this task twice, as quickly and efficiently as they could, that is, they were told to avoid colliding with the columns and walls, as well as avoid pausing/interrupting their movement. Upon completing this task, they were then instructed to continue to move through the environment until they reached area 2, where the second moment of this session was to take place.

   In area 2, participants were asked to position themselves in the center of the corridor (between the two large platforms which leant on opposite walls of the VE) and align their "bodies" with the first black box (numbered 6.5), and then face the safety sign at the end of the corridor (see Figure 18.3 for an example). At this distance (i.e., which was, respectfully, 6.5 m away from the safety sign), participants were asked to describe what they could view/read. Subsequently, they were then told to move forward, slowly, until they could view/read another of the sign's elements. When the participants could view/read another of the safety sign's visual features, they were asked to stop/pause their movement and press the gamepad's button. These actions were then repeated until the participants could distinguish each of the safety sign's key components (i.e., signal word; hazard identification text; safety symbol; and risk/consequence and/or action statements). When all of the safety signs' individual components were identified, the participants were asked to position themselves at the distance at which they could view/read all of the safety signs' information comfortably, without any image/pixel distortion (i.e., the *preferred viewing distance*). After having established this last position, the participants were told to press, once again, the gamepad's button.

After completing this first visual exercise, the participants were told to leave area 2 and to enter area 3. From areas 3 to 5, the participants repeated the process for each of the remaining safety signs. When all three signs were evaluated, they were then instructed to continue to move through the environment. When they reached area 6, they were told to circulate through this area, as efficiently and as quickly as they could (i.e., with temporal pressure), and most importantly, to simulate a natural, fluid, and pause-free movement, as well as to avoid, at all costs, collisions with the VE's architectural elements. Once the participants reached area 1, once again, the experimental session ended.

3. *First posthoc/follow-up questionnaire phase.* After completing the first experimental session, the participants had a 5 min break and then filled out the *simulator sickness questionnaire*, to check for any preliminary indications of VRISE, and then the *demographic questionnaire*, to collect data regarding their individual characteristics.

4. *Second experimental session.* With the same VE prototype, in the second experimental session participants had to repeat the same path that they had done in the former session. However, in this session, they were told not to repeat the whole process of viewing/reading each of the safety signs' components, nor press the gamepad's button. They were told, once again, to circulate through these areas as efficiently and as quickly as they could (i.e., with temporal pressure), and most importantly, to simulate a natural and fluid movement, with no pauses, and to avoid at all costs collisions with the VE's architectural elements. The main reason behind this approach was to evaluate whether the participants' dexterity in using the system's devices had improved in comparison to their performance in the first experimental session.

5. *Second posthoc/follow-up questionnaire phase.* At the end of the second experimental session, the participants had another 5 min break and then filled out three questionnaires, namely the: (1) *simulator sickness questionnaire*, once again, to assess if there was an increase in VRISE, due to having been exposed twice to the simulation, and over a time period of approximately 30 min; (2) *VE's interaction quality questionnaire*, to evaluate the usability of the VE prototype's VR system set-up; and (3) *VE's visual quality questionnaire*, to analyze the prototype's visual quality.

18.3   RESULTS

18.3.1   Discussion of the First Test Bed's Findings

For the first test bed, the following quantitative data were obtained for each of the experiment's sessions: (1) *number of collisions (NC)*; (2) *area duration (AD)*; and (3) *experimental session duration (ESD)*. Descriptive statistics for such measures, for both experimental sessions, are depicted in Tables 18.3 and 18.4, and explained below.

In Table 18.3, the study's different measures (i.e., *NC, AD,* and *ESD*), regarding the first test bed, for both experimental sessions, are depicted. Since each session differed in the instructions provided to the participants, as well as in the number and type of tasks/exercises, a direct comparison between both sessions, regarding the three types of measurements, cannot be done. Nevertheless, since the conditions regarding the travel/navigation tasks/techniques in areas 1 and 6 were the same, for both experimental sessions, such interactions/performances can be compared.

TABLE 18.3

Descriptive Statistics (Mean and Standard Deviation Values) Regarding the First Test Bed's Overall Measurements in Areas 1 and 6, for Both Experimental Sessions

		Area 1		Area 6		Total	
		NC	AD	NC	AD	NC	ESD
1st session	Mean	29.67	09'47	3.50	01'48	45.17	25'40
	SD	27.30	03'04	7.39	00'29	48.40	07'39
2nd session	Mean	20.33	04'22	0.83	01'12	31.33	07'43
	SD	18.58	01'12	0.69	00'35	31.75	02'03

TABLE 18.4

Descriptive Statistics (Mean and Standard Deviation Values) of the First Test Bed's Measurements, for Both Experimental Sessions, Regarding the Number of Times Participants Collided with the VE's Columns, when Performing the Chicane Task in Area 1

	1st Session	2nd Session
Mean	3.50	2.67
SD	2.81	1.37

These results concerning only the two common denominators, that is, *NC* and *AD*, are presented in the subsequent section.

1. *NC.* In area 1, in the first experimental session, the participants collided with the VE's different architectural elements (i.e., walls, columns, and platforms), on average, 29.67 times; whereas, in the second session, they collided, on average, 20.33 times. Therefore, the total number of collisions in area 1 decreased by 31.5% between sessions. Among the overall number of impacts, Table 18.4 demonstrates that an average of 3.50 collisions was accounted for the number of times the participants collided with the VE's columns, while performing the chicane task, in the first session; and an average of 2.67 times in the second session; thereby resulting in a 23.7% decrease. In area 6, the participants collided, on average, 3.50 times in the first session; while, in the second session, they collided, on average, 0.83 times. Therefore, the total number of collisions in area 6 decreased by 76.3% between sessions.

2. *AD.* In the first experimental session, the participants spent, on average, 9 min and 47 s to complete the defined tasks in area 1; whereas, in the second session, they spent, on average, 4 min and 22 s. Therefore, the total time spent in area 1 decreased by 55.4% between sessions. In what concerns area 6, the participants spent, on average, 1 min and 48 s to pass through it, in the first experimental session; and in the second session, they spent, on average, 1 min and 12 s; thereby resulting in a 24.3% decrease between sessions.

Such results reveal that there were significant differences between the two experimental sessions, for both areas. This is, in the second experimental session, the participants had (1) collided less with the VE's obstacles, in both areas as well as (2) spent less time to complete the different tasks, in both areas. By comparing the two session's data, in both areas, one can assume that the participants had better executed each of the area's specific tasks, in the second experimental session. In what regards the *NC*, one can infer that between sessions and areas, the participants had learnt how to interact with the VE, as well as had acquired the necessary skill/ability to manipulate/control/use the VE prototype's system devices/equipment (i.e., gamepad).

18.3.2 Discussion of the Second Test Bed's Findings

For the second test bed, the distances at which the participants could view/read each of the four safety signs' individual visual components were calculated. Such observation distances included, as previously stated, the (1) *signal word viewing distance (SWVD)*; (2) *hazard identification viewing distance (HIVD)*; (3) *safety symbol viewing distance (SSVD)*; (4) *message panel viewing distance (MVD)*; and lastly, (5) *preferred viewing distance (PVD)*. Descriptive statistics for these distances are depicted in Tables 18.5 and 18.6.

As shown in Tables 18.5 and 18.6, the observation distances at which the participants could view/read each of the safety signs' individual visual components (which had distinct typographic fonts and sizes) inside the VE, were different to those defined for favorable reading conditions, in real-life settings, for users with 20/20 (or corrected 20/20) vision. Such results reveal that there are significant differences between the maximum viewing distances defined for real-life environments, and the actual viewing/reading distances regarding this particular VE prototype.

TABLE 18.5

Descriptive Statistics (Mean and Standard Deviation Values) of the Second Test Bed's Measurements, Regarding the Observation/Viewing Distances for Each of the Safety Signs' Individual Visual Components

	Danger Sign *"PERIGO"*					Warning Sign *"AVISO"*				
	SWVD	HIVD	SSVD	MVD	PVD	SWVD	HIVD	SSVD	MVD	PVD
Mean	4.15	3.07	2.65	1.99	1.83	4.55	2.87	2.91	1.17	1.00
SD	0.55	0.15	0.41	0.07	0.10	0.68	0.50	0.91	0.10	0.09
	Caution Sign *"CUIDADO"*					Notice Sign *"ATENÇÃO"*				
	SWVD	HIVD	SSVD	MVD	PVD	SWVD	HIVD	SSVD	MVD	PVD
Mean	4.32	3.22	2.67	2.05	1.84	5.10	3.08	2.24	1.14	1.05
SD	0.20	0.20	0.53	0.25	0.09	0.54	0.23	0.62	0.12	0.03

TABLE 18.6

Mean Values Regarding the Safety Signs' Overall Observation/Viewing Distances, for Each of the Four Key Visual Components, and the *Preferred Viewing Distance*

Observation/Viewing Distances	Maximum Viewing Distances for Real-Life Settings (m)	Mean Viewing Distances in the VE (m)
SWVD	9.00	4.53
HIVD	5.10	3.06
SSVD	5.58–8.28	2.24–2.91
MVD	2.10	1.59
PVD	–	1.43

This is, inside the VE, each graphical component was viewed at shorter/closer distances: (1) *SWVD*—all four signal words, which were 3.0 cm in height, were viewed/read at an average distance of 4.5 m, instead of at a maximum viewing distance of 9.0 m (i.e., a 49.7% decrease); (2) *HIVD*—the four hazard descriptions, which were 1.7 cm in height, were viewed/read at an average distance of 3.1 m, instead of at a maximum viewing distance of 5.1 m (i.e., a 40% decrease); (3) *SSVD*—the four safety pictorials/symbols, which varied between 9.3 and 12.4 cm in height, were viewed at an average distance between 2.2 and 2.9 m, instead of at a maximum viewing distance between 5.6 and 8.3 m (i.e., a decrease between 59.9% and 64.9%); and (4) *MVD*—the four risk/consequence and/or action statements, which were 0.7 cm in height, were viewed/ read at an average distance of 1.6 m, instead of at a maximum viewing distance of 2.1 m (i.e., a 24.3% decrease).

By analyzing the *preferred viewing distances (PVD)*, it becomes clear that the observation/viewing distances of safety signs, when comparing the recommendations for real-life settings and VEs, are not the same. In light of this fact, one can infer that the prototype's system specifications may have limited the quality of the image's projection, and thereby, distorted specific details of the VE's visual cues/stimuli. Therefore, one can only assume that such image distortions made it harder for the middle-aged participants to view/read different types of graphical information from farther distances.

18.3.3 Discussion of User Experience Findings

As for the study's more qualitative and subjective measures, the most important qualitative data was gathered with the following methods:

1. *Self-reporting protocol.* The most important findings gathered from this method were related to the participants' ability to detect and discern different types of graphical information, with distinct typographic and pictorial sizes, inside the VE. Although the correct viewing/reading of the VE's safety signs' individual components was impacted between 24.3% and 64.9%, the participants' ability to identify the four signs' shapes and colors were not affected. Subsequently, when positioned/aligned 6.5 m away from the safety signs, the participants were able to accurately describe the four safety signs' overall compositions/layouts, namely the different colored shapes (red, orange, yellow, blue, black, and white rectangles). Most importantly, they could identify the danger, warning and caution signs' yellow triangular-shaped symbols. Thus, when the participants were asked to describe what they could view/read when entering areas 2–5, they immediately recognized, understood, and reported that they were analyzing warning type signs. In light of such fact, one can assume that the participants had some knowledge regarding hazard communication and warning systems which in turn enabled them to draw inferences on what they were viewing/reading.

   As for the quality of the interaction, the most interesting findings gathered were concerned with the gamepad's control sensitivity, namely the speed set for the head movements. Most participants felt that the speed, at which their virtual head rotated, when directing their viewpoints to certain points inside the VE, was too fast, and to some extent unnatural/unrealistic, as well as difficult to control.

2. *Simulator sickness questionnaire (SSQ).* The results from both SSQs (applied at the end of each experimental session) reveal that there were no occurrences of ocular/visual disturbances, disorientation, and nausea, both during, or after the experiment. When analyzing the *experimental sessions duration (ESD)*, as shown in Table 18.3, one can infer that even after a lengthy (i.e., an average of 25.40 min in the first session; plus an average of 7.43 min in the second session; which therefore sums up to a total average of 33.23 min) exposure to the VE, the experimental prototype's design (i.e., the study's system set-up,

TABLE 18.7

Descriptive Statistics (Mean and Standard Deviation Values) Regarding the Quality of VE Interaction Questionnaire, and the VE's Visual Quality Questionnaire

	Quality of VE Interaction Questionnaire						VE's Visual Quality Questionnaire			
	Q 1	Q 2	Q 3	Q 4	Q 5	Total	Q 1	Q 2	Q 3	Total
Mean	4.80	4.80	5.20	5.00	4.80	4.90	5.30	5.00	4.80	5.00
SD	4.50	4.50	5.00	4.50	4.50	4.60	5.50	4.50	5.00	5.00

VE layout/floor plan, and methodological procedure) did not affect the preretirement workers' overall well-being.

3. *Quality of VE interaction questionnaire.* As depicted in Table 18.7, all five inquiries (which evaluated the VE's ease of navigation; navigation control; vision naturalness; viewpoint control; and VE performance categories) attained an average rating of 4.90 (SD = 4.60). That is, the participants rated the VE prototype's usability as "average/moderate" to "fairly easy/good." Thus, with such data, one can conclude that the participants believed to have had an above-average interaction/experience with the VE, and that they were able to control their movement with ease.

4. *VE's visual quality questionnaire.* As shown in Table 18.7, all three questions (which assessed the VE's *object identification, sign visibility,* and *safety signs' observation* categories) scored an average rating of 5.00 (SD = 5.00). Therefore, the participants rated the VE's overall graphical quality as "fairly easy/good." Consequently, with such data, one can assume that the participants believed that they were able to view/read and identify different types of visual information, present in the VE, reasonably well.

18.4 CONCLUSION

This chapter presents the structure and findings of a pilot study which aimed to assess usability matters concerning an experimental VE prototype that was explicitly designed for undergoing ergonomic and design studies with older populations/users (preretirement middle-aged adults, between 50 and 70 years old) and workplace safety signs. Such a pilot study sought to validate the proposed VE prototype for safety sign research. In order to conduct such an evaluation, the study comprised of three key moments, namely to (1) examine to what extent and/or quality such preretirement workers could perform certain interactions inside the VE, by determining their ability/dexterity in using/manipulating/controlling the prototype's system devices/equipment; (2) analyze whether they could view/read the VE's visual cues/stimuli, by assessing the observation/viewing distances for each of the VE's graphical information (i.e., different safety signs); and (3) analyze their overall user experience with the prototype. Consequently, two system usability test beds were performed, as well as four qualitative questionnaires applied.

By analyzing the study's results regarding the first usability test bed, which aimed to determine whether the quality with which the middle-aged workers could perform certain interactions inside the VE, as well as their skill/dexterity in using/manipulating/controlling the prototype's VR system devices/equipment (i.e., the gamepad), one can conclude that, in general, the participants were able to successfully interact with the VE, as well as its system's set-up. By comparing data between both sessions, one can infer that in the first experimental session the participants were unfamiliar with the VE prototype's VR system devices/equipment (i.e., the gamepad), as well as the interaction techniques (i.e., using both of the device's controls to simulate a natural and fluid move-ment) required to perform inside the VE, and for these two reasons, they collided more often with the VE's architectural elements and spent more time to complete the different tasks. Since there were significant differences (i.e., decreases across the test bed's measurements) between the experi-ment's sessions, one can observe that from one session to the other, and in-between sessions, the participants had learnt/trained how to (1) travel/move/navigate from one location to another, while simultaneously changing their viewpoint accordingly; (b) avoid/contour architectural elements (e.g., walls, columns, and platforms) and pass through narrow corridors; and (c) circulate through the VE's course with temporal pressure, that is, as quickly and accurately as possible.

In light of the study's second usability test bed, which sought to evaluate whether the partici-pants could view/read the VE's visual cues/stimuli, by assessing the observation/viewing distances for each of the VE's graphical information (i.e., different safety signs), one can conclude that, in general, the participants were able to successfully detect and discern different types of graphical information, with distinct typographic fonts and sizes, in the VE. However, in what concerns the results gathered for the observation/viewing distances at which the participants could view/read each of the VE's safety signs' individual visual components, one can conclude that the observation/viewing distances of safety signs, when comparing the recommendations for real-life settings and VEs, are not the same. With such data, it becomes clear that the ANSI/ISO safety standards' criteria cannot be applied/generalized to VEs which have the same specifications and system set-up as this study's prototype.

In what concerns the study's user experience analysis, which aimed to assess the participants' overall experience with the VE, namely their perceptions on how they had viewed/visualized and felt while they interacted inside the VE, one can conclude that such a prototype provided its users with a sickness-free and an above-average experience. This is, it can be inferred that overall the participants found the VE prototype's system set-up fairly easy to use and learn, and that its graphi-cal information was easily identified and viewed/read. Subsequently, such subjective results indi-cate that the participants were believed to have acquired the necessary skill/ability to perform and interact inside the VE.

In short, such a study concludes that the participants had an adequate and satisfactory interaction with the proposed VE prototype. The attained results, across all measures, indicate that, overall, the preretirement workers were able to perform certain interactions inside the VE; view/read the VE's visual cues/stimuli (i.e., different types of safety signs); and have an acceptable experience. In con-clusion, one can infer that VEs (which have the same system set-up such as this study's prototype) may be successfully used by older working populations, and subsequently, that they can be used to conduct safety sign research with such samples.

As final considerations, when compared to conventional and/or traditional evaluation methods (i.e., laboratory- and/or field-based research), the potential benefits of using VE prototypes, par-ticularly for the purpose of conducting safety communications and warning sign studies with older populations/users, are manifold: such technology provides researchers with the means to simulate interactive and quasi-real scenarios (i.e., in which hazardous situations can be studied in a safe manner) with an enhanced control, as well as ecological validity over the experimental conditions. However, researchers who wish to conduct warning interaction or behavioral compliance studies must be aware that certain VE system set-ups, which have the same specifications (i.e., certain reso-lution/pixel limitations regarding its image projection/display) such as this study's prototype, may impact the correct viewing/reading of safety signs.

By evaluating the maximum observation/viewing distances for each of the safety signs' indi-vidual graphical information, this pilot study highlights which of the safety signs' visual features/characteristics, as well as typographic sizes can be easily viewed/read by middle-aged adults. Therefore, in order to guarantee the correct viewing/reading of safety signs inside VEs such as this one, researchers in these fields of study should consider this pilot study's findings.

However, in order to completely understand such distances, and to provide effective VE criteria, further usability testing is required, for example: (1) tests using younger working adult populations (i.e., 20–35 years old), in order to evaluate if the same visual decreases, regarding the observation viewing distance references for each of the safety signs' individual components, also occur; (2) tests in order to assess to what extent the participants' knowledge regarding hazard communication and warning systems may have influenced their task of actively viewing/reading each of the safety signs' distinct graphical information, that is, understand if the participants guessed/assumed and/or inferred (instead of accurately identifying or discriminating each word) the safety signs' contents; (3) tests to compare the observation/viewing distances attained in the VE prototype are the same as those gathered in a real-life environment and with printed paper-based versions of the safety signs; and in addition; (4) tests with other VR system set-ups and image specifications in order to assess if the same image distortions and visual decreases also occur.

In light of the larger research project which proposes to highlight the use of technology-based safety signs and/or warning systems as inclusive solutions for compensating and/or assisting age-related deficits that may characterize older, yet still active, populations/users, future work will be dedicated to the definition of more effective and inclusive VE criteria measures for warning inter-action or behavioral compliance studies. Given the lack of such usability standards for this area of research, such a project seeks to design and implement a number of VE prototypes, and subse-quently evaluate the impact of using different interaction techniques and devices, as well as levels of engagement, on older population performances. Since such an analysis has not yet been conducted, we hope to create a body of work which will promote VEs as feasible research tools for enhancing the fields of human factors and ergonomics, inclusive design, and most importantly, safety commu-nications and warning signs.

ACKNOWLEDGMENTS

This chapter is a full and revised version of the paper published in the book of the *AHFE 2014: 5th International Conference on Applied Human Factors and Ergonomics*. Reference: Lara Reis, Emília Duarte, and Francisco Rebelo. 2014. Evaluation of an experimental virtual environment prototype for older population warning studies. In *Advances in Ergonomics in Design, Usability & Special Populations (Part I)*, edited by M. Soares and F. Rebelo, pp. 543–554. AHFE Conference copyright 2014. ISBN: 978-1-4951-2106-7. Furthermore, a PhD scholarship (SFRH/BD/79622/2011) granted to Lara Reis, from FCT: *Fundação para a Ciência e Tecnologia* (the *Portuguese Science and Technology Foundation*), supported this study.

REFERENCES

American National Standards Institute. 2007. *Environmental and Facility Safety Signs (ANSI Z535.2-2007)*. Rosslyn, Virginia: National Electrical Manufactures Association.

Barroso, M.P., P.M. Arezes, L.G. da Costa, and A.S. Miguel. 2005. Anthropometric study of Portuguese workers. *International Journal of Industrial Ergonomics* 35: 401–410. doi:10.1016/j.ergon.2004.10.005.

Bohannon, R.W. and A.W. Andrews. 2011. Normal walking speed: A descriptive meta-analysis. *Physiotherapy* 97(3): 182–189. doi:10.1016/j.physio.2010.12.004. http://dx.doi.org/10.1016/j.physio.2010.12.004.

Bowman, D., J.L. Gabbard, and D. Hix. 2002. A survey of usability evaluation in virtual environments: Classification and comparison of methods. *Presence: Teleoperators and Virtual Environments* 11(4): 404–424.

Bowman, D.A., D.B. Johnson, and L.F. Hodges. 1999. Testbed evaluation of virtual environment interaction techniques. *Presence: Teleoperators and Virtual Environments* 10: 26–33. http://dl.acm.org/citation.cfm?id=323667.

Cobb, S.V.G. and P.M. Sharkey. 2007. A decade of research and development in disability, virtual reality and associated technologies : Review of ICDVRAT 1996 – 2006. *The International Journal of Virtual Reality* 6(2): 51–68.

Czaja, S.J. and C.C. Lee. 2007. The impact of aging on access to technology. *Universal Access in the Information Society* 5(4): 341–349. doi:10.1007/s10209-006-0060-x. http://dx.doi.org/10.1007/s10209-006-0060-x.

Duarte, E., F. Rebelo, J. Teles, and M.S. Wogalter. 2014. Behavioral compliance for dynamic versus static signs in an immersive virtual environment. *Applied Ergonomics* 45(5): 1367–1375.

Duarte, E., F. Rebelo, and M.S. Wogalter. 2010. Virtual reality and its potential for evaluating warning compliance. *Human Factors and Ergonomics in Manufacturing & Service Industries* 20(6): 526–537. doi:10.1002/hfm.20242. http://dx.doi.org/10.1002/hfm.20242.

Gabbard, J.L. 1997. *A Taxonomy of Usability Characteristics in Virtual Environments*. Blacksburg, Virginia: Faculty of the Virginia Polytechnic Institute and State University. http://scholar.lib.vt.edu/theses/available/etd-111697-121737/.

Gabbard, J.L., D. Hix, and J.E. Swan II. 1999. User-centered design and evaluation of virtual environments. *Computer Graphics and Applications, IEEE* 19: 51–59. doi:10.1109/38.799740.

Hix, D. and J.L. Gabbard. 2002. Usability engineering of virtual environments. In *Handbook of Virtual Environments: Design, Implementation and Applications*, edited by K. Stanney, pp. 681–699. Mahwah, New Jersey: Lawrence Erlbaum Associates. http://people.cs.vt.edu/jgabbard/publications/hvet02.pdf.

International Organization for Standardization. 2004. *Graphical Symbols—Safety Signs—Safety Way Guidance Systems (SWGS) (ISO 16069:2004)*. Geneva, Switzerland: International Organization for Standardization.

International Organization for Standardization. 2011. *Graphical Symbols—Safety Colours and Safety Signs—Part 1: Design Principles for Safety Signs and Safety Markings (ISO 3864-1:2011)*. Geneva, Switzerland: International Organization for Standardization.

Ishihara, S. 1988. *Test for Colour-Blindness*. 38th edn. Tokyo: Kanehara & Co., Ltd.

Kennedy, R.S., N.E. Lane, K.S. Berbaum, and M.G. Lilienthal. 1993. Simulator sickness questionnaire: An enhanced method for quantifying simulator sickness. *The International Journal of Aviation Psychology*. 3(3): 203–220. doi:10.1207/s15327108ijap0303_3.

Liu, C.-L. 2009. A neuro-fuzzy warning system for combating cybersickness in the elderly caused by the virtual environment on a TFT-LCD. *Applied Ergonomics* 40(3): 316–324. doi:10.1016/j.apergo.2008.12.001. http://www.ncbi.nlm.nih.gov/pubmed/19144322.

Machado, S., E. Duarte, J. Teles, L. Reis, and F. Rebelo. 2012. Selection of a voice for a speech signal for personalized warnings: The effect of speaker's gender and voice pitch. *Work* (Reading, Massachusetts) 41(Suppl 1): 3592–3598. doi:10.3233/WOR-2012-0670-3592. http://www.ncbi.nlm.nih.gov/pubmed/22317268.

Mayhorn, C.B. and K.I. Podany. 2006. Warnings and aging: Describing the receiver characteristics of older adults. In *Handbook of Warnings*, edited by M.S. Wogalter, pp. 355–361. Mahwah, New Jersey: Lawrence Erlbaum Associates.

McLaughlin, A.C. and C.B. Mayhorn. 2014. Designing effective risk communications for older adults. *Safety Science* 61: 59–65. doi:10.1016/j.ssci.2012.05.002. http://www.sciencedirect.com/science/article/pii/S0925753512001129.

Moffat, S.D. A.B. Zonderman, and S.M. Resnick. 2001. Age differences in spatial memory in a virtual environment navigation task. *Neurobiology of Aging* 22(5): 787–796. http://www.ncbi.nlm.nih.gov/pubmed/11705638.

Nichols, S. and H. Patel. 2002. Health and safety implications of virtual reality: A review of empirical evidence. *Applied Ergonomics* 33(3): 251–271. http://www.ncbi.nlm.nih.gov/pubmed/12164509.

Pacheco, C., E. Duarte, F. Rebelo, and J. Teles. 2010. Using virtual reality for interior colors selection and evaluation by the elderly. In *Advances in Cognitive Ergonomics*, edited by D. Kaber and G. Boy, 5234, pp. 784–792. *Advances in Human Factors and Ergonomics Series*. Boca Raton, Florida: CRC Press. doi:10.1201/EBK1439834916. http://www.crcnetbase.com/doi/book/10.1201/EBK1439834916.

Rizzo, A. and G.J. Kim. 2005. A SWOT analysis of the field of virtual reality rehabilitation and therapy. *Presence: Teleoperators and Virtual Environments* 14(2): 119–146. doi:10.1162/1054746053967094. http://www.mitpressjournals.org/doi/abs/10.1162/1054746053967094.

Rousseau, G.K., N. Lamson, and W.A. Rogers. 1998. Designing warnings to compensate for age-related changes in perceptual and cognitive abilities. *Psychology & Marketing* 15(7): 643–662.

Stanney, K.M., M. Mollaghasemi, L. Reeves, R. Breaux, and D.A. Graeber. 2003. Usability engineering of virtual environments (VEs): Identifying multiple criteria that drive effective VE system design. *International Journal of Human-Computer Studies*. 58: 447–481. doi:10.1016/S1071-5819(03)00015-6. http://linkinghub.elsevier.com/retrieve/pii/S1071581903000156.

Wilson, J.R. 1999. Virtual environments applications and applied ergonomics. *Applied Ergonomics* 30(1): 3–9. http://www.ncbi.nlm.nih.gov/pubmed/10098812.

Witmer, B.G. and M.J. Singer. 1998. Measuring presence in virtual environments: A presence questionnaire. *Presence* 7(3): 225–240. http://www.mitpressjournals.org/doi/abs/10.1162/105474698565686.

19 Percept Walk
Promoting Perception Awareness in the Elderly with Low Vision

Miguel de Aboim Borges and Fernando Moreira da Silva

CONTENTS

19.1 INTRODUCTION

As part of an ongoing research project—Percept Walk, developed at CIAUD—Research Center in Architecture, Urbanism and Design at the Faculty of Architecture of the University of Lisbon, this research has as its final objective the promotion of an independent way for elderly people with a low vision condition to act autonomously in hospital environments using a sensorimotor wayfinding system, combining a kinaesthetic haptic perception system applied on the floor with an adjusted low vision pathologies wayfinding system.

This PhD research project overlaps different research areas, using literature review and case study methodology, using relevant international case studies, and has an experimental phase developed in two Portuguese hospitals, using participatory design.

As main result, it is expected that the developing of new tools and the creation of an information design model supported by a user-centered design process for an orientation system that, besides the most important act of orientation, will also promote independency and, consequently, self-esteem in people with visual impairment. It is also expected that, giving nurses and medical professionals more time to do their job and not having to help users who cannot perceive and interpret the space, by showing the destination required at all times, their work will be more valuable for the institution where they work.

A number of researchers have commented on the limited number of studies dealing with person–environment fit in hospitals (Devlin and Arneill, 2003, p. 667). Steinfeld proposes, within a universal design approach, that in order to build an environment that eliminates unnecessary expenditure of people's effort, it is necessary to organize space and design devices that simplify the task of using it (i.e., a building), eliminating useless movements, and the use of color and textures contrasts contributing to this simplification of the environment (Imrie and Hall, 2004, p. 15). Ostroff states that the designer can learn a great deal from the experience of the potential consumers, where the needs and limitations of users may be unfamiliar (Imrie and Hall, 2004, p.15). In this matter RNIB reinforces the need of breaking down barriers and exclusiveness through a design approach that sets out to include as many people as possible, without denying the need for solutions to meet the needs of specific type of impairments (RNIB—Royal National Institute for the Blind in Barker et al., 1995, p. 13).

19.2 CONTEXT

A new paradigm has significantly transformed the world's demographic organization in the last century, as a result of a rapid increase in the older population, as a reflex of the decline of birth rate and the increase of life expectancy. Today the elderly have more longevity, reaching old age later and with a better quality of life. But this better quality of life does not reflect possible problems related to health. Aging is related with the state of health and implicitly related with the diminishing of general functioning (Kent and Butler, 1988 apud Fontaine, 2002, p. 55) and it is of common knowledge that these problems grow in a constant form (Birren and Shaie, 2006) (Figure 19.1).

Some biological changes occur in the human organism with aging and may influence human behavior, such as a decline in mobility and dexterity, a difficulty in balance handling, a decrease in strength, and also a reduction in sensory acuity. These disabilities related to aging affect the way elderly people perceive and interact with the built environment. Some changes in how we communicate visual information in buildings, mainly hospitals, ought to be reviewed and redesigned in response to this ederly population increase, who in reality are their main users.

19.3 WAYFINDING

The word wayfinding exists in the English language since the sixteenth century, with the sense of traveling (Berger, 2009). A generation of designers emerged after the economic boom from the United States in the 1950s, with important names in graphical material production emigrating to the United States, such as Paul Rand, Saul Bass, William Golden, who recognized the need for the

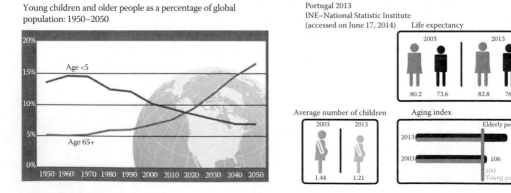

FIGURE 19.1 World population prospects: The 2010 review United Nations (WHO, 2011; available at: http://www.who.int/ageing/publications/global_health/en/) (left image) and INE, National Statistic Institute and demographic statistics for Portugal 2003/2013 (available at: https://www.ine.pt) (right image). (By author Aboim Borges.)

introduction of signage and orientation systems in buildings, companies, and cities in response to the multiracial and linguistic public (Berger, 2009), a reflex of the emigration between 1880 and 1930 (Massey, 1995). In the period referred, 28 million emigrants, from different countries, entered the United States as described by Handlin in *The uprooted: The epic story of the great migrations that made the American people* (Handlin, 1951). In result of this massive emigration and consequent multiculturalism and profusion of languages, cities became bigger, needing a restructuration and the development of visual information and orientation systems so they could be used by all citizens, both existing and new ones. In the 1960s the word *wayfinding* was used for the first time in Kevin Lynch's (1960) published book *The Image of the City* that was based on spatial orientation and its precondition the cognitive map, what Lynch defined as the "image," that was the element of recognition of some already visited points that act as reference and localization. In his book the city was characterized through paths, edges, references, nodes, and districts designations that are today still used in information design theory. Lynch transmits us the idea that we are not simple observers, but in a stage, in conjunction with other participants, in the observation of this references (images) that we are absorbing from the city, even if the city is never an accomplished project but something in constant evolution. In the 1970s was created the SEGD—*Society of Environmental Graphic Design* in response to the need of the creation and sharing of technical information and counseling of this new specialization area. The wayfinding system, this new discipline, had in 1968 an important role on the definition of accessibility in the Mexican Olympic Games, where Lance Wyman developed a program of information design covering the subway distribution scheme for Mexico City, the brand for the event, and the public visual information system composed by pictograms of all the modalities involved (Berger, 2009). This representative way of communicating had a more universal understanding by a public that came from different parts of the globe.

This design generation applied principles of different valences of design, architecture, color theory, and symbols in the information construction that would deconstruct and would make the built environment accessible to all (Golledge, 1999).

Charles Hilgenhurst stated in the sequence of the cities, increase into a scale where new tools would be necessary in orienting people and for the appropriation feel (Berger, 2009, p. 19): *"Today we are strangers in our own towns. We do not know how and cannot see it works. Our support systems … are remote. The information supplied in the environment is largely irrelevant to our immediate purposes or to an understanding of the world in which we live."*

Wayfinding is the process of following a path or route between an origin and a destination (Golledge, 1999, p. 6) and can be applied to a variety of situations and environments, using for this effect a compromise between a pictogram of a broad interpretation and a text as an identification of the destination. It assumes an approach of a deconstruction of a space, converting it into being understandable and usable through a study, a diagnosis, and an evaluation of a tenet set, so that a specific space performs that primary objective: orientation through graphic and textual information and the support of three-dimensional graphic elements, a spatial language that permits a more realistic space analogy (Larkin and Simon, 1987 apud Freksa, 1999). Wayfinding is a natural process that people learn since childhood (Piaget and Inhelder, 1967 apud Freksa, 1999) and it is a purposive, directed, and motivated activity (Golledge, 1999, p. 6).

Wayshowing is proposed by Mollerup (2005) as the understanding for what needs to be introduced in a wayfinding system so that it produces the objectives identified, that is, leads a person to its destiny through a comprehensive structure of information with text and pictograms, previewing that this comprehension includes also people with disabilities.

In most public buildings, in particular hospitals, most wayfinding systems do not reflect some of the limitations identified, which in fact need to be adapted to information with effective color contrasts, dimensionally rewritten textual and pictogram/infogram information, or even haptic perception devices to more efficiently promote legibility and readability, orientation and mobility, and above all autonomy and well-being. A wayfinding system should aid design features that could reinforce multisensory cues to accommodate people with sensory limitations (Steinfeld and Maisel,

2012, p. 276), and also for those who are visiting a building for the first time, or occasional visitors for whom the environment is an unknown place.

Nelson-Shulman (1983–1984) found that patients who have the benefit of an information system are more self-reliant and make fewer demands on staff (Rooke et al., 2009, p. 5). Carpman et al. suggest that directional signs should be placed at or before every major intersection, at major destinations, and where a single environmental cue or a series of such cues (e.g., changes in flooring material) convey the message that the individual is moving from one area into another (Rooke et al., 2009, p. 5). Heulat links good wayfinding with good patient flow, and asserts that applying simple organizational architecture and graphic principles not only reduce patient stress and anxiety, but also can lead to improved health (Rooke et al., 2009, p. 5).

19.4 INCLUSIVE DESIGN

"The need for design change is not limited to consumer products. It should also be a priority for designers involved in the public services. From design of printed matter and communications and information technology, to the design of transport, housing and public buildings, a better understanding of users' needs, can dramatically improve the independence and quality of life of the vast number of older users. But for these endeavors to be most effective, we need to go beyond the numbers, to understand the lifestyles of today's older adults, as well as their physical and mental capabilities" (Huppert, 2003, p. 31).

Inclusive design also called Ddsign for all or universal design has of recent years become synonymous with a designed world enabling everybody to participate in life and the activities taking place in our society on equal terms (Kenning and Ryhl, 2002, p. 6) and emerged out of the disabilities rights movement ensuring equal opportunity and eliminating discrimination based on disabilities (Kenning and Ryhl, 2002, p. 15). If we do not implement universal design now, the economic burden of an aging society will even be greater in the future (Steinfeld and Maisel, 2012, p. XII). Design is an active, purposeful adaptation method that people use to adjust their world to their needs (Steinfeld and Maisel, 2012, p. 1). Inclusive design constitutes a framework and growing body of practice within which business decision makers and design practitioners can understand and respond to the needs of diverse users, with the ultimate aspiration of developing products and services that can meet the needs of the whole population within the context of a consumer society (Coleman et al., 2003, p. 10).

The design council refers that an inclusive environment does not attempt to meet every need, but by considering people's diversity can break down barriers and exclusion and will achieve solutions that benefit everyone (The principles of inclusive design, 2006). Places need to be designed so that they can adapt to changing uses and demands (The principles of inclusive design, 2006). Everyone will at some time experience limited mobility or visual problems.

As referred, people are living longer and the lifespan has increased. Potential consumers or users of the design who may be functionally limited by age or disability are increasing at a dramatic rate, representing a significant population (Story et al., 1998, p. 14). An understanding of human diversity is critical to designing effectively, and a successful application of universal design principles requires an understanding of how abilities vary with age, disability, the environment, or the circumstances (Story et al., 1998, p. 17).

Buildings should provide appropriate design features and navigational aids to enable people with a range of sensory impairments to move around with confidence and ease (Imrie and Hall, 2004, p. 3). The principles of inclusive design are important and potentially progressive in seeking to restore disabled people's self-esteem, dignity, and independence, while encouraging the development and implementation of user-friendly design (Imrie and Hall, 2004, p. 16).

Salmen and Ostroff suggest that designers cannot get information from books, databases, or design criteria alone, they must involve the future users, the consumers of the design, and develop a process which is broadly representative, user responsive, and participatory (Imrie and Hall, 2004, p. 15).

19.5 USER-CENTERED DESIGN

Park (Norman, 1990, p. 188) refers to user-centered design as a philosophy (POET—*The psychopathology of everyday things*) based on the needs and interests of users, with an emphasis on making products usable and understandable, on which he refers the main principles for evaluation:

- Make it easy to determine what actions are possible at any moment
- Make things visible, including the conceptual model of the system, the alternative actions, and the results of actions
- Make it easy to evaluate the current state of the system
- Follow natural mappings between intentions and the required actions; between actions and the resulting effect; and between the information that is visible and the interpretation of the system state

These recommendations besides placing the user at the center of the evaluation process, focuses on the role of the designer, which is assuring the usage of that particular product with a minimum effort. The design process has to involve the user by an investigation and identification of their needs by performing the tasks and analysis thereof. User-centered design is not giving the user what he wants, but identifying and mapping all necessary data through a participatory design methodology. When developing a hospital wayfinding system for elderly people with visual disabilities, knowing how they perceive space and act within it will contribute in a more effective way in design decisions. The evaluation of needs is being developed with a battery of questionnaires, visual and haptic tests, subdivided into four main phases (see Section 19.11). The importance of knowing how users sense and perceive the environment is a key issue for the design process.

19.6 DISABILITIES AMONG ELDERLY PEOPLE

"Our senses are the instruments of communication that facilitate our relationship with the environment" (Meerwein et al., 2007, p. 13).

19.6.1 Visual Disabilities

Increased longevity due to the progress of medical science and socioeconomic conditions has brought an apparent well-being for a longer period of time. But reality shows that a large part of older people present a major dependency, in general, related to health problems. This situation includes the population that as a result of aging have declining visual capabilities (Tielsch et al., 1995 apud Gohar, 2009) with the gradual aging of the functioning of the eyes, with changes in the retina and the eyes' nervous system affecting acuity, accommodation, speed of adaptation to change, and a variety of perceptual disabilities such as blurring, decrease spatial abilities, loss of color discrimination, etc. (Wijk and Sivik, 1995 apud Gohar, 2009). We all can suffer from low vision, with major incidence in the older age-groups (CEBV, s.d.). Most of these visual losses are related to low vision, in people over 60 years, and it provokes intensive medical care needs in hospital ophthalmology services. With age, people change physically, mentally, and psychologically and these changes involve multiple, minor impairments in eyesight, hearing, dexterity, mobility, and memory (Haigh, 1993 apud Coleman et al., 2003, p. 121).

The effects on aging in the visual system appears at 40 years on the optical structure and from 60 years on the retinal structure (Fontaine, 2002, pp. 72–74) producing modifications in both optical structure and retinal system. In the optical structure, it affects the transmissibility and its accommo-dation capacity, translated into problems of the distance perception of objects, depth, and sensibility to obfuscation and colors. This happens due to modifications in the four optical structures:

- Cornea opacifies becoming thick and rigid provoking an augmenting of astigmatism and blurred vision
- Atrophy of ocular muscles linked to the crystalline (the lens that permits the projection of images on the retina) provoking a decrease in the capability of eye accommodation, result-ing in difficulty of seeing objects when near
- Crystalline becomes rigid causing a diminishing in the accommodation capability, and yellows, modifying the light composition projected on the retina and sometimes becomes opaque, causing cataract in the elderly
- Posterior chamber, containing the vitreous body liquid, liquefies and has a tendency with aging to become clearer and gelatinous and the result is a augmentation to sensibility and obfuscation

On the retinal system the receptor cells (cones responsible for seeing in colors and rods respon-sible for seeing in black and white) are located in the fovea (the central part of the macula) and suffer chemical alterations by lighting excitation, which results in the generation of nervous influxes that reach the occipital areas of the brain through the optic nerve. With aging the probability of macula cells' degeneration is augmented resulting in the loss of fine detail vision. These effects on vision and also in hearing are general cognitive functioning sensorial deficits and of a central nature, asso-ciated to neuronal deteriorations, as seen in some research works on elderly people aged over 60–70 years (Werner et al.,1990; Fozard, 1990 apud Fontaine, 2002, p. 75), although other researchers do not see a significant association in these deficits (Schaie et al., 1964; Horn, 1980; Raz et al., 1990 apud Fontaine, 2002, p. 75).

There is considerable diversity in visual capabilities within the older population, although in general, prevalence of visual impairment accelerates after age 65 (Fisk et al., 2004, p. 49). The diversity is due in part to variability in aging processes, in the increased use of assistive devices for age-related changes compensation, and also the increased frequency of surgical interventions that modify the visual system (Fisk et al., 2004, p. 50).

19.6.2 Low Vision Pathologies

Few eyes that have survived 65 or more years of life are free from at least some slight sign of dete-rioration, degeneration, or past or present disease (Jackson and Wolffsohn, 2007, p. 78). Low vision could be defined as vision that, when corrected by optimal refractive correction, is not adequate for the patient's needs (Jackson and Wolffsohn, 2007, p. 10). Also it has to be seen as the attention to the retained residual vision and not that which has been lost (Figure 19.2).

The most prevalent etiologies in low vision are AMD (age-related macular degeneration) respon-sible for the majority of cases in severe loss of central vision, cataract symptoms include decrease in visual acuity, color perception, and contrast sensitivity, glare disability and gradual onset of blurred vision in distance, glaucoma with either visual field (peripheral) loss and/or, difficulty in functioning in dim light, decrease in contrast sensitivity, glare disability, and decrease in dark/light adaptation, diabetic retinopathy characterized as blurred vision (Jackson and Wolffsohn, 2007, pp. 77–92; Rosenberg and Sperazza, 2008). Sighted persons explore the environment with the sensory apparatus available, and so does the blind and the visually impaired primarily through tactile, auditory, olfactory, and kinesthetic information gathering (supplemented by any residual vision) (Barker et al., 1995, p. 14). For the creation of buildings or environments that respond effectively to the needs of the visually impaired people, it is important that there should be some understanding of the nature of visual loss (Barker et al., 1995, p. 21). Any design change must be useful, practical,

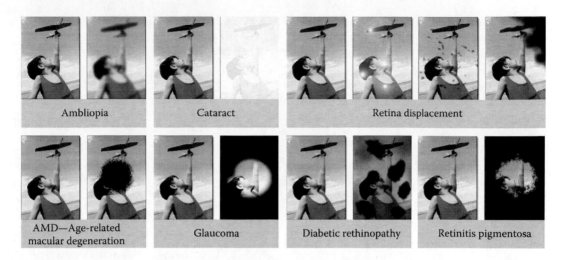

FIGURE 19.2 Low vision pathologies adapted. (From http://www.cebv.pt, accessed May 20, 2011.)

empowering or enabling, and the individual user is always the best judge of any building modification, therefore the participation of the visually impaired in the design process being essential (Barker et al., 1995, p. 18). The elderly visually impaired are the group that optometric practitioners deal more with, and most of them react well with the situation, although a small minority has never accepted the situation and isolate themselves in their homes.

19.7 SELF-ESTEEM AND INDEPENDENCE

Self-esteem refers to a person's sense of value and worth, of competence and adequacy, of self-satisfaction (Tuttle and Tuttle, 2004, p. 6). "Self-esteem is the lived status of one's competence in dealing with the challenges of living in a worthy way over time" (Mruk, 1999 apud Tuttle and Tuttle, 2004, p. 6). Success in independent travel tasks, even at the beginning levels, can bring about an improved self-concept, a greater sense of independence, and improved motivation for other tasks, but the opposite may lead to an increase of dependency, isolation (Welsh, 1980 apud Tuttle and Tuttle, 2004, p. 28). Self-esteem and self-satisfaction result from independence acting on the decision of doing things by ourselves and not relying or expecting from others to accomplish that specific task. Chapman describes the acquisition of independence as a gradual process of decision making, an act by itself and responsible in nature (Tuttle and Tuttle, 2004, p. 59). Research is beginning to show that self-esteem may be an important factor in health behaviors, coping, and well-being (Birren, 2007, p. 461). Welsh (1980 apud Tuttle and Tuttle, 2004, p. 28) points that the lack of independent travel tasks may lead to increased dependency, isolation, and hopelessness but also to the opposite, success in independent travel tasks, even at the beginning levels, can cause an improved self-concept, a greater sense of independence, and improved motivation for other tasks. One of the key ingredients of better adjusting patterns is a healthy self-concept or positive self-esteem, the affective component of self-concept (Tuttle and Tuttle, 2004, p. 56).

Information design oriented to this target group can act as a way of promoting this self-esteem through independent mobility. By analyzing through a participatory design process users' usage of the hospital (identifying main circulation axes, existing navigational signage, space materials, and lighting characteristics), and a space existence characterization, the information obtained in both fields will constitute an important tool in diagnosing the essentials for building an effective action plan for that determined space. Although the research seeks the relation of patients with a space, it proposes overall a simplification of the orientation information. Simplifying a wayfinding system in a determined space will without doubt improve its comprehension and promote self-esteem by promoting an independent mobility to its users.

When walking, most of the elderly with low vision defend themselves from unbalance or falling by dragging their feet in order to feel the ground. By this defensive way of walking they can react when perceiving eventual irregularities on the ground, because falling is feared by the elderly representing a long time in hospital or at home in bed due to extended time for bone recovery, a factor related to aging. Klatzky and Lederman (2003, p. 147) describe this action of exploratory movements as an active modality for perceiving information from the world.

19.8 VISUAL AND HAPTIC PERCEPTION

19.8.1 Perception

Perception is the process of interpreting information received from the senses (Goldstein, 2010 apud Steinfeld and Maisel, 2012, p. 108).

Perceptions are a set of physiological and psychological mechanisms whose general functions are to gather information in the environment or in their own organisms (Lieury, 1990 apud Fontaine, 2002) and are polysensorial because they are product of an integration of diverse sensorial messages. We all use more than one perception to sense a determined space or an action. In everyday perception, touch and vision operate together (Klatzky and Lederman, 2003, p. 147). Perception is the awareness of more complex characteristics of the stimuli, the activation of the sensation cells, and the interpretation of that information by calling on stored memory (Fisk et al., 2004, p. 13). However, limitations associated with perception, cognition, and the control of movements increase in prevalence as one ages.

In order to avoid falling, elderly people with low vision walk in a defensive way so they can react when perceiving irregularities on the ground. Klatzky and Lederman (2003, p. 147) describe this action as exploratory movements as an active modality for perceiving information from the world. But our ability to cope effectively with the environment begins with our capacity to process the sensory input (Schieber apud Birren and Shaie, 2006, p. 129). Our senses have been carefully crafted by the forces of nature to effortlessly extract critical information from the world around us, although advancing adult aging brings with it systematic reductions of efficiency of the sensory systems (Birren and Shaie, 2006).

People over 65 years of age are more subject to develop disabilities related to aging, showing difficulties with vision and balance, that can condition the capacity of acting independently and performing daily tasks. The ability to cope effectively with the environment begins with the capacity to process sensory input but advancing adult aging brings with it systematic reductions in efficiency of our sensory systems (Schieber, 2003 apud Birren and Shaie, 2006, p. 129). This perception of the environment was defined by Lieury (1990) as a set of psychological and physiological mechanisms whose general function is environmental information gathering, not being a passive reception of messages proceeding from the environment, but a set of complex activities of reception and analysis, where it can be distinguished in the first phase as a reception of signals (sensation) proceeding from the environment and its transformation in nervous influxes (Fontaine, 2002, pp. 61–62). Our perceptions are plurisensorial for they result from diverse sensorial messages integration (Fontaine, 2002, p. 61). These later-life scenarios in the increasing demands placed on elders by shifts on their personal resources and changes in their near and far environments are the object of study of environment–aging relations,[*] supported by different disciplines (e.g., psychologists, sociologists, health professionals architects, community planners, social policy makers, designers) dedicated to the understanding of the behavioral and psychological implications of encounters between elders and their environment (Scheidt and Windley apud Birren and Shaie, 2006, p. 105). The environmental

---

[*] Environmental gerontology.

gerontology can benefit from action research, a collaborative knowledge building involving clients, practitioners, consultants, and researchers to produce knowledge useful to the everyday lives of people (Senge and Scharmer, 2001 apud Birren and Shaie, 2006, p. 114). The primary purpose in action research in environmental gerontology is quality improvement for older persons through a collaborative research building, contributing to better design and environmental modification information of social and physical living arrangements (Scheidt and Windley apud Birren and Shaie, 2006, p. 114). A wider view integrating theory and practice, environmental gerontology theorists and environmental design practitioners, as well as health professionals, can improve the quality of life of older people (Birren and Shaie, 2006). The ultimate goal of science and practice of human factors is to ensure that human–system and human–environment interactions will be safe, efficient, and effective (Fisk et al., 2004, p. 13).

19.8.2 HAPTIC PERCEPTION

During the early-to-mid-twentieth century, only three individuals David Katz (1925/1989), Geza Révész (1878/1955), and James J. Gibson (1904/1979) stand out for their uncommon emphasis on the critical importance of active, voluntary manual exploration.

Pallasmaa (2005) in his book *The Eyes of the Skin—Architecture of the Senses* underlines the importance of the tactile sense for our experience and understanding of the world, over the visual sense. Referring Ashley Montagu's medical evidence on confirming the primacy of haptic realm he writes, "touch is the parent of our eyes, ears, nose and mouth. It is the sense which became differentiated into the others, a fact that seems to be recognized in the age-old evaluation of touch as the mother of the senses" (Pallasmaa, 2005, p. 11).

We should not adapt ourselves to the environment, but it is the environment that should be adapted to us, is one of the inclusive design principles (Herssens and Heylighen, 2008, p. 103).

In both haptic and visual perception, the stimuli rely on material (texture, temperature, density) and space characteristics (form, place, orientation, length) (Hatwell, 2003 apud Herssens and Heylighen, 2008). Information from movement output thus plays an important, and probably crucial, role in tactile recognition (Millar, 1994 apud Herssens and Heylighen, 2008). The elderly walk in a defensive and secure mode, feeling their way by dragging their feet. This situation permits them feeling the floor, avoiding the unlevelled and/or irregularities of the ground. The use of a haptic system (on the floor) in a wayfinding system in a hospital environment can act as a means to lead people in finding the information required and so promote better and more detached information and thus, assist those with low vision condition to find their way. This can be referred as the practical intelligence (Vygotsky apud Tijus, 2001, p. 25) where the-knowing-what-to-do is not expressed by language but by the action itself, and by Gibson's (1986) affordances theory which, by feeling the haptic texture on the floor, people will know what they are meant to do. Révèsz also suggests that in space perception the haptic space is centered on the body whereas vision is centered on external coordinates (Millar, 2006, p. 27).

The haptic system uses combined inputs from both the cutaneous and kinaesthetic systems (Klatzky and Lederman, 2003, p. 148). Haptics is commonly viewed as a perceptual system mediated by two afferent subsystems, cutaneous and kinaesthetic, that most typically involves active manual exploration (Lederman and Klatzky, 2009, p. 1439). The sensing modality of touch can be categorized into three main channels: *kinaesthetic* describing our bodily perception when moving, sensing the orientation and rotation of our muscles, joints, and tendons; *tactile stimuli* are applied in our skin, when we are passive; and a *haptic sensation* is produced during the exploration of an object (Loomis and Lederman, 1984 apud Pohl and Loke, 2012, p. 2). Touch is the parent of our eyes, ears, nose, and mouth. It is the sense that becomes differentiated from the others, a fact that seems to be recognized in the age-old evaluated of touch as the "mother of the senses" (Pallasmaa, 2005, p. 2). All the senses, including vision, are extensions of the tactile sense; the senses are specializations of the skin tissue, and all sensory experiences are modes of touching and thus related to tactility (Pallasmaa, 2005, p.10). Bachelard proposes polyphony of the senses as the way our senses interacts in space perception, a multisensory experience (Pallasmaa, 2005, p. 41).

19.9 PERCEPT WALK

Perception (Kerzel apud Binder et al., 2009, p. 3098) is the conscious experience of sensory stimulus and reflects the stimulation of the sensory system (e.g., eyes, ears, skin), and is also determined by higher cognitive processes, such as attention and memory. Perception is our mind's window on the world (Binder et al., 2009, p. 3098) and the awareness of more complex characteristics of stimuli (Fisk et al., 2004, p. 13).

The research project, Percept Walk, can be defined as the sensorial capacity that allows a human being to collect explicit information of the physical environment, in particular spatial organization, in order to act in conformity, which in the present research is of visual and haptic touch order. Within the project different phases of sensorial evaluations (see Section 19.11) will be held with low vision patients, users, and medical staff (Figure 19.3).

The use of a sensory-motor wayfinding systems joining a correct color use for legibility and readability of written information, associated with simplified pictograms, also applied in effective contrasts, and supported by a haptic system (foot touch) on the floor, may result in an effective way of displaying information in hospital environments. Older people with a low vision condition need a wayfinding system that can be seen by them at a relatively small distance.

The ability to detect and recognize objects in the visual environment varies considerably as a function of target size, contrast, and spatial orientation (Olzak and Thomas, 1985 apud Birren and Shaie, 2006, p. 141). Numerous studies reveal a consistent pattern of age-related change in the CSF (contrast sensitivity function) collected under photopic conditions (foveal presentation at moderate-to-high luminance levels). Contrast sensitivity declines by approximately 0.3 log units across the latter half of the adult life span (Birren and Shaie, 2006, p. 141).

The theme "color promotes wayfinding" has been largely studied and is still being studied by different authors (Carpman et al., 1983; Mahnke, 1996; Arthur and Passini, 2002; Mollerup, 2005; 2009; Meerwein et al., 2007; Gibson, 2009; Helvacıoğlu and Olguntürk, 2010; Zingale, 2010; Katz, 2012; Moreira da Silva, 2013).

What is of real importance is that the colors and the system used are effective for all persons that use that specific building (Figure 19.4).

A wayfinding system, using effective colors, contrasting the written information with the background color, associated with a haptic touch texture on the floor, as a tactile physical help in defining

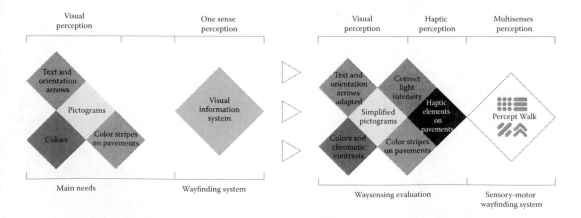

FIGURE 19.3 (**See color insert.**) Wayfinding system and Percept Walk wayfinding system. (By author Aboim Borges.)

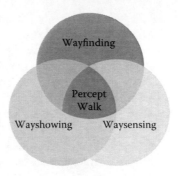

FIGURE 19.4 (**See color insert.**) Percept Walk diagram. (By author Aboim Borges.)

and signalizing the existing orientation information, offers the visually impaired user access for reading the information at his most convenient distance. There is an acceptability by people for touching floor surfaces, via the shoes, in contrast to people being averse to touching certain objects in public spaces (Visell et al., 2009, p. 149). The introduction of a sensorimotor task on the floor does not imply an increased sense since it is implicitly engaged in normal walking (Visell et al., 2009, p. 150). Active touch, what is ordinarily called touching, is an exploratory sense and may be termed tactile scanning, by analogy with ocular scanning (Gibson, 1962, p. 477). Visually impaired people as all humans compensate for deficiencies or weaknesses in one sensory modality by relying on another sense, without necessary being aware of it (Krippendorff, 2006, p. 143). We use all our senses to perceive and use the environment. We do not use one specific sense to do it, but as normal vision people sight is the first sense used, although leaving other senses aware. When using a building, in particular a hospital, we do not have to adapt ourselves to this environment, but it is the environment that has to be adapted to us (Herssens and Heylighen, 2007). There is a need to rethink and adapt the functionality of a particular building to its users' needs or special needs. Those are inclusive design objectives, aiming at usability and comfort for as many people as possible regardless of age, ability, or circumstance (Herssens and Heylighen, 2007). Our ability to cope effectively with the environment begins with our capacity to process sensory input, although advancing adult aging brings with it systematic reductions in the efficiency of our sensory systems (Birren and Shaie, 2006, p. 129). The Society of Technical Communication (STC) defines the discipline of information design as the translating of complex, unorganized, or unstructured data into valuable, meaningful information (Baer, 2008, p. 12).

People are living longer, remaining more active into older age, and preferring to stay in their homes longer before feeling the need for "assisted living" arrangements. Aging brings with it changes in perception, cognition, and movement control (Fisk et al., 2004, p. 4). The ultimate goal of science and practice of human factors is to ensure that human–system and human–environment interactions are safe, efficient, and effective (Fisk et al., 2004, p. 13). The ability to function independently is much related to our mental capabilities as to our physical capabilities (Huppert, 2003, p. 41).

19.10 CASE STUDY

The first phase of the experimental research is being held in a public ophthalmology hospital (serving around 5000 patients/month) offering a variety of services, such as consultations, exams, eye surgeries, and treatments.

This hospital unit, *IOGP—Instituto de Oftalmologia Dr. Gama Pinto* is the only public ophthalmology hospital existing in the country and its creation dates back to the nineteenth century when the kings D. Luiz and later D. Carlos I ruled. The building is a successful conversion from a palace into a hospital, and is located in the old part of Lisbon.

This hospital is exclusively dedicated to problems related to vision, beginning with identification through an initial screening, with different pathologies follow-up and control in specialized consultancies, surgeries when necessary, to the reinsertion of patients with extreme low vision conditions into active life through the teaching, and use of the white cane as an auxiliary mean of dislocation.

The attending structure disposes of a vast set of consulting, examination, and medical treatment rooms, offering a personalized reception environment on the attendance of general consultancy and subspecialties like retina, glaucoma, strabismus, refractive/external ocular surface surgery, ocular genetics, and low vision. The subvision (low vision) area offers full support to users through a multidisciplinary team. The number of medical consultations was around 47,000 people in 2010, ophthalmology being responsible for 89%, distributed by retina 20%, estrabismus–genetical–pediatry 7%, glaucoma 7%, ophthalmology general consultancy 52%, anterior segment 4%, and low vision 2%.

A general observation of the natural and artificial light conditions will permit the establishing and parameterizing the optimal condition for interpreting the orientation information for all pathologies and minimize the effects of shade in some areas. Through direct observation of the patients' and users' usage of the building, the main axes of circulation and the most used areas will be evaluated so that the localization of the necessary information for an effective wayfinding/waysensing system will be implemented. The importance of the patients' sensorial evaluation is, in this particular case, visual and haptic touch oriented. A research held in another environment with different special needs users may result in an approach to the evaluation of other senses, but still aiming for the promotion of a safe, inclusive, and structured usage of the space.

19.11 METHODOLOGY

The research is supported by a participatory design methodology applied to the hospital's low vision elderly users collected through interviews, questionnaires, and sensorial tests (visual and haptic) of their environment perception limitations information. The visual tests will evaluate the visual acuity, color, and contrasted color perception, legibility and readability in the different patients' low vision pathologies; the haptic tests will measure the foot haptic perception to different textures, and through the observation of their mobility behavior, locomotion and dexterity capabilities will also be evaluated.

The visual tests are supported by printed plates with different color contrasts, standard and simplified pictograms and texts with different typographies and scales shown in different lighting conditions, while haptic tests are based on different shapes and thicknesses to evaluate the most perceived and interpreted within the different patients' pathologies (Figure 19.5).

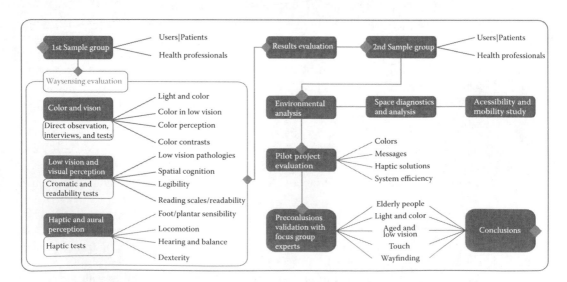

FIGURE 19.5   Waysensing: Research evaluation process. (By author Aboim Borges.)

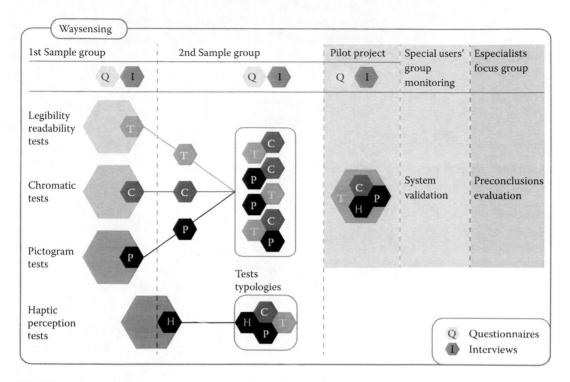

FIGURE 19.6   (**See color insert.**) Waysensing: Steps in evaluation process. (By author Aboim Borges.)

Through different phases of evaluation of the target group perceptions the research will seek the necessary answers building efficient information for the design project. All the data acquired will permit the obtaining of some preconclusions of the research phase that will lead to the development of the design process of creating a pilot project for a specific part of this hospital. This area will be a lab of experiences and tests with patients and users to validate the preconclusions and recommendations obtained in the prior phases (Figure 19.6).

Within the different phases of the waysensing process, the evaluation of the target group sensorial perceptions will seek the necessary answers, building efficient information for the design project. The behavior of users and medical staff will be analyzed and evaluated in both areas, the intervened and the actual one, and compared in order to validate that the proposed wayfinding system effectively responds to users' independent mobility, thus promoting self-esteem and confidence in the safe usage of the hospital.

In order to diagnose effective wayfinding needs, an analysis of the building accessibility and mobility, materials and lighting, users' flows, and existing visual communication will be carried out (Figure 19.7).

19.12 WORK IN PROGRESS AND FUTURE WORK

The Percept Walk sensory-motor wayfinding system represents a more sensorial way involving the deconstruction of the built environment, in particular ophthalmology hospitals for people with a low vision condition or other special needs. By promoting an independent mobility in these patients it will also promote their self-confidence and self-esteem. Having patients and users acting by themselves in the building is expected to leave more effective working time for doctors, auxiliary teams, and nurses who actually spent a great deal of time showing and helping users to find their way around, is the research hypothesis expected to be achieved.

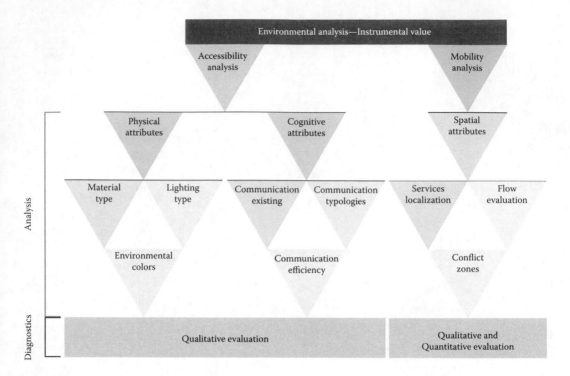

FIGURE 19.7   **(See color insert.)** Environmental analysis and diagnostics. (By author Aboim Borges.)

The Percept Walk system concept is perfectly applicable to situations with absence of light even if they are for people of normal vision, for instance auditoriums or exhibition places. There is always information that can be conveyed through haptics, especially when an excess of information or visual pollution is present.

The expected future work is in obtaining relevant data from the methodology purposed for the empirical phase developed in hospital, and the preparation for the pilot project installation and its implementation. The wayfinding system that will be achieved will result from the participatory design methodology involving all users of the hospital, the evaluation by a focus group of specialists, and the validation by users. As final result it is expected to develop a wayfinding system that will simplify and communicate effectively with all users, by promoting an easier, safer, uncomplicated way of using this space. This study will compile important data that can help information designers to understand the limitations of the elderly especially those with low vision, and develop sustainable efficient wayfinding systems.

REFERENCES

Arthur, P. and Passini, R. 2002. *Wayfinding—People, Signs and Architecture* (Vol. 1). Ontario: Focus Strategic Communications Incorporated.

Baer, K. 2008. *Information Design Workbook: Graphic Approaches, Solutions, and Inspiration+ 30 Case Studies*. Massachusetts: Rockport Publishers.

Barker, P., Barrick, J., and Wilson, R. 1995. *Building Sight—A Handbook of Building and Interior Design Solutions to Include the Needs of Visually Impaired People*. London: Royal National Institute for the Blind.

Berger, C. M. 2009. *Wayfinding: Designing and Implementing Graphic Navigational Systems*. Mies: RotoVision.

Binder, M. D., Hirokawa, N., and Windhorst, U. 2009. *Encyclopedia of Neuroscience*. Springer. Retrieved from http://library.wur.nl/WebQuery/clc/1243349.

Birren, J. E. 2007. *Encyclopedia of Gerontology*. Oxford: Elsevier.

Birren, J. E. and Shaie, K. W. (Eds.). 2006. *Handbook of the Psychology of Aging* (6th ed., Vol. 2). London: Elsevier Academic Press.

Carpman, J. R., Grant, M. A., and Simmons, D. A. 1983. Wayfinding in the hospital environment: The impact of various floor numbering alternatives. *Journal of Environmental Systems*, 13(4), 353–364.

CEBV. (s.d.). CEBV—Centro especializado da baixa visão. Retrieved from http://www.cebv.pt/index.php.

Coleman, R., Lebbon, C., Clarkson, J., and Keates, S. 2003. *Inclusive Design: Design for the Whole Population*. London: Springer Science & Business Media.

Devlin, A. S. and Arneill, A. B. 2003. Health care environments and patient outcomes a review of the literature. *Environment and Behavior*, 35(5), 665–694. http://doi.org/10.1177/0013916503255102.

Fisk, A. D., Rogers, W. A., Charness, N., Czaja, S. J., and Sharit, J. 2004. *Designing for Older Adults: Principles and Creative Human Factors Approaches*. Boca Raton: CRC press.

Fontaine, R. 2002. *Psicologia do Envelhecimento* (Vol. 1). Lisboa: Climepsi Editores.

Freksa, C. 1999. Spatial aspects of task-specific wayfinding maps. In: J. S. Gero and B. Tversky (eds.), *Key Centre of Design Computing and Cognition*, University of Sydney, Sidney, pp. 15–32.

Gibson, D. 2009. *The Wayfinding Handbook: Information Design for Public Places*. New York: Princeton Architectural Press.

Gibson, J. J. 1962. Observations on active touch. *Psychological Review*, 69(6), 477–491. Retrieved from http://wexler.free.fr/library/files/gibson%20(1962)%20observations%20on%20active%20touch.pdf.

Gibson, J. J. 1986. *The Ecological Approach to Visual Perception*. New Jersey: Lawrence Erlbaum Associates, Inc.

Gohar, N. 2009. The application of colour and colour contrast in the home environment of the elderly and visually impaired individuals. [http://www.homemods.info/]. Retrieved October 1, 2013, from http://www.homemods.info/resource/bibliography/application_colour_and_colour_contrast_home_environment_elderly_and_visually_impaired_individuals.

Golledge, R. G. 1999. *Wayfinding Behavior—Cognitive Mapping and Other Spatial Processes* (Vol. 1). Baltimore: The John Hopkins University Press. Retrieved from www.press.jhu.edu.

Handlin, O. 2002. *The uprooted: The epic story of the great migrations that made the American people.* University of Pennsylvania Press. (Originally published in 1951.)

Helvacıoğlu, E. and Olguntürk, N. 2010. Colour and wayfinding, pp. 464–468. Retrieved from http://rice.iuav.it/247/1/03_helvacioglu-olgunturk.pdf.

Herssens, J. and Heylighen, A. 2007. Haptic architecture becomes architectural hap. Retrieved from https://uhdspace.uhasselt.be/dspace/handle/1942/13897.

Herssens, J. and Heylighen, A. 2008. Haptics and vision in architecture—Designing for more senses. In R. Lucas and G. Mair (Eds.), (pp. 102–112). *Presented at the Sensory Urbanism Proceedings, University of Strathclyde*, Glasgow: The Flâneur Press. Retrieved from https://doclib.uhasselt.be/dspace/handle/1942/13900.

Huppert, F. 2003. Designing for older users. In J. Clarkson, R. Coleman, S. Keates, and C. Lebbon (Eds.), (pp. 30–49). *Inclusive Design: Design for the Whole Population*. London: Springer-Verlag London Ltd.

Imrie, R. and Hall, P. 2004. *Inclusive Design: Designing and Developing Accessible Environments*. London: Taylor & Francis.

Jackson, A. J. and Wolffsohn, J. S. 2007. *Low Vision Manual*. Butterworth Heinemann/Elsevier. Retrieved from http://www.elsevierhealth.com/us/product/toc.jsp?isbn=9780750618151.

Katz, J. 2012. *Designing Information: Human Factors and Common Sense in Information Design*. New Jersey: John Wiley & Sons, Inc.

Kenning, B. and Ryhl, C. 2002. *AaOutils: Teaching Universal Design*. Retrieved January 20, 2003.

Klatzky, R. L. and Lederman, S. J. 2003. Touch. *Handbook of Psychology—Experimental Psychology* (Vol. 4, pp. 147–176). Retrieved from http://www.psy.cmu.edu/~klatzkyfaculty/Touch.prepub.pdf.

Krippendorff, K. 2006. *The Semantic Turn: A New Foundation for Design*. Philadelphia: CRC Press.

Lederman, S. J. and Klatzky, R. L. 2009. Haptic perception: A tutorial. *Attention, Perception, & Psychophysics*, 71(7), 1439–1459. Retrieved from http://link.springer.com/article/10.3758/APP.71.7.1439.

Lynch, K. 1960. *The Image of the City*. Cambridge: MIT press.

Mahnke, F. H. 1996. *Color, Environment and Human Response*. New York: Van Nostrand Reinhold.

Massey, D. S. 1995. The new immigration and ethnicity in the United States. *Population and Development Review*, 631–652. Retrieved from http://www.jstor.org/stable/10.2307/2137753.

Meerwein, G., Rodeck, B., Mahnke, F. H. et al. 2007. *Color: Communication in Architectural Space*. Birkhauser Verlag. Retrieved from http://www.degruyter.com/view/product/201873.

Millar, S. 2006. *Processing Spatial Information From Touch and Movement: Implications From and for Neuroscience*. In M. A. Heller and S. Ballesteros (Eds.) (Vol. Touch and blindness: Psychology and neuroscience). London: Lawrence Erlbaum Associates Publishers. Retrieved from http://www.lavoisier.fr/livre/notice.asp?id=OSOWASAKL3XOWO.

Mollerup, P. 2005. *Wayshowing* (pp. 112–114). Baden: Lars Muller Publishers.

Mollerup, P. 2009. Wayshowing in hospital. *Australasian Medical Journal*, 1(10), 112–114. Retrieved from http://researchbank.swinburne.edu.au/vital/access/services/Download/swin:15045/SOURCE2.

Moreira da Silva, F. 2013. *Cor e inclusividade—Um projecto de Design de comunicação visual com idosos (Colour and Inclusivity—A visual Communication Design Project with Older People) (Caleidoscópio)*. Lisboa: Caleidoscópio.

Norman, D. A. 1990. *The Design of Everyday Things* (Vol. 1). New York: Library of Congress-in-Publication Data.

Pallasmaa, J. 2005. *The Eyes of the Skin: Architecture and the Senses*. London: Wiley.

Pohl, I. M. and Loke, L. 2012. Engaging the sense of touch in interactive architecture. In *Proceedings of the 24th Australian Computer-Human Interaction Conference*. University of Sidney, Australia (pp. 493–496). Retrieved from http://dl.acm.org/citation.cfm?id=2414611.

Rooke, C. N., Tzortzopoulos, P., Koskela, L. J., and Rooke, J. A. 2009. Wayfinding: Embedding knowledge in hospital environments. *HaCIRIC*, Imperial College Business School, London, UK (pp. 158–167). Retrieved from http://usir.salford.ac.uk/id/eprint/3411.

Rosenberg, E. A. and Sperazza, L. C. 2008. The visually impaired patient. *American Family Physician*, 77(10), 1431–6. Retrieved from http://schindlermd.com/visually%20impaired%20article.pdf.

Steinfeld, E. and Maisel, J. 2012. *Universal Design: Creating Inclusive Environments*. New Jersey: John Wiley & Sons.

Story, M., Mueller, J., and Mace, R. 1998. The universal design file: Designing for people of all ages and abilities. *Design Research and Methods Journal*, 1(1). Retrieved from http://design-dev.ncsu.edu/openjournal/index.php/redlab/article/view/102.

The Principles of Inclusive Design. 2006. *Design Council*. Retrieved from http://www.designcouncil.org.uk/knowledge-resources/guide/principles-inclusive-design.

Tijus, C. 2001. *Introdução à Psicologia Cognitiva*. In J. N. de Almeida (Trans.). Paris: Climepsi Editores.

Tuttle, D. W. and Tuttle, N. R. 2004. *Self-Esteem and Adjusting with Blindness: The Process of Responding to Life's Demands*. Charles C Thomas Publisher, Illinois.

Visell, Y., Law, A., and Cooperstock, J. R. 2009. Touch is everywhere: Floor surfaces as ambient haptic interfaces. *Haptics, IEEE Transactions on*, 2(3), 148–159. Retrieved from http://ieeexplore.ieee.org/xpls/abs_all.jsp?arnumber=5166445.

WHO. 2011. *Global Health and Aging*. World Health Organization. Retrieved from http://www.who.int/en/.

Zingale, S. 2010. Wayfinding using colour: A semiotic research hypothesis. In *Design and Semantics of Form and Movement* (pp. 22–32). Lucerne, Switzerland.

20 Designing Interfaces for Seniors in the Context of Healthcare

Audrey Abi Akle, Katarzyna Borgieł, Stéphanie Minel, and Christophe Merlo

CONTENTS

20.1 INTRODUCTION

With the constant aging of the world's population (Christensen et al., 2009), the increasingly widespread use of information and communication technologies (ICT) (Hart et al., 2008), and the growing interest in patient-centered healthcare (Davis et al., 2005), the use of computers by seniors in the context of healthcare is of utmost interest. As stated by Koch and Hägglund (2009): "in the light of an aging society, effective delivery of healthcare will be more dependent on different technological solutions supporting the decentralization of healthcare, higher patient involvement and increased societal demands."

Senior-centered design is a vast field of research in human factors and ergonomics (Denno et al., 1992; Fisk et al., 2009). More particularly, the design of interfaces for seniors has attracted special attention in this field. One of the subjects studied is the impact of aging on abilities with regard to the use of human–machine interfaces (HMI). For example, Hawthorn (2000) provides an in-depth analysis of age-impacted abilities and provides a set of suggestions to be considered when designing interfaces for older users. He lists numerous characteristics that need to be taken into account: vision, speech and hearing, psychomotor abilities, attention and automated responses, memory and learning, and finally intelligence and expertise. In the same direction, Demiris et al. (2001) provide a set of guidelines for a web-based system for the elderly. They list functional impairment and a lack of computer skills as the two main reasons for the need of senior-specific guidelines for HMI design.

The use of computers and the Internet by seniors is another field of research. Older adults mostly use computers for the following: communication and social support, leisure and entertainment, finding information about health-related and education-related subjects, as well as productivity (Wagner et al., 2010). Other authors also mention e-banking or e-shopping, and argue that the use of computers by seniors does not differ greatly from that of younger users (Vuori and Holmlund-Rytkönen, 2005).

Nevertheless, the use of ICT by seniors in the healthcare context has only been studied from the perspective of patient-centered healthcare provision. The most popular service categories addressed by ICT and identified in the research literature are: handling adverse conditions, assessing their state of health, consulting and learning, motivation and feedback, ordering services, and finally social inclusion (Ludwig et al., 2012).

In this chapter, we present a perspective on seniors as computer-users in two different contexts of healthcare services: home care (case 2) and thermal resorts (case 1). More particularly, we pay attention to the behavior of seniors while using the websites and to their needs with regard to the content of healthcare-oriented interfaces. The constitution of the World Health Organization (WHO, 1946) states "Health is a state of complete physical, mental and social well-being and not merely the absence of disease or infirmity." The social dimension of well-being is equally important to both physical and mental health. In this context, it seems important to consider seniors' expectations and their use of computers not only as directly considering medical information (doctor's diagnosis, analysis results, etc.), but also possibly as regards the social context of healthcare delivery. For the purposes of our research, we define seniors (or older adults, elderly users) as people aged 65 and over and without a professional activity.

Our initial problem is derived from case 2, concerning the replacement of the paper patient record (located at patients' homes) by a technological tool. In this context, the population of end-users is very heterogeneous and the tool will be merely addressed to medical professionals. However, since seniors represent the largest number of home care service users and since the tool is to be placed in their households, they need to be considered as the potential users of the ICT. This is why we decided to refer to another healthcare structure, case 1, concerning the reengineering of an information tool (website) for a thermal health resort. In this case, the end-users are mainly seniors who are readily available (on the spot). Our two industrial cases are comparable within HMI design for elderly people in a health context. Indeed, the tools (i.e., HMI) are used by seniors in the private sphere (at home). In addition, although case 2 deals with hospitalization and case 1 with well-being, they both address the use of HMI for access to care.

In order to identify the expectations of seniors and "HMI behavior," we conducted two experiments. Firstly, we performed a campaign of user tests on HMI in case 1 to analyze seniors' HMI behavior. Secondly, we carried out semi-structured interviews within the population of seniors from case 2. This second phase helped us to draft generic recommendations, based on the expectations of seniors in the context of healthcare.

The contents of this chapter are organized as follows. We begin with a brief presentation of selected research work concerning senior users and technology for healthcare. Next, we introduce the scope of our research and the different issues addressed, pertaining to our two case studies. Then, the context is described, and the method used to analyze them is presented in three steps. After the results section, we propose operational-oriented design recommendations. We finish with a general discussion and draw conclusions about the transfer of our recommendations to other contexts and for other user profiles.

20.2 THE SENIOR AS A USER

Senior-centered design is a vast field of research in human factors and ergonomics (Denno et al., 1992; Fisk et al., 2009). Here, we briefly recall the definition of who seniors are and we identify three different approaches to seniors in service and product design: "senior as the disabled user profile," "senior as another user profile," and "senior as the innovative user profile."

20.2.1 DEFINING SENIORS

Today very powerful information is conveyed enabling us to classify other people as members of a particular group of society in everyday life. Age is the most popular characteristic to define an "older person," a "senior," and the "elderly" (we will use these three terms interchangeably in this paper). For example, in medical contexts over 65s are defined as aged, whereas in social contexts, a senior citizen is someone who has passed retirement age. Furthermore, Fisk et al. (2009) makes the distinction between the "younger-old" (aged 60–75) and the "older-old" (aged over 75).

With the ongoing extension of life expectancy it is, however, becoming increasingly difficult to assume that everyone in the "old" category will share the same characteristics as individuals, users, and clients. Although different authors agree about the impact of aging on diverse perceptual, motor, cognitive, and physical capabilities, we argue that these are not the only ones that will impact the use of a product or service.

First, not all people aged over 65 will experience the aging of their body in the same way. We agree with Newell (2008), who highlights the importance of distinguishing three subgroups of older people: fit older people, frail older people, and disabled people who grow older.

Second, age and age-impacted abilities are not the only characteristics to take into account when talking about the use of technology. For example, according to Czaja et al. (2006), factors impacting the use of computers among seniors are: age, education, fluid intelligence (abstract problem-solving ability), crystallized intelligence (cultural knowledge), computer efficacy (belief about ability), computer anxiety, and prior technology experience.

20.2.2 SENIOR USER PROFILE

User-centered design depends highly on the perception that the designer (or more broadly the researcher) has about the user. Looking beyond user characteristics identified by experts (we refer to designers and researchers as experts) it is primarily the definition of categories that enables users to be classified: the user profiles.

Older adults are perceived by experts in very different ways. We observed three main approaches, independently of age-related classifications in the literature: the "disabled" approach, the "as another" approach, and the "innovative" approach.

The first approach defines seniors as people with reduced capacities, that is, the "disabled" user profile. Rogers and Fisk (2010) state that design for aging involves understanding the unique capabilities and limitations of older adults. Obviously, experts who see the elderly as "disabled" do not only characterize users by their lower capabilities. For example, Fink et al. (1998) propose building their model (i.e., profile) based on four characteristics: motor and sensory abilities, interests and preferences, domain knowledge, and competence. Thus, the definition of the "disabled" profile is represented by only a quarter of their characteristics.

In the second approach, seniors are seen as just one of several user profiles. Thus, experts try to better understand the barriers that exist between the product/service and their target user (i.e., older adults) to improve design. For instance, Adams et al. (2005) conclude that an increase in Internet advertising would be an improvement. Indeed, in their study, it was found that "computer or Internet experience increased perceptions of ease of use and efficacy of the Internet and reduced perceived complexity of navigation."

Likewise, in their study, O'Brien et al. (2008) are interested in understanding the barriers. The authors show that older adults interact with a variety of technologies in the course of their daily activities and overall they are Internet users (mainly for communication and shopping). Thus, they show that one of the primary barriers is the education of older adults.

In the same approach, we can mention another interesting example which deals with packaging design usable by elderly people (Chavalkul et al., 2011). We highlight this study because it is close to our own vision.

Indeed, for the "as another" approach, we distinguish two types of motivation:

• Elderly user-centered design in order to gain new users (i.e., consumers)
• Elderly user-centered design because seniors are already users

We deal with the second case because in our two industrial cases, seniors were already users.

Finally, the third approach is rarer: the "innovative" approach. Here, seniors are seen as the source of innovation. Essén and Östlund (2011) question the concept of older users lagging behind and show how they can contribute to service design early in the innovation process (see also Östlund, 2011; Lee et al., 2013).

The importance for us to distinguish between these different approaches is not only to understand the design path. In fact, it is essential to point out the impact of these approaches on the final product/service. Indeed, the way the product is characterized is hardly dependent on the upstream definition of user characteristics.

20.3   USE OF TECHNOLOGY FOR HEALTHCARE

The important contribution of human factors and ergonomics to the design of new technology-supported services for seniors relies on describing and understanding how they use technology. Another major part of our work consisted of evaluating the interface design and different interaction paradigms in order to identify those which best suit particular needs. Finally, there is a great deal of interest in the study of current ICT uses in order to understand the place of technology in users' lives.

20.3.1   USE OF TECHNOLOGY BY SENIORS

Design for aging is about identifying the needs of older adults, their preferences, and desires for technology in their lives (Rogers and Fisk, 2010). We could also add users' motivation. As stated by Tsai et al. (2012), older adults would like to spent time using a product and learning about a product from a manual. Obviously, in our work, we focus on the human–machine-interface (HMI) for the elderly (designed with the "as another" approach). Therefore, learning from a manual is difficult when we deal with HMI. However, this indication is important when we look at elderly user-centered design. For instance, older adults show high menu disorientation when using hyperlinks (Ziefle et al., 2007). In our opinion, this means guidance is required.

Through observation and analysis, identifying needs, preferences, and desires is about understanding use. In a social networking use case, we can see that older adults are different in use (in comparison to younger adults). In this case, the use of social networking by older adults is more formal, within a restricted network. Moreover, their use of additional features is limited and seems merely focused on basic components (Pfeil et al., 2009).

HMI design for aging is also about promoting adoption by the elderly. O'Brien et al. (2008) demonstrate that the usefulness, compatibility, complexity, technology generation, and relative advantage of a technology are important characteristics that influence adoption. For the authors "compatibility of the technology with existing goals and lifestyles may also facilitate increased adoption." Adams et al. (2005) advocate more simple and uniformly designed Internet pages, as well as more user-friendly online and error message terminology.

20.3.2 SENIORS AND TECHNOLOGY FOR HEALTHCARE

Interest in the use of ICT in the field of healthcare by seniors is a result of major evolutions in the healthcare system, diversification of the healthcare needs of seniors, and the constant evolution of technological tools, including mobile and tactile technologies. Nowadays, many countries are tending toward the adoption of patient-centered care provision and patients are increasingly demanding greater access to the best healthcare available and ascertaining their rights as decision makers (Fieschi, 2002). The use of technology is also seen as a particularly cost-effective solution for the distant monitoring of chronic diseases and diverse health problems linked with aging. The constant growth of the accessibility of ICT is a major facilitator for such evolutions (Boulos et al., 2011).

In this context, many industrial and research projects are implemented in order to create adequate solutions, that both satisfy healthcare needs and promote the adoption of technology. We can thus identify two most important subjects: the application and the device.

Ludwig et al. (2011) have conducted a literature review in order to identify the principal services addressing health issues through technology for the elderly. Their results show that the most popular services address:

- Handling adverse conditions (manual emergency call; automated detection of deviant behavior, falls, or cardiac emergencies; and handling potentially dangerous situations)
- Assessing state of health (recognition of unknown diseases and medical conditions, monitoring of known diseases, and monitoring of therapeutic interventions)
- Consultation and education
- Motivation and feedback
- Ordering services
- Social inclusion

Regarding the device, significant work within the human factors and ergonomics community has been carried out in order to evaluate and guide the design of new ICT depending on the nature of the interface (graphic interface, touch interface, and ambient interface) or a particular ICT device (e.g., mobile phone or website) in order to promote their adoption and thus promote the accessibility of healthcare services. Clearly, the choice of a particular interaction paradigm must be made based on the context of the task and the experience of the user, including the senior user (Wood et al., 2005), since every input device has its advantages and limits (Taveira and Choi, 2009), even if according to some authors, touch interfaces seem to be particularly adapted for the elderly (Holzinger, 2003; Leonardi et al., 2010; Piper et al., 2010; Caprani et al., 2012). Finally, prior use of a particular technology type could impact the adoption of the future application, for example, a mobile phone application (Kurniawan, 2008), a web site (Demiris et al., 2001; Kaufman et al., 2006; Or et al., 2011), or a tablet PC application.

A large amount of work has thus been done in order to design useful and easy to use ICT's in the context of healthcare where seniors are the final users. In our work, we want to highlight another subject, underdeveloped in our opinion in today's research: the use by seniors of general public ICT-supported healthcare services and thus the consideration of seniors in (re)design as one of diverse user profiles.

20.4 SCOPE OF WORK

We based our work on two different case studies linked by the context (healthcare) and the target HMI users (seniors). Figure 20.1 presents the scope of our work.

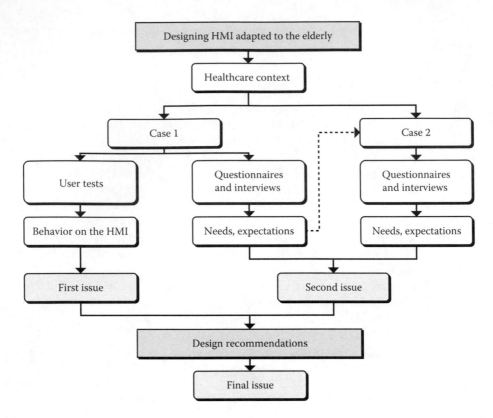

FIGURE 20.1   Scope of the work.

Our research focuses on two aspects of user-centered design in human–machine interfaces for elderly people in the field of healthcare services. As mentioned above, both these cases are comparable in terms of using HMI in the private sphere and in care accessibility.

The first issue concerns the behavior of elderly people with regard to human–machine interfaces: How do elderly people explore web pages? In this part, we focus on click locations, on areas of interest, and on gaze paths (eye tracking).

The second issue concerns the needs and expectations of seniors related to healthcare services, accessible through the interfaces. More particularly, we want to find answers about the information considered important in the context of thermal care and home care. Our observations during the experiment in case 1 allow us to identify the needs which we refer to as information "in the periphery of health." Indeed, besides information about medical care services, seniors show great interest for information such as "planning," "location," "activities," etc. linked with care services. Through interviews and questionnaires, we verified the importance of both types of information (medical and "peripheral") in the context of case 2.

The final issue, and the heart of this research project, is to propose design recommendations for HMI in the context of healthcare, based on our findings from both case studies. In this work, our recommendations are operational oriented and we see them as design levers (action parameters) for other contexts and fields.

20.5   INDUSTRIAL CONTEXT

To achieve our objectives, we conducted experiments in two distinct healthcare structures. The first case (case 1) is a specific short-term study project, that is, 1 month and senior users (we obtain a mean of 65 years for the subjects). The second case (case 2) is part of a multi-level long-term research project, that is, 3 years and a multi-actor project (Borgiel et al., 2013).

20.5.1 Case 1: Specific Short-Term Study

Industrial case 1 (see Figure 20.2) is a thermal resort. It proposes 3-week courses of medical treatment. The care services depend on the type of treatment (rheumatology or phlebology). The thermal doctor prescribed the treatment (72 sessions in total). The types of care varied, so we are in a multi-actor context. In this project, we are interested solely in patients whose average age is 65.

In this case, the industrial demand is the ergonomic analysis of the homepage of the thermal resort's website. For us, it mainly involves studying the behavior of the elderly on an HMI in the context of healthcare to define design recommendations. We want to identify the appropriation of web tools by seniors, the image they have, and how they explore a web page, locate clicks, and areas of interest.

20.5.2 Case 2: Multi-Level Long-Term Research Project

Industrial case 2 (see Figure 20.3) is a French home care structure offering two kinds of distinct services: hospital at home and nursing at home. The first one, hospital at home, is an alternative to traditional hospitalization and allows people with serious, acute, or chronic diseases to stay in their family environment. The second one addresses elderly or disabled people who need coordinated nursing services.

The home care structure takes care of about 500 patients every day, with a level of 20% within the hospital at home service. The hospital at home service is aimed at all ages of patients; however, the population of patients of the studied structure is represented mostly by elderly people. Indeed, for the years 2010–2013, 70% of patients in the hospital at home service were over 65 years old. The home care structure has recently initiated a project to replace the paper health record, located at the patient's home, by a tablet PC with healthcare traceability software. This project is aimed at the hospital at home service and is motivated by different issues that will not be presented here (cf. Borgiel et al., 2013).

Given the hypothesis that the patient's medical record is seen as a tool mostly for healthcare professionals, the initial project did not take into account the patients' use of the future system. However, patients and their family circle (mostly represented by the family) is the central actor in home care activities. They are not simply "customers" but participate actively in the care process.

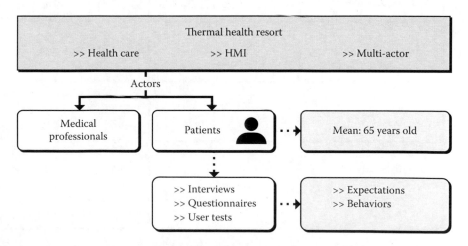

FIGURE 20.2 Illustration of case 1—specific short-term study.

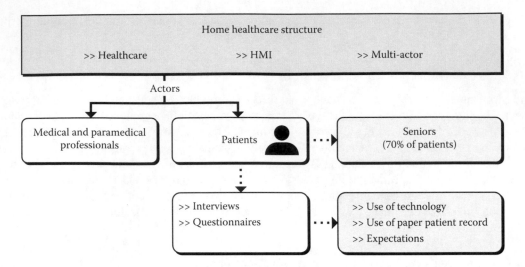

FIGURE 20.3   Illustration of the case 2—general long-term study.

They can be responsible for some of the care tasks and are very often in charge of medication delivery and sometimes—administration. Their constant presence makes them the best source of knowledge about the evolution of care. In the context of the new project, it therefore seems obvious to include patients and their families in the group of future users. We consequently decided to approach the definition of the future use of tools from the patient's perspective.

20.6   METHOD

Our experiment was divided into two distinct phases represented by our two industrial cases. The first phase conducted in industrial case 1 consists of two steps: "pretest" and "test," and the second phase conducted in industrial case 2 consists of one step. In this chapter, we call this step "posttest" (see Figure 20.4). As described in the scope of work section, we reinject in case 2 the needs and

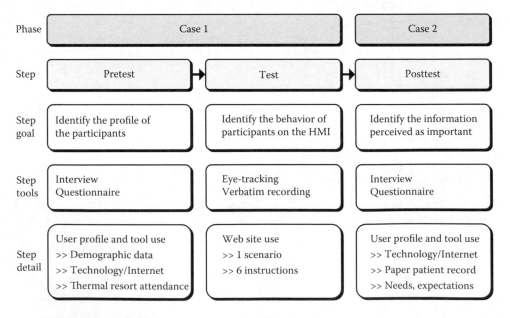

FIGURE 20.4   Experiment procedure.

expectations identified and observed in case 1. This enables us to control the genericity of needs and expectations. Indeed, step 3 consolidates our "qualitative" results. It is worth remembering that both industrial cases deal with the use of technology by seniors in the private sphere for access to care. Our research focuses on this aspect. Thus, we could couple the results from both industrial fields.

20.6.1 STEP 1: THE PRETEST

The first step (pretest) is used to identify the user profile of participants (seniors). Here, we mainly seek to identify the relationship of the elderly with technology and the Internet. To do so, we conducted semi-structured interviews with a questionnaire for support. We used an "hourglass" structure to conduct the interviews (Figure 20.5). We started with very open questions at first, semi-open question afterwards to narrow down the answers, and reopen the issues at the end. Reopening the questionnaire is mainly useful for increasing the interviewees' feeling of freedom.

20.6.2 STEP 2: THE TEST

Step 2 consists of getting patients to test the homepage of the thermal resort's website. To do this, we developed a scenario with six instructions. The scenario and instructions during the test on HMI are presented in Table 20.1. After the participant has selected the item in the web page containing information about rheumatology, the scenario continues with the following instruction. Thus, each time an instruction is given, the participant explores the web page and interacts with it until clicking on the expected item (according to the instruction).

The material used for the test is a screen, a keyboard, a mouse, and an eye-tracking system: Tobii X2 (see Annex for an example of the heatmap result).

During this step, we measure the time taken to complete the task, the number of "incorrect" and "correct" clicks and their location (different possibilities), as well as the time elapsed between first noticing the item and clicking on it. Also, we observe areas of interest on the web page and the path of participant eyes on the web page.

It is worth noting that during the test we kept a verbatim record to identify "invisible" information. And, step 3 consolidates our "qualitative" results.

20.6.3 STEP 3: THE POSTTEST

To define the future use of tools by Patients and their Family circle (PF), we decided to analyze the present context from different perspectives. First, we found it important to define the PF's use of technological tools in general. Second, we decided to analyze present activities regarding the paper medical record. Third, we decided to analyze the patients' needs related more generally to the

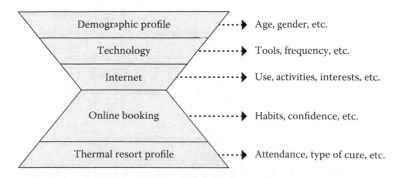

FIGURE 20.5 Structure of questionnaire in the pretest step.

TABLE 20.1
Scenario from the User Test of the Thermal Resort Website

Scenario Element	Content
Introduction	"Imagine that your doctor prescribes a course of treatment in Dax (France) whose orientation is rheumatology"
Instruction 1	"We are going to pretend you are on the thermal resort's website looking on the web for information about the course of treatment prescribed by your doctor"
Instruction 2	"Now we are going to pretend you are looking for activities and additional treatments"
Instruction 3	"Now that you have chosen your course of treatment as well as your additional treatments, we are going to imagine that you want to come to the thermal resort by train in May/June. As you come by train, we are going to imagine you are looking to see how far Dax's railway station is from the thermal resort"
Instruction 4	"Now let's imagine you want to stay in a studio for one person at Thermotel and you are looking for rental rates"
Instruction 5	"We are going to pretend that you are about to book your stay at the thermal resort and you want the list of things to expect before the start your course of treatment"
Instruction 6	"Finally, we are going to imagine the text is too small and difficult to read, and you want to increase the text size"

organization of care activities, including rounds of healthcare professionals. The data were gathered through semi-structured interviews and questionnaires. The evaluation of the frequenc*y of use of technological tools is based on a* five-point Likert scale, from "never" to "daily." The evaluation of the importance of access to medical and organizational data are also based on a five-point Likert scale, from "very important" to "not important at all" (1—not important at all, 2—little important, 3—relatively important, 4—important, and 5—very important).

We gathered data from eight households where patients had been taken into care under the hospital at home service. Participation is voluntary, thus the patients were previously contacted by a nurse manager. The interviews were organized at homes and depending on the situation, both patients and/or their family members were interviewed.

20.7 RESULTS

In this part, we present results obtained from the research relevant to our industrial case studies. Given our method (Figure 20.4), we first describe the results obtained from industrial case 1 and then from industrial case 2.

20.7.1 RESULTS FROM CASE 1

The sample of subjects who participated in the user tests consists of 11 women and 5 men, whose average age is 65.0 (SD = 13.0). The questionnaire carried out during the pretest indicates that 80% of participants are daily Internet users and 100% of the sample uses the Internet through a computer (compared to a tablet or a smartphone).

Figure 20.6 presents the website's homepage (screenshot). The colored rectangles correspond to clickable items (defined by the HMI designer) and the rectangles without fillings correspond to items that we propose making clickable.

As specified in the previous section, the participant interacts with the web page for each instruction from the scenario. To simplify the presentation of results (Figure 20.7), we gave a title to each instruction: instruction 1 = "TREATMENT" information, instruction 2 = "CARE" information, instruction 3 = "LOCATION" information, instruction 4 = "RATES" information, instruction 5 = "BEFORE TREATMENT" information, and instruction 6 = "ZOOMING."

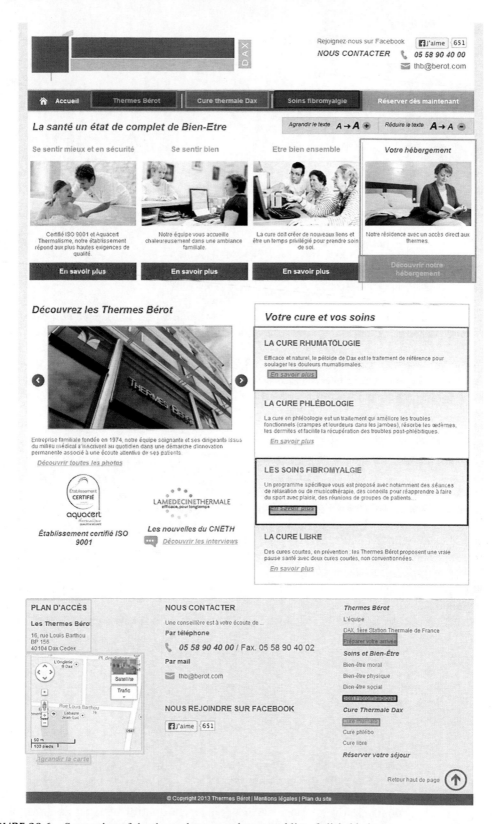

FIGURE 20.6 Screenshot of the thermal resort webpage and list of clickable items.

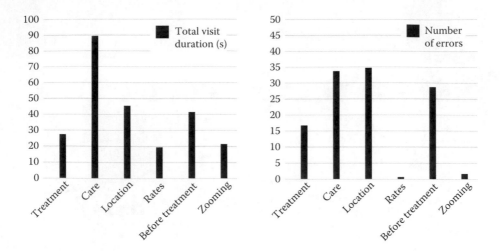

FIGURE 20.7   Total duration of the visit and number of errors for each instruction given.

We observed that the information in "CARE" is long to find on the homepage at nearly 90.0 s. We also measured the time needed to find the location of a thermal resort ("LOCATION" information) and the list of things to expect before arriving ("BEFORE TREATMENT" information) at respectively 46.5 and 42.0 s. The time needed to find the information for rheumatology treatments ("TREATMENT" information), thermotel rates ("RATES" information), and the zooming button ("ZOOMING" information) are within acceptable limits (<30 s).

We noted that for "RATES" and "ZOOMING," errors are almost nonexistent (0.8%). Errors when looking for information about "TREATMENT" amount to 17, which corresponds to about one "bad" click per person, and is quite acceptable. On the other hand, errors relating to finding information on further care ("CARE"), the location of thermal resorts ("LOCATION"), and the list of things to expect before treatments ("BEFORE TREATMENT") are too important and highlight confusion and a lack of clarity among participants.

Clicks in block correspond to clicks on the block + on the title in the block + on the link in the block. Only the click on the link (or in the menu bar) allows the user to finish and pass onto the next instruction (because these are the clickable items defined by the HMI designer).

The clicks result for each instruction is

- Instruction 1 (TREATMENT): 26 clicks in block (with 10 on the link) versus 6 clicks in the menu bar
- Instruction 2 (CARE): 26 clicks in block (with 12 on the link) versus 4 clicks in the menu bar
- Instruction 3 (LOCATION): 28 clicks on the map versus 3 clicks on the address of the resort (textual)
- Instruction 4 (RATES): 16 clicks on the button (in a block) (there is 1 error due to the label)
- Instruction 5 (BEFORE TREATEMENT): 11 clicks in the footer page versus 5 clicks in the menu bar
- Instruction 6 (ZOOMING): no real mistakes (two errors due to label)

We thus observed that for instructions 1 and 2, participants (who are seniors) tend to interact with thematic blocks. For instruction 3, we can also consider "block-clicking" behavior because we obtained 28 clicks on the map that can be seen as a thematic block. In addition, the resulting clicks with regard to instruction 5 clearly indicate that seniors interact very little with the menu bar on the web page. Indeed, we obtained 11 clicks in the footer of the web page compared to 5 in the menu bar.

Also, during tests, we are able to identify certain information expressed by participants regarding their needs and expectations. We decided to check these points more quantitatively through industrial case 2.

20.7.2 RESULTS FROM CASE 2

In case 2, we studied eight different households. From all the households, only one patient lived alone, the rest of them lived with their spouses. Three patients were unable to express their opinion due to their pathologies and one patient did not take part in the interview, although he was present in the same room. The age of participants varied between 60 and 81 (mean = 71.6). Given the small sample size, we shall present here selected descriptive statistics for the results obtained.

Table 20.2 presents results about the use of technological tools and the Internet. The results show that computers and the Internet are popular technological tools among the group studied. Only one in eight households does not have a computer or the Internet. The remaining seven use computers mostly for Internet access and the most popular activities are checking mail, reading the news, and looking for information.

The results from the questionnaire about the use of the paper patient record (Table 20.3) show that in only 37.5% of interviewed households the inhabitants read the paper patient record. However, more detailed interviews show that this reading is not regular and is limited to only a few documents, particularly prescriptions. Similar conclusions may be drawn about others consulting the paper record, that is, the patients' children or friends. Finally, the fact that nobody writes any information on the record confirms how the document is perceived by patients and their families: they see it as a caregiver's tool and do not dare to add any comments about medical data.

TABLE 20.2
Use of Technological Tools and the Internet among Eight Study Participants

| | | Use of Technological Tools | | | Use of the Internet | |
| | | Frequency of Use | | Frequency | | |
	Tools Used	Computer	Tablet PC	of Use	Activities
1	TV, computer, tablet PC, and mobile phone	Daily	Weekly	Daily	Mail, weather, bank, and bourse
2	TV, computer, and mobile phone	Annual	Never	Annual	Songs, speech, and mail
3	TV, computer, and mobile phone	Daily	Never	Daily	News and journals
4	TV and mobile phone	Never	Never	Never	–
5	TV, computer, and mobile phone	Weekly	Never	Weekly	News, journals, and information research
6	TV, computer, and mobile phone	Weekly	Never	Weekly	Mail, weather, and sales
7	TV, computer, tablet PC, and mobile phone	Daily	Weekly	Daily	News and journals
8	TV, computer, and mobile phone	Daily	Annual	Daily	News, journals, and yellow pages

TABLE 20.3
Use of Paper Patient Record

Potential User	Type of Use	Yes	No
Patient and spouse (inhabitants)	Reading	3	5
	Writing	0	8
Family and friends (nonresidents)	Reading	2	6
	Writing	0	8

TABLE 20.4

Descriptive Statistics for the Evaluation of Importance of Access to Medical and Organizational Data

Data Type	Data Item	Min	Max	Median	Mean	SD	SE
Medical	Prescriptions	4.0	5.0	5.0	4.75	0.46	0.16
	Health evolution	4.0	5.0	4.0	4.25	0.46	0.16
	Analyses results	3.0	5.0	4.0	4.25	0.71	0.25
Organizational	Caregivers' timetable	3.0	5.0	4.0	3.75	0.71	0.25
	Caregivers' names	2.0	5.0	3.0	3.25	1.04	0.37
	Caregivers' delays	2.0	5.0	3.5	3.5	0.93	0.33

TABLE 20.5

Descriptive Statistics for the Evaluation of Access to Medical versus Organizational Data

Data Type	Min	Max	Median	Mean	SD	ED
Medical	3.0	5.0	4.0	4.42	0.58	0.12
Organizational	2.0	5.0	3.5	3.5	0.88	0.18

Table 20.4 presents descriptive statistics for results about the perception of importance of access to diverse data linked to patients staying within the home care structure, both medical and organizational.

The results obtained show that access to prescriptions is of utmost importance to patients and their families in the context of home care. Other medical data like the global evolution of health or laboratory test results are also important. On the other, information about the organization of health seems to be important, but not as much as medical data. With reference to the initial issue, it is important to specify that today access to most of the data listed in Table 20.4 is not provided in written form, except for prescriptions and laboratory tests results. Prescriptions are written by family General Practitioners (GPs) at home and the results are sent by laboratories directly to the patient's address. The remaining information is given to patients and their families orally, directly at the moment of care (health evolution, care givers' names, delays, etc.).

Table 20.5 presents descriptive statistics for the results of the evaluation of the importance of access to medical versus organizational data.

The results obtained suggest that even if access to organizational information is seen as important in the context of home care, access to medical information is even more important.

To summarize, our case study about the use of ICT by seniors in the context of home care allows us to state the following facts. Patients and their spouses are active computer and Internet users in their daily lives, so they will probably be interested in using the new tool. Even if their use of actual paper medical records is not very frequent, the binder contains very important data that need to be made freely accessible, especially prescriptions. Thus, the future system needs to provide an easy and direct access to these data. Finally, the organizational data, or the information "in the periphery of health" are seen as important. The future system could support the provision of this information.

20.7.3 RECOMMENDATIONS

On the basis of the results obtained in our two industrial cases, we propose recommendations for elderly user-centered HMI design in the field of healthcare. Our recommendations concern two aspects of HMI: the structure and selecting information to display (HMI content).

TABLE 20.6

Needs and Expectations

Applicability	Needs and Expectations
Generic	• Planning/scheduling
	• Names of actors/healthcare staff
Specific for case 1	• Additional activities proposed and possible
	• Location/city/reception
Specific for case 2	• Delays in "rounds"

First, we propose "hybrid" interfaces, that is, mosaic type (e.g., Windows 8 interface), but equipped and preserving the menu bar for more "traditional" browsing. Indeed, we observed that the HMI-related behavior of seniors is exploratory, they "surf." They tend to operate and click on thematic blocks rather than on links or a menu bar. In view of this, we recommend reducing the visual emphasis of links in order to make the buttons stand out more and clearly visually distinguish thematic blocks, for example, using visual variables such as separation, continuity, etc. (Khöler, 1964; Bertin, 1983; Card et al., 1999) or using preattentive variables such as color, size, orientation, etc. (Tidwell, 2006).

Our second level of recommendation concerns selecting information to display. Our work highlights the need for information "in the periphery of health." In our study, seniors have expectations going beyond the access to medical information (doctor's diagnosis, analysis results, etc.) and they express a need either for social projection or to be reassured (cf. Table 20.6).

The first part of observed expectations is common for both case studies; we call them "generic." The second part is specific to each of the case studies, given the specificities of the healthcare context.

20.8 DISCUSSION

The objective of our study is to propose HMI design recommendations for seniors in the context of healthcare. We base our findings on two industrial case studies which cover a large scope of HMI use by seniors with regard to the accessibility of care in the private sphere (thermal resort and home care) in an attempt to provide a generic framework. The proposed recommendations are divided into two parts. The first part is formed by recommendations regarding HMI structure, whereas the second part relates to the HMI content, with a particular interest in information "in the periphery of health."

Our research does not pretend to offer solutions to all elderly people. Here, we want to consider a user panel as any other with its characteristics, experiences, etc. because seniors are not only a research topic related to "reduced capabilities" or handicaps for HMI design (Wildroither et al., 2015).

To promote the recommendations in a more generic framework, we plan to continue our work by implementing our recommendations in the design of the future tool in case 2 and to experiment on tablets with more user profiles (seniors, active people, young people, etc.). Indeed, our results do not guarantee that our recommendations apply to other ICT tools, such as tablets. In addition, it would be interesting to verify the transfer of our recommendations to a context other than that of health, such as shopping, or more precisely—booking. In fact, our industrial case 1 is very close to this type of context.

These design recommendations will subsequently be used for research work in case 2, in reference to the future use of the electronic health record by home care patients and their families. We believe that the adoption of the new tool by patients and their families will promote new relationships between the home care structure and its customers, and thus will be a source of organizational innovation for all the actors. We assume that our work presented is just a part of a project with a larger span that will lead to the redesign and implementation of a tool dedicated to home care services and we plan to verify the validity of our results with other actor profiles in the context of home care, mainly different healthcare professionals. Even if health is at the center of their activity, they could also express needs "in the periphery of health."

20.9   CONCLUSION

Interest in the use of ICT for healthcare services by seniors is a result of major evolutions in the healthcare system, the diversification of the healthcare needs of seniors, and the constant evolution of technological tools, including mobile and tactile technologies.

In this chapter, we present design recommendations, based on findings from two different case studies, where seniors are only one amongst several different final user profiles. The example of the thermal hotel (case study 1) shows that older adults are keen to use the web when they need it and that they tend to see it as an important source of information. In home care services the technology can be both a support for organization and an understanding between the healthcare structure and its patients. We believe that the information "in the periphery of health" will become increasingly important as patients become more and more empowered and active in their care.

If both user and technology characteristics are inseparable parts of the technology we use, within every project they have to be applied to context, to specific demands, needs, and constraints. We believe that the most important thing when considering the design and evaluation of ICT for seniors, for example, healthcare services and applications, is to simply approach them as fully valuable user profiles. Next, depending on the application and the technology (both interaction and device), the senior user profile should be specified in a more detailed way. And, it is with this specification that the different design recommendations and guidelines could have meaning.

We assume that our recommendations do not provide a full solution to the problem. Senior-centered design is a research area that needs to evolve. It is essential we continue along this path because technology evolves very quickly and we are obviously the elderly of tomorrow.

ANNEX

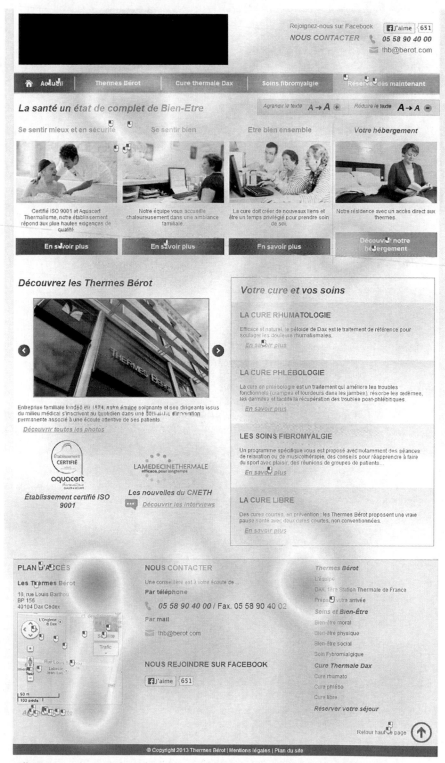

Illustration of eye-tracking data captured for Instruction 5 ("before treatment" information).

REFERENCES

Adams, N., D. Stubbs, and V. Woods. 2005. Psychological barriers to Internet usage among older adults in the UK. *Informatics for Health and Social Care* 30 (1): 3–17. doi:10.1080/14639230500066876.

Bertin, J. 1983. *Semiology of Graphics*. Madison, Wisconsin: University of Wisconsin Press.

Borgiel, K., X. Latortue, S. Minel, and C. Merlo. 2013. Holistic approach to management of innovation: A home care case study. Paper presented at CONFERE 2013, Biarritz, France, July 4–5. https://hal.archives-ouvertes.fr/hal-01015818/document.

Boulos, M., S. Wheeler, C. Tavares, and R. Jones. 2011. How smartphones are changing the face of mobile and participatory healthcare: An overview, with example from eCAALYX. *Biomedical Engineering Online* 10 (1): 24. doi:10.1186/1475-925x-10-24.

Caprani, N., O'Connor, N.E., and C. Gurrin. 2012. Touch screens for the older user. In F. A. Auat Cheein (ed.), *Assistive Technologies*, 1st edn., pp. 95–118. InTech Open Access Publisher. doi:10.5772/38302.

Card, S. K., J. D. Mackinlay, and B. Shneiderman. 1999. *Readings in Information Visualization*. San Francisco, California: Morgan Kaufmann Publishers.

Chavalkul, Y., A. Saxon, and R. N. Jerrard. 2011. Combining 2D and 3D design for novel packaging for older people. *International Journal of Design* 5 (1): 43–58. http://www.ijdesign.org/ojs/index.php/IJDesign/article/view/852/328.

Christensen, K., G. Doblhammer, R. Rau, and J. W. Vaupel. 2009. Ageing populations: The challenges ahead. *The Lancet* 374 (9696): 1196–1208. doi:10.1016/s0140-6736(09)61460-4.

Czaja, S. J., N. Charness, A. D. Fisk, C. Hertzog, S. N. Nair, W. A. Rogers, and J. Sharit. 2006. Factors predicting the use of technology: Findings from the center for research and education on aging and technology enhancement (create). *Psychology and Aging* 21 (2): 333–352. doi:10.1037/0882-7974.21.2.333.

Davis, K., S. C. Schoenbaum, and A.-M. Audet. 2005. A 2020 vision of patient-centered primary care. *Journal of General Internal Medicine* 20 (10): 953–957. doi:10.1111/j.1525-1497.2005.0178.x.

Demiris, G., S. M. Finkelstein, and S. M. Speedie. 2001. Considerations for the design of a web-based clinical monitoring and educational system for elderly patients. *Journal of the American Medical Informatics Association* 8 (5): 468–472. doi:10.1136/jamia.2001.0080468.

Denno, S., B. A. Isle, G. Ju, C. G. Koch, S. V. Metz, R. Penner, L. Wang, and J. Ward. 1992. Human factors design guidelines for the elderly and people with disabilities. Cs.Cmu.Edu. http://www.cs.cmu.edu/~khaigh/ILSAEXTERNALWEBSITE/content/publications/.

Essén, A. and B. Östlund. 2011. Laggards as innovators? Old users as designers of new services & service systems. *International Journal of Design* 5 (3): 89–98. http://www.ijdesign.org/ojs/index.php/IJDesign/article/view/915/368.

Fieschi, M. 2002. Information technology is changing the way society sees health care delivery. *International Journal of Medical Informatics* 66 (1–3): 85–93. doi:10.1016/s1386-5056(02)00040-0.

Fink, J., A. Kobsa, and A. Nill. 1998. Adaptable and adaptive information provision for all users, including disabled and elderly people. *New Review of Hypermedia and Multimedia* 4 (1): 163–188. doi:10.1080/13614569808914700.

Fisk, A. D., Rogers, W. A., Charness, N., Czaja, S. J., and J. Sharit. 2009. *Designing for Older Adults: Principles and Creative Human Factors Approaches*, 2nd edn. Boca Raton, Florida: CRC Press.

Hart, T. A., B. S. Chaparro, and C. G. Halcomb. 2008. Evaluating websites for older adults: Adherence to "senior-friendly" guidelines and end-user performance. *Behaviour & Information Technology* 27 (3): 191–199. doi:10.1080/01449290600802031.

Hawthorn, D. 2000. Possible implications of aging for interface designers. *Interacting with Computers* 12 (5): 507–528. doi:10.1016/s0953-5438(99)00021-1.

Holzinger, A. 2003. Finger instead of mouse: Touch screens as means of enhancing universal access. In N. Carbonnell and C. Stephadinis (eds.), *User Interfaces for All*, 1st edn., pp. 387–397. Berlin, Heidelberg: Springer-Verlag.

Kaufman, D. R., J. Pevzner, C. Hilliman, R. S. Weinstock, J. Teresi, S. Shea, and J. Starren. 2006. Redesigning a telehealth diabetes management program for a digital divide seniors population. *Home Health Care Management & Practice* 18 (3): 223–234. doi:10.1177/1084822305281949.

Koch, S. and M. Hägglund. 2009. Health informatics and the delivery of care to older people. *Maturitas* 63 (3): 195–199. doi:10.1016/j.maturitas.2009.03.023.

Kohler, W. 1964. *Psychologie De La Forme*. Paris, France: Gallimard.

Kurniawan, S. 2008. Older people and mobile phones: A multi-method investigation. *International Journal of Human-Computer Studies* 66 (12): 889–901. doi:10.1016/j.ijhcs.2008.03.002.

Lee, C., R. Myrick, L. A. D'Ambrosio, J. F. Coughlin, and O. L. de Weck. 2013. Older adults' experiences with technology: Learning from their voices. In C. Stephanidis (ed.), *HCI International 2013—Posters' Extended Abstracts*, pp. 251–255. Berlin, Heidelberg: Springer. doi:10.1007/978-3-642-39473-7_51.

Leonardi, C., A. Albertini, F. Pianesi, and M. Zancanaro. 2010. An exploratory study of a touch-based gestural interface for elderly. *Proceedings of the 6th Nordic Conference on Human-Computer Interaction: Extending Boundaries, Reykjavik*, Iceland, pp. 845–850. New York, NY, USA: ACM. doi:10.1145/1868914.1869045.

Ludwig, W., K.-H. Wolf, C. Duwenkamp, N. Gusew, N. Hellrung, M. Marschollek, M. Wagner, and R. Haux. 2012. Health-enabling technologies for the elderly—An overview of services based on a literature review. *Computer Methods and Programs in Biomedicine* 106 (2): 70–78. doi:10.1016/j.cmpb.2011.11.001.

Newell, A. 2008. User-sensitive design for older and disabled people. In A. Helal, M. Mokhtari, and B. Abdulrazak (eds.), The Engineering Handbook of Smart Technology for Aging, Disability, and Independence, pp. 785–802. Hoboken, New Jersey: John Wiley & Sons. doi:10.1002/9780470379424.ch43.

O'Brien, M. A., K. E. Olson, N. Charness, S. J. Czaja, A. D. Fisk, W. A. Rogers, and J. Sharit. 2008. Understanding technology usage in older adults. Paper presented at *the 6th International Conference of the International Society for Gerontechnology,* Pisa, Italy, June 4–7. http://www.gerontechnology.info/Journal/Proceedings/ISG08/papers/014.pdf

Or, C. K. L., B.-T. Karsh, D. J. Severtson, L. J. Burke, R. L. Brown, and P. F. Brennan. 2011. Factors affecting home care patients' acceptance of a web-based interactive self-management technology. *Journal of the American Medical Informatics Association* 18 (1): 51–59. doi:10.1136/jamia.2010.007336.

Östlund, B. 2011. Silver age innovators: A new approach to old users. In F. Kohlbacher and C. Herstatt (eds.), *The Silver Market Phenomenon: Marketing and Innovation in the Ageing Society*, pp. 15–26. Berlin, Heidelberg: Springer. doi:10.1007/978-3-642-14338-0_2.

Pfeil, U., R. Arjan, and P. Zaphiris. 2009. Age differences in online social networking—A study of user profiles and the social capital divide among teenagers and older users in MySpace. *Computers in Human Behavior* 25 (3): 643–654. doi:10.1016/j.chb.2008.08.015.

Piper, A. M., R. Campbell, and J. D. Hollan. 2010. Exploring the accessibility and appeal of surface computing for older adult health care support. *Proceedings of the SIGCHI Conference on Human Factors in Computing Systems*, Atlanta, Georgia, pp. 907–916. New York, NY, USA: ACM. doi:10.1145/1753326.1753461.

Rogers, W. A. and A. D. Fisk. 2010. Toward a psychological science of advanced technology design for older adults. *The Journals of Gerontology Series B: Psychological Sciences and Social Sciences* 65B (6): 645–653. doi:10.1093/geronb/gbq065.

Taveira, A. D. and S. D. Choi. 2009. Review study of computer input devices and older users. *International Journal of Human-Computer Interaction* 25 (5): 455–474. doi:10.1080/10447310902865040.

Tidwell, J. 2006. *Designing Interfaces,* 1st edn. Beijing: O'Reilly.

Tsai, W.-C., W. A. Rogers, and C.-F. Lee. 2012. Older adults' motivations, patterns, and improvised strategies of using product manuals. *International Journal of Design* 6 (2): 55–65. http://jodesign.org.tw/ojs/index.php/IJDesign/article/view/1028.

Vuori, S. and M. Holmlund-Rytkönen. 2005. 55 + people as Internet users. *Marketing Intelligence & Planning* 23 (1): 58–76. doi:10.1108/02634500510577474.

Wagner, N., K. Hassanein, and M. Head. 2010. Computer use by older adults: A multi-disciplinary review. *Computers in Human Behavior* 26 (5): 870–882. doi:10.1016/j.chb.2010.03.029.

WHO 1946. Constitution of the World Health Organization. *American Journal of Public Health and the Nation's Health* 36 (11): 1315–1323.

Wildroither, H., L. Hagenmeyer, S. Breker, and M. Panou. 2015. On designing automotive HMIs for elderly drivers: The AGILE initiative. In C. Stephanidis and J. Jacko (eds.), *Human-Computer Interaction: Theory and Practice* (Part II), 1st edn., pp. 323–327. Mahwah, New Jersey: Lawrence Erlbaum Associates.

Wood, E., T. Willoughby, A. Rushing, L. Bechtel, and J. Gilbert. 2005. Use of computer input devices by older adults. *Journal of Applied Gerontology* 24 (5): 419–438. doi:10.1177/0733464805278378.

Ziefle, M., U. Schroeder, J. Strenk, and T. Michel. 2007. How younger and older adults master the usage of hyperlinks in small screen devices. *Proceedings of the SIGCHI Conference on Human Factors in Computing Systems*, San Jose, California, pp. 307–316. New York, NY, USA: ACM. doi:10.1145/1240624.1240676.

Section IV

Design Development

21 Human Body–Sleep System Interaction in Young Adult Residence

A Methodology and Tool for Measure and Evaluation of Interaction Patterns Using a Software iSEE with Observation of Postural Behaviors during Sleep

Gustavo Desouzart, Ernesto Filgueiras, Rui Matos, and Filipe Melo

CONTENTS

21.1 INTRODUCTION

Sleep disruption is a growing problem that may have serious health effects (Wright et al., 2007). In many Western societies, decreased time available for sleep and/or increased sleep disturbance is often associated with a demanding life style and is a growing problem (Rajaratman and Arendt, 2001; National Sleep Foundation, 2005; Soares, 2005; Wright et al., 2007).

Sleep and circadian rhythms, one of several biological rhythms found in humans, are produced jointly by the action of various structures of the nervous system and are influenced by various environmental factors. The importance of sleep as a restorative and homeostatic agent has evident influence on the waking state of the individual. Sleep disorders can bring various effects to humans, causing loss of quality of life, autonomic dysfunction, and decreased professional or academic performance (Danda et al., 2005).

Many young adults have occasional sleep disorders and pain may be one of the factors that cause them. However, in some cases, these problems can become chronic, causing serious consequences in their behavior and their quality of life (Pter, 1990).

Humans spend approximately one-third of their lives in bed, while a synergy of psychological, physiological, and physical conditions affects the quality of sleep. An insufficiently adapted sleep system (i.e., mattress + support structure + head cushion) or an incorrect sleeping posture may cause back pain (BP) or sleep disorders in general (Haex, 2005). The comfort and support of the sleep surface are related to problems of sleep quality and efficiency. Certain sleep surfaces have resulted in complaints of lower back discomfort, pain, or stiffness (Addison et al., 1986; Jacobson et al., 2009). The risk of BP has a multi-factorial nature and is one of the most compelling problems in the industrialized world, being that poor posture is one of these factors (Vieira and Kumar, 2004; Haex, 2005; Geldhof et al., 2007; Silva et al., 2009). Posture, according to Silva et al. (2009), is considered to be the biomechanical alignment and the spatial arrangement of body parts in relation to their segments.

BP is a leading cause of disability. It occurs in similar proportions in all cultures, interferes with quality of life and daily performance, and is the most common reason for medical consultations. Few cases of BP are due to specific causes; most cases are nonspecific. Acute BP is the most common presentation and is usually self-limiting, lasting less than 3 months regardless of treatment. Chronic BP is a more difficult problem, which often has a strong psychological overlay (Ehrlich, 2003).

BP is one of the most common forms of chronic pain and is a significant cause of disability and cost in society (Andersson et al., 1993; Mantyselka et al., 2001; Walker et al., 2004). Chronic BP substantially influences the capacity to work and has been associated with the inability to obtain or maintain employment and productivity lost (Stang et al., 1998; Stewart et al., 2003). About 80% of the population have experienced BP at some time in their lives (Schmidt and Kohlmann, 2005).

For example, musculoskeletal BP is the most common reason for medical evacuation in the military with return to occupation being uncertain. BP is also a common reason for long-term disability (Lincoln et al., 2002; Cohen et al., 2009, 2010).

Despite continuous research and development of new interventions, BP remains a clinical challenge because it is a condition with high incidence and prevalence (Koes et al., 2010; Mayer et al., 2010) and a condition that has a considerable cost in health care, with negative socioeconomic impact in industrialized societies (Apeldoorn et al., 2010; Van Middelkoop et al., 2011).

Pain results from either exacerbated noxious impulses, or lack of them. In addition, pain cannot be conditioned by peripheral nociceptive stimuli, and psychogenic pain a good example of these exacerbated impulses (World Health Organization, 2001).

Maintaining an ideal posture and exercise are essential for a healthy body, free of pain, especially the spine.

Gross et al. (2000), often claim that the occurrence of BP precedes or is concomitant with changes in body posture. This association can be explained by the fact that many body postures adopted in day to day are inadequate to the anatomical structures, increasing the total stress on the body elements, especially on the spine, which may cause discomfort, pain, or functional disability.

Current research and international clinical guidelines recommend people with BP to take a more active role in their recovery, to prevent pain chronicity (Moffett, 2002; van Tulder et al., 2005; Liddle et al., 2007). A postural intervention program is often used for people with complaints of BP, predominantly based on postural recommendation approaches, which enables a more active participation of this population in their rehabilitation process (Moffett, 2002; Casserley-Feeney et al., 2008; Liddle et al., 2009; Hurley et al., 2010).

However, little or nothing is referenced in the literature on the effectiveness of the physical therapy approach using postural recommendations for aspects such as pain in the region of the spine or related disorders in the sleep period (Hurley et al., 2010). According to the American College of Physicians and the American Pain Society, one of the recommendations made to patients with complaints of BP is postural intervention, such as cognitive behavioral therapy (recommendations of postural behavior) (Chou et al., 2008).

Sleep and rest are as essential for the musculoskeletal system as they are for the central nervous system. It would be illogical for the musculoskeletal system to remain fully operational during the body's rest periods (Gracovetsky, 1987). The behavioral and postural habits and sleep rhythm can be changed depending on the type of daily activity (work or academic activities) or other types of events, but this has rarely been reported in the literature, which would allow an analysis and evaluation of this behavior through sleep disorders. The decreased time available for sleep and/or increased sleep disturbance is often associated with a demanding life style and other challenges imposed by modern society. Perhaps, this is related to the fact that the evaluation of this behavior is complex and the observation of these postural behaviors in the environmental context is needed.

Curcio et al. (2006) reported that many young adults of different educational levels and different type of work suffer from sleep disorders. It was found that higher cognitive functions such as attention, memory, or performance of complex tasks are compromised when there are changes in sleep patterns. On the same theme, Ban and Lee (2001) reported that the deficiency or sleep disorders are known to have serious consequences in various ways, particularly causing problems such as decreased concentration, memory, decreased ability to perform daily tasks, decreased willingness, and decreased interpersonal relationships.

Young adults are recognized as having insufficient sleep on weekdays and sleeping long hours during the weekends. For example, two-thirds of young adult university students reported occasional sleep disturbances, and about a third of these reported suffering regularly from severe sleep difficulties. These disorders are marked by gradual late waking up or more absences from classes, leading to poor academic performance and excessive sleepiness during the week (Brown et al., 2002).

Brown et al. (2002) reported that university students are known for their very variable schedules. Such schedules along with other student practices (e.g., alcohol and caffeine), are associated to poor sleeping habits. Sleeping schedules, anxiety to go to bed, environmental noise, and concern about falling asleep contribute to poor quality of sleep.

Perhaps, this is related to the fact that the evaluation of this behavior is complex and the observation of these postural behaviors in the environmental context is needed. However, the observation methodology based on iSEE software (Filgueiras et al., 2012) allows the classification and registration of postural behaviors for long periods of time and can be applied in this context (Figure 21.1).

Although sleep research has been around for almost a hundred years, there is still a huge need for studies that look at the influence of environmental effects on sleep, both from a physiological as well as a psychological point of view. In general, ergonomic sleep studies benefit from long-term monitoring in the home environment to cope with daily variations and habituation effects (Willemen et al., 2012). In sleep research and clinical settings, both subjective and objective measures are used widely to evaluate patients with sleeping problems (Devine et al., 2005; Kushida et al., 2005; Morgenthaler et al., 2007).

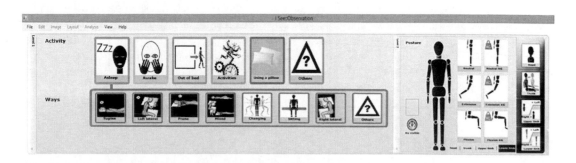

FIGURE 21.1   Functional areas of the iSEE software interface.

The complexity of some newer product interactions in complex context systems demands a higher level of user performance and involves risk that may possibly negatively impact the user's safety and health. For this reason, the evaluation or design of new products used in complex systems requires extensive knowledge of human interaction, including the operation and vulnerabilities of the whole system. Therefore, taking this into consideration, the use of video analysis increases the capability to collect more detailed information on human activity during the interaction of the user with a product–environment system. With this data comes increased understanding of user strategies and awareness of possible safety and health issues as well as system dysfunctions (Rebelo et al., 2011).

Video analysis has been used in many areas. This approach is also used in sleeping posture analysis (Kazmierczak et al. 2006; Spielholz et al. 2001; Strauss and Corbin 1990; Penzel and Conradt, 2000; Kuo et al., 2004; Rebelo et al., 2011). Most methods are used only for a small set of postures. There are other general approaches to posture recognition (Wu and Aghajan, 2007; Liao and Yang, 2008).

In a general way, data regarding sleep analysis are collected in simulated laboratory conditions. Although these kinds of studies interfere with the tasks and with natural behaviors in the sleep period, they have some advantage such as: an accurate control of variables, a high potential to collect physiological measures, and the accuracy of data collected, mainly the quantitative data (Engstrom and Medbo, 1997; De Bruijn et al., 1998; Forsman et al., 2002).

The adoption of a systemic approach of an activity through the analysis of all possibilities of interactions in a real context is the main element for good ergonomic research (Hendrick and Kleiner, 2001).

Recently, with the technological advances of digital video equipment and computers, associated with low costs, video analysis is being routinely used in human behavior research. Video analysis usage makes multiple revisions possible, thereby allowing the collection of detailed information that would be impossible to collect in field studies involving only the researcher's visual memory. In this case, the use of a single source of observation (visual memory) may cause losses due to memory lapses and potential interpretation difficulties. It is, however, important to point out that for the ergonomist, the exclusive use of video analysis is not a substitute for traditional tool usage in ergonomic analysis. In addition, some aspects, such as user interpersonal relations, environmental issues, and macro-ergonomic data, are also important in analyzing product quality (Rebelo et al., 2011).

In order to minimize the difficulty in applying these experimental methods in the real context, researchers combine some objective with subjective techniques, which generally are qualitative such as questionnaires, interviews, and direct activity observation, making it possible to minimize the difficulties in applying these experimental methods in the real context. Usually, this approach is also related to the interpretation and evaluation of the comfort or discomfort, that is, experienced through users' testimony and the understanding of the real activity through self-reporting (Bergqvist, 1995; Straker et al., 1997; Fenety and Walker, 2002; Filgueiras et al., 2012).

In this sense, the main goal of this study was to investigate the human interaction with postural behaviors during sleep in the residences' bedroom of university students and air force military (sleeping positions, head posture, trunk posture, upper limb, and lower limb posture) during the periods in which the subjects were asleep, awake, out of bed, doing activities, using a pillow (interaction categories [ICs]), in different time periods and with ecological validation, through digital video recording using a methodology proposed by Rebelo et al. (2011) and Desouzart et al. (2014a,b).

This knowledge will allow: (a) the understanding of the possible origin of incidence of BP referred in day-to-day activities and (b) the elaboration of more specific recommendations to the changes in postural behaviors and to the development of products. However, in this chapter, we present the results for interaction patterns during the use of a set of specific IC (Rebelo et al., 2011).

21.2 METHODOLOGY

This chapter presents a part of a larger study which aims to analyze the relationship between the perception of BP and the sleeping position and is based on the observation of the human interaction (with visual display terminals) with postural behaviors in bed during the night period (8 h/night) in the residences' bedrooms of female university students and male air force military. We used the same group of equipment and the same model of bed, in order to analyze if there are similar patterns of interaction between users. This study started in April 2013 and finished in October 2013. Data were collected from Portuguese university students and Portuguese air force military.

21.2.1 STUDY SITE AND RESIDENCE BEDROOMS' PROPERTIES

Data were collected from 24 young adults aged between 18 and 25 years old (mean = 20.96 $\pm$ 1.899) of the Portuguese air force military and Portuguese students. Twelve male soldiers, of different categories (first corporal, second corporal or soldier) and 12 female university students, studying in the healthcare domain (physiotherapy, occupational therapy, speech therapy, nursing, and dietary therapy), residing in dormitories of the air base and the university were selected. The bedrooms were in dormitories with two beds in each bedroom with the same type of bed, mattress, and pillow.

21.2.2 SUBJECTS AND NIGHT ACTIVITIES

At the beginning, 89 students of the Polytechnic Institute of Leiria in Portugal aged between 18 and 25 years, residing in dormitories answered a questionnaire about the perception of quality of life according to the abbreviated questionnaire on quality of life (WHOQOL-Bref) and 12 female (mean = 19.75 years old $\pm$ 1.138), studying in the healthcare domain (physiotherapy, occupational therapy, speech therapy, nursing, and dietary therapy) were volunteers in this study.

On the other hand, were 134 soldiers (112 male and 22 female), aged between 18 and 25 years, belonging to the air base no. 5 of the Portuguese air force when the research began. Of these, 66 soldiers answered a questionnaire about the perception of quality of life according to the abbreviated questionnaire on quality of life (WHOQOL-Bref) and 12 male (mean = 22.17 years old + 1.749) were volunteers in this study. The soldiers worked in the area of mechanical aviation material ($n = 6$), hospitality services and sustenance ($n = 2$), car driving ($n = 1$), mechanical, electrical, and flight instrument work ($n = 1$), weaponry and equipment mechanics ($n = 1$), and health service ($n = 1$).

The participants were informed about the study's objective and procedures through a group meeting and an individual approach on the day before each video recording. Finally, the participants were instructed to perform their tasks as usual and not to change their schedule due to the presence of cameras.

All video collection was authorized by the participants through a consent form and all procedures in this project are in line with national and international guidelines for scientific research involving human subjects, including the Declaration of Helsinki in 2013 on Ethical Principles for Medical Research Involving Human Subjects, and the 1997 Convention on Human Rights and Biomedicine (the "Oviedo Convention"). The ethics committee of the Faculty of Human Kinetics, University of Lisbon, approved the experimental procedures with no. 13/2014.

21.2.3 RECORDING PROCEDURE AND FEATURES

Participants' interactions with the bedroom equipment were video recorded on a normal rest period day and were assessed using: (a) one infrared digital camera (Wireless AEE Weatherproof—2,5 GHz—color); (b) one multiplexer video recorder (ACH MPEG-4 Realtime DVR); and (c) DVD recorder HD (LG recorder). All device lights were turned off or hidden and participants were informed about the placement of all cameras. However, they did not know the real video recording time.

The digital video cameras turned on automatically from midnight to 8:00 a.m. and during the periods in which the subjects were asleep, awake, out of bed, doing activities, using a pillow, they

FIGURE 21.2   Images of the first plan (frontal superior) of the bed observations.

were filmed using one plan (frontal superior) considering that it provided the best visualization of the participant and activity (Figure 21.2).

In order to ensure similar interaction times in the bedroom and not to interfere with evening activity and sleep period, all volunteers were filmed for 3 days for 10 h continuously (starting at 12:00 a.m.). After the filming period for each participant, a quick analysis of the video was done in order to select the best two days, according to the following criteria:

- Longer stay of students in the bed (preferred > 5 h)
- More than 60% of the video had a good visualization of the postural behaviors during sleep times

21.2.4   DATA COLLECTION AND ANALYSIS METHOD

Data collection and analysis methodology by observing video was based on the methodology developed for the observation of the interaction human/equipment/environment proposed by Rebelo et al. (2011), who analyzed the postural behavior in real situation bedroom residences using the software iSEE (Figure 21.1).

The fundamental aspect of this analysis was the development of behavioral ICs that will be quantified later. Following the analysis of the results of the previous steps and observation of the collected videos, categories were defined.

For this analysis, the software iSEE was classified to (a) evaluate the behaviors of interaction in a real environment and for long periods of time; (b) allow sorting an impossible number of observables in other techniques at the same event; (c) observe activities, actions, means of interaction (equipment), and postural behaviors in the same event; (d) create hierarchies to allow for observables; (e) to question the events in greater depth and detail and be able to sort all visible behaviors and test their viability during analysis by category OTHERS; among others. Following the analysis of the results of the previous phases and of the observation of the collected videos, the categories were defined.

According to Filgueiras et al. (2012), as mentioned, the analysis was done using software developed for this purpose. It allows classifying the ICs, through video analysis, in levels. Six categories of behaviors were defined, that represent the night activity or posture behaviors in this residences' bedrooms, divided into three base categories, two ICs, and one other nonspecific category (Figures 21.3 through 21.6).

Importantly, these ICs correspond to the information provided by the reference studies for the realization of these ICs (Saad et al., 2004; Haex, 2005; Huang et al., 2010; Verhaert et al., 2012; Desouzart et al., 2016).

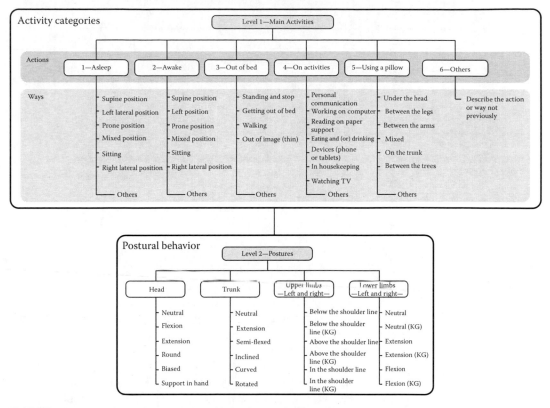

FIGURE 21.3 Level 1—six categories of behaviors and postural behaviors.

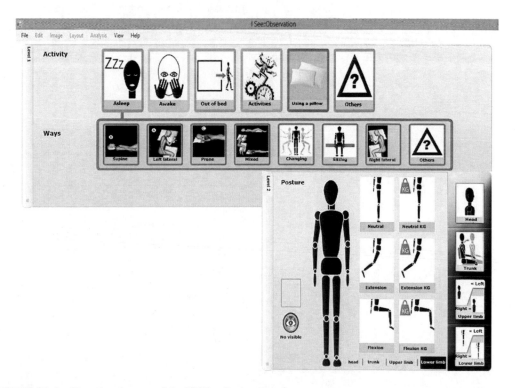

FIGURE 21.4 Functional areas of the iSEE software interface.

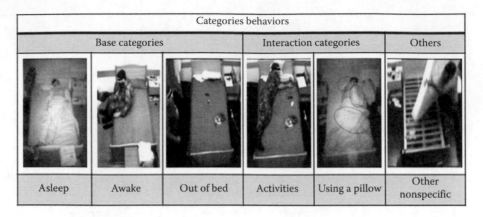

FIGURE 21.5   Level 1—six categories of behaviors.

The equipment for the sleep period (i.e., the combination mattress, bed base, and pillows) is an important factor in the sleep environment as it influences the physical recovery during sleep because it corresponds to the body support surface. However, several factors influence the interaction between the human body and the equipment to the sleep period. The factors which contribute to the physical recovery include body size, body weight, and body postures distribution used during sleep involving the entire mattress surface (Verhaert et al., 2012).

The ergonomic analysis of the environment and the activities that are involved, is a way to structure the person–task–environment system for the defining and ranking of ergonomic problems, whether postural, interaction, movement, operational, spatial, or physical environmental. The person–task–environment involves understanding the difficulties that users find in the execution of activities in daily life and the environment in which this task is accomplished. Thus, the ergonomic analysis can be characterized as a record of the activities in real situations implementing tasks, in order to promote improvements in the conditions of its implementation and the user's health (Moraes and Mont'Alvão, 2009; Cantalice and Tinoco, 2013).

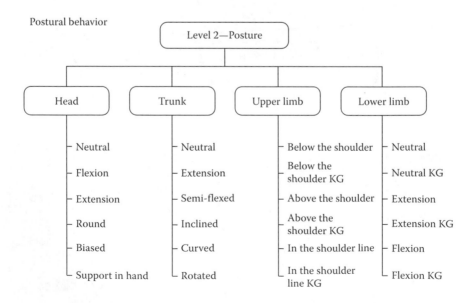

FIGURE 21.6   Level 2—postural behaviors.

21.2.4.1 Level 1: Observation Categories Refer to Activities

Tables 21.1 through 21.6 show the codes and description for ICs "asleep," "awake," "out of bed," "activities," "using a pillow," and "others" groups.

21.2.4.2 Observation Categories Related to Postures

The analysis and ecological validity of a postural study must reproduce the methods and conditions of real life situations that are under investigation. Furthermore, laboratory systems are expensive and complex with limited capacity to provide postural feedback, time consuming for everyday tasks, which is significant given the potential role of postural awareness (Kapandji, 2009; O'Sullivan et al., 2011).

TABLE 21.1
Categories, Ways, and Description for Asleep Group

Category	Ways	Description
Asleep		Period of time when the participant did not have any activity associated to the bed
	Supine	They were in the supine sleep position with their back in contact with the mattress
	Prone	They were in the prone sleep position with their frontal trunk in contact with the mattress
	Left lateral decubitus	They were in the left lateral sleep position with their trunk and hip left side in contact at mattress
	Right lateral decubitus	They were in the right lateral sleep position with their trunk and hip right side in contact at mattress
	Mixed position	They were at least two different sleep positions at same time with their trunk, hip, and shoulder in different contact at mattress
	Position change	They were a change at sleep position
	Sitting	They were in the sitting sleep position with their back in contact at pillow and your hip in contact at mattress
	Other	Any posture activity that represent some kind of specific category which was not anticipated

TABLE 21.2
Categories, Ways, and Description for Awake Group

Category	Ways	Description
Awake		Period of time when the participant had activity associated to the bed
	Supine	They were in the supine sleep position with their back in contact with the mattress
	Prone	They were in the prone sleep position with their frontal trunk in contact with the mattress
	Left lateral decubitus	They were in the left lateral sleep position with their trunk and left side hip in contact with the mattress
	Right lateral decubitus	They were in the right lateral sleep position with their trunk and right side hip in contact with the mattress
	Mixed position	They were in at least two different sleep positions at the same time with their trunk, hip, and shoulder in different contact with the mattress
	Position change	There was a change in sleep position
	Sitting	They were in the sitting sleep position with their back in contact with the pillow and their hip in contact with the mattress
	Other	Any posture activity that represents some kind of specific category which was not anticipated

TABLE 21.3
Categories, Ways, and Description for Out-of-Bed Group

Category	Ways	Description
Out of bed		Period of time in which the participants are not in bed
	Standing	They are standing
	Getting out of bed	They are getting out of bed
	Walking	They are walking
	Out of the picture	They are absent from the picture
	Other	Any activity that represents some kind of specific category which was not anticipated

Based on the above, the second level was organized to contain four groups representing the main body segments (e.g., head, trunk, upper limbs, and lower limbs). Each group has six categories of observation representing some of the key positions for each segment. Tables 21.7 through 21.10 show the codes and description for postures.

As mentioned, the analysis was done using software developed for this purpose. It allows classifying the IC (through video analysis) in levels. According to Filgueiras et al. (2012), although the system allows the observation and registering of categories in continuous time, the high number of categories for this analysis represents a cognitive overload to the observer and may contribute to a significant increase in classification errors.

Thus, the classification of systematic activity sequences was done using samples controlled by the software (10 s of analysis for each 100 s of activity). Each one of these activity sequences represents an "event" which remained in looping (10 s) until all ICs were registered (Figure 21.7).

Postural control is no longer considered simply the sum of static reflexes, but rather a complex skill based on the interaction of dynamic sensorimotor processes. Postural control involves the

TABLE 21.4
Categories, Ways, and Description for Activity Group

Category	Ways	Description
Activity		It includes all the activity/action behaviors of a specific task, which interacts with other categories
	Personal communication	The category "personal communication" must be activated whenever the participant observes a facial articulation characteristic of an oral or gestural communication, for a period of time greater than or equal to 5 s
	Using the computer	The category "using the computer" should be considered whenever there is contact of the participant with a computer system
	Reading	The category "reading" where the position of the participant's head is facing a readable medium (e.g., paper, book), for a period of time greater than or equal to 5 s
	Eating/Drinking	The category "eating/drinking" records all behavior related to eating or handling liquids (drink) or solids (food) for a period of time greater than or equal to 5 s
	Using mobile devices (phone or tablet)	The category "using mobile devices" must be activated whenever the participant observes a set of actions relating to the handling of equipment (e.g., tablet and mobile phone), for a period of time greater than or equal to 5 s
	In housekeeping	The category of "housekeeping" aims to identify all the situations in which the participant was engaged in activities for the organization, cleaning, or adjusting of their bed or bedroom (e.g., organizing or repositioning equipment and making the bed), for a period of time greater than or equal to 5 s
	Watching TV	The category "watching TV" where the position of the participant's head is turned to the television screen, for a period of time greater than or equal to 5 s
	Other	Any activity that means some kind of specific category which was not anticipated

TABLE 21.5

Categories, Ways, and Description for Using a Pillow Group

Category	Ways	Description
Using a pillow		Means for using pillow (s) in all the behaviors that the participants have any pillow contact with a body part associated with the bed, and the recorded image is viewable for a period of time greater than or equal to 5 s for each event
	Under the head	The category "under the head" must be activated whenever the participant used one pillow under the head
	Between the legs	The category "between the legs" must be activated whenever the participant used one pillow between the legs
	Between the arms	The category "between the arms" must be activated whenever the participant used one pillow between the arms
	Mixed	The category "mixed" must be activated whenever the participant used two or more pillows at the same time in some place of the body (head, leg, arm, feet, and/or trunk)
	On the trunk	The category "on the trunk" must be activated whenever the participant used one pillow under the trunk
	Between the feet	The category "between the feet" must be activated whenever the participant used one pillow between the feet
	Without support	The category "without support" must be activated whenever the participant did not use any pillow
	Other	Any activity that means some kind of specific category which was not anticipated

TABLE 21.6

Categories, Ways, and Description for Others Group

Category	Ways	Description
Others	Other	Any posture activity that represent some kind of specific category which was not anticipated

TABLE 21.7

Categories, Ways, and Description for Head Posture

Category	Ways	Description
Head		The postures qualified for the head-neck joint reflect all movements and anatomical axes made at the cervical level
	Neutral	Period where the subject is in the neutral position of the head-neck. Admits a slight rotation or slope ($<10°$)
	Flexion	Period where the subject is with the head-neck in flexion, with head tilted forward or down, with flexion in excess of $10°$
	Extension	Period where the subject is with the head-neck in extension, with the head tilted back or up, with any extension beyond $10°$
	Rotation	Period where the subject is with the head-neck in rotation to one side, sometimes the category requires identification of a note to the sagittal and horizontal planes, however, when a rotation angle greater than $40°$ is visible in any of the planes
	Side slope	Period where the subject is with the head-neck in lateral tilt to one side, sometimes the name of that category requires a note to the sagittal and horizontal planes, however, when a slope with angle greater than $40°$ is visible in any of the planes
	Head with support in hand or arm	Period where the subject is with the head-neck in one of the positions described above, normally the headrest is made with one or both upper limbs

TABLE 21.8
Categories, Ways, and Description for Trunk Posture

Category	Ways	Description
Trunk		Six categories of observation were identified to represent the main postures of the trunk
	Neutral	Period where the subject is in neutral position of the trunk. Admits a slight rotation or tilt (<10°)
	Extension	Period where the subject is extending the upper trunk to 10°
	Semi-flexion	Where the period is subject to the upper body bent slightly forward, in flexion
	Side slope	Where the period is subject to the trunk side tilt to one side, sometimes the name of that category requires observation and being horizontal, however, when a slope with an angle greater than 40° is visible in any of the planes
	Full flexion	Period where the subject is with the trunk in a previous full flexion. This position is also called "fetal" position
	Rotation	Period where the subject is in the trunk rotation to one side, this rotation can be performed by the shoulder girdle over the pelvic girdle or vice versa

active alignment of the trunk, limbs, and head in relation to gravity. The somatosensory information, vestibular, and visual systems are integrated in postural control, and are dependent upon the movement task goals and the environmental context (Horak, 2006).

The posture while sleeping reveals important information for the health of the individual (Huang et al., 2010), to provide an adequate postural support to allow each part (head, trunk, and limbs) of the human body to recover to perform daily activities being the most important function of the sleep period (Verhaert et al., 2012).

Postural analysis by applying the observation shows advantages related to the observations of postural changes for long periods of time in order to check which constraints posture can generate complaints, discomfort, or musculoskeletal pain in the spine, and that may be related to the

TABLE 21.9
Categories, Ways, and Description for Upper Limb (Left or Right)

Category	Ways	Description
Upper limb		The categories of these groups are related to placements made by the segments arm, elbow, forearm, and hand, displayed according to macro analysis and classified under the position of the glenohumeral joint (shoulder), left or right
	Arm below the shoulder line	Period where the subject is with the upper limb in position below the ipsilateral shoulder line
	Arm below the shoulder line with weight (kg)	Period where the subject is with the upper limb in position below the ipsilateral shoulder line, which has the weight or support any part of the body (e.g., having the body on top of the arm) or objects (e.g., hold a computer)
	Arm above the shoulder line	Period where the subject is with the upper limb in position above the ipsilateral shoulder line
	Arm above the shoulder line with weight (kg)	Period where the subject is with the upper limb in position above the ipsilateral shoulder line, which has the weight or support any part of the body (e.g., having the body on top of the arm) or objects (e.g., holding a computer)
	Arm at shoulder line	Period where the subject is with the upper limb in the line position of the ipsilateral shoulder to 90°. Admits a slight lower or upper slope (<10°)
	Arm at shoulder line with weight (kg)	Period where the subject is with the upper limb in the line position of the ipsilateral shoulder to 90°. Admits a slight lower or upper slope (<10°), which has the weight or support any part of the body (e.g., having the body on top of the arm) or objects (e.g., holding a computer)

TABLE 21.10

Categories, Ways, and Description for Lower Limb

Category	Ways	Description
Lower limb		The categories classified for the lower limbs correspond to postural situations to the lower left and right members. The categories of these groups are related to placements made by leg, knee, and tibiotarsal, displayed according to macro analysis, and classified by virtue of the position of the hip joint
	Lower limb in neutral position	Period where the subject is with the lower limb in neutral. Admits a slight flexion or extension (<10°)
	Lower limb in neutral position with weight (kg)	Period where the subject is with the lower limb in ipsilateral neutral position, which has the weight or support any part of the body (e.g., have the other leg on top of this) or be in the standing position. Admits a slight flexion or extension (<10°)
	Lower limb in extension	Period where the subject is with the lower limb in the position extent of ipsilateral limb
	Lower limb in extension with weight (kg)	Period where the subject is with the lower limb in the position extent of ipsilateral limb, which has the weight or support any part of the body (e.g., have the other leg on top of this) or be in the standing position
	Lower limb in flexion	Period where the subject is with the lower limb in ipsilateral flexion
	Lower limb in flexion with weight (kg)	Period where the subject is with the lower limb in ipsilateral flexion, which has the weight or support any part of the body (e.g., have the other leg on top of this) or be in the standing position

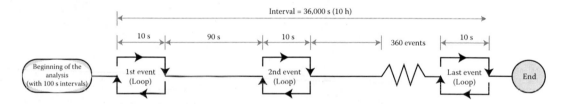

FIGURE 21.7 Flowchart with the systematic observation stages used by the software.

difficulties felt with daily activities. A tool or systemic observation method must register the postural behavior of each segment individually (Guérin et al., 2007).

For the development of this study, based on the study of Fransson-Hall et al. (1995), we found that less than 10° variations with reference to the anatomical position, are not sufficiently visible to most observers to alter postural rating. Following this principle, all the behaviors of supination, pronation, abduction, adduction, deviations, inversion, eversion, and some push-ups and extensions of the upper limbs (elbow and wrist) and lower (knee and hock), admittedly difficult to identify by image video, were eliminated.

Figure 21.8 shows the rotation angle of the main body joints produced based on the study of Fransson-Hall et al. (1995) and known to those skilled in this study. The gray area represents the angular range where the classification errors occur more than once, since it is often difficult for a less experienced observer to clearly distinguish between the neutral position and movements of flexion, extension, and rotation.

21.3 RESULTS

A sample of 13,928 observations, which corresponds to 392 sleep-hours of 12 university students and 12 air force military, was classified into six ICs. The results can be seen in Figure 21.9.

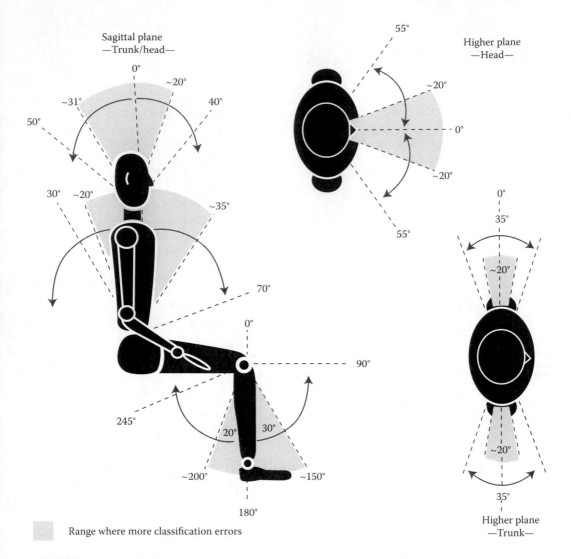

FIGURE 21.8 Angular areas difficult to identify in observational studies. The gray areas indicate the angular range, approximately, where there are more classification errors in observational studies.

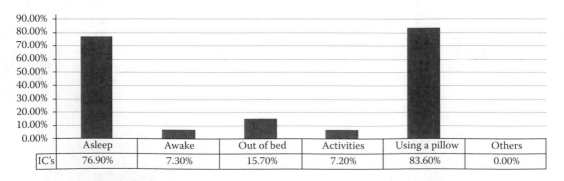

	Asleep	Awake	Out of bed	Activities	Using a pillow	Others
IC's	76.90%	7.30%	15.70%	7.20%	83.60%	0.00%

FIGURE 21.9 Results for ICs groups.

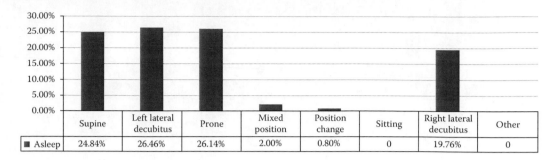

FIGURE 21.10 Results for "asleep" category.

	Supine	Left lateral decubitus	Prone	Mixed position	Position change	Sitting	Right lateral decubitus	Other
■ Asleep	24.84%	26.46%	26.14%	2.00%	0.80%	0	19.76%	0

The result of the ICs groups showed the group "using a pillow" with 83.6% ($N = 11,641$ records) as the main image registration, followed by the group "asleep" with 76.9% ($N = 10,710$ records), "out of bed" with 15.7% ($N = 2190$ records), "awake" with 7.3% ($N = 1020$ records), and "activities" with 7.2% ($N = 1002$ records).

Group 1 represents the categories of observation of the postures adopted during the period in which (a) participant is "asleep." Data on activities that occurred during this period allowed verifying that 46.2% of the participants presented the lateral position (26.46% on the left and 19.76% on the right) as the most common postural behavior during sleep (Figure 21.10).

The category "awake" presented the sitting position as the most common postural behavior, with 25.49% of postural behavior, followed by the supine position, with 23.14% (Figure 21.11).

In the "out of bed" category, the most common observation was the "out of the picture," with 96.07% of observation (Figure 21.12).

When the participant stood in the "activity" category during the video capture, the most common activity was "using a computer," with 48.30% of observation, which corresponds to approximately 30 min of computer use per participant per night (Figure 21.13). The use of any electronic device

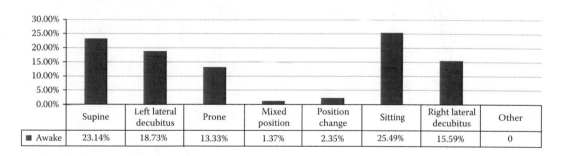

FIGURE 21.11 Results for "awake" category.

	Supine	Left lateral decubitus	Prone	Mixed position	Position change	Sitting	Right lateral decubitus	Other
■ Awake	23.14%	18.73%	13.33%	1.37%	2.35%	25.49%	15.59%	0

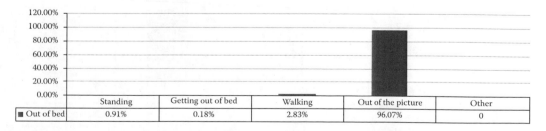

	Standing	Getting out of bed	Walking	Out of the picture	Other
■ Out of bed	0.91%	0.18%	2.83%	96.07%	0

FIGURE 21.12 Results for "out of bed" category.

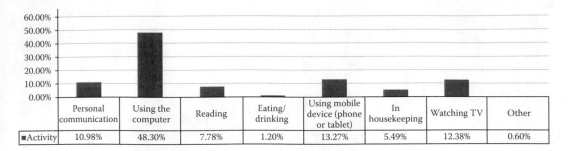

FIGURE 21.13   Results for "activity" category.

(using the computer and mobile device) in the rest period is more than 60% of the activities occurring in residence rooms.

Concerning the "using a pillow" category, 93.89% of participants used only one pillow under their head (Figure 21.14).

In the postural behaviors, which corresponds to iSEE software level 2 (Figure 21.6), it shows the head posture, trunk posture, left or right upper limb posture, and left or right lower limb posture (Figure 21.15). The maximum observation for head posture was round, with 34.79%, the trunk posture was semi-flexed, with 26.99%, the upper limb posture was below the shoulder, with a mean of 61.17% between right and left (61.34% and 61.00%, respectively), and the lower limb posture was flexed, with a mean of 35.65% between right and left (35.13% and 36.16%, respectively), as shown in Figure 21.15.

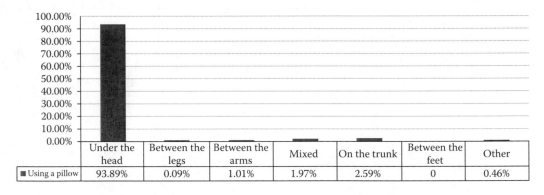

FIGURE 21.14   Result for "using a pillow" category.

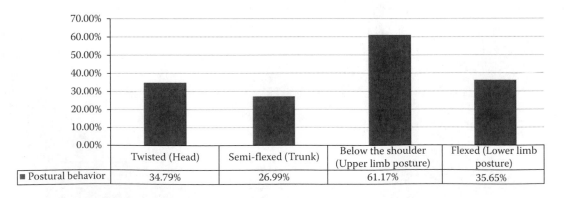

FIGURE 21.15   Maximum postural behaviors in each body classification.

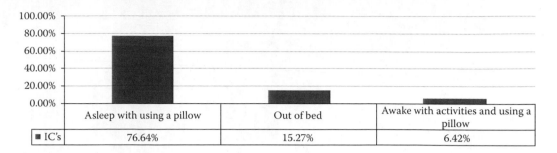

FIGURE 21.16 Results for IC and their interactions.

In the ICs, the most common interactions were "asleep using the pillow" with 76.64%, followed by category "out of bed" with 15.27%, and further followed by the category "awake with activities and using the pillow" with 6.42% (Figure 21.16).

The category "asleep and using a pillow" are the most common IC with 76.64% of all classifications. In the ICs of most interaction, prone using a pillow under the head occurs 22.25%, left lateral decubitus using a pillow under the head corresponds to 22.21%, supine using a pillow under the head occurs 21.93%, and right lateral decubitus using a pillow under the head occurs 17.48% (Figure 21.17).

The combination of the categories of interaction and body postures suggests that the main motor behaviors performed during the sleep period were right lateral decubitus using a pillow under the head with head in flexed position, semi-flexion trunk, upper limb bellow the shoulder line, and lower limb in flexion, corresponding to 4.34% (Figure 21.18).

During the image capture, participants answered a questionnaire about the perception of pain in the spine according to the visual analogue scale. 100% of university students and air force military indicated complaints in BP; 37.5% referred to the evening as the period in which the pain was more intense; 37.5% of the participants reported that pain disrupted their sleep at least one night per week and; the biggest indication of average of pain was 4.46 ± 1.382, and the lumbar region was the place where there was a higher incidence of pain (with mean 3.33 ± 0.771), value that corresponds to moderate pain.

When we analyze the average of BP indications in the group of participants, the students ($N = 12$) showed 4.42 ± 1.379, 41.7% referred to the evening as the period in which the pain was more intense, and 25% of the participants reported that pain disrupted their sleep. The military ($N = 12$) showed that 4.50 ± 1.446 was the average of BP indications, 33.3% referred to the evening as the period in which the pain was more intense, and 50% of the military reported that pain disrupted their sleep.

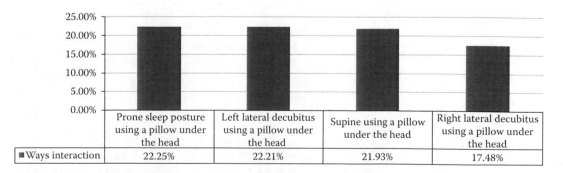

FIGURE 21.17 Results for ways interaction.

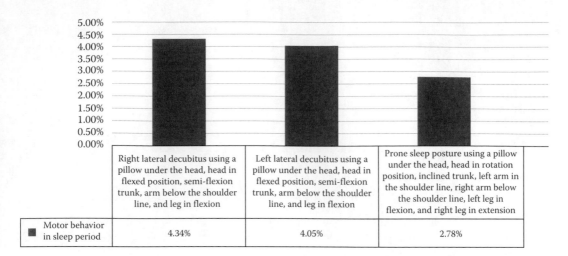

FIGURE 21.18   Results for motor behaviors during the sleep period.

21.4   CONCLUSION

The ICs of "asleep" (with more or less 6 h and 30 min) and "using a pillow" (with more or less 7 h) are the largest periods of the postural behaviors in bed during the rest period. The influence of the sleep position on physiological damage in the period of rest with or without activities in bed is not well known.

The results obtained with this method of analysis of postural behaviors for long periods of continuous time are important to understand their influence on musculoskeletal conditions. This data can be associated with the increase of musculoskeletal problems, which can be found among young Portuguese university students and air force military when they remain in apparently bad postures in bed for long periods of time.

Finally, this iSEE methodology was considered efficient for the proposed objectives and the findings offer new challenges for future research. Our findings allow us to suggest what graphical interface designers must seek as new strategies and solutions for behavior change in posture in bed, exploring other peripheral equipment for the position of laptops; or, at least, to improve the posture of the participants when using the laptop in bed, and if these ergonomic changes influence the reduction of BP indications.

These data are essential for health care professionals who can use this information to enable a reduction factor of complaints of BP, to make recommendations to schools and universities to change the demands of academic activities by distributing them throughout the semester and not at the end of each period.

However, the iSEE software methodology defines the categories of observation, called categories of interactions, and the software to quantify those (Filgueiras et al., 2012). This software is essential to analyze data. Without it, it would be necessary to directly observe the video with notes on paper or a computer record of every change without the application of categories for the purpose and with a loss of important information.

ACKNOWLEDGMENTS

The authors would like to thank the Portuguese Polytechnic Institute of Leiria and in particular its School of Education and Social Sciences, and School of Health Sciences and Social Services. The authors would also like to thank the Portuguese Air Force and in particular its air base no. 5 as well as the general staff of the Portuguese Air Force. The authors also thank the commander of the air base for assisting in the sample collection process and the research assistants, particularly Wing Commander Ismael Alves, Aspiring Officer Madalena Carvalho, and Wing Commander Fernando Damásio, for their help during the data collection phase. We would also like to thank the university students and air force military participants for their contribution of time and effort to the research. Without them, this study would not be possible.

REFERENCES

Addison, R., M. Thorpy, and T. Roth. A survey of the United States public concerning the quality of sleep. *Journal of Sleep Research* 16, 1986: 244.

Andersson, H., G. Ejlertsson, I. Leden, and C. Rosenberg. Chronic pain in a geographically defined general population: Studies of differences in age, gender, social class, and pain localization. *The Clinical Journal of Pain* 9, 1993: 174–182.

Apeldoorn, A., R. Ostelo, H. van Helvoirt, J. Fritz, H. de Vet, and M. van Tulder. The cost-effectiveness of a treatment-based classification system for low back pain: Design of a randomised controlled trial and economic evaluation. *BMC Musculoskeletal Disorders* 11, 2010: 58.

Ban, D. and T. Lee. Sleep duration, subjective sleep disturbances and associated factors among university students in Korea. *Journal of Korean Medical Science* 16, 2001: 475–480.

Bergqvist, U. Visual display terminal work—A perspective on long-term changes and discomforts. *International Journal of Industrial Ergonomics* 16(3), 1995: 201–209.

Brown, F., W. Buboltz, and B. Soper. Relationship of sleep hygiene awareness, sleep hygiene practices, and sleep quality in university students. *Behavioral Medicine* 8(1), 2002: 33–38.

Cantalice, J. and M. Tinoco. Ergonomic analysis of proposed and dimensioned furniture space in social housing. *VI PROJETAR 2013*. Salvador: UFBA. 2013: E3008–E3033.

Casserley-Feeney, S., G. Bury, L. Daly, and D. Hurley. Physiotherapy for low back pain: Differences between public and private healthcare sectors in Ireland. *Manual Therapy* 13(5), 2008: 441–449.

Chou, R., A. Qaseem, V. Snow, D. Casey, J. Cross, and P. Shekelle. Diagnosis and treatment of low back pain: A joint clinical practice guideline from the American College of Physicians and the American Pain Society. *Annals of Internal Medicine* 148(3), 2008: 247–248.

Cohen, S., C. Nguyen, S. Kapoor, V. Anderson-Barnes, L. Foster, C. Shields, B. Mclean, T. Wichman, and A. Plunkett. Back pain during war: An analysis of factors affecting outcome. *Arch Intern Med* 169, 2009: 1916–1923.

Cohen, S., C. Brown, C. Kurihara, A. Plunkett, C. Nguyen, and S. Strassels. Diagnoses and factors associated with medical evacuation and return to duty for service members participating in operation Iraqi freedom or operation enduring freedom: A prospective cohort study. *Lancet* 375, 2010: 301–309.

Curcio, G., M. Ferrara, and L. De Gennaro. Sleep loss, learning capacity and academic performance. *Sleep Medicine Reviews* 10(5), 2006: 323–337.

Danda, G., G. Rocha, M. Azenha, C. Sousa, and O. Bastos. Standard sleep–wake cycle and excessive daytime sleepiness in medical students. *Jornal Brasileiro de Psiquiatria* 54(2), 2005: 102–106.

De Bruijn, I., J. Engels, and J. Van Der Gulden. A simple method to evaluate the reliability of OWAS observations. *Applied Ergonomics* 29(4), 1998: 281–283.

Desouzart, G., E. Filgueiras, F. Melo, and R. Matos. Human body–sleep system interaction in residence for university students: Evaluation of interaction patterns using a system to capture video and software with observation of postural behaviors during sleep. In Soares, M. and Rebelo, F. (eds.), *Design, Advances in Ergonomics In Design, Usability & Special Populations: Part I*. pp. 169–178. Poland, Kraków: AHFE Conference 2014a. (ISBN: 978-1-4951-2106-7).

Desouzart, G., E. Filgueiras, F. Melo, and R. Matos. Human-bed interaction: A methodology and tool to measure postural behavior during sleep of the air force military. In Marcus, A. (ed.), *DUXU 2014*, Vol. Part III, pp. 662–674. Switzerland: Springer International Publishing, 2014b.

Desouzart, G., R. Matos, F. Melo, and E. Filgueiras. Effects of sleeping position on the back pain in physically active seniors: A controlled pilot study. Work: *A Journal of Prevention, Assessment and Rehabilitation* 53, 2016: 235–240.

Devine, E., Z. Hakim, and J. Green. A systematic review of patient reported outcome instruments measuring sleep dysfunction in adults. *Pharmacoeconomics* 23, 2005: 889–912.

Ehrlich, G. Low back pain. *Bulletin of World Health Organization* 81, 2003: 671–676.

Engström, T. and P. Medbo. Data collection and analysis of manual work using video recording and personal computer techniques. *International Journal of Industrial Ergonomics* 19(4), 1997: 291–298.

Fenety, A. and J. Walker. Short-term effects of workstation exercises on musculoskeletal discomfort and postural changes in seated video display unit workers. *Physical Therapy* 82(6), 2002: 578–589.

Filgueiras, E., F. Rebelo, and F. Moreira da Silva. Support of the upper limbs of office workers during a daily work. Work: *A Journal of Prevention, Assessment & Rehabilitation* 41(1), 2012: 676–682.

Forsman, M., G. Hansson, L. Medbo, P. Asterland, and T. Engström. A method for evaluation of manual work using synchronised video recordings and physiological measurements. *Applied Ergonomics* 33(6), 2002: 533–540.

Fransson-Hall, C., R. Gloria, A. Kilbom, and J. Winkel, A portable ergonomic observation method (PEO) for computerized on-line recording of postures and manual handling. *Applied Ergonomics* 26(2), 1995: 93–100.

Geldhof, E., D. Clercq, I. Bourdeaudhuij, and G. Cardon. Classroom postures of 8–12 year old children. *Ergonomics* 50(10), 2007: 1571–1581.

Gracovetsky, S. The resting spine: A conceptual approach to the avoidance of spinal reinjury during rest. *Physical Therapy* 67, 1987: 549–553.

Gross J., J. Fetto, and E. Rosen. *Musculoskeletal Examination. Porto Alegre: Artmed*, 2000, pp. 85–128.

Guérin, F., A. Laville, F. Daniellou, and J. Duraffourg. *Understanding and Transforming Work. The Pratice of Ergonomics*. Lyon: ANACT Network Editions, 2007.

Haex, B. *Back and Bed: Ergonomic Aspects of Sleeping*. Routledge: Taylor & Francis Group, 2005.

Hayden, J., M. van Tulder, A. Malmivaara, and B. Koes. Exercise therapy for treatment of non-specific low back pain. *The Cochrane Database System of Reviews* 20(3), 2005: CD000335.

Hendrick, H. and B. Kleiner. *Macroergonomics: An Introduction to Work System Design*. Santa Monica, CA: Human Factors and Ergonomics Society, 2001.

Horak, F. Postural orientation and equilibrium: What do we need to know about neural control of balance to prevent falls? *Age Ageing* 35 (suppl. 2), 2006: ii7–ii11.

Huang, W., A. Wai, S. Foo, J. Biswas, C. Hsia, and K. Liou. Multimodal sleeping posture classification, 2010, *Pattern Recognition, International Conference on 2010*, Istanbul, Turkey: ICPR, pp. 4336–4339.

Hurley, D. et al. Physiotherapy for sleep disturbance in chronic low back pain: A feasibility randomised controlled trial. *BMC Musculoskeletal Disorders* 11, 2010: 70.

Jacobson, B., A. Boolani, and D. Smith. Changes in back pain, sleep quality, and perceived stress after introduction of new bedding systems. *Journal of Chiropractic Medicine* 8, 2009: 1–8.

Kapandji, I. *Articular Physiology: Trunk and Head*. São Paulo: Guanabara Koogan, vol. 3. 2009.

Kazmierczak, K., S. Mathiassen, P. Neumann, and J. Winkel. Observer reliability of industrial activity analysis based on video recordings. *International Journal of Industrial Ergonomics* 36, no. 3, 2006: 275–282.

Koes, B. et al. An updated overview of clinical guidelines for the management of non-specific low back pain in primary care. *European Spine Journal* 19(12), 2010: 2075–2094.

Kuo, C., F. Yang, M. Tsai, and M. Lee. Artificial neural networks based sleep motion recognition using night vision cameras. *Biomedical Engineering—Applications, Basis & Communications* 16, no. 2, 2004: 79–86.

Kushida, C. et al. Practice parameters for the indications for polysomnography and related procedures: An update for 2005. *Sleep* 28(4), 2005: 499–521.

Liao, W. and C. Yang. Video-based activity and movement pattern analysis in overnight sleep studies. In *Proceedings of the 21st Pattern Recognition,* Florida, USA: ICPR, 2008, pp. 1–4.

Liddle, S., G. Baxter, and J. Gracey. Physiotherapist's use of advice and exercise for the management of chronic low back pain: A national survey. *Manual Therapy* 14, 2009: 189–196.

Liddle, S., J. Gracey, and G. Baxter. Advice for the management of low back pain: A systematic review of randomised controlled trials. *Manual Therapy* 12, 2007: 310–327.

Lincoln, A., G. Smith, P. Amoroso, and N. Bell. The natural history and risk factors of musculoskeletal conditions resulting in disability among US army personnel. *Work* 18, 2002: 99–113.

Mantyselka, P., E. Kumpusalo, R. Ahonen, A. Kumpusalo, J. Kauhanen, H. Viinamaki, P. Halonen, and J. Takala. Pain as a reason to visit the doctor: A study in Finnish primary health care. *Pain* 89, 2001: 175–180.

Mayer, J., S. Haldeman, A. Tricco, and S. Dagenais. Management of chronic low back pain in active individuals. *Current Sports Medicine Reports* 9(1), 2010: 60–66.

Moffett, J. Back pain: Encouraging a self-management approach. *Physiotherapy Theory and Practice* 18(4), 2002: 205–212.

Moraes, A. and C. Mont'Alvão. *Ergonomics: Concepts and Applications* (4th ed). Rio de Janeiro: 2AB, 2009.

Morgenthaler, T. et al. Practice parameters for the use of actigraphy in the assessment of sleep and sleep disorders: An update for 2007. *Sleep* 30, 2007: 519–529.

National Sleep Foundation, NSF. *Sleep in America Poll.* National Sleep Foundation, 2005, pp. 1–54. Available: https://sleepfoundation.org/sleep-polls-data/sleep-in-america-poll/2005-adult-sleep-habits-and-styles.

O'Sullivan, K., L. Galleotti, W. Dankaerts, P. O'Sullivan, and L. O'Sullivan. The between-day and inter-rater reliability of a novel wireless system to analyse lumbar spine posture. *Ergonomics* 54(1), 2011: 82–90.

Penzel, T. and R. Conradt. Computer based sleep recording and analysis. *Sleep Medicine Reviews* 4(2), 2000: 131–148.

Pter, J. Sleep apnea and cardiovascular diseases. In C. Guilleminault and M. Partinen (eds.), *Obstructive Sleep Apnea Syndrome: Clinical No Treatment*, New York: Raven Press, 1990, pp. 81–98.

Rajaratman, S. and J. Arendt. Health in a 24-h society. *Lancet* 358, 2001: 999–1005.

Rebelo, F., E. Filgueiras, and M. Soares. Behavior video: A methodology and tool to measure the human behavior: Examples in product evaluation. In W. Karwowski, M. Soares, and N. Stanton (eds.). *Human Factors and Ergonomics in Consumer Product Design: Methods and Techniques.* Boca Raton, USA: CRC Press, Taylor & Francis, 2011, p. 320.

Saad, M., D. Masiero, A. Lourenço, and L. Battistella. Proposed a method of quantitative evaluation of the lying posture based on photography. *Acta Fisiátrica* 11(2), 2004: 60–66.

Schmidt, C. and T. Kohlmann. What do we know about the symptoms of back pain? Epidemiological results on prevalence, incidence, progression and risk factors. *Zeitschrift fur Orthopadie und Ihre Grenzgebiete* 143(3), 2005: 292–298.

Silva, A., D. Punt, P. Sharples, J. Vilas-Boas, and M. Johnson. Head posture and neck pain of chronic nontraumatic origin: A comparison between patients and pain-free persons. *Archives of Physical Medicine and Rehabilitation* 90, 2009: 669–674.

Soares, C. Insomnia in women: An overlooked epidemic? *Archives Women's Mental Health* 8, 2005: 205–213.

Spielholz, P., B. Silverstein, M. Morgan, H. Checkoway, and J. Kaufman. Comparison of self-report, video observation and direct measurement methods for upper extremity musculoskeletal disorder physical risk factors. *Ergonomics* 44(6), 2001: 588–613.

Stang, P., M. Korff, and B. Galer. Reduced labor force participation among primary care patients with headache. *Journal of General Internal Medicine* 13, 1998: 296–302.

Stewart, W., J. Ricci, E. Chee, D. Morganstein, and R. Lipton. Lost productive time and cost due to common pain conditions in the US workforce. *JAMA* 290, 2003: 2443–2454.

Straker, L., C. Pollock, and J. Mangharam. The effect of shoulder posture on performance, discomfort and muscle fatigue whilst working on a visual display unit. *International Journal of Industrial Ergonomics* 20(1), 1997: 1–10.

Strauss, A. and J. Corbin. *Basics of Qualitative Research: Grounded Theory, Procedures, and Techniques.* Newbury Park, New Jersey: Sage, 1990.

van Middelkoop, M., S. Rubinstein, T. Kuijpers, A. Verhagen, R. Ostelo, B. Koes, and M. van Tulder. A systematic review on the effectiveness of physical and rehabilitation interventions for chronic non-specific low back pain. *European Spine Journal* 20(1), 2011: 19–39.

Verhaert, V., H. Druyts, D. Van Deuna, V. Exadaktylos, J. Verbraecken, M. Vandekerckhove, B. Haex, and J. Sloten. Estimating spine shape in lateral sleep positions using silhouette-derived body shape models. *International Journal of Industrial Ergonomics* 42(5), 2012: 489–498.

Vieira, E. and S. Kumar. Working postures: A literature review. *Journal of Occupational Rehabilitation* 14(2), 2004: 143–159.

Walker, B., R. Muller, and W. Grant. Low back pain in Australian adults: Prevalence and associated disability. *Journal of Manipulative and Physiological Therapeutics*, 27, 2004: 238–244.

Willemen, T. et al. Automatic sleep stage classification based on easy to register signals as a validation tool for ergonomic steering in smart bedding systems. Work: *A Journal of Prevention, Assessment & Rehabilitation* 41(s1), 2012: 1985–1989.

World Health Organization, World Health Report, 2001. Available at: www.who.int/whr/index.htm.

Wright, C., H. Valdimarsdottir, J. Erblich, and D. Bovbjerg. Poor sleep the night before an experimental stress task is associated with reduced cortisol reactivity in healthy women. *Biological Psychology* 74, 2007: 319–327.

Wu, C. and H. Aghajan. Model-based human posture estimation for gesture analysis in an opportunistic fusion smart camera network. AVSS 2007. *IEEE Conference*, London, UK, 2007, pp. 453–458.

22 Applications of Haptic Devices and Virtual Reality in Product Design Evaluation

Christianne Falcão, Marcelo M. Soares, and Tareq Ahram

CONTENTS

22.1 INTRODUCTION

Virtual environment is an instance of a virtual world, where objects are presented in a simulated interactive field, known as virtual reality (VR) (Sherman and Craig, 2003). Virtual environment is an interactive environment that is associated with multisensory inputs and synthesized by a computer. The environment enables users to immerse themselves in a computer simulation (Barfield and Nash, 2006).

The research community points out the importance of usability in the product development life-cycle and stressed the need for usability testing to be performed throughout the product development process, starting from prototype testing based on various techniques.

When evaluating user product interaction, user activities are coordinated with a complex combination of sensory stimuli that is recorded via different sensory channels, such as touch, vision, and hearing. In order for results to be validated, it is necessary that the prototypes simulate the actual behavior and the appearance of the product being developed, and this ability is closely linked to the performance capability of simulation technologies, both hardware and software (Ferrise et al., 2013). The technological development in virtual environments, specifically virtual and augmented reality, has led to the development of more intuitive interfaces for design (Shen et al., 2010), enabling the implementation of users testing activities from the interaction with virtual prototypes.

Virtual prototyping is based on simulating realistic and three-dimensional (3D) virtual environment of the functionality expected from the actual product. Virtual prototyping technologies can reduce the time and costs associated with the development and construction of new products and bridge the disciplinary and distance gaps among different collaborators working on the same project (Lu et al., 1999). From the development of multimodal and multisensory environments, it is possible for the user to interact with the digital model in a way that is similar to how the user interacts with a physical model (Ferrise et al., 2013).

In this context, the sense of touch, or haptic sense, appears to be fundamental to assess product properties. This need became evident as a result of knowing that vision and/or auditory senses are not sufficient to permit the recognition of objects. In this case, it is necessary to use devices linked to haptic VR systems for identifying objects. Despite the large number of research studies in usability, there are few studies that focus on virtual environments and haptics. This chapter outlines several haptic devices and demonstrates the potential benefits and application of this technology on usability evaluation of product design in a virtual environment.

22.2 BACKGROUND

The definition of usability is sometimes shortened to "ease of use," but this offers poor information about the user interface. From a better known concept, usability is defined as the ability of a product or system to be used in an effective, efficient, and enjoyable way by a specific range of users, for tasks that need specific tools within a given environment (Falcão and Soares, 2012).

In order to measure the level of user satisfaction, as well as effectiveness and efficiency of the product, usability testing is a required process. Usability evaluation allows designers to see what users actually do when they use the product, what works best for them and their preferences. Usability testing refers to activity that focuses on observing users using a product, performing tasks that are real and meaningful to them (Barnum, 2011).

Simulation technology in virtual environment corresponds to simulate a realistic environment, and the best virtual environment is the one that most closely matches this situation. According to Gutierrez et al. (2008), for most people, simulation technology corresponds to the dominant perception of senses and the main means for information acquisition. However, a full simulation in virtual environment is not limited to what we can see, but also involves other components that are part of perceptual experience such as hearing, tactile feedback, smell, and taste. Despite the fact that hearing (auditory), smell (olfactory), and taste (gustatory) represent important functions in this experiment; they are unexplored due to the complex technological application (Gutiérrez et al., 2008; Fortineau et al., 2013).

In this sense, the virtual environment interface allows the evaluation of user skills and knowledge while using the virtual product in two situations: interactions with systems and interactions with virtual objects. In interactions with systems, multimodal interfaces should be integrated, including vision, sounds, voice input, gesture recognition, etc. While the interaction with virtual objects supports a realistic visual feedback, while movements and behavior is simulated (Wan et al., 2004). During the usability testing in a virtual environment, the virtual prototype should be viewed and interacted with by all the actors involved in its design, including the potential users, as if it was a real physical product. This is where VR can play a significant role, since it allows evaluating various alternative solutions and compares them in a realistic and dynamic way, not only visually, but also considering other interaction aspects such as sound and force feedback.

VR is a high-quality user–computer interface that involves real-time simulation and interactions through multiple sensory channels. These sensory modalities are visual, hearing, touch, smell, and taste (Burdea and Coiffet, 2003), and the visual channel is commonly the main aspect used in VR applications (Kirner and Siscoutto, 2007). Of the several peripheral devices, such as motion capture and haptic interfaces, VR offers an immersive work environment with different forms of interaction between the user and the system.

In product design, researchers introduced VR applications in almost every stage of development, and presented intuitive 3D interaction, enabling users to exploit their creativity in developing concepts from the 3D immersion (Shin et al., 2000; Lee et al., 2004; Bordegoni et al., 2011).

22.2.1 HANDS INTERACTION IN A VIRTUAL ENVIRONMENT

Interaction with 3D virtual prototypes in a virtual environment involves mainly human hands and provides a realistic simulation of movements and behaviors, such as those realized in the real world

(Wan et al., 2004; Han and Wan, 2010). Along with increased realism of the movements, the sense of immersion in the virtual environment, as well as the visual realism, will be felt better. Thus, important information about user behavior can be obtained from the 3D object simulation. However, objects have various textures and shapes, which makes the simulation of holding an object with the hands and moving it one of the most important and challenging tasks in the research field of virtual environments (Wan et al., 2004; Gutiérrez et al., 2008).

In the real world, people hold an object with one or both hands depending on the shape and weight, while the interaction in a virtual environment is based on the user ability to select and directly manipulate the virtual object with the hands, and the virtual hand used to visualize user inputs. The hand position in the virtual environment is obtained from 3D trackers, while the movements of the fingers and the bending angle are mapped by a data glove. To select an object, the user performs movements like in the real world that are repeated by the virtual hand, which is used as an "avatar" of the user's hand.

The human hand is a very sophisticated and complex organ of the human body, it features a structure with extra joints that allow precise manipulations and forces exertion on a variety of objects (Cushman and Rosenberg, 1991). Simulating how the human hand grasps objects is difficult due to the high number of degrees of freedom (DOF) that it has, and further the several seizure styles regarding the shape of the object and purpose (Kyota and Saito, 2012). Therefore, it is challenging to build a realistic and geometric model of the human hand (Gutiérrez et al., 2008; Han and Wan, 2010).

Kyota and Saito (2012) proposed a system that allows the user to choose the holding style from a taxonomy of postures with his/her hands. The system generates various postures of apprehension in accordance with various 3D object forms, since each joint is calculated separately. The postures are generated interactively and presented as a thumbnail table and can be used in the product evaluation.

Virtual models of human hands are most commonly used in VR applications which are made from a combination of separate parts for easier kinematic control over the visual realism (Wan et al., 2004). The *VirtualHand SDK* software developed by CyberGlove is an example of such system. The software adopts a simple model of the hand and has a solution that adds movement, interaction, and feedback force for application simulations (CyberGlove, 2015).

The construction of a virtual human hand includes geometric modeling and kinematics modeling, considering various factors like shape, motion, force feedback, and collision detection (Wan et al., 2004; Han and Wan, 2010). The haptic interaction with the virtual hand plays a fundamental role in many VR applications, particularly in the analysis of usability of products, since they provide an intuitive way for users to interact with virtual objects.

The virtual hand geometric and shape is represented using a polygon mesh and modeled in a way that is accurate enough to reflect the shape of the human hand, it should not be too complicated to not impede the simulation of hand movements in real time (Wan et al., 2004). To meet this requirement, Han and Wan (2010) suggest the construction of the virtual hand geometry with a triangular mesh based on the hand shape knowledge and its anatomical structure.

Hand movements are restricted by the joints and often one articulation limits another when in motion, and it is difficult simulate the movements realistically and the muscular deformations with simple kinematic models (Wan et al., 2004). Thus, for the kinematic model of the virtual hand, it is usually used a model of three layers consisting of the skeletal layer, the muscular layer, and skin layer, as shown in Figure 22.1 (Wan et al., 2004; Han and Wan, 2010).

The skeletal layer is built on the anatomical structure of the hand. Each finger is captured by a chain and the angle of rotation of each set of joints is captured by a CyberGlove (2015). The muscle in the human hand is between the skeleton and the skin and helps in stretching/shrinking skeleton motion, resulting in deformation of the skin. Therefore, the muscular layer is configured as an intermediate layer to drive the geometric deformation of the skin. The skin layer corresponds to the polygonal mesh used for the purpose of presentation and its deformation is conducted by the other two layers (Wan et al., 2004).

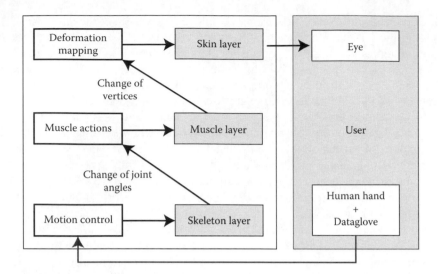

FIGURE 22.1   Kinematic model of virtual hand. (Adapted from Wan, H. et al. 2004. Realistic virtual hand modeling with applications for virtual grasping. *Proceedings of ACM SIGGRAPH International Conference on Virtual Reality Continuum and its Applications in Industry.* New York, pp. 81–87. ISBN: 1-58114-884-9.)

The feedback perception involves both the physiological and psychological issues that should be taken into account in haptic interaction. Han and Wan (2010) proposed that the touch model should fully respect the contact function in the area between the fingers and the virtual object in order to present a more realistic force feedback performed.

Real-time collision detection is used to automatically identify if there is interference between the virtual hand and virtual objects. In general, collision detection requires an overhead of calculations, especially when it involves deformable models, and should be effective to ensure a good tactile feedback in real time (Han and Wan, 2010). Han and Wan (2010) proposed a virtual hand model to facilitate haptic interaction. As shown in Figure 22.2, data from the user's hand movements are captured by a 3D tracker and a data glove to drive the virtual hand.

As indicated in Figure 22.2, the motion model, which includes in one of the blocks the virtual hand building, is directly driven by the user's hand movement from the use of a data glove and 3D trackers. The collision detection model is used to check the interference between the virtual hand and other objects, the haptic model is used in order to stabilize the force feedback, and the shape model is used for the virtual presentation.

22.2.2   TAXONOMY OF POSTURES WITH VIRTUAL HAND

The action taken by the hand, according to object shape, forms the hands in different postures. The observation of handle postures allows investigating the behavior and reaction of users with products, facilitating the capture of active human responses from the various scenarios (Rusák et al., 2008). In order to determine the positions taken in a simulation, researchers investigated a number of taxonomies developed by several researchers such as Cushman and Rosenberg (1991), Wells and Greig (2001), Wan et al. (2004), Rusák et al. (2008), Feix et al. (2009), and Kyota and Saito (2012).

Many of these taxonomies have been developed in order to understand what types of handles people often use in daily tasks and thus apply them in product design projects. For a better understanding, in this study, the types of handles correspond to each static posture of the hand with an object that can be grasped securely with one hand (Feix et al., 2009).

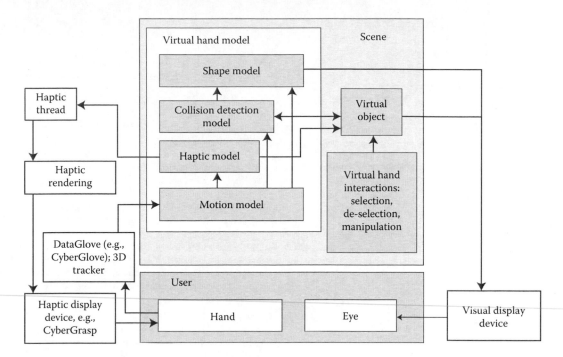

FIGURE 22.2 Haptic interaction of virtual hand framework. (From Springer Science+Business Media: *Transactions on Edutainment IV, LNCS 6250*, A framework for virtual hand haptic interaction, 2010, pp. 229–240, Han, X. and Wan, H.)

Kamakura et al. (1980) proposed a taxonomy of hand movements according to the position of the fingers and the contact areas for application in the field of occupational therapy. Fourteen static standards used in daily life were classified into four categories. Feix et al. (2009) conducted a review of the current literature to find the maximum number of types of handles and carried out a comparative study. As a result, the authors developed a taxonomy classifying postures according to the requirements of force and accuracy, as well as the hand palm position and fingers.

Wimmer (2011) proposed the *GRASP* model. This is a descriptive model of the handle meanings and how objects are apprehended from five factors: objective, relation between the user and the object, anatomy, object configurations, and properties.

By observing these approaches, researchers can identify important factors, such as user hand positions taking into account not only the shape of the object but also the purpose of the task. Therefore, it is important to classify these positions to better define the movements performed in the virtual environment simulation.

A taxonomy of hand movements was developed for this study from the following authors, Cushman and Rosenberg (1991), Wells and Greig (2001), Wan et al. (2004), and Feix et al. (2009), as shown in Figure 22.3. The hands posture were classified according to the hands palm and fingers, being force or precision grips, apprehension/nonapprehension, or contact. The force and precision grips are different from the contact points on the hand palm and thumb (Rusák et al., 2008). If the object is safe, or even partially, the hand palm and with the fingers support exerting force to grab an object, provides a handle apprehension force. The accuracy handle involves only the use of the fingers and thumb, and is a nonapprehension handle (Cushman and Rosenberg, 1991; Wan et al., 2004). Finally, the contact does not involve apprehension but only the action of touching objects, and may also press them with the fingertips or thumb (Wells and Greig, 2001).

To identify the hand positions taken by the user in virtual simulation, it must be based on the measurement of the real-time movement. According to Rusak et al. (2008), three steps can be taken:

	Apprehension handle	
	Handle prism	This force grip allows the application of full force of the wrist and the upper arm
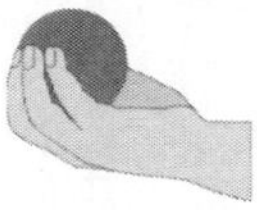	Palm-finger included	This force handle allows involve an object with the fingers and the palm
	Nonapprehension handle	
	Palm handle fingers	This precision handle corresponds to a gripping force or a pull with fingers, being a hanger grip
	Handle thumb tip	Corresponds to a precision handle and extremity that involves the tip of the thumb and another, being used to pick up and manipulate small objects (pincer grip)
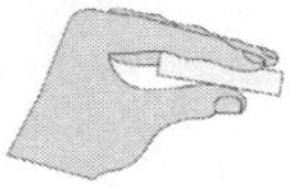	Handle palm thumb	Corresponds to a precision handle with pressure and involves taking an object with the thumb and another finger, the object is grasped with the lateral support of finger and thumb
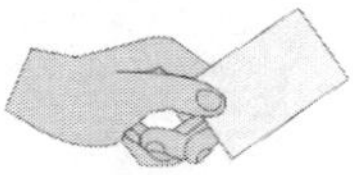	Side handle index finger-thumb Flat surfaces	Corresponds to a lateral precision grasped and involves manipulation with intermediate forces and moderate accuracy
	Side handle index finger-thumb Volumetric objects	
	Two fingers and thumb handle	Corresponds to accurately handle known as the writing handle
	Fingertips-thumb included	Corresponds to a precision disc handle and involves pick up an object with the tips of all fingers
	Contact	
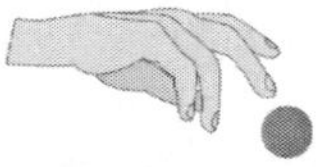	Touch of fingers	Action of touching objects with the fingertips
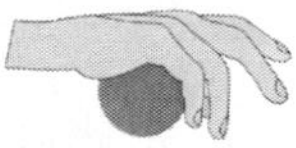	Touch of the palm	Action of touching objects with the palm
	Pressing	Action of touch and press the objects with fingertips or thumb

FIGURE 22.3 Posture of the hands taxonomy. (Adapted from Wells, R. and Greig, M. 2001. *Ergonomics* 44(15):1392–1402.)

(a) approach the object when there is no further contact between the hand and object; (b) touch the object when there is contact between the hand and the object, but the points of contact configuration do not form an apprehension posture; and (c) apprehension of the object, when the contact points configuration form a posture. If any apprehension posture is presented, it is recognized only as a contact posture.

22.3 HAPTIC DEVICES

Haptic devices are used to understand the tactile sense of the user in the virtual environment, providing the user a feel of the geometry and other properties of virtual objects (Han and Wan, 2010). The word derives from the Greek "haptesthai" (or touch) and is related to "being able to come into contact with" (Aziz and Mousavi, 2009), or the sensory information from touch or physical contact. The human action of touching corresponds to a bi-directional sense, that is, a force is applied to an object and a response is perceived by those who applied. This relation can be best understood from the example cited by Hayward et al. (2004), which a conventional mouse is compared to a haptic mouse enabled with programmable mechanical properties, as shown in Figure 22.4.

According to the figure above, a conventional mouse is limited to a one-way entrance and the user receives almost no information of his/her movements, only visually by the screen, although the properties of inertia and attrition can assist the user in performing movements. The haptic mouse, on the other hand, can provide the user with a programmable feedback based on the sense of touch or force, allowing a faster and more intuitive interaction with the computer. The arrow represents the information flow direction.

As pointed out earlier, the simulation in virtual environment involves interaction through multiple sensory channels (visual, auditory, tactile, olfactory, and gustatory) (Burdea and Coiffet, 2003). The visual and auditory feedback was researched the most. On the other hand, feedback associated with the touch (or haptic) represents a challenging research problem due to the unavailability of equipment at full capacity to support tactile systems, due to the complex structure of the underlying physiology of these processes, hardware and software costs, system complexity, and limited workspace (Boud et al., 2000; Lecuyer et al., 2000; Liu et al., 2004).

However, it is important to consider that in evaluating the properties of a product the sense of touch is crucial. This sensory channel complements the visual and auditory feedback mode, which are generally used in the present simulations in virtual environments (Burdea, 2000). Thus, it is evident to support the necessity of using haptic devices capable of allowing users to touch and feel the simulated objects.

The haptic interface or haptic feedback, corresponds to mechanical devices configured to mediate the communication between the user and the computer system, allowing the user to touch, feel, and manipulate 3D objects in virtual environments (Berkley, 2003; Tideman et al., 2006).

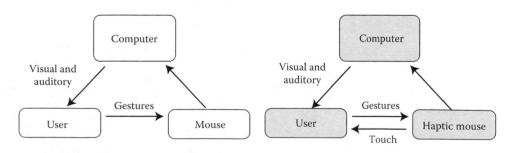

FIGURE 22.4 Comparative example between a conventional mouse and a haptic mouse. (Adapted from Hayward, V. et al. 2004. *Sensor Review* 24(1):16–29.)

TABLE 22.1
Definitions of Haptic Terminology

Term	Definition
Haptic	Relating to the sense of touch
Proprioceptive	Relating to sensory information about the state of the body (including cutaneous, kinesthetic, and vestibular sensations)
Vestibular	Pertaining to the perception of head position, acceleration, and deceleration
Kinesthetic	Meaning the feeling of motion. Relating to sensations originating in muscles, tendons, and joints
Cutaneous	Pertaining to the skin itself or the skin as a sense organ. Includes sensation of pressure, temperature, and pain
Tactile	Pertaining to the cutaneous sense but more specifically the sensation of pressure rather than temperature or pain
Force feedback	Relating to the mechanical production of information sensed by the human kinesthetic system

Source: Oakley, I. et al. 2000. Putting the feel in "look and feel." In: Press, A. (ed.). *Proceedings of the SIGCHI Conference on Human Factors in Computer System*, The Hague, Netherlands, pp. 415–422.

In the same way as a video interface allows the user to view a scene generated by the computer, the haptic interface allows the user to "feel" the object generated by the computer. The haptic system is the interaction of the components sensory, motor, and cognitive brain body system, which is more commonly described as "proprioceptive" (Oakley et al., 2000). The sense of touch is related to three components: tactile, kinesthetic, and skin. In humans, these feelings cannot be separated, but the same does not happen with the equipment developed for simulation. For this reason, there are devices that provide both or only one of these sensations (Oakley et al., 2000; Aziz and Mousavi, 2009).

For interactions with 3D objects, the motion and behavior with virtual objects should be simulated with a realistic touch feedback. Burdea (2000) divides the haptic feedback modality in two groups: tactile feedback and force feedback. Tactile feedback allows users to feel the rugosity of virtual surfaces, their edges, temperature, or slippage. Force feedback reproduces the weight of grasped virtual objects, their mechanical compliance, inertia, as well as motion constraints. Table 22.1 shows the definitions of each term related to haptic sensation. According to Oakley et al. (2000), using these definitions, devices can be categorized and understood by the sensory system that they primarily affect.

Currently, haptic devices vary in sophistication and fidelity and have been developed with the aim of offering ways to recognize and manipulate objects in virtual environments. These can be simple joysticks with force feedback which allows limited interactivity, on the other hand, a haptic device with a higher level of sophistication provides greater level of interaction and allows higher DOF in translation and rotation movement. The research literature indicates that haptic devices are divided into three main categories, according to their use: for hands, arms, or legs, and body.

22.3.1 Capturing Hands Motion

The hand movements captured by simulation devices usually provide the sense of touch to the user fingers or hand, as well as pressure, heat, or vibration. Among these, gloves were shown as having more natural interaction with objects. The gloves are provided with sensors which measure the movement of each finger and some models also have sensors that measure the angles of some or all fingers, as well as have a 3D tracker to indicate the user's hand position. The CyberGrasp, for example, consists of the glove CyberGlove (Figure 22.5a) with the addition of an exoskeleton (Figure 22.5b) (Fisch et al., 2003).

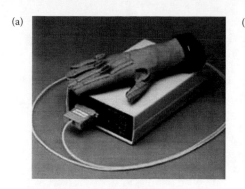
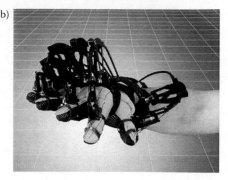

(a)   (b)

FIGURE 22.5   (a) CyberGlove and (b) CyberGrasp. (From http://cyberglovesystems.com.)

During the simulation in virtual environment, CyberGlove reads the user's hand and transmits the data to the central computer, which together with the data from the 3D tracker fixed on the wrist, control a virtual hand (Burdea and Coiffet, 2003). Another example of adapting CyberGlove corresponds to CyberTouch (Figure 22.6), which were added by six vibration actuators (one at the back of each finger and the palm) to provide a vibratory tactile feedback to the user. This glove model is best suited for handling tasks with hands where contact is with the fingertips, as it has the ability to provide feedback to the fingers individually (Burdea and Coiffet, 2003).

The CyberGrasp glove can be used for tasks that require greater skills and necessary force feedback on each finger. The CyberGlove interface box transmits to a force control unit (FCU) all CyberGrasp data resulting from finger positions. The same FCU receives the wrist position data from a 3D scanner, according to the previous paragraph. The information with the positions resulting from the 3D hand is sent to a central computer running the simulation via an Ethernet line (local area network). Then, the host computer performs the collision detection, and inputs the forces resulting from contact of the fingers in the FCU. The exoskeleton structure is connected to CyberGlove through the fingers rings, a rear plate, and Velcro strips. This arrangement allows only unidirectional forces to be applied at the fingertips, on the opposite hand closing direction (Burdea and Coiffet, 2003), and the user can have the feeling of grabbing an object instead of just touching it.

Maciel et al. (2004) investigated the potential of CyberGrasp to improve the perception in the simulation in VR. Researchers identified a number of disadvantages and limitations for the CyberGrasp. The main disadvantage of this system corresponds to its complexity and the inability to simulate the weight and inertia of the virtual object, also mentioned by Burdea and Coiffet (2003). Also, the

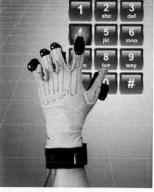

FIGURE 22.6   CyberTouch. (From http://cyberglovesystems.com.)

lack of procedures for reducing the force implemented in hardware and direction of applied forces, which is limited to the direction of the hand opening (Maciel et al., 2004). In addition, other factors like tactile and visual feedback appear to play an important role, beyond the force feedback. Maciel et al. (2004) referencing that many applications are not possible, especially those that require very high frequencies as vibrations or contact with very irregular and deformable objects. For these objects the device known as PHANToM™, illustrated in this chapter was found to be the most suitable.

22.3.2 CAPTURING ARMS AND LEGS MOVEMENT

The arm and leg motion capture device aims to offer force feedback, this interface differs from the tactile interfaces because of the need to provide substantial forces to restrict user movement, which implies large actuators, heavier structures (to ensure mechanical rigidity), and higher complexity (Burdea and Coiffet, 2003). One reason that capture devices are not focused on the sense of touch is that because the hand has more sensitivity and is used often for this function. Some devices offer touch and force feedback with the primary function to restrict the user's actions while using them.

A commercially available solution uses an arm to provide the mechanical sense of touch and force feedback, and has the advantage to allow movement with six DOF, which support the applications with stereoscopic displays. The PHANToM—personnel haptic interface mechanism, of the SensAble company corresponds to one example of these devices (Figure 22.7).

This PHANToM device has a base which is connected to a mechanical arm which ends with a component similar to a pen. The orientation of the pen is passive and does not allow the application of torques in the user's hand, and its shape allows the simulation of similar instruments such as brushes, screwdrivers, surgical tools, etc. From the six DOF arm, three of them are active and provide translational force feedback. The resulting workspace arm is used by the user's wrist with the forearm resting on a support (Burdea and Coiffet, 2003). The device for hands is equipped with a support structure and a force feedback. Figure 22.8 shows the CyberGrasp glove, which when added to a mechanical arm allows the simulation of the weight and inertia of the object, and begins to correspond to an arm device called CyberForce. However, this type of device is not portable, since they must be supported on a base. The most portable devices, such as force feedback gloves is supported on the user's forearm and allow greater freedom of movement and a more natural interaction with the simulation (Burdea and Coiffet, 2003). Another factor to consider is the reduced freedom of movement within reach of the mechanical arm, which is significantly smaller than the CyberGrasp, followed by the increased complexity of the system with the addition of mechanical arm.

(a)
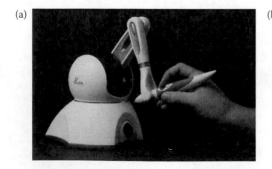
(b)

FIGURE 22.7 Haptic interface mechanisms developed by SensAble—(a) (b) PHANToM Desktop. (From http://www.sensable.com.)

FIGURE 22.8   CyberForce glove. (From PHANToM Omni e http://cyberglovesystems.com.)

22.3.3   BODY MOTION CAPTURE

Haptic devices can be used to capture body movements simulating a projected platform to provide vibration or interaction with the user's sense of balance. In some cases, these are treated as haptic devices that do not allow the perception of objects, but adjust the balance of the body, and are commonly known as mobile platforms and are widely used by the entertainment industry to develop innovative video games. The aviation and automotive industries utilized these platforms to build novel simulators (Moreau et al., 2004; Jimeno and Puerta, 2007; Guerlesquin et al., 2012).

22.3.4   APPLICATION OF HAPTIC DEVICES TO SUPPORT PRODUCT DESIGN

The development of virtual prototypes for product evaluation is based on visualization technologies which have reached a high level of quality in the reproduction of information, while searching for solutions that are close to market needs at reduced cost (DeFanti et al., 2011; Nesbitt, 2013). On the other hand, haptic technology has not kept pace with the advancement of product development. With respect to virtual prototype testing, technology has been limited to display aspects (Ferrise et al., 2013). However, evaluation is best suited for assessing the aesthetic aspects of the product and does not apply to aspects of user interaction.

In evaluating the usability of a product, it is necessary to record the feeling of touch and how the user physically manipulates the object as the task is performed. To meet this requirement, simple haptic control devices have been developed for testing the user interaction with products. For example, the study by Bruno and Muzzupappa (2010) cited a 3D joystick for selecting virtual objects, and included several usability tests with VR in consumer products.

By observing the examples of more complex application, such as the use of CyberForce, researchers and designers appreciate the ability for the user to hold and manipulate virtual objects from tactile interfaces with multiple fingers, which help to increase realism and haptic perception. However, as pointed out by Maciel et al. (2004), some applications are not compatible, and add to the cost of acquiring the equipment needed. Wang and Lu (2009) developed a real-time modeling and simulation system incorporating the PHANToM device for the design and evaluation of mobile devices. The product was evaluated at the initial stage of design, using a virtual template for modeling steps, interaction and evaluation, providing savings of cost and time.

Other studies cited a combination of devices as means to improve the performance of the haptic interface. Nikolakis et al. (2004) proposed the combination of PHANToM and CyberGlove to provide better feedback. Gao et al. (2012) used a dual interface with PHANToM to output the internal force and allow the elasticity feel of the products.

Thus, the overall decision regarding which device is best suited for use and how it should be applied, must take into account the complexity of the 3D environment and the task the user will be performing. The use of these devices still have many limitations. Despite the fact that haptic devices provide a good kinesthetic feedback, the lack of tactile sensation, which is not available through the pen or the glove, reduces the realism of the simulation. As pointed out by Ferrise et al. (2013), the representation fidelity of haptic behavior is high, but user interaction tests with the product are limited to press buttons, whereas other gestures used in general are not considered.

22.4   CONCLUSIONS

At the beginning of the digital era when computers were first introduced to product development activity, few believed in the important role machines could play in real-world practice. The systems were quite complex, time-consuming to setup, expensive, and complicated to use. However, driven by the strong belief in the potential of these technologies, many researchers have worked to break down technological barriers that led to the current achievements presented in this chapter.

In this view, the chapter aimed to point out the recent enhancements in user interaction and the potential of haptic devices applied in evaluating product design. Because of the need to work more closely with the user's domain in usability testing, 3D tasks have been suggested as tools and applications, which allow the same work in a more realistic setting.

Research has indicated the need to provide feedback through the manipulation of physical input devices representing virtual objects. The development of haptic devices and adapting this technology to research requirements is an important step to fill the gap between knowing what we want to do and how to achieve the intended uses with the product.

There is enthusiasm around haptic devices, and much work has been done to advance this technology and create useful systems. However, major innovations are still needed for the intended application in virtual environment for product design. Progress in haptic interfaces research is lagging behind research in the visual arena (i.e., displays and image generation), however, the current state of hardware for the manipulation of the product prototype in virtual environment is not satisfactory in most cases. Thus, from the development of technology identified, it is expected that the practice of the designer be reconsidered. VR and haptic technologies provide a natural interface that will allow the evolution of product development processes in new paradigms. We are witnessing the beginning of major changes that will lead to many excellent possibilities in the near future, and the design community has an inescapable duty to take a leadership role in exploring these great opportunities.

REFERENCES

Aziz, F. and Mousavi, M. 2009. A review of haptic feedback in virtual reality for manufacturing industry. *Journal of Mechanical Engineering* ME 40(1):68–71.

Barnum, C. 2011. *Usability Testing Essentials: Ready, Set … Test!* Elsevier: Burlington.

Barfield, W. and Nash, E.B. 2006. Virtual environments. In: Karwowisk, W. (ed.). *International Encyclopedia of Ergonomics and Human Factors*. 2nd edn. Taylor & Francis Group: Boca Raton, Florida, pp. 1332–1335.

Berkley, J. 2003. Haptic devices. White Paper. http://www.hitl.washington.edu/people/tfurness/courses/inde543/READINGS-03/BERKLEY/White%20Paper%20-%20Haptic%20Devices.pdf   (accessed September, 2015).

Bordegoni, M., Ferrise, F., and Lizaranzu, J. 2011. The use of interactive virtual prototypes for products specification in the concept design phase. *Virtual Reality Conference (VR) Proceedings IEEE*, Singapore, pp. 197–198.

Boud, A.C., Baber, C., and Steiner, S.J. 2000. Virtual reality: A tool for assembly? *Presence* 9(5):486–496.

Bruno, F. and Muzzupappa, M. 2010. Product interface design: A participatory approach based on virtual reality. *International Journal of Human-Computer Studies*, 68:254–269.

Burdea, G. and Coiffet, P. 2003. *Virtual Reality Technology*. 2nd edn. John Wiley & Sons: Hoboken, New Jersey.

Burdea, G. 2000. Haptics issues in virtual environment. *Proceedings of Computer Graphics International*, Piscataway, NJ. pp. 295–302.

Cushman, W.H. and Rosenberg, D.J. 1991. *Human Factors in Product Design*. Elsevier: Amsterdam.

CyberGlove. http://cyberglovesystems.com (accessed September, 2015).

DeFanti, T. et al. 2011. The future of the CAVE. *Central European Journal of Engineering* 1(1):13–37. doi: 10.2478/s13531-010-0002-5.

Falcão, C. and Soares M. 2012. Ergonomics, usability and virtual reality: A review applied to consumer product. In: Rebelo, F. and Soares, M.M. (eds.). *Advances in Usability Evaluation. Part II*. CRC Press: Boca Raton, Florida, pp. 297–306.

Feix, T., Pawlik, R., Schmiedmayer, H., Romero, J., and Kragic, D. 2009. A comprehensive grasp taxonomy. *Robotics Science and Systems Conference: Workshop on Understanding the Human Hand for Advancing Robotic Manipulation*. http://www.csc.kth.se/grasp/taxonomyGRASP.pdf

Ferrise, F., Bordegoni, M., and Cugini, U. 2013. Interactive virtual prototypes for testing the interaction with new products. *Computer-Aided Design & Applications* 10(3):515–525.

Ferrise, F., Bordegoni, M., and Graziosi, S. 2013. A method for designing users' experience with industrial products based on a multimodal environment and mixed prototypes. *Computer-Aided Design & Applications* 10(3):461–474.

Fisch, A., Mavroidis, C., Bar-Cohen, Y., and Melli-Huber, J. 2003. Haptic devices for virtual reality, telepresence and human-assistive robotics. In: Bar-Cohen, Y. and Breazeal, C. (eds.). *Biologically Inspired Intelligent Robots*. SPIE: Bellingham, Washington, pp. 4/1–4/24.

Fortineau, E., Arnaud, S., and Nahon, S. 2013. Simulations immersives en réalité virtuelle: une innovation technologique au service de l'intervention ergonomique? *Proceedings of SELF 2013 – 48ème Congrès de la Société d'Ergonomie de Langue Française*, August 28–30, 2013, Panthéon-Sorbonne, Paris.

Gao, Z., Wang, J., and Jiang, Z. 2012. Haptic perception evaluation for the virtual prototyping of elastic handheld product designs. *Virtual and Physical Prototyping* 7(2):117–128.

Guerlesquin, G., Mahdjoub, M., Bazzaro, F., and Sagot, J. 2012. Virtual reality as a multidisciplinary convergence tool in the product design process. *Systemics, Cybernetics and Informatics* 10 (1). ISSN: 1690–4524. http://www.iiisci.org/journal/CV$/sci/pdfs/HFA101DF.pdf (accessed September, 2015).

Gutierrez, M., Vexo, F., and Thalmann, D. 2008. *Stepping into Virtual Reality*. Springer-Verlag: London.

Han, X. and Wan, H. 2010. A framework for virtual hand haptic interaction. In: Pan, Z. et al. (eds.). *Transactions on Edutainment IV, LNCS 6250*. Springer-Verlag: Berlin, Heidelberg. pp. 229–240.

Hayward, V., Astley, O., Cruz-Hernandez, M., Grant, D., and Robles-De-La-Torre, G. 2004. Haptic interfaces and devices. *Sensor Review* 24(1):16–29.

Jimeno, A. and Puerta, A. 2007. State of the art of the virtual reality applied to design and manufacturing processes. *The International Journal of Advanced Manufacturing Technology*, 33:866–874. doi: 10.1007/s00170-006-0534-2.

Kamakura, N., Matsuo, M., Ishii, H., Mitsuboshi, F., and Miura, Y. 1980. Patterns of static prehension in normal hands. *The American Journal of Occupational Therapy* 34(7):437–445.

Kirner, C. and Siscoutto, A. 2007. Fundamentos de Realidade Virtual e Aumentada. In: Kirner, C., Siscoutto, R. (eds.). *Realidade Virtual e Aumentada: Conceitos, Projeto e Aplicações*. Porto Alegre, Brazil: SBC. pp. 2–21.

Kyota, F. and Saito, S. 2012. Fast grasp synthesis for various shaped objects. *EUROGRAPHICS 2012* 31(2):765–774. doi: 10.1111/j.1467-8659.2012.03035.x.

Lecuyer, A., Coquillart, S., Kheddar, A., Richard, P., and Coiffet, P. 2000. Pseudo-haptic feedback: Can isometric input devices simulate force feedback? *IEEE International Conference on Virtual Reality*, New Brunswick, New Jersey, pp. 83–90.

Lee, S., Chen, T., Kim, J., Kim, G., Han, S., and Pan, Z. 2004. Affective property evaluation of virtual product designs. *Proceedings of IEEE Virtual Reality*, Chicago, IL, pp. 207–292.

Liu, X., Dodds, G., McCartney, J., and Hinds, B.K. 2004. Virtual design works—Designing 3D CAD models via haptic interaction. *Computer-Aided Design* 36(12):1129–1140.

Lu, S.C.-Y., Shpitalni, M., and Gadh, R. 1999. Virtual and augmented reality technologies for product real-ization. *Annals of CIRP* 48, pp. 1–25. http://wireless.ucla.edu/gadh/pdf/99k.pdf (accessed September, 2015).

Maciel, A., Sarni, S., Buchwalder, O., Boulic, R., and Thalmann, D. 2004. Multi-finger haptic rendering of deformable objects. *Proceedings of Eurographics Symposium on Virtual Environments*, June 8–9, Grenoble, France.

Moreau, G., Fuchs, P., and Stergiopoulos, P. 2004. Applications of Virtual Reality in the manufacturing industry: From design review to ergonomic studies. *Mecanique & Industries* 5(2):171–180. doi: 10.1051/meca:2004018.

Nesbitt, A. 2013. Virtual prototyping. appliance design. May, 2013. http://www.appliancedesign.com/articles/93590-virtual-prototyping (accessed September, 2015).

Nikolakis, G., Tzovaras, D., Moustakidis, S., and Strintzis, M. 2004. CyberGrasp and PHANTOM integration: Enhanced haptic access for visually impaired users. *Proceedings of the SPECOM 2004 9th Conference Speech and Computer*. St. Petersburg, Russia.

Oakley, I., McGee, M., Brewster, S., and Gray, P. 2000. Putting the feel in "look and feel." In: Press, A. (ed.). *Proceedings of the SIGCHI Conference on Human Factors in Computer System*, The Hague, Netherlands, pp. 415–422.

Rusák, Z., Antonya, C., van der Vegte, W., and Horváth, I. 2008. Implementing real time grasping simulation based on anthropometric data: A work in progress report. In: Talabā, D. and Amditis, A. (eds.). *Product Engineering: Tools and Methods Based on Virtual Reality*. Dordrech, The Netherlands: Springer Science, pp. 523–540.

Shen, Y., Ong, S.K., and Nee, A.Y.C. 2010. Augmented reality for collaborative product design and develop-ment. *Design Studies* 31:118–145.

Sherman, W. and Craig, A. 2003. *Understanding Virtual Reality: Interface, Application and Design*. San Francisco, CA: Elsevier Science.

Shin, J., Joo, J., Choi, I., Han, S., and Cho, H. 2000. A prototype virtual reality system through IDEF model-ing for product configuration and analysis. *International Journal of Industrial Engineering* 7(1):15–25.

Tideman, M., van der Voort, M.C., and van Houten, F. 2006. Haptic virtual prototyping for design and assess-ment of gear-shifts. In: ElMaraghy, H.A. and ElMaraghy, W.H. (eds.). *Advances in Design*. Springer: Berlin, pp. 461–472.

Wan, H., Luo, Y., Gao, S., and Peng, Q. 2004. Realistic virtual hand modeling with applications for virtual grasping. *Proceedings of ACM SIGGRAPH International Conference on Virtual Reality Continuum and Its Applications in Industry*. New York: VRCAI'04, pp. 81–87. ISBN: 1-58114-884-9.

Wang, J. and Lu, W.F. 2009. Development of a haptic modeling and simulation system for handheld product design and evaluation. *Proceedings of the ASME 2009 International Design Engineering Technical Conferences & Computer and Information in Engineering Conference*. San Diego, California, DETC2009, 2009.

Wells, R. and Greig, M. 2001. Characterizing human hand prehensile strength by force and moment wrench. *Ergonomics* 44(15):1392–1402.

Wimmer, R. 2011. Grasp sensing for human-computer interaction. *Proceedings of the Fifth International Conference on Tangible, Embedded and Embodied Interaction*. Funchal, Portugal, pp. 221–228.

23 Hazard Perception of 3D Household Packages
A Study Using a Virtual Environment

*Hande Ayanoğlu, Emília Duarte, Paulo Noriega,
Júlia Teles, and Francisco Rebelo*

CONTENTS

23.1 INTRODUCTION

A well-designed product is one that presents a reduced risk of injury. Usually this is achieved by designing out the hazard. Unfortunately, it is not always possible to eliminate hazards without compromising the product's function or performance, as is the case of many household chemicals. Although such chemicals are intended for domestic usage, they may have substances which put the user at risk (e.g., they may be poisonous to drink, toxic to inhale, cause skin irritation). According to the hazard-control hierarchy (Wogalter and Laughery, 2001), in these cases, other safety measures should be implemented, such as guarding against hazards (e.g., child-resistant caps, dosing dispensers) and/or warning about hazards. However, these solutions may be ineffective due to many reasons (e.g., hazards may be hidden, and/or their nature and impact misunderstood). Since the packages are the first thing that users come into contact with before using the product (Serig, 2001), such problems can be aggravated by package similarity, which may pose some problems for the users, namely, users may have difficulties in distinguishing the contents and determining their hazardousness.

In this context, in order to ensure users safety, one may assume that an effective package should not only allow the users to correctly identify the content before using it, but also to communicate, through visual cues, the meaning and function of the product (Radford, 2007).

Generally, warning labels are employed to inform users and increase their safety, since they provide a method to convey the required safety information (e.g., hazard, consequences, and recommended behavior). However, people do not always search for or read labels (e.g., Laughery and Wogalter, 1997). In some cases, some packages have such small dimensions, which make it difficult to place all the necessary information on the warning label. Also, there are findings in the literature which suggest that warnings are read more carefully, and adhered to more often, when placed on products that are perceived as more hazardous and less familiar (e.g., Wogalter et al., 2001). Moreover, other studies suggest that the degree of caution that people are willing to take is largely determined by their hazard perceptions (e.g., Wogalter et al., 1991).

Many efforts have been made to improve the effectiveness of warnings (e.g., Johnson, 2006; Cunningham, 2007; Lesch, 2008; Hammond and Parkinson, 2009; Brannstrom et al., 2015), however, relatively few studies have examined other package variables such as shape (e.g., Wogalter et al., 2001), and/or pairs of variables such as color and shape (e.g., Ngo et al., 2011; Serig, 2001), and color and words (e.g., Luximon et al., 1998). The packages' openings is another topic commonly studied for safety reasons (e.g., Winder et al., 2002; Kozak et al., 2003; Galley et al., 2005; Langley et al., 2005; Yoxall et al., 2006; Caner and Pascall, 2010). However, these studies mostly explored the reasons behind the injuries that were caused when food packages were opened.

Research (e.g., Ratneshwar and Chaiken, 1991) indicates that some situational variables (e.g., limited time to perform an action, and/or concurrent tasks) or limited knowledge of the product/context may reduce the users' systematic processing and lead them to rely on heuristics. Systematic processing is assumed to be more effective than heuristic processing. However, according to the heuristic–systematic model, systematic processing will only occur when an individual possesses adequate levels of both cognitive capacity and motivation (Zuckerman and Chaiken, 1998). Thus, the package's shape, which can be, to some extent, familiar to users and can contain implicit information (e.g., perceived affordances/signifiers) that is relevant for a given judgment, as well as requires less cognitive capacity to be processed, can play a significant role in facilitating a correct risk perception and promoting compliance with warnings. Gibson (1986) introduced the term affordances to refer to the actionable properties between the world and an actor. According to Gibson's ecological perception theory, affordances are a part of nature and they are there even if they are not seen, known, or desirable. Norman (2010) states that when users fail to notice the affordance, designers should add visible signs of its existence with what he calls signifiers. In other words, signifiers make the affordances more salient. By manipulating the signifiers, or making them more or less suitable according to the hazardous content, the users' safety may be enhanced even before they handle a package (Ayanoğlu et al., 2013).

In sum, if a package is poorly designed, users may make wrong assumptions about its content and, as a result, injuries related to its hazardous contents may take place (e.g., Desai et al., 2005). In order to promote safer actions, better communication strategies between users and packages should be implemented so as to induce correct hazard perceptions. In this context, this study's objective was to examine the effect of packages' shape on hazard-related perceptions (i.e., levels of content hazardousness and awareness of consequences).

Hazard perception is related to behavioral intentions such as an intended carefulness and willingness to comply (Wogalter et al., 1998). Furthermore, Wogalter et al. (2001) denote that perceptions, whether they are correct or not, are probably based on previous experiences with familiar products. Familiarity is closely related to prior experience and frequency of use. Nonetheless, product (i.e., content) familiarity is not always associated to product experience. People may be familiar with products that they seldom or never use (Dejoy, 1999).

Different methodologies to evaluate hazard perception are described in the literature; for example, two-dimensional (2D) images and questionnaires are the most common tools (e.g., Serig, 2001; Wogalter et al., 2001). However, 2D images have a limited ability to display all of the properties of an object, and for that reason, may affect users' comprehension and, therefore, influence their judgments (e.g., Landabaso, 2006; Ayanoğlu et al., 2013). Furthermore, users' opinions are strongly affected by the context in which they find and/or use the product, as well as by having, or not having, the opportunity to interact with it (e.g., manipulate, observe from different viewpoints). Therefore, virtual reality (VR), as a tool, can benefit the evaluation of products and its associated user experience (UX) (e.g., Rebelo et al., 2012). Furthermore, research has shown that VR simulations can be an effective way to present scenarios that facilitate effective interaction between users and products (e.g., Rosson and Carroll, 2002; Ayanoğlu et al., 2013), thus increasing the study's validities. In this manner, a virtual environment (VE) was used in this study, in which the participants could observe the three-dimensional (3D) packages from diverse viewpoints.

23.2 METHOD

23.2.1 SAMPLE

The sample consisted of 20 undergraduate design students, aged between 18 and 24 years, equally distributed by gender (mean age = 20.35, SD = 1.87).

23.2.2 DESIGN OF STUDY

The study used a within-subjects design in which the type of package is the main factor, with four levels (HF = hazardous familiar, NHF = nonhazardous familiar, HUF = hazardous unfamiliar and NHUF = nonhazardous unfamiliar). Hazard perception and awareness of consequences were the dependent variables. Familiarity was used as a control variable.

23.2.3 MATERIALS

23.2.3.1 Stimuli

The stimuli were eight 3D packages, that is, the same ones used in a previous study by Ayanoğlu and colleagues (2013), which were designed in Rhinoceros® and then exported to Unity 3D. The initial selection process of the packages was carried out through a focus group session in which experts in ergonomics made the selection, from a large set of 2D (i.e., silhouettes) packages, according to the following criteria: (a) familiarity (familiar or unfamiliar); (b) content hazardousness (hazardous or nonhazardous); and (c) shape (rectilinear and curvilinear).

23.2.3.2 The VE

The VE was a closed room (with no doors or windows), measuring 6.6 m × 6.6 m, and contained a table (260 cm length, 30 cm depth, and 90 cm height) in the middle of the room. When the simulation began, the participants' view was as if they were standing 1 m away from the table. Also, the environment was designed to be minimalist, only considering aspects such as accessibility (e.g., dimensions and space layout) and visibility (e.g., light, shadows, and contrast). There was no sound in the VE.

The eight packages were placed on top of the table, as well as placed 20 cm away from each other. Each package was associated to a letter, from A to H, in order to facilitate their correct identification (see Figure 23.1). All extra details beyond the packages' shape, such as colors, textures, labels, and brands, were removed so as to not influence the participants' judgments.

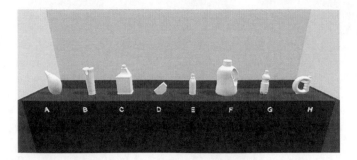

FIGURE 23.1 Screen shot of the VE showing the eight 3D packages associated with a letter. *Note*: HF, hazardous familiar (Packages C and F), NHF, nonhazardous familiar (Packages E and G), HUF, hazardous unfamiliar (Packages A and B), and NHUF, nonhazardous unfamiliar (Packages D and H).

23.2.3.3 Experimental Settings

Participants performed the experiment in a Virtual Reality Laboratory in which a computer and a video projector were used for data collection. During both training and experimental sessions, participants remained seated and viewed the VE egocentrically. Participants were free to visually explore the 3D packages, using a mouse to navigate inside the VE. By pressing the mouse's left button, participants were able to move forward, whereas the right button was used to move backwards. By moving the mouse completely, the participants were able to control their point of view. The participants' direction, when moving forward, was given by the direction of their point of view.

23.2.3.4 Forms

Three forms were used: a consent form, a demographics questionnaire, and a two-part questionnaire which assessed the hazard-related perceptions.

1. *Part I—Packages' hazard-related ratings:* The first part of the questionnaire was intended to evaluate the participants' hazard-related perceptions regarding the different household packages (see Table 23.1). Twelve questions were asked, which were adapted from the questionnaires used by Wogalter et al. (1999, 2001). Each package was rated for each question. The questions were organized according to the following scales: *hazard perception* (questions 1–8, and 10), *awareness of consequences* (questions 9, 11, and 12), and *familiarity* (question 6). Each question was associated with a 9-point Likert-type scale, from 0 to 8, where 0 indicates the minimum and 8 indicates the maximum.

2. *Part II—Packages' content classification:* In the second part of the questionnaire, participants were asked to write what the content of the liquid household packages could be. They were not given any information or restrictions, but they were told that they could skip the package if they did not know and/or recognize it. The packages' real contents were as follows (see Figure 23.2): Package A = toilet bowl cleaner; Package B = abrasive cleaner; Package C = toilet bowl cleaner; Package D = juice; Package E = recovery drink; Package F = laundry detergent; Package G = water; and Package H = energy drink.

23.2.4 Procedure

Upon arrival, participants received brief information about the study and were requested to sign a consent form in which they agreed to participate in the study voluntarily. They were told that the study was about package design and that it was mainly concerned with users' safety. At this point, the researcher stressed the fact that they could stop the experiment at any time without any prejudice. They were instructed about the study's procedure, which consisted of two sessions: training and experimental.

TABLE 23.1
Questions and Scales of Hazard-Related Perceptions to Rate 3D Household Packages

Questions and Scales

1. **Hazardous contents**:
    Based on this package's shape, how hazardous would you rate its contents?
    (0) not at all hazardous, (2) slightly hazardous, (4) hazardous, (6) very hazardous, (8) extremely hazardous

2. **Hazardous package**:
    Based on this package's shape, how hazardous would you rate the package itself?
    (0) not at all hazardous, (2) slightly hazardous, (4) hazardous, (6) very hazardous, (8) extremely hazardous

3. **Hazardous to children**:
    Based on this package's shape, how hazardous would it be if children came into contact with it?
    (0) not at all hazardous, (2) slightly hazardous, (4) hazardous, (6) very hazardous, (8) extremely hazardous

4. **Flammable/combustible hazard**:
    Based on this package's shape, how likely is it for it to be containing a flammable/combustible substance?
    (0) never, (2) unlikely, (4) likely, (6) very likely, (8) extremely likely

5. **Read label**:
    If this product had a label, how likely would it be that you would read it?
    (0) never, (2) unlikely, (4) likely, (6) very likely, (8) extremely likely

6. **Familiarity**:
    How familiar are you with this package?
    (0) not at all familiar, (2) slightly familiar, (4) familiar, (6) very familiar, (8) extremely familiar

7. **Hazardous to drink**:
    Based on this package's shape, how hazardous would its contents be when/if drunk?
    (0) never, (2) unlikely, (4) likely, (6) very likely, (8) extremely likely

8. **Hazardous to inhale**:
    Based on this package's shape, how hazardous would it be to inhale its contents?
    (0) not at all hazardous, (2) slightly hazardous, (4) hazardous, (6) very hazardous, (8) extremely hazardous

9. **Hazardous to skin contact**:
    Based on this package's shape, how hazardous would it be if it contacted your skin?
    (0) not at all hazardous, (2) slightly hazardous, (4) hazardous, (6) very hazardous, (8) extremely hazardous

10. **Easiness of use**:
    Based on this package's shape, how easy would it be to properly dispose its contents?
    (0) very easy, (2) easy, (4) fair, (6) hard, (8) very hard

11. **Cautious intent**:
    Based on this package's shape, how cautious would you be when using this package?
    (0) not at all cautious, (2) slightly cautious, (4) cautious, (6) very cautious, (8) extremely cautious

12. **Hazardous in closed spaces**:
    Based on this package's shape, how hazardous would it be if used in a closed/confined place?
    (0) not at all hazardous, (2) slightly hazardous, (4) hazardous, (6) very hazardous, (8) extremely hazardous

13. **Likelihood of injury**:
    Based on this package's shape, how likely are you to be injured by this package?
    (0) not at all hazardous, (2) slightly hazardous, (4) hazardous, (6) very hazardous, (8) extremely hazardous

14. **Severity of injury**:
    Based on this package's shape, to what extent (i.e., degree or magnitude) may you be severely injured by this
    package?
    (0) not at all hazardous, (2) slightly hazardous, (4) hazardous, (6) very hazardous, (8) extremely hazardous

Note: The scale of question 10 was reversed (lowest value is better).

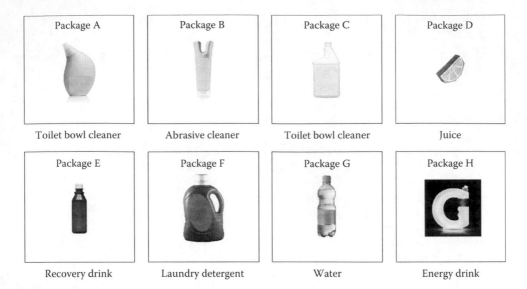

FIGURE 23.2   Images of the real packages used in this study.

23.2.4.1   Training Session

All participants were requested to perform a training stage in which they explored a VE which was designed to get them acquainted with the setup and also to allow them to practice navigating with the mouse, as well as to make a preliminary check for simulator sickness.

The training VE was a closed room in which different 3D objects (e.g., sphere, cylinder, cone, cube) were placed on a table. Participants were asked to visually explore the 3D objects while navigating around the environment and to reply to questions (i.e., similar to the ones that would be part of the experimental session). If they were able to accomplish the task (i.e., examining the objects while navigating and replying to the questions) without showing any symptoms of simulator sickness, they were considered able to do the experimental session.

23.2.4.2   Experimental Session

After declaring that they were able to start the experimental session, the researcher gave them the hazard-related perceptions questionnaire and a scenario (cover story):

> Imagine that your friend is moving to a new house and he/she asks you to help unpack and to organize a group of liquid household products' packages according to their level of hazard (e.g., how poisonous can the content be when drunk, how toxic can it be when inhaled, or how irritant/harmful can it be if it comes into contact with skin).

After being informed about the scenario, participants were given the following task: observe the packages and complete the questionnaire. They were told that they could take as much time as they needed to examine the packages and reply to the questions. Once the simulation started, no dialogue between the participant and the researcher occurred. The experimental session ended when the participants replied to all of the questions or when they declared that they wanted to quit the session.

23.3   RESULTS

The statistical analysis was performed with the software IBM SPSS Statistics, version 21, and a significance level of 5% was considered.

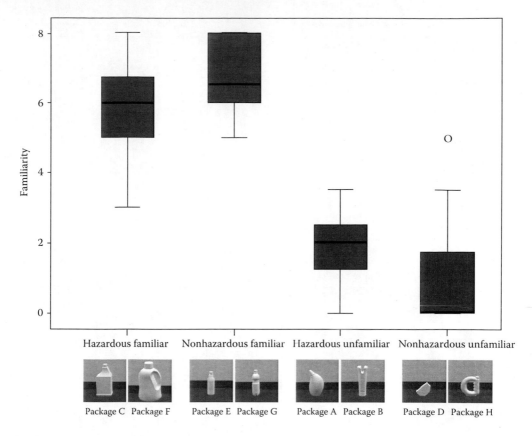

FIGURE 23.3 Box-plots of familiarity scores by package.

23.3.1 FAMILIARITY

Question 6 asked participants to state how familiar they were with the packages. As seen in Figure 23.3, the packages that were previously classified by the researchers as being familiar were also familiar to the participants, and the same can be found for the other two packages, which the participants considered unfamiliar.

23.3.2 PACKAGES' HAZARD-RELATED RATINGS

23.3.2.1 Assessment of Agreement

The intra-class correlation coefficient (ICC) was used to assess whether the participants classified, in the same manner, the two packages of the same type regarding hazard perception and awareness of consequences scores. These results are displayed in Table 23.2. Scatter plots (and the concordance line) of hazard perception scores for the pairs of HF, NHF, HUF, and NHUF packages are displayed in Figures 23.4 and 23.5 for hazard perception and awareness of consequence scores, respectively.

TABLE 23.2
ICC for Hazard Perception and Awareness of Consequence Scores

Type of Package (Packages)	HF (C and F)	NHF (E and G)	HUF (A and B)	NHUF (D and H)
ICC for hazard perception scores	0.825	0.796	0.375	0.337
ICC for awareness of consequences scores	0.822	0.706	0.610	0.637

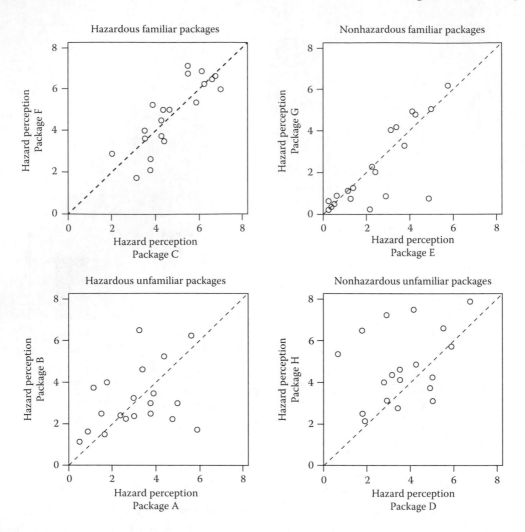

FIGURE 23.4   Scatter plots (and the concordance line) of hazard perception scores for the pairs of HF, NHF, HUF, and NHUF packages.

The values that were obtained, ranging from 0.337 to 0.825, allows one to consider the mean score for each pair of packages to represent the score of the corresponding type of package, for both hazard perception and awareness of consequence scores.

23.3.2.2   Type of Package Effect

Table 23.3 shows descriptive measures of (a) hazard perception and (b) awareness of consequences presented by the type of package.

The results of Friedman tests revealed that the type of package (i.e., HF, NHF, HUF, NHUF) has a significant effect on both hazard perception ($X^2(3) = 20.27$, $p < 0.001$) and awareness of consequences scores ($X^2(3) = 15.47$, $p = 0.001$).

1. *Hazard perception.* Concerning the hazard perception scores, post hoc comparisons revealed that there were significant differences between HF (mean rank = 3.38) and both NHF (mean rank = 1.78; $p = 0.001$) and HUF type of package (mean rank = 2.00; $p = 0.005$). The chart in Figure 23.6 presents a box-plot of the hazard perception subscale for each package.

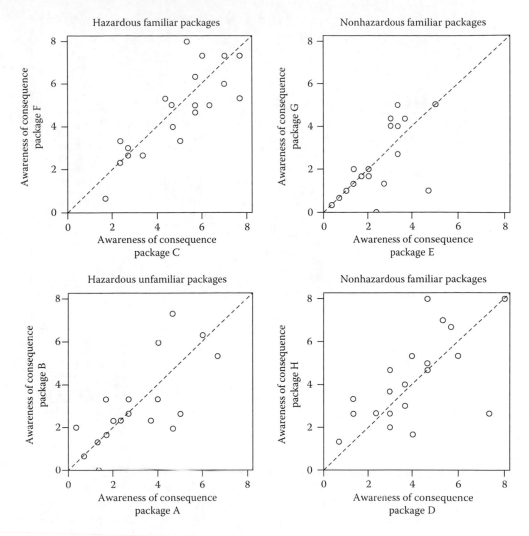

FIGURE 23.5 Scatter plots (and the concordance line) of awareness of consequences scores for the pairs of HF, NHF, HUF, and NHUF packages.

TABLE 23.3
Minimum, Maximum, Median, and the 25th and 75th Percentiles of Hazard Perception and Awareness of Consequence Presented by the Type of Package

Type of Package	Hazard Perception				Awareness of Consequences			
	HF	NHF	HUF	NHUF	HF	NHF	HUF	NHUF
Minimum	2.44	0.25	0.81	2.00	1.17	0.00	0.67	1.00
Maximum	6.69	5.94	5.94	7.31	7.50	5.00	6.17	8.00
25th Percentile	3.61	0.81	2.40	3.02	2.87	1.04	1.42	2.58
Median	4.63	2.04	3.13	4.06	5.00	1.92	2.59	3.83
75th Percentile	6.30	3.70	3.95	4.95	6.50	3.63	3.79	5.50

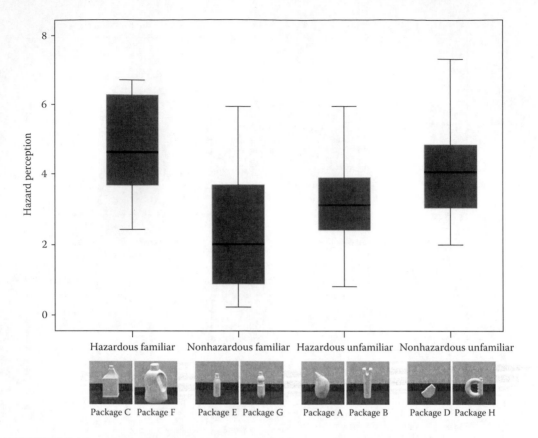

FIGURE 23.6   Box-plots of hazard perception scores by the type of package.

2. *Awareness of consequences.* Regarding the awareness of consequences scores, post hoc comparisons also revealed that there were significant differences between HF (mean rank = 3.28) and both NHF (mean rank = 1.90; p = 0.005) and HUF type of package (mean rank = 2.03; p = 0.013). The box-plots of the awareness of consequences subscale, for each package, are presented in Figure 23.7.

23.3.3 Packages' Content Classification

Regarding the packages' content, the participants' evaluations were first grouped into five categories (i.e., alimentary, household chemicals, car products, personal care, and others, unrelated replies), and second, the correct identification/classification of its category of contents were assessed (see Figure 23.8).

For familiar packages, the percentages of the correct classification of the package's category of contents were (i) for the hazardous ones: 65.0% (Package C) and 89.5% (Package F) and (ii) for nonhazardous ones: 55.0% (Package E) and 70.0% (Package G). Cochran's test revealed that there were no significant differences in the percentages of correct classification (Q = 6.000, df = 3, p = 0.125).

Concerning unfamiliar packages, the percentages of correct classification were: (i) for hazardous ones: 95.0% (Package A) and 22.2% (Package B) and (ii) for nonhazardous ones: 20.0% (Package D) and 0.0% (Package H). Cochran's test revealed that there were significant differences in the percentages of correct classification (Q = 24.522, df = 3, p < 0.001). Pairwise comparisons using Bonferroni correction showed that package A attained a higher percentage of correct classification than packages B (p = 0.007), D (p = 0.002), and H (p < 0.001).

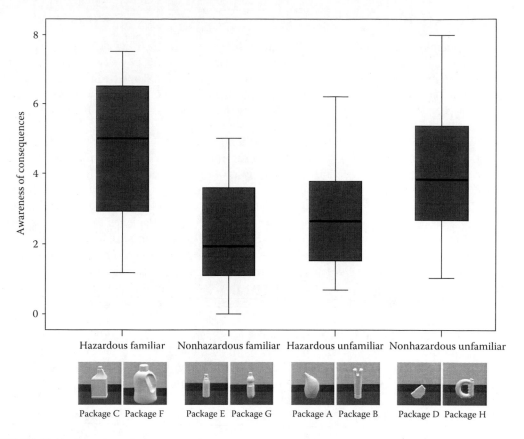

FIGURE 23.7 Box-plots of awareness of consequences scores by type of package.

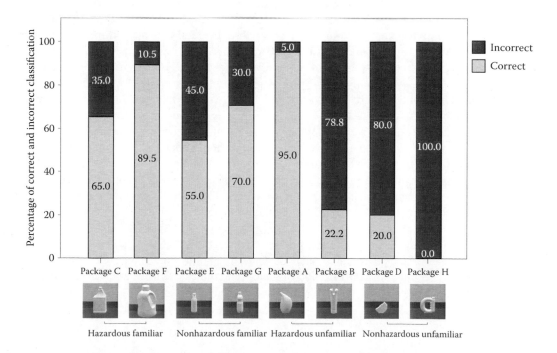

FIGURE 23.8 Percentages of correct classification of the package's category of content.

23.4   DISCUSSION

The main objective of this study was to examine the extent to which the packages' shape can be used to communicate, in an implicit manner, the hazard level of its contents and, therefore, contribute to increase users' safety. For such, the effect of packages' shape (from eight 3D household packages) on hazard-related perceptions (i.e., content hazardousness and awareness of consequences) was examined using a simulator-based methodology. Moreover, the familiarity of the package and the correct identification of its contents were analyzed.

The results showed that the participants were able to perceive diverse levels of hazard just by analyzing the packages, even though they were examining 3D prototypes which had minimal details. However, there were differences according to the familiarity level. The hazard-related perceptions were more accurate (i.e., perceiving which ones were hazardous and which were nonhazardous) for familiar packages than for unfamiliar ones. When considering only the hazardous packages, the hazard-related perceptions were higher for familiar ones than for unfamiliar ones. This finding is not surprising, since it indicates that the users' might face a higher level of risk when dealing with new designs that do not follow well-known patterns. However, interestingly, package A, which is a hazardous unfamiliar package, attained the highest value of correct content identification/classification. One reason for this can be that the participants considered content identification and familiarity as different concepts. Further investigation is required to clarify this finding.

In general, the current results suggest that a package's shape plays an important role in communicating risk information, by providing the users' with cues about the content's hazardous nature and associated level of hazardousness. These results support previous studies (Wogalter et al. 2001; Serig, 2001) in which researchers found that shape might serve as a cue to communicate the type and the level of hazard associated to its contents.

The results also indicate that the VR simulation used in the experiment, including both the devices and the VE, did not negatively affect the task performance. The participants did not report any simulator sickness symptoms. These results suggest that simulator-based methodologies can be successfully used to assess users' perceptions about packages' hazardousness, and embrace important advantages when compared to other approaches (e.g., easy manipulation of the packages' and environment's features, allowing good control over all extraneous variables and accuracy on the measures taken). However, further investigation is required to understand the extent to which this methodology has more advantages when compared to using static 2D images.

A main implication of this study is that it would be inappropriate to deliver a hazardous content in a familiar nonhazardous content package. Because of universal patterns of use, when designing a package for hazardous content, it is crucial to take into account whether the same package was used elsewhere for nonhazardous products, since it will increase the probability of people perceiving it as nonhazardous.

Informing users about the correct level of hazard associated with hazardous products can be one of the most important measures to help promote safety. Through the manipulation of different shape features, designers might weaken or strengthen the signifiers that change desirable action possibilities.

This study is an initial research concerning users' hazard-related perceptions of 3D virtual package prototypes. Additional research shall be done to examine the effects of other package features (e.g., color, texture, material) in order to determine the extent to which such aspects may affect hazard-related perceptions, as well as promote safe behaviors.

ACKNOWLEDGMENT

A part of this study was published in the *AHFE 2014 Conference Proceedings*.

REFERENCES

Ayanoğlu, H. 2013. *Implications of Package Design on Hazard Perception.* Unpublished doctoral dissertation, Second University of Naples.

Ayanoğlu, H., E. Duarte, P. Noriega, L. Teixeira, and F. Rebelo. 2013. The importance of integrating perceived affordances and hazard perception in package design. In: Rebelo, F. and M. Soares (eds.), *Advances in Usability Evaluation Part I*, pp. 627–636. CRC Press/Taylor & Francis, Ltd: Boca Raton, FL.

Ayanoğlu, H., F. Rebelo, E. Duarte, P. Noriega, and L. Teixeira. 2013. Using virtual reality to examine hazard perception in package design. In: Marcus, A. (ed.), *Design, User Experience, and Usability. User Experience in Novel Technological Environments SE - 4*, Vol. 8014, pp. 30–39. Lecture Notes in Computer Science. Springer: Berlin.

Brannstrom, C., H. L. Brown, C. Houser, S. Trimble, and A. Santos. 2015. You can't see them from sitting here': Evaluating beach user understanding of a rip current warning sign. *Applied Geography Elsevier Ltd.*, 56 61–70. http://linkinghub.elsevier.com/retrieve/pii/S0143622814002483.

Caner, C. and M. A. Pascall. 2010. Consumer complaints and accidents related to food packaging. *Packaging Technology and Science Wiley Online Library*, 23(7): 413–422.

Cunningham, R. 2007. *Package Warnings: Overview of International Developments. Canadian Cancer Society.* http://editor.annenbergpublicpolicycenter.org/wp-content/uploads/Release_WarningLabels_200703202.pdf.

Dejoy, D. M. 1999. Attitudes and beliefs. In: Wogalter, M.S., D. Joy, and K. Laughery (eds.), *Warnings and Risk Communication*, pp. 189–219. Taylor & Francis: London.

Desai, S. P., B. C. Teggihalli, and R. Bhola. 2005. Superglue mistaken for eye drops. *Archives of Disease in Childhood*, 90(11):1193.

Galley, M., E. Elton, and V. Haines. 2005, October. Packaging: A box of delights or a can of worms? The contribution of ergonomics to the usability, safety and semantics of packaging. Paper presented at *the FaraPack Briefing 2005.* New Technologies for Innovative Packaging: Loughborough, UK (unpublished presentation).

Gibson, J. J. 1986. *The Ecological Approach to Visual Perception.* Taylor & Francis: New York.

Hammond, D. and C. Parkinson. 2009. The impact of cigarette package design on perceptions of risk. *Journal of Public Health (Oxford, England)*, 31(3): 345–353.

Johnson, D. A. 2006. Practical aspects of graphics related to safety instructions and warnings. In: Wogalter, M. S. (ed.), *Handbook of Warnings.* Lawrence Erlbaum Associates: Mahwah, New Jersey, pp. 463–476.

Kozak, G. R., F. Terauchi, M. Kubo, and H. Aoki. 2003. Food packaging analyzed through usability: User survey about ways of opening, using and discarding packages. *Idemployee.id.tue.nl.*

Landabaso, J. L., M. Pardas, and J. R. Casas. 2006. Reconstruction of 3D shapes considering inconsistent 2D silhouettes. In: *IEEE International Conference on Image Processing*, pp. 2209–2212. http://ieeexplore.ieee.org/xpls/abs_all.jsp?arnumber=4107003.

Langley, J., R. Janson, J. Wearn, and A. Yoxall. 2005. Inclusive' design for containers: Improving openability. *Packaging Technology and Science*, 18(6):285–293.

Laughery, K. R. and M. S. Wogalter. 1997. Warnings and risk perception. In: Salvendy, G. (ed.). *Handbook of Human Factors and Ergonomics*, 2nd edn., Vol. 2. New York: Wiley, pp. 1174–1197.

Lesch, M. F. 2008. A comparison of two training methods for improving warning symbol comprehension. *Applied Ergonomics*, 39(2): 135–143.

Luximon, A., L. W. Chung, and R. S. Goonetilleke. 1998. Safety signal words and color codes: The perception of implied hazard by Chinese people. In: *Proceedings of the 5th Pan-Pacific Conference on Occupational Ergonomics*, Japan, Vol. 486, pp. 30–33. http://www-ieem.ust.hk/dfaculty/ravi/papers/safety.pdf.

Ngo, M. K., B. Piqueras-Fiszman, and C. Spence. 2011. On the colour and shape of still and sparkling water: Insights from online and laboratory-based testing. *Food Quality and Preference Elsevier Ltd*, 1–9. http://linkinghub.elsevier.com/retrieve/pii/S0950329311002400.

Norman, D. A. 2010. *Living with Complexity. Signals.* London: The MIT Press.

Radford, S. K. 2007. *Have You Seen the New Model? Visual Design and Product Newness.* University of Missouri, Colombia.

Ratneshwar, S. and S. Chaiken. 1991. Comprehension's role in persuasion: The case of its moderating effect on the persuasive impact of source cues. *Journal of Consumer Research*, 18: 52–62.

Rebelo, F., P. Noriega, E. Duarte, and M. Soares. 2012. Using virtual reality to assess user experience. *Human Factors: The Journal of the Human Factors and Ergonomics Society*, 54(6): 964–982.

Rosson, M. B. and J. M. Carroll. 2002. Usability engineering: Scenario-based development of human–computer interaction interface. *Morgan Kaufmann Series in Interactive Technologies.* San Francisco: Academic Press.

Serig, E. M. 2001. The influence of container shape and color cues on consumer product risk perception and precautionary intent. In: Wogalter, M. S., S. L. Young, and K. R. Laughery (eds.), *Human Factors Perspective on Warnings, Vol. 2: Selections from Human Factors and Ergonomics Society Annual Meetings, 1994–2000*, pp. 185–188. Santa Monica, CA: Human Factors and Ergonomics Society.

Winder, B., K. Ridgway, A. Nelson, and J. Baldwin. 2002. Food and drink packaging: Who is complaining and who should be complaining. *Applied Ergonomics*, Elsevier, 33(5): 433–438.

Wogalter, M. S., J. W. Brelsford, D. R. Desaulniers, and K. R. Laughery. 1991. Consumer product warnings: The role of hazard perception. *Journal of Safety Research*, 22: 71–82.

Wogalter, M. S., M. J. Kalsher, L. J. Frederick, A. B. Magurno, and B. M. Brewster. 1998. Hazard level perceptions of warnings components and configurations. *International Journal of Cognitive Ergonomics*, 2(1–2): 123–143.

Wogalter, M. S. and K. R. Laughery. 2001. Warnings. In: W. Karwowski (ed.), *International Encyclopedia of Ergonomics and Human Factors*, Vol. 2. Taylor & Francis: New York.

Wogalter, M. S., K. R. Laughery, and D. A. Barfield. 2001. Effect of container shape on hazard perceptions. In: *Human Factors Perspective on Warnings, Vol. 2: Selections from Human Factors and Ergonomics Society Annual Meetings, 1994–2000*, pp. 231–235.

Wogalter, M. S., S. L. Young, J. W. Brelsford, and T. Barlow. 1999. The relative contributions of injury severity and likelihood information on hazard-risk judgments and warning compliance. *Journal of Safety Research*, 30(3): 151–162.

Yoxall, A., R. Janson, S. R. Bradbury, J. Langley, J. Wearn, and S. Hayes. 2006. Openability: producing design limits for consumer packaging. *Packaging Technology and Science Wiley Online Library*, 19(4): 219–225.

Zuckerman, A. and S. Chaiken. 1998. A Heuristic-systematic processing analysis of the effectiveness of product warning labels. *Psychology & Marketing*, 15(7): 621.

24 Contributions to the Design of Knowledge-Based Development Tools for Intelligent Systems

Mário Simões-Marques and Isabel L. Nunes

CONTENTS

24.1 INTRODUCTION

"Intelligent systems" is a generic designation referring to computer-based tools used to provide support to decision-making or problem solving processes using artificial intelligence. Such tools are particularly relevant when the decision factors are complex or the timeliness of the decision is critical, making the decision-makers' task difficult, for instance due to the amount of information to process, because of the uncertainty and vagueness involved, or because of the stressful pressure resulting from the environment or the impact of the decision. Human experts are a high value and costly asset which is not always available, and even the experts may feel the need to have aids to support their decision-making process. Therefore, there is an ongoing effort to make available tools that replicate experts' reasoning process offering support to human experts or, in the case of lack of experts, becoming an alternative means to ensure the access to the expertise required to deal with specific problems. Expert systems (ES) are a particular type of intelligent systems which, as Turban et al. (2007) note, through the use of applied artificial intelligence techniques, aim to reach a level of performance comparable to a human expert, mimicking them in a particular area. The artificial intelligence methodologies used by intelligent systems are diverse, including fuzzy reasoning, rule-based, case-based, evolutionary algorithms, machine learning approaches, to name a few. When compared to natural intelligence-based decision support in a specific domain, an artificial intelligence system offers some advantages since (Turban et al., 2007): (i) it is more permanent, (ii) is easy to duplicate and disseminate, (iii) can be less expensive, (iv) is consistent and thorough, (v) can be documented, (vi) can execute certain tasks much faster than a human, and (vii) can perform certain tasks better than most people. However, artificial intelligence lacks the creativity and the ability to learn which is intrinsic to natural intelligence. In fact, as Turban et al. note human experts are capable of (i) recognizing and formulating a problem, (ii) solving this new problem quickly and correctly, (iii) explaining the solution, (iv) learning from experience, (v) restructuring knowledge, (vi) breaking rules if necessary, (vii) determining relevance, and (viii) degrading gracefully (i.e., being aware of one's limitations).

Irrespective of their limitations, current intelligent systems, and particularly ES, are able to engage in complex inference processes, necessary for evaluating alternative options and offering good quality conclusions and advice, and also to offer explanations about the rationale that led to such conclusions.

A key component of any intelligent system is the knowledge base, since this is where the knowledge elements about a specific subject are stored. The process of transferring the expertise of humans into computers can be quite complex and challenging when the type of decision-making problem is unstructured or semi-structured, since they are often based on human intuition.

Developing knowledge bases for this type of application is a quite difficult task, since there is the need to figure out and map, among others, the knowledge elements, organization, context of use, composition and representation, relations, importance, and the reasoning processes used to feed the inference process, combining the inputs coming from real world data with such knowledge in order to present the desired outputs, ranging from structured (e.g., models) to unstructured (e.g., metaheuristic algorithms).

In this chapter, we propose to address the issues involved in defining the requirements for designing a knowledge-based development tool (KBDT), which supports cooperative and participatory processes of knowledge elicitation, which are, despite the eventual complexity of the problem at hand, intuitive and easy to implement. This calls for an approach that carefully considers the principles and methodologies proposed by user-centered design.

This work builds on the experience in the fields of knowledge engineering and knowledge management (KM) that the authors gathered in the development of three independent ES in the areas of ergonomic assessment of work places, occupational risk assessment, and emergency management.

The structure of the chapter includes the present introduction, which sets the general context, followed by three sections. Section 24.2 characterizes some relevant concepts related with (i) knowledge, (ii) ES, (iii) KM, and (iv) user-centered design. Section 24.3 discusses issues related with developing a KBDT from a user-centered design standpoint. The final section presents some conclusions.

24.2 RELEVANT CONCEPTS

24.2.1 KNOWLEDGE

For this work, it is important to characterize the concept of knowledge in the context of ES. A common and useful approach is the one of the data–information–knowledge–understanding–wisdom (DIKUW) hierarchy, since this is a central model in the domain of KM, contributing to define the nature and relationships among each of these entities (Rowley, 2007). As Rowley notes, there is some debate on the authorship of the concepts underlying the model, however, Ackoff's paper "From data to wisdom" (Ackoff, 1989) is often cited when the DIKUW hierarchy is referred. Ackoff's definitions for the entities of the hierarchy are (Ackoff, 1989, 1999):

- Data are defined as symbols that represent properties of objects, events, and their environment. They are the products of observation. But are of no use until they are in a useable (i.e., relevant) form. The difference between data and information is functional, not structural.
- Information is contained in descriptions, answers to questions that begin with such words as who, what, when, and how many. Information systems generate, store, retrieve, and process data. Information is inferred from data.
- Knowledge is know-how, and is what makes possible the transformation of information into instructions. Knowledge can be obtained either by transmission from another who has it, by instruction, or by extracting it from experience.

- Understanding is ability to explain why things are as they are and why the means chosen produce the outcomes they do. It requires diagnosis and prescription, that is, requires detecting error, knowing why it was made and how to correct it.
- Wisdom is the ability to perceive and evaluate the long-run consequences of behavior. Wisdom adds value, which requires the mental function that we call judgment. The ethical and aesthetic values that this implies are inherent to the actor and are unique and personal.

Ackoff (1999) states that data, information, knowledge, and understanding presuppose each other, however, they are acquired and develop interdependently. This author argues that even though these entities form a hierarchy with respect to value, none is more fundamental than the others.

Ackoff further noted that knowledge and understanding focus on efficiency, while wisdom focuses on effectiveness, and, later on, in an allusion to a Peter Drucker's sentence,* stated that knowledge and understanding is about "doing things right" (efficiency) while wisdom is about "doing the right things" (effectiveness). In fact, this author clarifies that it takes understanding to be aware of the functions a system serves, but it takes wisdom to know whether these functions are good or bad (Ackoff, 1999). It is common to present the DIKUW hierarchy graphically as a pyramid. Figure 24.1 presents this pyramid adding to it some of the above presented concepts.

When considering the characteristics of computerized support systems, it is possible to map the levels of the DIKUW hierarchy with typical categories of computer-based information systems: the data level maps with transactions processing systems, the information level maps with management information systems, and the knowledge level maps with decision support systems (Turban et al., 2007). Considering that intelligent systems not only are expected to support complex decisions, by performing inference processes based on symbolic manipulations, but also to provide advice and explanations, these type of systems map into the understanding level.

The European Committee for Standardization issued, in March 2004, the document "European Guide to Good Practice in Knowledge Management—Part 1: Knowledge Management Framework," which states as a working definition of knowledge (CEN, 2004):

> Knowledge is the combination of data and information, to which is added expert opinion, skills and experience, to result in a valuable asset which can be used to aid decision making. Knowledge may be explicit and/or tacit, individual and/or collective.

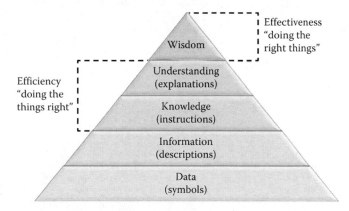

FIGURE 24.1 DIKUW hierarchy. (From Ackoff, R.L. 1999. *Re-Creating the Corporation: A Design of Organizations for the 21st Century.* Oxford University Press.)

---

* "There's a difference between doing things right and doing the right thing."

This definition brings together not only the concepts related with the DIKUW hierarchy, but also the ones presented in 1995 by Nonaka and Takeuchi in the book *The Knowledge-Creating Company* (Nonaka and Takeuchi, 1995), where these authors argue that knowledge is initially created by individuals and later becomes organizational knowledge through a process illustrated in Figure 24.2. The authors consider two dimensions of organizational knowledge creation: the epistemological, where two types of knowledge tacit and explicit are identified; and the ontological, considering individuals as the lower level, and teams, groups, and organizations as increasingly higher levels. Explicit knowledge is the knowledge that can be expressed in words and numbers and readily shared (for instance in as data, scientific formulae, specifications, and manuals) between individuals in a formal and systematic way (Nonaka and Konno, 1998). Tacit knowledge is the one which is more difficult to transfer since it is highly personal and hard to formalize, often resulting from experience. Nonaka and Takeuchi state that "A spiral emerges when the interaction between tacit and explicit knowledge is elevated dynamically from a lower ontological level to higher levels." This spiral is translated in the socialization, externalization, combination, internalization (SECI) model (Figure 24.2) that addresses the four modes of knowledge conversion (Nonaka and Konno, 1998):

- *Socialization (from tacit-to-tacit knowledge)*. Tacit knowledge is exchanged through joint activities, such as being together, spending time, living in the same environment, rather than through written or verbal instructions.
- *Externalization (from tacit to explicit knowledge)*. The articulation of tacit knowledge, that is, the conversion of tacit into explicit knowledge, involving: (i) techniques to express one's ideas or images as words, concepts, figurative language (such as metaphors, analogies, or narratives), and visuals and (ii) translating the tacit knowledge of customers or experts into readily understandable forms.
- *Combination (from explicit to explicit knowledge)*. Conversion of explicit knowledge into more complex sets of explicit knowledge. The key issues are communication and diffusion processes and the systemization of knowledge, relying on three processes: (i) capturing and integrating new explicit knowledge (e.g., public data); (ii) dissemination of knowledge directly by using presentations or meetings; and (iii) editing or processing explicit knowledge to make it more usable (e.g., documents such as plans, reports, and market data).
- *Internalization (from explicit-to-tacit knowledge)*. Conversion of explicit knowledge (embodied in action and practice, or using virtual situations) into tacit knowledge, for instance through learning-by-doing, training, and exercises.

Figure 24.3 represents the DIKUW hierarchy, and SECI model within it, with regard to programming complexity versus decision-making value. In fact, computer systems designed to manipulate data are the simplest to program but are the ones that add less value in terms of decision making.

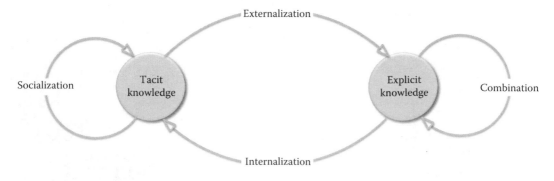

FIGURE 24.2   SECI model.

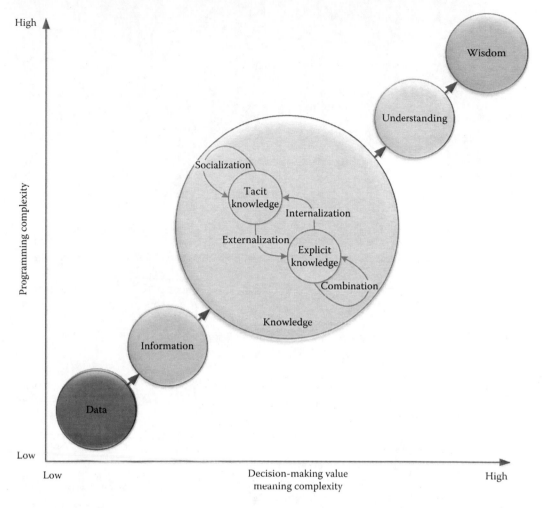

FIGURE 24.3 DIKUW hierarchy and SECI model representation with regard to programming complexity versus decision-making value.

The complexity increases as the focus evolves to information and to knowledge; this corresponds, however, to increased levels of decision-making value and meaningfulness of systems output.

In the following sections, we will address the issues of representing and manipulating knowledge in intelligent systems, and particularly in ES.

24.2.2 Expert Systems

As mentioned in the introduction, an expert system is a particular type of intelligent system which tries to replicate the reasoning of a human expert. An ES is expected to perform fast, providing replies to users' needs in a specific domain, extending their level of expertise in an easy, intuitive, and controllable way.

The generic architecture of an ES is illustrated in Figure 24.4. An ES is composed by four core building blocks that perform the following functions:

- *Knowledge base.* This component stores the knowledge required for the specific problem-solving application.

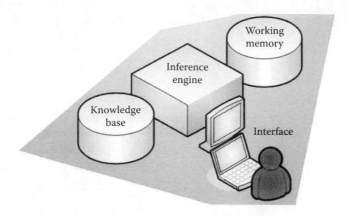

FIGURE 24.4 Generic architecture of ES.

- *Working memory.* This component stores data about the particular problem context to analyze.
- *Inference engine.* This component runs the knowledge against the data, assessing the concrete situation and generating the conclusions and advice, as well as any explanations required by the user.
- *User interface.* This component offers the human–computer interaction means necessary to input data, insert requests, and obtain system outputs.

The fields of application, and the number of ES proposed and implemented in growing steadily. Turban et al. (2007) identified some factors which are required, but not sufficient, to guarantee success in the implementation of ES:

- The problem must be important and difficult enough to warrant development of an ES
- The problem must be sufficiently narrow in scope
- The problem to be solved must be mostly qualitative (fuzzy)
- The level of knowledge must be sufficiently high
- Expertise must be available from at least one cooperative expert
- ES shell must be of high quality and naturally store and manipulate the knowledge
- The user interface must be friendly for novice user
- ES must have a favorable impact, resulting in end-user job improvement
- Management support must be cultivated

The same authors recognize that some problems have slowed down the commercial spread of ES (Turban et al., 2007):

- Knowledge is not always readily available
- It can be difficult to extract expertise from humans
- The approach of each expert to a situation assessment may be different yet correct
- It is hard, even for a highly skilled expert, to abstract good situational assessments when under time pressure
- Users of ES have natural cognitive limits
- ES work well only within a narrow domain of knowledge
- Most experts have no independent means of checking whether ES conclusions are reasonable
- The vocabulary experts use to express facts and relations is often specific and not understood by others

- Knowledge engineers are rare and expensive, a fact that can make ES construction costly
- Lack of trust on the part of end-users may be a barrier to ES use
- Knowledge transfer is subject to a host of perceptual and judgmental biases

All considered the pros largely outweigh the cons of intelligent systems and this technology is here to stay, dressed in many different forms. The Internet is a good example of fertile ground for a wide range of intelligent systems applications, from advanced search engines to commercial advertisement and product offers that try to adapt to users' preferences and interests, learning from their habits or guessing from recent activity.

The evolution and innovation of technological solutions led to an explosive offering of different alternative devices with high computational and communication capabilities that offer the basis for cooperative work and collaborative decision making. Naturally, this is another field that offers many opportunities for the development of ES. Figure 24.5 presents an example of such applications, illustrating the use of a distributed expert to support inter-agency cooperation in the context of emergency management (Simões-Marques and Nunes, 2014). As the authors note, in response to critical situations (for instance in disaster relief operations), it is normal to involve several entities that cooperate to improve the effectiveness and efficiency of their collective effort. This calls for common

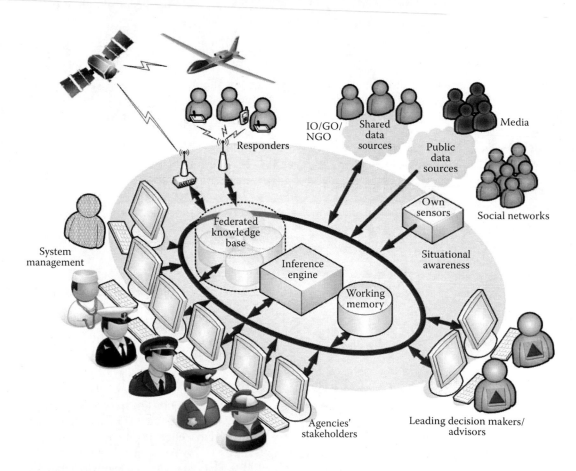

FIGURE 24.5 Example of a complex distributed expert system to support inter-agency cooperation in the context of emergency management. (Adapted from Simões-Marques, M. and I.L. Nunes. 2014. Contributions to the design of emergency management intelligent systems, In: Arezes, P. et al. (eds.), *Occupational Safety and Hygiene II—Selected Extended and Revised Contributions from the International Symposium Occupational Safety and Hygiene, SHO 2014*, Guimarães, Portugal, February 13–14, CRC Press/Balkema, Boca Raton, pp. 781–786.)

tools (like the one illustrated in the figure) for generating and sharing a common operational picture that ensures situational awareness, and that offers common evaluation criteria and advice which help anticipating the courses of action of other stakeholders involved in the process, while also helping to break cultural and procedural barriers, contributing to strengthen the maturity of the interaction among groups of responders, and to elevate the level of global preparedness and response. This is a particularly relevant goal since the management of disaster relief operations is a complex process that requires the coordination of different actors, with different cultures, aims, and views of the world.

The structure and contents of all these blocks have to be designed according to the type and specificities of the decision support required. Naturally, designing an ES for complex broad scope decision-making problems is more challenging than designing one for dealing with more narrow problems. Examples of ES designed using the basic architecture for stand-alone application are FAST ERGO_X (Nunes, 2009), which is an ES that performs ergonomic assessment of workplaces and offers advice on potential lines of action to correct the identified inadequacies and supply chain decision making fuzzy decision support system (SCDM FDSS) (Nunes et al., 2012) is a supply chain disturbance management fuzzy decision support system, to support decision-making processes aimed at improving the performance/resilience of supply chains. An example of a more complex ES, designed to be used as a distributed and collaborative decision-making platform is SINGRAR (Simões-Marques, 1999; Simões-Marques and Pires, 2003; Simões-Marques et al., 2000), which is a system developed to support crisis management activities on navy ships.

24.2.3 KNOWLEDGE MANAGEMENT

Independently of the scope and complexity of the application, developing an ES requires performing KM activities. KM supports the process of transferring expertise from human experts to computers and back to humans. Turban et al. (2007) identify four activities in such process:

- Knowledge acquisition (from experts to other sources) is the activity related with the externalization and combination phases of the SECI model, since in this stage, it is required to make explicit knowledge that eventually is still tacit or to combine knowledge which is already explicit.
- Knowledge representation (in the computer) requires dealing with abstract concepts (e.g., events, time, physical objects, and beliefs) and their relations, which constitute the content of the knowledge base, involving a new scientific field called ontological engineering (Russell and Norvig, 2010).
- Knowledge inferencing is the reasoning capabilities that can build higher-level knowledge from existing heuristics. This reasoning consists of inferencing from facts and rules using heuristics or other search approaches.
- Knowledge transfer (to the user) is the activity of delivering knowledge to ES users, namely to nonexpert recipients, using adequate knowledge presentation and visualization interface or environment.

A topic which is concurrent with knowledge representation is the one of knowledge visualization which is defined as "the use of complementary visual representations to transfer and create knowledge between at least two persons" (Burkhard, 2005). Knowledge visualization exploits new ways of graphically representing insights, experiences, gathered evidence, and know-how in order to share knowledge, create new knowledge, or apply knowledge to decision making (e.g., lists, trees, semantic networks, and schemas translating the know what/where/how/when/why), making use of the whole spectrum of graphic representations, ranging from simple hand-drawn sketches to immersive virtual three-dimensional worlds (Eppler and Pfister, 2013). Figure 24.6 illustrates some alternative standardized representations used in the depiction of elements of information and their relations which can help in conveying knowledge.

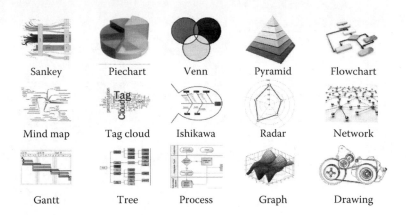

Sankey   Piechart   Venn   Pyramid   Flowchart

Mind map   Tag cloud   Ishikawa   Radar   Network

Gantt   Tree   Process   Graph   Drawing

FIGURE 24.6 Example of standardized representations used in the depiction of elements of information and their relation. (From Springer Science+Business Media: *Knowledge and Information Visualization*, Towards a framework and a model for knowledge visualization: Synergies between information and knowledge visualization, LNCS 3426, 2005, pp. 238–255, Burkhard, R.A.)

The diagram presented in Figure 24.7, adapted from Simões-Marques and Nunes (2013), illustrates an example of relations that can be mapped in the knowledge base (KB) of the emergency management ES previously presented in Figure 24.5. As the authors note, this diagram refers to knowledge required for solving an inter-agency responders' assignment problem on disaster relief operations, and depicts relationships between responders and incident types, considering factors such as incident impact type, geographical context, or response coordination. The same authors say that the completeness and the complexity of the model depend on the amount of attributes considered in these relations, while the accuracy of the advice depends on how well the knowledge-based relations reflect real-world relationships.

In Nunes (2007) and Simões-Marques and Pires (2003), the authors described the knowledge acquisition (or elicitation) and knowledge representation activities, respectively, for FAST ERGO_X and SINGRAR ES previously referred.

Developing an expert system KB is a quite challenging task, since there is the need to figure out and map, among others, the knowledge elements, organization, context of use, composition and representation, relations, importance, and the reasoning processes used to feed the inference process, combining the inputs coming from real world data with such knowledge in order to present the desired outputs. Furthermore, the acceptance of an ES depends greatly on the quality of the outcome of these developing activities, making it a critical success factor. This was observed by Hecht et al. (2011), who analyzed the factors influencing the adoption, acceptance, and assimilation of the design of KM systems. These authors argue that some factors cannot be directly influenced by design (e.g., environment, technological infrastructure, resources, organizational characteristics, social influence, attitude toward technology use, management characteristics, and institutional characteristics), while others can be directly influenced by design (e.g., innovation characteristics, fit, expected results, communication characteristics, effort expectancy, performance expectancy, social system characteristics, and process characteristics). Among the factors that can be influenced by design, the ones classified as effort and performance expectancy, are intimately related with usability (e.g., complexity, ease of use, usefulness, results demonstrability, and job-fit).

Eliciting the contents of KB is a big task on its own; therefore, it is desirable to have an ES shell available which avoids the burden on developers of designing the entire ES for each application. This is particularly true for the KB component. In fact, having KBDTs available which are "user friendly" help reducing the application implementation effort to a minimum. This assertion implies the adoption of a user-centered design approach in the project and creation of a KBDT. We will proceed reviewing some basic concepts related with the topic user-centered design.

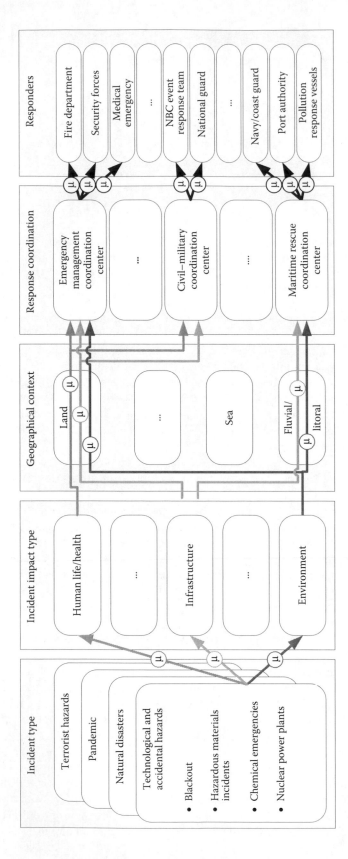

FIGURE 24.7 Example of fuzzy relations elicited for the KB of an ES to support inter-agency cooperation in the context of disaster relief operations. (From Simões-Marques, M. and I.L. Nunes. 2013. A fuzzy multicriteria methodology to manage priorities and resource assignment in critical situations. In: Zeimpekis, V., S. Ichoua, and I. Minis (eds.), *Humanitarian & Relief Logistics: Research Issues, Case Studies and Future Trends.* Springer under Series Operations Research/Computer Science (ORCS), pp. 129–153.)

24.2.4 User-Centered Design

User-centered design is a structured development methodology to attain software usability, focused on the needs and characteristics of users, which should be applied from the beginning of the development process in order to make applications more useful and easy to use (Averboukh, 2001; Nunes, 2006; Nunes and Simões-Marques, 2013; Simões-Marques and Nunes, 2012).

There are different perspectives about the principles that developers should adopt when designing a product to achieve an appropriate usability. For instance, the European Union's Council Directive, 90/270/EEC, of May 29, lists usability principles that should be taken into account when designing, selecting, commissioning, and modifying software. Jordan identified a list of 10 general principles to observe: consistency, compatibility, consideration of user resources, feedback, error prevention and recovery, user control, visual clarity, prioritization of functionality and information, appropriate transfer of technology, and explicitness (Jordan, 1998). Reiss presents usability techniques that help improve product design regarding functionality, responsiveness, and clarity (including its visibility, understandability, logicalness, consistency, and predictability) making it ergonomic and foolproof (Reiss, 2012). Despite the fact that the breadth, depth, and terminology may vary, the core principles are equivalent.

Gerhardt-Powals identified a set of heuristics to improve performance, which includes automating unwanted load, reducing uncertainty, condensing data, presenting new information with meaningful ways to support their interpretation, using names that are conceptually related to functions, limiting data-oriented tasks, including only information on the screens that the user needs at any given time, providing multiple coding of data (where appropriate), and practicing a judicious redundancy (Gerhardt-Powals, 1996).

ISO 9241 provides requirements and recommendations for human-centered design principles and activities throughout the life cycle of computer-based interactive systems. It is intended to be used by those managing design processes, and is concerned with ways in which both hardware and software components of interactive systems can enhance human-system interaction. ISO 9241-210:2010 identifies four key activities related with the user-centered design[*] approach (see Figure 24.8), which should be planned and implemented in order to incorporate the requirements of usability in the process of software development: understand and specify context of use; specify the user and organizational requirements; produce design solutions; and evaluate design against requirements. These activities are performed iteratively, with the cycle being repeated until the requirements have been achieved.

24.3 KNOWLEDGE-BASED DEVELOPMENT TOOL

A KBDT is an environment designed to support KM activities, adapted to the ES context of use and to the application requirements (Figure 24.9). Regarding the knowledge acquisition phase, it must support the elicitation of knowledge from experts (e.g., through structured and unstructured interviews) and the import of or link to explicit knowledge from different sources (e.g., other KB, multimedia, and web-based). Regarding representation, the KBDT has to support alternative ways (e.g., graphical, tabular, and tree) of presenting and relating the knowledge (e.g., ontologies, models, heuristic rules, semantic networks, and description logics). The knowledge has to be coded in formats that can be used during the knowledge inferencing and knowledge transfer phases.

These very generic statements unveil the need for an edition and visualization environment which is quite complex, since there is a great variety of alternative ways and formats of collecting and representing knowledge that must be made compatible as much as possible and, whenever possible, presented in multiple formats. In fact, it is long recognized that, according to their level of proficiency and personal likings, whenever there are different ways of doing the same task, users tend

---

[*] In fact, the ISO 9241 standard uses term "human-centered design" is used rather than "user-centered design" in order to emphasize that this part of standard also addresses impacts on a number of stakeholders, not just those typically considered as users. However, in practice, these terms are used synonymously.

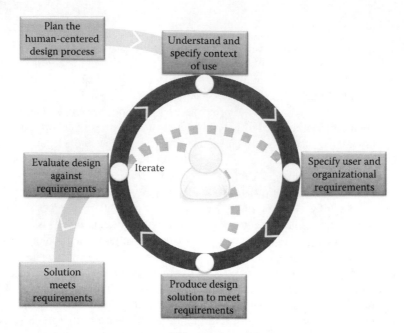

FIGURE 24.8 Activities of user-centered design. (Adapted from ISO 9241. 2010. *Ergonomics of Human-System Interaction—Part 210: Human-Centred Design for Interactive Systems.* International Organization for Standardization, Geneva.)

to choose the alternative that performs best in a way that the user can control. An obvious example is the use of keyboard shortcuts to perform functions that otherwise would force the user to take the hands off the keyboard, use a pointing device to make multiple clicks to navigate on menus or on button panes. Naturally, the option of using shortcuts is only controlled by individuals that are proficient in the use of the application and when the environment offers such functionality. Of course if you do not know a specific software you can always try to use the commonly used shortcuts (e.g., CTRL + C, CTRL + V) and hope the developers of such application adhere to the "standard." If you succeed, this means that the development of such software followed the some basic usability principles, making its use predictable. A new challenge imposed by the overwhelming progress of technology and the increasing number of computer device types, is finding the corresponding interaction alternatives when you use different devices (e.g., desktop or portable computer, tablets, and smartphones). Note that the interaction with touch and multitouch devices follows paradigms that do not match many of the standards established for conventional human–computer interaction.

Resuming our previous discussion about proficiency and preference, the scope is much wider than just the issue of shortcuts. For instance, there are many ways of creating, accessing, and editing databases. "Beginners" tend to prefer an environment where they visually interact with objects that represent the structure and the relations of a particular instance of a database, hiding the cryptic language of scripts and queries. "Intermediate" level users probably risk switching to the database management system (DBMS) text editor environment and fine tuning the code automatically generated in the visual environment. "Experts" will probably start using the text editor environment to quickly reutilize code already developed, change it, and eventually at the end switch to the visual environment to check in a glance if everything looks good. The readers can find identical type of examples in their own field of expertise. In fact, experts (e.g., engineers and scientists) usually fill the need to adopt formats for dealing with the details of a particular domain of knowledge that are not well suited to be the media of interfacing with nonexperts (e.g., clients, workers, and students).

Therefore, a KBDT has to offer means to make possible editing and visualizing knowledge in different formats. A self-sustained tool that contains all types of the required functionalities might

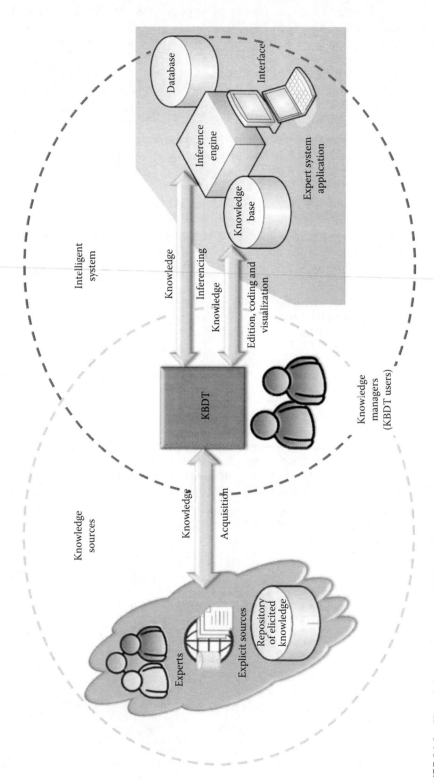

FIGURE 24.9 Knowledge management activities required to develop Intelligent Systems and supported by the KBDT.

be a desirable solution to offer a coherent and uniform environment. However, developing and keeping updated such a product seems almost impossible. Another fact to consider is that there are already many products available that partially solve the problem.

A pragmatic approach might be the KBDT to act as mediator between a KB (or a federation of KB) that stores the knowledge elements and different applications (e.g., open-source, commercial) that offer the required functionalities for edition and visualization. Such an approach would allow a modular growing of the solution and would require minimum effort for keeping the KBDT up with new interaction and technological standards. Finally, it would also make it possible to build the KBDT as a collaborative endeavor.

From a user-centered design standpoint, an advantage of this approach would be the opportunity to build the KBDT based on solutions that comply with usability standards. According to Nielsen (1993), the characteristics that influence the usability of a system are

- *Ease to learn.* The system must be intuitive, allowing even an inexperienced user to be able to work with it satisfactorily
- *Efficiency of use.* The system must have an efficient performance, that is, the resources spent to achieve the goals with accuracy and completeness should be minimal
- *Memorability.* The use of the system must be easy to remember, even after a period of interregnum
- *Errors frequency.* The system must provide adequate conditions to users for achieving specific objectives with high accuracy and completeness
- *Satisfaction.* The attitude of the user toward the system should be positive

Thus, we advocate that a KBDT should be an environment complying with the following requirements. The KBDT must

- Provide a user interface that adheres to the commercial standards, is easy to learn, intuitive, and easy to remember
- Provide the means that support performing collaborative, participatory, and distributed KM, allowing the combination of the work of multiple users
- Provide access to alternative edition environments (e.g., text and graphical) which offer different ways of creating, editing, and representing the knowledge, ensuring an efficient performance for users with different skill levels and preferences, making it satisfactory to users
- Provide the means to import knowledge from compatible KB and from standard file formats (e.g., office and multimedia software packages), allowing the reuse and combination of already available knowledge
- Provide knowledge acquisition support, namely through the use of methodologies for dealing with alternative knowledge sources, namely humans, deterministic, stochastic and heuristic models, and big data (e.g., questionnaires, analytical methods, simulations, fuzzy logics, and data mining)
- Provide the means to support different ES output types, both for advice and for explanation (e.g., text, tabular, graphical, georeferenced, and actionable)
- Provide the means to support different inference processes (e.g., deterministic, stochastic, and heuristic)
- Provide the means to support systems exploitation, as well as the means to recover from errors

The implementation of the KBDT should be done using a spiral and incremental approach where typical users (i.e., knowledge managers) are involved to validate and refine the requirements, to evaluate design against requirements, and to assess the usability of the solutions. Figure 24.10 illustrates this user-centered perspective of the development of a KBDT.

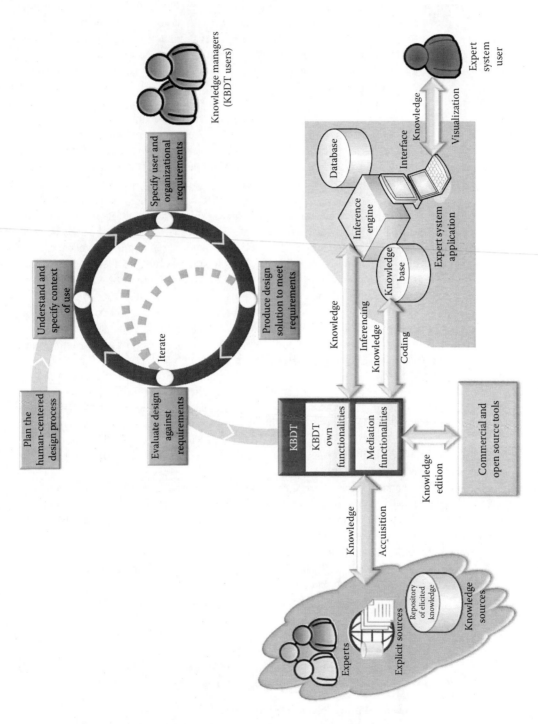

FIGURE 24.10 User-centered design perspective of KBDT.

24.4 CONCLUSIONS

Intelligent systems is a term used to refer to computer-based tools which provide decision support using artificial intelligence. Natural intelligence is capable, among other qualities, of recognizing and formulating problems, solving new problems, explaining the solutions, learning from experience, restructuring knowledge, breaking rules, determining relevance, and being aware of its own limitations. Artificial intelligence is not able to match such qualities; however, for a specific domain, intelligence systems may offer some advantages, when compared with human advice, since the reasoning they provide is more permanent, is easy to duplicate and disseminate, can be less expensive than human expertise, is consistent and thorough, can be documented, can execute certain tasks much faster than a human, and ultimately can perform certain tasks better than most people.

ES are a particular type of intelligent systems developed to analyze and offer recommendations and explanations about a specific problem domain, providing support when a human expert is not available, or helping experts in dealing with very demanding and critical problems, usually because of the complexity of the problem, the volume of information processed, and the pressure for short time answers.

A KB is a component of an ES, together with the inference engine, the working memory and the user interface. KBs are used to store structured and unstructured knowledge about a specific subject. Developing KBs for this type of application is a quite difficult task, since there is the need to identify and map, among others, the knowledge elements, organization, context of use, composition and representation, relations, importance, and the reasoning processes used to feed the inference process. The inference process combines the inputs coming from real world data with the knowledge stored in the KB in order to present the desired outputs.

A working definition of knowledge proposed by the European Committee for Standardization (CEN, 2004) is "Knowledge is the combination of data and information, to which is added expert opinion, skills and experience, to result in a valuable asset which can be used to aid decision making. Knowledge may be explicit and/or tacit, individual and/or collective."

A common and useful approach to contextualize knowledge is the one provided by Ackoff's DIKUW hierarchy, which also offers a perspective regarding the programming complexity and decision-making value of computer systems designed to address the different levels of this hierarchy. In fact, computer systems designed to manipulate data are the simplest to program but are the ones that add less value in terms of decision making. The complexity increases as the focus evolves to information and to knowledge; this corresponds, however, to increased levels of decision-making value and meaningfulness of systems output.

In this chapter, we presented the basic concepts related with knowledge, ES, and user-centered design, setting the stage to address the issues involved in defining the requirements for designing a KBDT, which follows basic usability principles. The proposed requirements envisage a KBDT that supports cooperative and participatory processes of knowledge elicitation, which are, despite the eventual complexity of the problem at hand, intuitive and easy to implement.

This work was built on the experience in the fields of knowledge engineering and KM that the authors gathered in the development of two independent ES in the areas of ergonomic assessment of work places, and emergency management, which were referred to illustrate some of the discussed subjects.

REFERENCES

Ackoff, R.L. 1989. From data to wisdom. *Journal of Applied Systems Analysis* 16:3–9.
Ackoff, R.L. 1999. *Re-Creating the Corporation: A Design of Organizations for the 21st Century.* Oxford University Press, Inc., New York.
Averboukh, E.A. 2001. Quality of life and usability engineering. In: Karwowski, W. (ed.), *International Encyclopedia of Ergonomics and Human Factors.* Taylor & Francis, London and New York, pp. 1317–1321.
Burkhard, R.A. 2005. Towards a framework and a model for knowledge visualization: Synergies between information and knowledge visualization. In: Tergan, S.-O. and T. Keller (eds.), *Knowledge and Information Visualization.* LNCS 3426, Springer-Verlag, Berlin, Heidelberg, pp. 238–255.

CEN. 2004. *European Guide to Good Practice in Knowledge Management—Part 1: Knowledge Management Framework.* European Committee for Standardization, Brussels.

Eppler, M.J. and R. Pfister. 2013. Best of both worlds: Hybrid knowledge visualization in police crime fighting and military operations. In *Proceedings of the 13th International Conference on Knowledge Management and Knowledge Technologies,* Graz, Austria. ACM, New York, pp. 17:1–17:8.

Gerhardt-Powals, J. 1996. Cognitive engineering principles for enhancing human–computer performance. *International Journal of Human–Computer Interaction* 8(2):189–211.

Hecht, M., R. Maier, I. Seeber, and G. Waldhart. 2011. Fostering adoption, acceptance, and assimilation in knowledge management system design. In *Proceedings of the 11th International Conference on Knowledge Management and Knowledge Technologies,* Graz, Austria. ACM, New York, pp. 7:1–7:8.

ISO 9241. 2010. *Ergonomics of Human-System Interaction—Part 210: Human-Centred Design for Interactive Systems.* International Organization for Standardization, Geneva.

Jordan, P. 1998. *An Introduction to Usability.* Taylor & Francis, London.

Nielsen, J. 1993. *Usability Engineering.* Academic Press, Boston.

Nonaka, I. and N. Konno. 1998. The concept of "Ba": Building a foundation for knowledge creation. *California Management Review* 40(3):40–54. University of California, Berkeley.

Nonaka, I. and H. Takeuchi. 1995. *The Knowledge-Creating Company: How Japanese Companies Create the Dynamics of Innovation.* Oxford University Press, Oxford.

Nunes, I.L. 2006. Ergonomics & usability—Key factors in knowledge society. *Enterprise and Work Innovation Studies* 2:87–94.

Nunes, I.L. 2007. Knowledge acquisition for the development of an upper-body work-related musculoskeletal disorders analysis tool. *Human Factors and Ergonomics in Manufacturing & Service Industries* 17(2):149–162.

Nunes, I.L. 2009. FAST ERGO_X—A tool for ergonomic auditing and work-related musculoskeletal disorders prevention. *Work: A Journal of Prevention, Assessment, & Rehabilitation* 34(2):133–148.

Nunes, I.L., S. Figueira, V. Cruz-Machado. 2012. Combining FDSS and simulation to improve supply chain resilience. In: Hernández, J.E., P. Zaraté, F. Dargam, B. Delibasic, S. Liu, and R. Ribeiro (eds.), *Decision Support Systems—Collaborative Models and Approaches in Real Environments.* LNBIP 121, Springer-Verlag, Berlin, Heidelberg, pp. 42–58.

Nunes, I.L. and M. Simões-Marques. 2013. SINGRAR usability study. In: Marcus, A. (ed.), *Design, User Experience, and Usability. Design Philosophy, Methods, and Tools Second International Conference, DUXU 2013, Held as Part of HCI International 2013,* Las Vegas, NV, USA, July 21–26, 2013, *Proceedings* Part I. LNCS 8012, Springer-Verlag, Berlin, Heidelberg, pp. 359–368.

Reiss, E. 2012. *Usable Usability: Simple Steps for Making Stuff Better.* John Wiley & Sons, Inc., Indianapolis.

Rowley, J. 2007. The wisdom hierarchy: Representations of the DIKW hierarchy. *Journal of Information Science* 33(2):163–180.

Russell, S. and P. Norvig. 2010. *Artificial Intelligence: A Modern Approach,* 3rd edn. Pearson Prentice-Hall Inc., Upper Saddle River, New Jersey.

Simões-Marques, M. and I. L. Nunes. 2012. Usability of interfaces. In: Nunes, I. L. (ed.), *Ergonomics.* InTech., Rijeka. http://www.intechopen.com/books/ergonomics-a-systems-approach/usability-of-interfaces.

Simões-Marques, M. and I.L. Nunes. 2013. A fuzzy multicriteria methodology to manage priorities and resource assignment in critical situations. In: Zeimpekis, V., S. Ichoua, and I. Minis (eds.), *Humanitarian & Relief Logistics: Research Issues, Case Studies and Future Trends.* Springer under Series Operations Research/Computer Science (ORCS), pp. 129–153.

Simões-Marques, M. and I. L. Nunes. 2014. Contributions to the design of emergency management intelligent systems. In: Arezes, P. et al. (eds.), *Occupational Safety and Hygiene II—Selected Extended and Revised Contributions from the International Symposium Occupational Safety and Hygiene, SHO 2014,* Guimarães, Portugal, February 13–14, CRC Press/Balkema, Boca Raton, pp. 781–786.

Simões-Marques, M.J. 1999. *Sistema de Apoio à Decisão Difuso para a Gestão de Prioridades de Reparação de Equipamentos e Afectação de Recursos, em Navios, em Situação de Combate.* MSc thesis. FCT/UNL, Lisboa.

Simões-Marques, M.J. and F. Pires. 2003. SINGRAR—A fuzzy distributed expert system to support command and control activities in naval environment. *European Journal of Operations Research,* 145:343–362.

Simões-Marques, M.J., R.A. Ribeiro and A. Gameiro-Marques. 2000. A fuzzy decision support system for equipment repair under battle conditions. *Fuzzy Sets and Systems* 115(1):141–157.

Turban, E., J.E. Aronson and T.P. Liang. 2007. *Decision Support Systems and Intelligent Systems,* 7th edn, Prentice-Hall of India, Inc., New Delhi.

25 Certain Ergonomic Considerations and Design Solutions Connected with the Safety and Comfort of City Buses

Aleksandar Zunjic, Svetozar Sofijanic, and Evica Stojiljkovic

CONTENTS

25.1 INTRODUCTION

A certain number of studies from the anthropometric viewpoint was previously published in relation to the design of the workplace of a bus driver. But, it is very difficult to find research that has been conducted in relation to the anthropometric designing of the interior of buses for urban transport. However, the studies that have been recently published in the scientific literature (Albertsson and Falkmer 2005; Halpern et al. 2005; Palacio et al. 2009; Zunjic et al. 2012), which are related to the injuries of passengers during collision and noncollision situations involving buses in urban city areas indicate the importance of the aforementioned type of research. From the analysis of ways of being injured and types of injuries, it can be concluded that one of the reasons for injury was the incompatibility of the designed equipment, which is an integral part of the interior of a bus, with the anthropometric characteristics of passengers. Research that will be presented here from the viewpoint of anthropometric designing deals with the problem of the safety and comfort of passengers in city buses.

25.2 INTRODUCTORY THEORETICAL METHODOLOGICAL CONSIDERATION

Of particular importance is the determination of proper anthropometric dimensions, which should be taken into account when designing the mentioned buses. In order to select the anthropometric variables that are relevant for the design of city buses, the comprehensive literature dealing with various problems in the field of anthropometry was examined (Dewangan et al. 2008; DOD 1991; Gordon et al. 2013; Kroemer and Grandjean 1997; NASA 1978a,b; Parkinson and Reed 2010; Pheasant and Haslegrave 2005; Robinette et al. 2002; Roebuck 1995; Wagner et al. 1996). Based on the analysis of the mentioned literature, an initial set of 132 anthropometric dimensions was chosen for the additional analysis. This set of anthropometric variables was selected on the basis of appreciation of several criteria. Firstly, it was necessary that this initial set includes all body regions. Besides, it was necessary that the initial set includes the wider range of anthropometric dimensions. Using this approach, the probability for omitting some anthropometric dimension that is of importance for designing is reduced.

Each of the selected dimensions was additionally analyzed, in terms of its significance for the designing of some interior part of the bus. In connection with this, each of these 132 anthropometric dimensions was also classified as a dimension that is important for designing from the aspect of passengers, as a dimension that is potentially important, as a dimension of little importance, or as a dimension of marginal importance. On account of this analysis, a certain number of relevant anthropometric dimensions were selected for the designing of the interior part of buses, which are intended for the transport of passengers in the sitting and standing position.

The classification of the chosen anthropometric dimensions (Zunjic et al. 2014) was performed in the following way. In the group "of importance" were classified the anthropometric dimensions that almost without exception should be included, if we want to achieve a design solution of proper quality for the interior of a city bus. Each individual dimension from this group of anthropometric dimensions has a specific purpose in connection with designing (positioning) the equipment from an ergonomic standpoint. The anthropometric dimensions that we should take into account when we want to achieve a new or specific design solution with an increased level of comfort and safety are classified in the group "of potential importance." This group includes also the dimensions that should be considered in the case when we want to specify a more precise position which is required for the accommodation of the equipment, as well as when in the case of a certain user population, some of the anthropometric dimensions from the group "of importance" do not appropriately reflect the situation for which it is necessary to accommodate the design solution. In addition to the group "of importance" when designing the interior of the bus, this group of anthropometric dimensions should certainly be taken into consideration.

The group "of little importance" includes anthropometric variables that less accurately reflect a situation or occurrence that should be taken into consideration when designing, in relation to the dimensions included in any of the above-mentioned two groups. Basically, each of these dimensions has a more precise alternative, which is contained in the group "of importance" or in the group of dimensions "of potential importance." In the best case, some of the anthropometric dimensions from this group could be utilized for very specific solutions. The last group "of marginal importance" primarily includes anthropometric variables that do not have an impact on the design of the interior of the bus, or their use is almost impossible to predict. Many of the anthropometric dimensions from this group also have significantly more precise replacements in all three previously mentioned groups. Due to their characteristics and on account of decreased importance for the designing of the interior of buses, the groups of anthropometric dimensions "of little importance" and "of marginal importance" in the big majority of cases can be omitted from consideration when designing the interior of buses for public transport of passengers.

In the lists that follow, the classification of the initial 132 anthropometric dimensions into the aforementioned four groups has been performed, according to their importance for designing the interior of city buses:

a. *The group of importance*
   • Stature
   • Chest depth

- Foot length
- Vertical grip reach
- Maximum body breadth
- Popliteal height
- Knee height, sitting
- Elbow rest height
- Sitting height
- Buttock–knee length
- Buttock–popliteal length
- Hip breadth, sitting
- Forearm–forearm breadth
- Forward grip reach, sitting
- Vertical grip reach, sitting
- Hand length

b. *The group of potential importance*
  - Waist depth
  - Acromial height
  - Elbow–fingertip length
  - Acromial height, sitting
  - Cervical height, sitting
  - Bideltoid breadth
  - Functional leg length

c. *The group of little importance*
  - Abdominal extension depth, sitting
  - Lateral femoral epicondile height
  - Scye depth
  - Buttock depth
  - Waist back length (natural indentation)
  - Waist back length (omphalion)
  - Interscye I
  - Interscye II
  - Knee height, midpatella
  - Radiale–stylion length
  - Sleeve outseam
  - Shoulder–elbow length
  - Span
  - Acromion–radiale length
  - Hip breadth
  - Wrist-wall length
  - Wrist-wall length, extended
  - Overhead fingertip reach, sitting

d. *The group of marginal importance*
  - Biacromial breadth
  - Eye height, sitting
  - Thigh clearance
  - Waist height, sitting (natural indentation)
  - Wrist height
  - Waist height sitting (omphalion)
  - Axilla height
  - Bustpoint/thelion–bustpoint/thelion breadth
  - Suprasternale height

- Waist height (natural indentation)
- Waist height (omphalion)
- Wrist height, sitting
- Bispinous breadth
- Chest height
- Crotch length (natural indentation)
- Crotch length (omphalion)
- Iliocristale height
- Neck–bustpoint/thelion length
- Tenth rib height
- Waist breadth
- Calf height
- Cervical height
- Chest breadth
- Shoulder length
- Strap length
- Waist front length (natural indentation)
- Waist front length (omphalion)
- Waist (natural indentation) to waist (omphalion) length
- Axillary arm circumference
- Buttock circumference
- Calf circumference
- Chest circumference
- Chest circumference at scye
- Chest circumference below breast
- Elbow circumference
- Knee circumference
- Lower thigh circumference
- Neck circumference
- Neck circumference, base
- Overhead fingertip reach, extended
- Scye circumference
- Shoulder circumference
- Thigh circumference
- Waist circumference (natural indentation)
- Waist circumference (omphalion)
- Wrist circumference
- Biceps circumference
- Buttock height
- Crotch height
- Forearm circumference, flexed
- Trochanteric height
- Vertical trunk circumference (ASCC)
- Vertical trunk circumference (USA)
- Waist–hip length
- Crotch length, posterior (natural indentation)
- Crotch length, posterior (omphalion)
- Ear length
- Ear length above tragion
- Ear protrusion
- Sleeve length, spine–elbow

- Sleeve length, spine–scye
- Sleeve length, spine–wrist
- Bitragion chin arc
- Bitragion coronal arc
- Bitragion crinion arc
- Bitragion frontal arc
- Bitragion submandibular arc
- Bitragion subnasale arc
- Bizygomatic breadth
- Ear breadth
- Head breadth
- Head circumference
- Head length
- Interpupillary breadth
- Menton–sellion length
- Ankle circumference
- Ball of foot circumference
- Ball of foot length
- Bimalleolar breadth
- Foot breadth, horizontal
- Hand breadth
- Hand circumference
- Heel breadth
- Lateral malleolus height
- Thumb breadth
- Wrist–center of grip length
- Wrist–index finger length
- Wrist–thumbtip length
- Gluteal furrow height
- Midshoulder height, sitting
- Neck height, lateral

25.3 ADVANCING THE METHODOLOGY

Anthropometric dimensions that are presented in the lists (a–d) are dimensions that can be found in the literature. However, sole use of the stated dimensions cannot completely provide the comfort and safety of passengers. In this regard, as a result of observation of body positions and behavior of passengers during use of buses for public transportation, it was noted that the need for determining additional anthropometric dimensions exists. As a consequence of appreciation of that need, new anthropometric dimensions were created based on the observation of passengers in the standing and sitting positions, in different conditions of transport (when the bus has few passengers, and especially in situations when a big crowd was registered).

The majority of the anthropometric dimensions that are listed above are related to some maximum distance. For instance, knowing the dimension overhead grip reach is necessary, but not sufficient if we want to provide adequate passenger comfort during transportation. One cannot expect that a passenger holds a hand in the position of maximum reach during the entire transportation, and assume that he is completely satisfied with this way of transportation. In this regard, the passengers primarily expect handrails that are located in positions that are most comfortable for them to grip and hold during an extended period of time.

As a result of the large number of interactions with passengers who use buses for urban transport, as well as conducted interviews, pilot surveys, and testing, certain new anthropometric dimensions have been established (Zunjic et al. 2014), which are called comfortable anthropometric dimensions. The purpose of these anthropometric dimensions is, first of all, to enable comfort, and also the safe transportation of passengers, when we take these dimensions into consideration in the case of designing the individual segments of the bus interior. To the group of comfortable anthropometric dimensions belong:

- Universal comfortable grip height
- Overhead comfortable grip height
- Below head comfortable grip height
- Universal comfortable grip length
- Overhead comfortable grip length
- Below head comfortable grip length
- Comfortable grip height (sitting)
- Comfortable grip length (sitting)

In addition to the comfortable anthropometric dimensions, several anthropometric variables also exist that we could not find in the previously reviewed literature. However, their measurements are also important because they contribute to the design of the interior of buses on the principles of safety and comfort. These anthropometric dimensions are

- Height of forward grip reach (sitting)
- Minimum grip height
- Lower thoracic height
- Upper thoracic height

However, their measurement is also significant because it contributes to the design of the interior of buses on the principles of safety and comfort. The first group of importance for designing involves the following anthropometric dimensions:

- Universal comfortable grip height
- Overhead comfortable grip height
- Below head comfortable grip height
- Height of forward grip reach (sitting)

The second group of potential importance for designing includes the following anthropometric dimensions:

- Universal comfortable grip length
- Overhead comfortable grip length
- Below head comfortable grip length
- Comfortable grip height (sitting)
- Comfortable grip length (sitting)
- Minimum grip height
- Lower thoracic height
- Upper thoracic height

In the following, all anthropometric measures (from the literature and the new ones), from groups of importance for designing and of potential importance for designing will be brought in connection with the design of a specific part of the bus interior.

25.3.1 ANTHROPOMETRIC DIMENSIONS THAT ARE OF IMPORTANCE FOR DESIGNING THE BUS

Below are shown the aspects of the application (functions) of the anthropometric dimensions that are of importance for the design of the interior space of a bus, which is intended for urban passenger transport.

Stature
- Vertical clearance required for accommodation

Chest depth
- Clearance between seat backs and obstructions
- Determining the number of people who can stand in a passage
- Planning the total number of passengers

Foot length
- Predicting the necessary space for the feet (sitting position)
- Determining the width of the passage
- Determining the length of the stair

Vertical grip reach
- Determining the maximum height of the handrail

Universal comfortable grip height
- Defines the most comfortable position in the vertical direction where to provide the handrail

Overhead comfortable grip height
- Defines the most comfortable position where to provide the handrail above the top of the head

Below head comfortable grip height
- The most comfortable position where to provide the handrail below the head

Maximum body breadth
- Determining the width of the bus doors
- Determining the width of the aisle
- Planning the total number of passengers

Popliteal height
- Defines the maximum acceptable height of the seat pan

Knee height, sitting
- Determines the position that is not to be reached by the chair in front of a passenger

Elbow rest height
- Defines the position of the top of the arm rest

Sitting height
- Determines the minimum height for the accommodation of a passenger in the sitting position

Buttock–knee length
- Determines the horizontal clearance from the seat back rest to accommodate the upper leg

Buttock–popliteal length
- Defines the maximum acceptable seat pan depth

Hip breadth, sitting
- Determines the minimum width of the seat

Forearm–forearm breadth
- Determines the width for positioning of the chair armrests

Forward grip reach, sitting
- Defines the maximal distance from the back to the position of the handrail in front of the passenger

Height of forward grip reach, sitting
- Determines the maximum distance from the floor at which the handrail can be placed in front of the passenger

Vertical grip reach, sitting
- Defines the maximum height on which the handrail can be placed for the sitting position

Hand length
- Can be used when determining the diameter of handrails

Note: If a subject positions the hand above the head during determination of the universal comfortable grip height, then it is not necessary to additionally measure the dimension overhead comfortable grip height, because these two dimensions in this case coincide. However, if a subject positions the hand below the level of the head during determination of the universal comfortable grip height, then it is not necessary to measure additionally the dimension below head comfortable grip height, because these two dimensions in this case coincide.

25.3.2   ANTHROPOMETRIC DIMENSIONS THAT ARE OF POTENTIAL IMPORTANCE FOR DESIGNING THE BUS

Below are shown the aspects of the application of the anthropometric dimensions that are of potential importance for the design of the interior space of a bus, which is intended for urban passenger transport.

Waist depth
- Can be used for determining the width of a passage
- Planning space for seating
- Determining the maximum number of people that can be accommodated on the bus

Minimum grip height
- Defines the lowest position where it is possible to position the handrail for the standing position

Acromial height
- Can be used for planning the highest point for positioning the handrail, if a higher level of comfort during holding the handrail is expected

Universal comfortable grip length
- Can be used for planning the necessary distance of a passenger to the handrail

Overhead comfortable grip length
- Can be used for planning the necessary distance from the back to the handrail, which is positioned above the top of the head

Below head comfortable grip length
- Can be used for planning the necessary distance from the back to the handrail, which is positioned below the head

Elbow–fingertip length
- Can be used for determining the length of the armrests

Acromial height, sitting
- Can be used for planning the location of the handrail that a passenger uses from the sitting position

Cervical height, sitting
- Can be used for planning the position of the head restraint of certain versions of chairs

Bideltoid breadth
- Determining the width of the backrest of the chair

Functional leg length
- Can be used for planning the legroom of passengers seated

Comfortable grip height, sitting
- Can be used for determination of the vertical location of a handrail in front of a passenger

Comfortable grip length, sitting
- Can be used for determination of the horizontal location of a handrail in front of a passenger

Lower thoracic height
- Can be used for designing the backrest of the chair in accordance with the shape of the spine

Upper thoracic height
- Can be used for designing the backrest of the chair in accordance with the shape of the spine

25.4   CERTAIN PRACTICAL PROBLEMS OF THE COMFORT AND SAFETY

The results of the mentioned pilot research and experience related to the solving of the problem of transport that is connected with the city buses point to the importance of the availability of handrails for passengers. If the handrail is not accessible to a passenger at every moment, the unsafe state occurs that can cause injuries to the passengers, due to a speed change or change of a course of a bus. Passenger safety directly depends on the availability of handrails, because their usage can prevent the incidence of injuries or to reduce them significantly. The central part of this practical research directly focuses on the placement of handrails from the ergonomic aspect. When motion through the bus is hampered due to the great number of passengers, very frequent is the case that some passengers cannot reach the handrail in such situation. In this way, it has endangered not just their safety, but also the safety of other passengers. Besides the aspect of safety, city buses for the transportation of passengers also need to fulfill determined conditions of comfort, in relation to the positioning of the handrails.

If one designs the interior of city buses by placing the handrails in positions (heights) that cannot be reached by all passengers, then the safety of certain passengers is certainly endangered. In the following will be considered the problem of positioning of handrails located above the passenger's head on the bus. This also encompasses handrails that are positioned at a maximum height on the bus, as shown in Figure 25.1.

One of the rare documents that deals with the constructional aspects of the interior of buses is the E/ECE/324-E/ECE/TRANS/505 recommendation. This recommendation is in international use, and numerous bus producers use it as a reference for designing city buses. However, the E/ECE/TRANS/505 recommendation specifies only the interval of heights where the handrail should be located. According to this recommendation, a handrail should be set in the range from 80 to 190 cm. However, there are no precise data relating to the positioning of the handrail in height, in a way that does not reduce the comfort of passengers. The question about the height limit (specified in the recommendation) for a handrail is also open. The highest height of the handrail is primarily connected with the safety aspect of designing of city buses.

25.5   APPLICATION OF THE METHODOLOGY TO THE SPECIFIED PROBLEM

The solutions of previously described problems require the application of the developed methodology, which is previously described. For solving the safety problem, it is necessary to take into the consideration a known anthropometric dimension called the vertical grip reach. This dimension is shown in Figure 25.2. If the anthropometric dimension the vertical grip reach is smaller than the height that corresponds to the position of the handrail in the bus, it is clear that a handrail is then

FIGURE 25.1   Position of the highest handrail on a city bus.

FIGURE 25.2   Vertical grip reach.

inaccessible to a user. If the other handrail is also not available to such a passenger (which is not uncommon in practice when there is a crowd in a bus), the user will not be able to maintain balance in the event of sudden and intense changes in the acceleration of a bus.

For solving the problem of comfort, it is necessary to take into consideration the anthropometric dimension that has the name overhead comfortable grip height. This dimension is shown in Figure 25.3. It determines the most comfortable position where to provide the handrail above the top of the head. If the measurement of this dimension is conducted under laboratory conditions, before measuring a subject receives instructions to imagine that he/she is located in a bus for public transportation of passengers and to put the hand (in which he/she holds the cylinder

FIGURE 25.3 Overhead comfortable grip height.

that simulates the handrail) into the position above the top of the head where he/she considers most appropriate (in terms of minimum fatigue over a longer period and maintaining stability). Then the distance from the floor to the center of the cylinder that he/she holds in the hand should be measured. If a handrail is available to the passenger at the position corresponding to the height of the overhead comfortable grip height, it can be expected that the user will be satisfied due to the reduced fatigue during transportation, while safety also will be at the appropriate level.

Research conducted to determine the position at which height the handrail should be placed on the bus is not known. The above-mentioned recommendation E/ECE/TRANS/505 also does not point to the research, on the basis of which such recommendation was made. With that in mind, as well as the practical need for the realization of conditions of comfort and safety of passengers on buses, research was carried out in order to determine the aforementioned two anthropometric dimensions. Measurement of these anthropometric dimensions can enable the evaluation of justification of the values given in E/ECE/TRANS/505 recommendation, regarding the positioning of the handrail by height (interval from 80 to 190 cm). However, the research described below is primarily focused on the assessment of the upper limit of the aforementioned interval. This means that the objective is determining the positions of handrails that are located in the highest position on the bus (as in Figure 25.1), as well as those which are intended to be primarily placed above the heads of most travelers.

The research was carried out in the city of Belgrade (Serbia). The study included 500 subjects, 350 males, and 150 females. The average age of the subjects was 33.2 years (standard deviation 15.4 years). The age range was in the interval of 18–68 years. All subjects were current users of the system of public transport, or they had used buses for the public transportation of passengers in the past. For measuring, mechanical anthropometric measuring devices were used. One part of the measurement was conducted in laboratory conditions. The second part of the measurement was done on the ground, in locations that enabled accurate measurement.

25.6   RESULTS AND ANALYSIS OF RESULTS

The results primarily related to the anthropometric dimensions of the vertical grip reach and overhead comfortable grip height are presented and analyzed below. In addition to these two anthropometric dimensions, the anthropometric dimension stature was measured and shown. It will enable the performing of additional comparative analyzes. Table 25.1 provides basic statistical values that describe research results which relate to the sample of 350 male persons. Results are related to values without shoes.

Table 25.2 contains the basic statistical values that describe research results which relate to the sample of 150 female persons. The results are also related to values without shoes.

Some analyzes that need to be performed, as in the case of determination of percentiles, require the existence of a normal distribution of the observed dimensions, within a selected sample. For this reason, it is necessary to test the existence of the normal distribution of dimensions stature, vertical grip reach, and overhead comfortable grip height.

The normality test of the distribution for the dimension stature, for the sample of female persons, was performed using the Shapiro–Wilk test. The calculated value of the test statistic is W-statistic $= 0.992$ ($P = 0.583$). With regard to the result obtained, it can be considered that the data matches the pattern that is expected for the population with a normal distribution.

The normality test of the distribution for the dimension vertical grip reach, for the sample of female persons, was performed using the Shapiro–Wilk test. The calculated value of the test statistic is W-statistic $= 0.991$ ($P = 0.468$). With regard to the result obtained, it can be considered that the data matches the pattern that is expected for the population with a normal distribution.

The normality test of the distribution for the dimension overhead comfortable grip height, for the sample of female persons was also performed using the Shapiro–Wilk test. The calculated value of the test statistic is W-statistic $= 0.987$ ($P = 0.186$). With regard to the result obtained, it can be

TABLE 25.1

Basic Statistical Parameters of the Measured Anthropometric Dimensions, Which Refer to the Sample of Male Persons

Parameter	Stature (cm)	Vertical Grip Reach (cm)	Overhead Comfortable Grip Height (cm)
Mean	176.4	209.5	185.9
SD	8.3	12.0	8.4
Max.	203.0	247.5	209.5
Min.	148.5	169.5	155.5
Range	54.5	78.0	54.0

TABLE 25.2

Basic Statistical Parameters of the Measured Anthropometric Dimensions, Which Refer to the Sample of Female Persons

Parameter	Stature (cm)	Vertical Grip Reach (cm)	Overhead Comfortable Grip Height (cm)
Mean	165.4	191.4	175.9
SD	6.7	7.8	9.2
Max.	181.0	209.5	198.5
Min.	147.5	170.5	154.5
Range	33.5	39.0	44.0

considered that the data also matches the pattern that is expected for the population with a normal distribution.

However, the application of the Shapiro–Wilk test on all three measured anthropometric dimensions for the sample of male persons has shown that there is no normal distribution. The normality test for the stature, for the sample of male and female persons (the total), was performed using the Shapiro–Wilk test. The calculated value of the test statistic is W-statistic $= 0.995$ ($P = 0.107$). With regard to the result obtained, it can be considered that in this case the data matches the pattern that is expected for the population with a normal distribution. However, the application of the Shapiro–Wilk test to the sample of male and female persons on the other two anthropometric dimensions (vertical grip reach and overhead comfortable grip height) showed that there is no normal distribution.

Bearing in mind the results of testing of the existence of a normal distribution, it can be concluded that it is justified in this situation to calculate percentiles for all three anthropometric dimensions, for the sample containing female persons. In addition, the determination of the percentiles for the stature dimension may be justified, which relates to both samples (men and women together). In Table 25.3, the percentiles for the mentioned anthropometric dimensions are presented.

It is necessary to establish whether a statistically significant difference exists between the dimensions of the vertical grip reach and the overhead comfortable grip height. If between these two anthropometric dimensions, there is no statistically significant difference, it would mean that it is possible to set a handrail at the height corresponding to the height of the vertical grip reach and expect that this position will also be comfortable for passengers.

TABLE 25.3

Percentiles (in cm) for the Dimensions of Stature, Vertical Grip Reach, and Overhead Comfortable Grip Height

Percentiles	Stature (Women)	Vertical Grip Reach (Women)	Overhead Comfortable Grip Height (Women)	Stature (Men and Women)
1st	149.79	173.03	154.56	151.30
2.5th	152.27	175.94	157.96	154.76
5th	154.41	178.46	160.91	157.76
10th	156.82	181.29	164.22	161.13
15th	158.43	183.19	166.43	163.37
20th	159.77	184.76	168.27	165.24
25th	160.91	186.09	169.83	166.83
30th	161.92	187.28	171.21	168.24
35th	162.79	188.30	172.41	169.45
40th	163.73	189.40	173.69	170.76
45th	164.59	190.43	174.89	171.98
50th	165.40	191.37	175.99	173.10
55th	166.20	192.32	177.09	174.23
60th	167.08	193.34	178.29	175.44
65th	168.01	194.44	179.58	176.75
70th	168.88	195.47	180.78	177.97
75th	169.89	196.65	182.16	179.37
80th	171.03	197.99	183.72	180.96
85th	172.37	199.56	185.56	182.84
90th	173.98	201.45	187.77	185.08
95th	176.39	204.28	191.08	188.45
97.5th	178.53	206.80	194.02	191.44
99th	181.01	209.72	197.42	194.91

In this connection, the existence of the statistically significant differences between the vertical grip reach and the overhead comfortable grip height for women was tested first. The paired t-test was used for this analysis. The calculated value of the test statistics is $t = 28.008$. Accordingly, we conclude that the length of vertical grip reach is greater than the overhead comfortable grip height for women ($\alpha = 0.05$).

The existence of a statistically significant difference between the vertical grip reach and the overhead comfortable grip height for men was then tested. Since it was previously determined that there is no normal distribution of these two anthropometric dimensions in the male sample, the paired t-test will not be used for testing . In this instance, the Wilcoxon signed rank test was applied. Statistical parameters of the test are $W = -61046.0$, $T+ =14.5$, $T = -61060.5$. Since the Z-statistic (based on positive ranks) $= -16.187$, it is confirmed that a statistically significant difference exists.

It is also necessary to determine whether there is a statistically significant difference between the vertical grip reach and the overhead comfortable grip height on the level of the entire sample (men and women together). Since it was previously determined that there is no normal distribution of these two anthropometric dimensions in the mixed sample, the Wilcoxon signed rank test was applied for testing. Statistical parameters of the test are $W = -124579.0$, $T+ =85.5$, $T = -124664.5$. Since the Z-statistic (based on positive ranks) $= -19.332$, it is confirmed that a statistically significant difference exists.

Given that in all the tested cases a statistically significant difference between the dimensions of the vertical grip reach and the overhead comfortable grip height has been revealed, we can conclude that these two anthropometric dimensions differ in size. Height corresponding to the vertical grip reach is greater than the height that corresponds to the overhead comfortable grip height, which means that the positioning of the handrail on the position of the vertical grip reach decreases passenger comfort.

Further, it is necessary to establish whether a certain correlation between the new anthropometric dimension of the overhead comfortable grip height and some other anthropometric dimension exists. In the anthropometry, the selected anthropometric dimension compares almost without exception with the stature (as the basic dimension). Considering this, we will try to establish whether there is a certain dependency between the anthropometric dimensions of the overhead comfortable grip height and the stature. Figure 25.4 shows the relationship between the aforementioned two dimensions, for the sample of persons of the female gender.

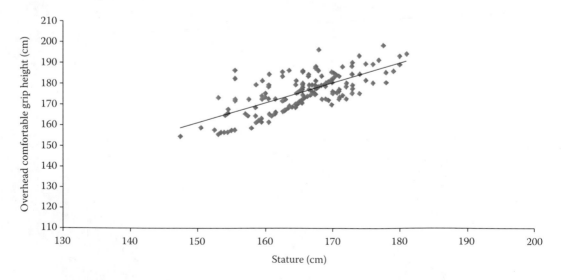

FIGURE 25.4 Graphical representation of dependency between anthropometric dimensions the overhead comfortable grip height and the stature, for the sample of persons of the female gender.

The magnitude (intensity) of dependency between these two dimensions can be determined using the Pearson's correlation coefficient. For the sample of persons of the female gender, this coefficient is $R = 0.703$. Although this value of Pearson's correlation coefficient deviates from 1, it still indicates the existence of a certain correlation between the dimensions that are compared (which can be regarded as statistically significant). Figure 25.5 shows the relationship between the aforementioned two dimensions, for the sample of persons of the male gender.

For the sample of persons of the male gender, the Pearson's correlation coefficient is $R = 0.776$. This value of the correlation coefficient indicates to the somewhat higher connectedness between the observed dimensions, compared to the sample of persons of the female gender. However, in both cases, a deviation exists from the absolute connectedness (when $R = 1$). The reason for this deviation probably arises from the nature of the dimension of the overhead comfortable grip height. As already mentioned, this dimension belongs to the group of comfortable anthropometric dimensions. As a result, it can be assumed that for the dimension of the overhead comfortable grip height there is a relatively narrow interval of values, for which an approximately equal probability to be selected by the subject exists (due to the existence of identical, or nearly the same comfort). Therefore, variations of this kind are likely the reason why the correlation coefficients deviate from 1.

However, for designers, the connectedness that exists between the two anthropometric dimensions is often important from the practical point of view. If a designer does not have information about a certain anthropometric dimension (e.g., if he did not perform its measurement for any reason), it is important to know the value of the respective other anthropometric dimension and the functional connection between the requested and the known dimension. In this way, the designer is able to determine the value of the missing anthropometric dimension.

It can be seen from Figures 25.3 and 25.4 that there is a potential linear relationship between the dimensions of the overhead comfortable grip height and the stature. In the case of the sample of female persons, based on the application of linear regression, the functional dependence of the form $y = 0.962x + 16.7$ is obtained. In the case of the sample of male persons, based on the application of linear regression, the functional dependence of the form $y = 0.784x + 47.52$ is obtained. In these equations, y is the dimension of the overhead comfortable grip height, and x refers to the stature. By the applying of the functional dependencies of this form, designers are able to determine the dimension of the overhead comfortable grip height, based on the knowledge of the dimension of the

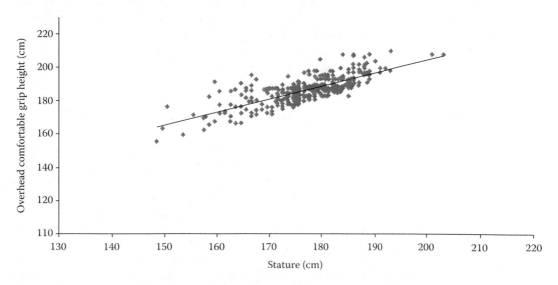

FIGURE 25.5 Graphical representation of dependency between anthropometric dimensions the overhead comfortable grip height and the stature, for the sample of persons of the male gender.

stature. However, in this case, one should bear in mind the limitation, which is associated with the specified correlation coefficients.

Now the safety aspect of the transport will be taken into consideration, which is associated with the availability of the handrail that has the highest position in the bus. As previously mentioned, the E/ECE/TRANS/505 recommendation specifies a height of 190 cm as the maximum position at which a handrail should be set. If we set a handrail at a height of 190 cm, from Table 25.3 (column for the dimension of the vertical grip reach), we can see that up to the 40th percentile this handrail will not be available to female passengers. The situation, however improves, if we predict an addition for the height of shoes. For women, in the average this amounts to 4.5 cm (Pheasant and Haslegrave 2005). When we apply this addition, on the basis of the aforementioned column of Table 25.3, it can be noticed that up to the 20th percentile the handrail on the mentioned position will also not be accessible for female passengers.

Certainly, a better situation can be expected in the case of a sample of males. However, given that the existence of a normal distribution was not confirmed in the male sample, percentage analysis will be used instead of an analysis based on the use of percentiles. In this way, taking into account the basic data relating to the vertical grip reach, the information was obtained that 6.57% of males in the sample could not reach a handrail that is set at a height of 190 cm. The situation becomes more favorable if we take into account the addition for the height of shoes. For men, in average, it amounts to 2.5 cm (Pheasant and Haslegrave 2005). Taking into account this addition for shoes, it is obtained the information that 5.14% of men in the sample could not reach a handrail positioned at a height of 190 cm.

25.7 CONCLUSION

Anthropometric dimensions that are presented constitute the basis for the design of individual elements of the interior of buses on ergonomic grounds. Considering the group of comfortable anthropometric dimensions will allow the design solution to be adequately aligned with the anthropometric characteristics of passengers. The focus of the research that was presented here has been placed on the assessment of the viability of positioning of the highest handrail on a bus for urban transport of passengers on the location that corresponds to the upper limit for positioning of the handrail of 190 cm, in accordance with the international recommendation E/ECE/TRANS/505. The issue of comfort regarding the positioning of handrails, which are placed over the heads of the passengers, has also been considered.

The results of this research show that the positioning of handrails at the upper limit of 190 cm should be avoided. Although for most of the men this height corresponds, for a large part of the population of women this height is not appropriate. The reason for that is because the population of women up to the 20th percentile has the dimension of the vertical grip reach smaller than 190 cm (taking into account the addition in height for women's shoes). If we do not take into account the addition for women's shoes, the women until the 40th percentile cannot reach the handrail. In this way, their safety is reduced.

Research has also shown that the overhead comfortable grip height greatly depends on the height of the subjects. Design solutions that are based on the application of this finding should take this fact into account in the proper way. Also, the anthropometric dimension of the overhead comfortable grip height is generally smaller than the dimension of the vertical grip reach. This is another reason why the positioning of handrails at the level of 190 cm should be avoided.

In order to achieve integrated improvement of the comfort and safety of city buses, one of the solutions is setting up the handrails to the position of the mean value of the dimension of the overhead comfortable grip height. It amounts to 176 cm for women. In this way, the comfort of passengers would be increased, and also the safety. The reason for that is because all women can reach this height (if one takes into account the addition for shoes). However, for passengers who are above average height, the aforementioned value of the height where a handrail would be positioned can be a problem in certain situations. Therefore, as an optimal solution the height of 183 cm can be proposed, as the upper limit for the positioning of a handrail. This value is almost identical to the average value of the dimension of the overhead comfortable grip height for men (when taking into account the addition for shoes). However, this height is accessible even to women of the 5th percentile (when taking into account the addition for footwear for women). From the anthropometric point of view, this is an acceptable solution. It can be expected that the implementation of these ergonomic solutions in the design of city buses will reduce the number of injuries during transportation, and also lead to the increase of the comfort of passengers.

REFERENCES

Albertsson, P. and T. Falkmer. 2005. Is there a pattern in European bus and coach incidents? A literature analysis with special focus on injury causation and injury mechanisms. *Accident Analysis and Prevention* 37(2):225–33.

Dewangan, K.N., Owary, C., and R.K. Datta. 2008. Anthropometric data of female farm workers from north eastern India and design of hand tools of the hilly region. *International Journal of Industrial Ergonomics* 38(1):90–100.

DOD. 1991. *Anthropometry of U.S. Military Personnel (DOD-HDBK-743A)*. Natick, MA: Department of Defense.

E/ECE/324-E/ECE/TRANS/505. 2002. Uniform provisions concerning the approval of large passenger vehicles with regard to their general construction. Rev.1/Add.35/Rev.2, Regulation No. 36. United Nations.

Gordon, C.C., Blackwell, C.L., Bradtmiller, B., Parham, J.L., Hotzman, J., Paquette, S.P., Corner, B.D., and B.M. Hodge. 2013. *2010 Anthropometric Survey of U.S. Marine Corps Personnel: Methods and Summary Statistics*. Natick, MA: Natick Soldier Research, Development and Engineering Center. http://oai.dtic.mil/oai/oai?verb=getRecord&metadataPrefix=html&identifier=ADA581918.

Halpern, P., Siebzehner, M.I., Aladgem, D., Sorkine P., and R. Bechar. 2005. Non-collision injuries in public buses: A national survey of a neglected problem. *Emergency Medicine Journal* 22(2):108–10.

Kroemer, K.H.E. and E. Grandjean. 1997. *Fitting the Task to the Human: A Textbook of Occupational Ergonomics*. Boca Raton, Florida: CRC Press/Taylor & Francis.

NASA. 1978a. *Anthropometric Source Book Volume I*: Anthropometry for Designers (NASA RP-1024). Houston, Texas: National Aeronautics and Space Administration.

NASA. 1978b. *Anthropometric Source Book Volume II: A Handbook of Anthropometric Data (NASA RP-1024)*. Houston, Texas: National Aeronautics and Space Administration.

Palacio, A., Tamburro, G., O'Neill, D., and C.K. Simms. 2009. Noncollision injuries in urban buses—Strategies for prevention. *Accident Analysis and Prevention* 41(1):1–9.

Parkinson, M.B. and M.P. Reed. 2010. Creating virtual user populations by analysis of anthropometric data. *International Journal of Industrial Ergonomics* 40(1):106–11.

Pheasant, S. and C.M. Haslegrave. 2005. *Bodyspace—Anthropometry, Ergonomics and the Design of Work*. Boca Raton, Florida: CRC Press/Taylor & Francis.

Robinette, K.M., Blackwell, S., Daanen, H., Boehmer, M., Fleming, S., Brill, T., Hoeferlin, D., and D. Burnsides. 2002. Civilian American and European Surface Anthropometry Resource (CAESAR) Final Report, Volume I: Summary (AFRL-HE-WP-TR-2002-0169), Wright-Patterson AFB, OH: Human Effectiveness Directorate Crew System Interface Division. http://www.humanics-es.com/CAESARvol1.pdf.

Roebuck, J.A. 1995. *Anthropometric Methods: Designing to Fit the Human Body*. Santa Monica, California: Human Factors and Ergonomics Society.

Wagner, D., Birt, J.A., Snyder, M.D., and J.P. Duncanson. 1996. *Human Factors Design Guide (HFDG) for Acquisition of Commercial Off-The-Shelf Subsystems, Non-Developmental Items, and Developmental Systems (DOT/FAA/CT-96/1)*. Atlantic City, New Jersey: Federal Aviation Administration.

Zunjic, A., Sofijanic, S., and E. Stojiljkovic. 2014. Anthropometric consideration of interior design of city buses. In Rebelo, F. and Soares, M. (eds.), *Advances in Ergonomics in Design, Usability & Special Populations Part II*, pp. 95–106. Danvers, Massachusetts: AHFE Conference.

Zunjic, A., Sremcevic, V., Sijacki Zeravcic, V., and A. Sijacki. 2012. Research of injuries of passengers in city buses as a consequence of non-collision effects. *Work* 41:4943–4950.

26 School Supplies Transportation System

An Ergonomic Approach between Two Distinct Realities

Ricardo Dagge and Ernesto Filgueiras

CONTENTS

26.1 INTRODUCTION

In the beginning of each new school year, teachers, health care professionals, and parents highlight the considerable amount of growth stage children reporting back pain and musculoskeletal disorders, pointing out the use of heavy scholar backpacks as one of the main contributors to this fact. This highlights the fact that backpack users still struggle to transport their school supplies in an efficient manner. This constant struggle may be related to extremely rapid market demands that tend to lead to scholar backpacks more geared to aesthetic appeal, disregarding design and research aspects that are essential to this product functionality (Dagge and Filgueiras, 2014).

Goodgold et al. (2002) and Ramprasad et al. (2010) assign the appearance of musculoskeletal disorders, and back pain in growth stage children, to the considerable amount of transported weight in the posterior area of their body.

Goodgold et al. (2002), studied behavioral motor patterns while transporting a backpack, and they proved that the body mass center shifts while transporting weight in the posterior area of the body. This shift generates an opposite unconscious reaction by the backpack's user that involves leaning his body into the opposite direction of the transported weight, in order to compensate for it. Goodgold et al. (2002) claim that this compensation may result in fatigue of the soft tissues and ultimately can lead to postural deformities.

Considered as a serious information gap, in the early stages of product development process, the non-observation of human factors and ergonomical aspects, may result in user's hazards. Dejean and Naël (2007) even claim that when product purposes are not satisfied, or are badly integrated accordingly to its user aims, product development could put at risk the safety of the user himself. Therefore, it is almost mandatory, that designers and development teams spend a considerable amount of time studying physical, cognitive, and sensory aspects that surround the context of use of products (Dagge and Filgueiras, 2014).

In this chapter, the application of a study in a developed European country, namely Portugal, similar to the one applied by Jayaratne (2012) in Sri Lanka, will be presented, with the purpose of understanding if the differences between these two realities, may interfere with ergonomic and human factors involved in the school supplies transportation task. Most of the results gathered with this study will then be compared with the ones achieved by Jayaratne (2012), in order to deepen the knowledge about the context of use surrounding school supplies transportation products, in two distinct economic and social realities.

In his study, conducted in a Sri Lanka district, Jayaratne (2012) had a sample of 1607 children attending 6th–8th school grades, to whom he applied a survey to gather classroom environmental ergonomic factors to understand its influence on their health.

It is expected, that this chapter will highlight whether economic and social differences, found between developing and developed countries, interfere with human and ergonomic factors surrounding the school supplies transportation task.

26.2 FIELD STUDY

26.2.1 METHOD

A survey, and a direct observation strategy, similar to the ones described by Chapanis (1965), were applied with the intent of gathering ergonomic and human factors concerning school the supplies transportation task and related products. Both strategies aimed to get qualitative and quantitative data, for users' interactional and behavioral strategies adopted, while transporting their school supplies to and from school.

This field study was conducted in a Portuguese public educational school, placed in the urban district of Oeiras, that accounts with a total amount of 1062 middle and upper-middle class students, mainly inhabitants of the area, with 491 attending the 3rd Basic Cycle of Education, and therefore falling within this field study population. Accordingly to the national ranking, this school tends to be present among the five best public educational schools in Portugal (DN, 2012).

This field study was only possible to implement after collecting the authorization and informed consents forms by the school principal, students' parents, and caretakers, respectively.

Carried out throughout a week of March 2012, this field study was inserted in 45 min in physical education classes. Its application was conducted in six similar classrooms, all of them with a 30 student capacity, each one with five large windows, allowing the entry of natural light.

26.2.2 SAMPLE

The school board allowed us to contact six classes, divided equally by the grades aimed for this study, two of 7th, two of 8th, and two of 9th grade, to whom informed consent forms were delivered, to be filled by the students' parents.

Of a total amount of 160 consent forms sent, 110 students were authorized to participate in this study. Therefore, the sample featured a total amount of 110 students, 56% of which were female, with an age range from 12 to 15 years old (M = 13.1; SD = 0.92). Within this sample, 38.18% of these were attending 7th grade, 32.73% the 8th grade, and 29.09% the 9th grade.

26.2.3 DATA COLLECTION

As mentioned earlier, this field study consisted of a survey and a direct observation strategy, based on the one described by Chapanis (1965), that had the purpose of gathering ergonomic and human factors data concerning the school supplies transportation task and related products. In order to pre-test both these strategies, they were previously applied to a group of students (Ghiglione and Matalon, 2001).

Based upon the survey applied by Jayaratne (2012), the survey presented to our sample was also composed of 12 main questions, with a set of pre-established answers. Among these, there was a question that figured a standardized Nordic questionnaire adaptation, that was based on the work by Kaewboonchoo et al. (1998), and in which the student was asked to highlight one, or more body areas, in which pain or discomfort occurred.

The themes for the 12 questions approached were

1. Product typology used, for the school supplies transportation task
2. Features, that the user would like to change in his product, intended for school supplies transportation
3. The most valued aspect of the product used
4. People that most likely influenced the buying decision process
5. The most common brands, and concepts associated to these, regarding the product used
6. Situations, outside school, where the same product was used
7. The standardized Nordic questionnaire adaptation, that included the period of the day, in which discomfort and pain emerged
8. A brief clinical history, to help in the distinction between chronic, and product use related, pain and discomfort
9. Quantity, and type of elements, carried inside school supplies transportation products
10. The life-time of the product in use
11. User's point of view, of his/her own habits and strategies adopted, while transporting his/her school supplies
12. And finally, transportation means used to travel to and from school

The direct observation strategy (Chapanis, 1965) was implemented as a complementary method to the survey, and it intended to verify, and further understand the answers gathered.

Survey answers, intended to be verified by the direct observation strategy, were

1. Product typology used in the school supplies transportation task
2. The most common brands and concepts associated to these, regarding the product used
3. Quantity and type of elements, carried inside the school supplies transportation products
4. User's point of view of his/her own habits and strategies adopted, while transporting his/her school supplies

However, the data that initially led to the use of the direct observation strategy intended to gather:

1. Product typology used, in order to classify them into different morphological classes
2. The most common adopted postures and behaviors, that were registered by photographic means, in order to understand user interaction with his/her school supplies transportation product

3. The weight capacity, against the effective weight carried in this kind of product, that involved a weighing of the user with and without the elected product
4. The quantity and typology of the items transported inside the product chosen by the student
5. And finally, the general anthropometrical measures, which were collected by placing a measurement tape on the wall of the observation room, in order to get minimal, mean, and maximal height, of the studied population

26.2.4 STUDY PROCEDURES

In order to apply this field study, physical educational teachers were asked to send a group, with a maximum of 10 students, to a previously arranged classroom. Upon students' arrival, the study aim was explained, and they were asked if anyone refused to participate in the study.

Then, survey application was carried out by the lead investigator, one student at a time. The survey was carried out in a two seat table, so that there would not be any students around listening, and therefore influencing themselves, and the subject under inquiry.

After completion of the survey, the student was asked to move to another classroom area where he/she was weighed, measured, and finally photographed, with and without, his/her school supplies transportation product.

26.3 RESULTS

In order to smooth the reading process, field study results were divided in two major groups: the first one regarding the results gathered with the applied survey and the second one regarding the data collected from the direct observation strategy.

26.3.1 SURVEY RESULTS

Survey results, are divided in three major categories, according to the main theme of each question, were as follows:

1. Product-related questions
2. School supplies transportation products and strategies used in their transport
3. And finally, the negative effects resulting from the school supplies transportation task

26.3.1.1 Product-Related Questions

The surveyed sample was distributed as follows: 82% of the users claim to transport their school supplies in backpacks; 12% stated to use a suitcase; 5% a shoulder bag; and just 1% of the surveyed sample claimed to use a trolley bag, to carry their school supplies to and from school.

When it comes to features, just 19% of the surveyed population stated that they would in fact change something in their product, intended for school supplies transportation. From these 19% (21 students), 42.9% claim that they would change the color of their product, 28.6% would change the shape of the product itself, and finally, 28.5% would like to change the way in which the product interacts with their body.

The surveyed sample revealed an 82% satisfaction with the shape of the product used for their school supplies transportation task, 14% do not consider that a relevant aspect, while 4% of the surveyed sample showed some kind of dissatisfaction with the shape of the product owned.

When it comes to the capacity of the school supplies transportation products, this study showed that while 33% of the students (30) claimed that their product did not have enough capacity to lodge their everyday supplies, most of the surveyed subjects (65%) considered that their product had just the ideal capacity. Despite this fact, it is important to highlight that 2% of the surveyed subjects considered their school supplies transportation product capacity bigger than they needed it to be.

Product-related results also revealed that 39% of the surveyed sample owned a product that had been acquired at the beginning of the current school year, while the same percentage claimed to have bought their product in the last 3 years, leaving 22% of the surveyed subjects stating that they owned a product more than 3 years old.

26.3.1.2 School Supplies Transportation Products, and Strategies Used in Their Transport

Major findings revealed that most of the surveyed sample (76%) tend to use their product only to carry their school supplies to and from school, claiming not to use them outside this routine.

The majority of the sample (55%), claimed to use only a motorized means of transportation, while 38% stated, not to use motorized means at all, while the remaining 7% of the surveyed sample claimed to use a combination of both methods mentioned before.

Within the universe of 82% of subjects that claimed to use a backpack to transport their school supplies, their perceived posture, while interacting with it, revealed the following distribution: 51% used both of backpack's carrying handles, and 31% used only one. This survey, also allowed to show that 13% of this particular universe tend to carry another product besides the backpack to carry their own school supplies. The most common referenced extra products were: a dossier and another bag. Meanwhile, this question also allowed us to see that just 5% of the surveyed sample claimed to make proper use of the backpack's whole features, transporting it using both of its carrying handles, while wearing its waist belt.

Regarding transported content, this survey found that 63% of the surveyed subjects carry five or more books every day to and from school. Besides that, 65% of the surveyed students claimed to carry their dossier to school, every day. Among the total amount of 110 students, 15% of them claimed to transport their laptop every day, in their school supplies transportation product to and from school.

26.3.1.3 Negative Effects Resulting from School Supplies Transportation Task

When it comes to negative health effects, caused by the school supplies transportation task, survey results showed a total amount of 55% of the inquired subjects claimed some kind of musculoskeletal discomfort, or even pain. From these, 45% classified it as minor discomfort, or slight pain, while the remaining 10% classified these episodes as moderate, or even acute, levels of pain.

From the 55% of surveyed subjects that reported musculoskeletal discomfort or pain, they distributed it according to the period of the day, in which these incidents tend to occur, as follows: 5% in the beginning of the school day; 31% throughout the school day; and the vast majority (64%) claimed that this type of episode, tends to occur in the end of each school day.

This survey also found that 20% of the inquired subjects presented some kind of musculoskeletal discomfort, or pain, in the cervical area of the spine. From these, 14.54% classified it as a minor pain or discomfort, while the rest 5.46% classified it, as a moderate or acute pain occurring in the same region.

When it comes to the shoulder area, 54% of the inquired subjects reported some kind of musculoskeletal discomfort or pain. Among these, 38% reported episodes of minor pain or discomfort, while the remaining 16% classified these episodes as moderate pain and discomfort.

When it comes to the upper back, 48% of the surveyed sample presented some type of pain or discomfort. Among these, 34% categorized these episodes as minor pain or slight discomfort, while the remaining 14% claimed to feel a moderate discomfort, or pain, in that same region.

This survey also found that 36% of the surveyed sample presented some kind of health problem in the month in which this field study was carried out, and 26% of the inquired subjects claimed to have felt some kind of musculoskeletal problem, within that same period of time.

26.3.2 Direct Observation Strategy Results

As mentioned earlier in this chapter, the direct observation strategy was carried out as a complementary method to the survey, in order to verify, and further understand, the answers gathered.

This direct observation strategy allowed to verify, from a total amount of 110 observed students, that only 3% of the capacity of the transported products seem to be insufficient; 25% presented a fairly capacity amount; while the vast majority (72%) of the observed products seemed to be oversized, when it comes to capacity. These results were obtained, by comparing the school supplies transportation products weight capacity with the effective quantity of items transported by the users in them.

When it comes to items carried by the students, in their school supplies transportation product, the direct observation strategy allowed us to find that 39% of those items were books (school and nonschool related); 16% were notebooks; 7% were dossiers; 12% school kit cases; 9% were calculators; 7% rulers; and finally, the remaining 10% were scattered, and labeled, as other products.

Direct observation strategy major findings allowed to find a mean weight of the product intended for school supplies transportation of 5.06 kg (SD = 1.99 kg) with a maximum and minimum of 10 and 0.2 kg, respectively.

This strategy also showed that the mean weight of the studied population, without the product used for school supplies transportation was 53.5 kg, which allowed us to find the mean percentage between the user's body and their school supplies transportation weight of 9.46%.

When it comes to the anthropometrical dimensions of the studied population, this strategy allowed to find that the mean height, was 163 cm (SD = 9 cm) with a maximum and minimum of 190 and 144 cm height, respectively.

26.4   DISCUSSION

As a guideline to understanding Sri Lanka's reality concerning school supplies transportation products, the article entitled "Inculcating the Ergonomic Culture in Developing Countries: National Healthy Schoolbag Initiative in Sri Lanka" written by Kapilla Jayaratne, served as a basis to understand the Portuguese reality for this matter.

Despite the social and economic differences found between these two countries, the major findings revealed that when it comes to the school supplies transportation task, the most divergent aspects were the carrying method of the product intended for school supplies transportation; and the negative effects that come from the users' body interaction.

Regarding the convergences found between both studies and summarized in Table 26.1, the major findings revealed:

1. Similar results, from the amount of the surveyed sample that stated the use of a backpack, to transport their school supplies to and from school. Which may be seen, through the small difference, between 82% and 79.6% of Portuguese and Sri Lankan students, respectively.
2. A mean percentage, of schoolbag relation to users' body weight, which can be seen in 9.46% for Portuguese students and 11.04% for Sri Lankan students.
3. And finally, the small difference found between the surveyed samples that reported musculoskeletal problems around the neck area, which can be seen by 15.1% of Portuguese students, against 20% of Sri Lankan ones.

TABLE 26.1
Convergences Found between Studied Realities

	Convergences Found	
	Portuguese Reality (%)	Sri Lankan Reality (%)
Backpack use	82	79.60
Weight proportion (bag vs. body)	9.46	11.04
Musculoskeletal problems (neck area)	15.10	20

TABLE 26.2
Divergences Found between Studied Realities

	Divergences Found	
	Portuguese Reality	**Sri Lankan Reality**
Shoulder bag use	17.90%	5.00%
Use of two carrying handles	51.00%	97.10%
Use of an extra product	49.00%	2.90%
Minor level of musculoskeletal problems	45.00%	10.00%
Moderate level of musculoskeletal problems	10.00%	62.00%
Mean weight carried	5.06 kg	3.72 kg

When it comes to observed divergences, for both studies results, the major findings revealed:

1. A much more common use for the shoulder bag by Portuguese students than by Sri Lankan ones, with 17.9% against 5% of the surveyed samples, respectively
2. That the carrying method of the backpack itself, had the higher differences observed, with a total amount of 51% of the Portuguese students, against 97.1% for Sri Lankan ones, that claimed to use both of backpacks' carrying handles
3. A larger number (45%) of the Portuguese surveyed sample that reported a minor level of musculoskeletal problems, against just 10% of the Sri Lankan one
4. That, when it comes to moderate level of musculoskeletal problems, the divergences were more pronounced, as it may be seen through 62% of Sri Lankan students, against 10% of Portuguese ones
5. And finally, a discrepancy of 1.34 kg, in the mean weight carried in school supplies transportation products (3.72 kg for Sri Lankan students and 5.06 kg for Portuguese ones) (Table 26.2)

26.5 CONCLUSION

Major findings allowed to demonstrate, that besides economic and social divergent situations of a developing Asian country and a developed European one, users' needs and methods of interaction with a product may vary.

Moreover, a mean percentage between body weight and the school supplies transportation product weight was found that leads us to assume that some of the differences found could be correlated with malnutrition, although further studies about this matter would be required. Another aspect, that could explain some of the results gathered, is that teaching and education appreciation may differ, between a developing country and a developed one.

Therefore, it is demonstrated here that a study applied to a single task, namely the carriage of school supplies to and from school, may originate divergences between both realities that may seem quite obvious. The most alarming aspect to address, should be the inadequacy from the mean body weight percentage recommended by international limits, of 10% and 15%, for the weight carried in school supplies transportation products, since it was shown that despite being near the lower limit (9.46% and 11.04%) of this range, there are still students claiming to suffer from musculoskeletal pain and discomfort.

These divergences may actively influence the health of growth stage children, attending both educational systems, which ultimately could influence these individuals in adulthood.

Therefore, this chapter tries to raise the awareness of all designers, product development teams, national education departments, and school board authorities, to the need of taking into consideration all the accessible ways to minimize the weight transported by children in their school supplies transportation products. Designers should have something to contribute, by the targeting of better fitted products for this task, leaving aside the typology development process solely based on the aesthetic aspect.

REFERENCES

Chapanis, A. 1965. Words, words, words, *Human Factors: The Journal of the Human Factors and Ergonomics Society*, 7(1), 1–17.

Dagge, R. and Filgueiras, E. 2014. Comparative analysis between two distinct realities concerning the transport of school material. In: Soares, M. and Rebelo, F. (eds.), *Advances in Ergonomics in Design, Usability & Special Populations: Part I*, AHFE Conference, USA, 16, p. 141.

Dejean, P. H. and Naël, M. 2007. Ergonomia e Projeto. In: Falzon, P. (ed.), *Ergonomia do produto*, São Paulo, Editora Blucher, pp. 56–58.

DN 2012. Veja o Ranking das escolas Secundárias e Básicas. Diário de Notícias Website: http://www.dn.pt/DNMultimedia/DOCS+PDFS/RankingEscolas2011/Secundarias.html.

Ghiglione, R. and Matalon, B. 2001. *O inquérito: teoria e prática*, Oeiras, Celta Editora.

Goodgold, S., Corcoran, M., Gamache, D., Gillis, J., Guerin, J., and Coyle, J. Q. 2002. Backpack use in children, *Pediatric Physical Therapy*, 14(3), 122–131.

Jayaratne, K. 2012. Inculcating the ergonomic culture in developing countries: National healthy schoolbag initiative in Sri Lanka, *Human Factors: The Journal of the Human Factors and Ergonomics Society*, 54(6), 908–924. doi: 10.1177/0018720812456870.

Kaewboonchoo, O., Yamamoto, H., Miyai, N., Mirbod, S. M., Morioka, I., and Miyashita, K. 1998. The standardized Nordic questionnaire applied to workers exposed to hand-arm vibration, *Journal of Occupational Health*, 40(3), 218–222.

Ramprasad, M., Alias, J., and Raghuveer, A. K. 2010. Effect of backpack weight on postural angles in preadolescent children, *Indian Pediatrics*, 47(7), 575–580. doi: 10.1007/s13312-010-0130-2.

27 Improving Bus Travel through Inclusive Service Design

Carlos Aceves-González, Sharon Cook, and Andrew May

CONTENTS

27.1 BACKGROUND

The aging population has highlighted the need to help people age well through the provision of enabling and supportive environments. Nevertheless, access to outdoor spaces and buildings is not fully guaranteed for people with varying needs and capacities. The World Health Organization (WHO, 2007, p. 15) points out that "in both developed and developing countries, people think that their city was not designed for older people" and they also report that the provision of commercial and public services presents problems in meeting older people's needs. Data from this organization suggested that along with the need for an accessible built environment, there is a compelling need for inclusive services that can be used by a broader range of users. Services, in which providers are able to understand how they can better respond to users irrespective their age or capabilities (BS 18477, 2010; WHO, 2007), therefore need to be developed. Public transport is one of the services that have been identified as especially relevant in supporting older people to remain healthy and active (WHO, 2002). A number of studies highlight the benefits of continued mobility and use of transport for older people, including health and well-being (Metz, 2003; Mollenkopf et al., 2005; Webber et al., 2010). Conversely, a lack of adequate public transport can result in increased isolation and "the denial of a range of human rights, including participation and equitable access to services" (United Nations Population Fund, 2012). Regardless of the benefits of public transport, the literature reports the presence of barriers relating to accessibility and use for all passengers and in particular for older and disabled people. The WHO (2007) specifically highlights how services such as transport have difficulty in meeting the needs of older citizens.

27.1.1 Inclusive Service Design Approach

Inclusive services are those which are available, usable, and accessible to all customers equally, regardless of their personal circumstances (BS 18477, 2010). The growing need for inclusive services represents a challenge for the design disciplines in terms of providing knowledge and tools for the evaluation, design, and improvement of such type of services. With this in mind, an inclusive service design approach, which comprises the integration of theory and methods of service design and inclusive design (both outlined below), has been the route proposed to guide the evaluation and design of inclusive services (Aceves-Gonzalez, 2014).

Service design is an emerging discipline that aims to innovate or improve services that are useful, usable, and desirable from the user perspective, and efficient and effective from the organization perspective (Moritz, 2005; Mager and Sung, 2011). The literature suggests that this emerging discipline provides several benefits to the end users' experience when applied to service sectors such as retail, banking, transportation, and healthcare (Stickdorn, 2010). In the public services arena, it has been pointed out that the approach is less about competition and contestability and more about reducing the gap between what organizations do and what users expect or need (Parker and Heapy, 2006).

Meroni and Sangiorgi (2011) highlight that service design, since its origins, has considered users as its main focus in the process of service delivery. These authors claim that this approach generally conceives users as a resource rather than a burden or a problem. Still, beyond being a user-centered approach, it is also considered as a human-centered approach that investigates or understands people's experiences (as users, service staff, communities, or humanity in a wider sense), interactions and practices "as a main source of inspiration for redesigning or imagining new services" (Meroni and Sangiorgi, 2011, p. 203). However, this design approach is limited in terms of evaluating and designing inclusive services, since there is not an explicit consideration that users are diverse and possess a variety of capabilities, needs, and desires.

Inclusive design "is a general approach to designing in which designers ensure that their products and services address the needs of the widest possible audience, irrespective of age or ability" (Design Council, 2008). One of the main objectives of this approach is avoiding design exclusion, which might be caused if the demands of the task exceed any of the corresponding user abilities (Clarkson et al., 2015). Furthermore, a key characteristic of inclusive design is to increase the target group of a product or service, but without compromising the business goals of profit and customer satisfaction (Coleman et al., 2003).

Inclusive design has been developed over the last 20 years in the UK. Among its research contributions are theoretical models, data on different user capabilities, methods and tools, as well as standards and guidelines (Clarkson and Coleman, 2015). Research from this design approach is mainly divided into two areas (a) understanding the capabilities and needs of end users and (b) understanding the needs of information of knowledge users (e.g., designers, policy makers, industries, etc.) (Dong et al., 2015). Overall, inclusive design offers several benefits when designing for inclusion. However, it appears that much research of this design approach has focused on the end users and on the design of products.

In essence, both approaches are closely related to the notion of developing better services. Whilst service design provides principles to deal with a human-centered and holistic perspective, inclusive design offers a strong focus on users' diversity. Consequently, the inclusive service design approach is underpinned by the idea that using philosophical principles, tools, and techniques coming from these design approaches, guided by a human factors perspective, can contribute to evaluating and designing more inclusive services.

27.2   AIM AND OBJECTIVES

Designing usable services is part of the aim of a service design approach (Mager and Sung, 2011), and usable environments have been recognized as a basic precondition for traveling (Carlsson, 2004). The concept of usability has been used to explore problems that people face using public transport and this concept is also very important to the provision of inclusive services (BS 18477, 2010). The overall aim of this chapter is to report on the use of the inclusive service design approach to evaluate (a) bus service use by younger and older people and (b) the role of other stakeholders in the provision of the bus service in Guadalajara, Mexico. In order to achieve this, the research was guided by the following objectives:

- To describe the main characteristics of the bus service
- To identify which elements of the bus system impose greater difficulty for younger and older people
- To determine if older people report different reasons for, and impacts from, those problematic elements than younger people
- To identify the key stakeholders who affect service implementation or who are affected by it
- To investigate the main issues that prevent delivery of an inclusive service

The research takes a broad and encompassing perspective with consideration of most of the elements of the bus journey experience and the influence of multiple stakeholders as well as transport conditions in the city. This approach enables a better understanding of the usability and accessibility barriers for younger and older people and the reasons behind the presence of those barriers. Such a level of understanding will provide some good insights for improving bus travel in the city.

27.3   METHODS

27.3.1   OVERVIEW

The research methodology comprised two complementary studies. The first study was designed to investigate the nature of the challenges facing younger and older users when using the bus service. In this study, participants were asked to identify those issues in the whole journey that impose difficulties to their use of the service and which would need to be redesigned to increase its uptake. The study helped to ascertain whether younger and older users face similar challenges in the use of the bus service or if each group has unique views and needs when traveling.

The second study was designed to obtain from stakeholders a broader understanding of the context of the service operation, and to explore how this might relate to the challenges identified in the first study. To better understand this context, the research needed to not only consider the users themselves but also extend beyond them by considering the contributions made by other stakeholders. Two methods were used to develop understanding in this area. The first related to consultations with a range of stakeholder groups in the form of personal interviews and group meetings whilst the second related to a document analysis of relevant items in the public domain. In this way, the design and operation of the service could be better understood at a system level from the context of legal requirements through to in-practice considerations.

27.3.2   FOCUS GROUP STUDY

In order to obtain information regarding how users think and feel (Barrett and Kirk, 2000) about the bus service, this study used a series of focus groups to identify and understand those problematic issues that impose greater difficulties to use the bus service to younger and older passengers. Discussions were aimed at ascertaining the true motivations and insights (Stickdorn and Schneider, 2010) behind the perceptions of each group of users.

The study recruited a convenience sample of older participants, via a general call in the Metropolitan Centre of the Elderly (CEMAM, according to its designation in Spanish). Twenty-six older people were selected based on the criteria of being aged 60 or over with appropriate language and cognitive abilities to participate in the group discussions and give informed consent. The study was particularly interested in those who had a desire to use public transport, but perceived that there were barriers to doing so. Therefore, a combination of frequent and nonfrequent travelers took part in the study.

A similar strategy was used to recruit 17 students in their first year at the University of Guadalajara. These younger participants were aged between 18 and 21 years, and they were all frequent users of the bus service. Appropriate language and cognitive abilities to participate in the group discussions and give informed consent were also required of these participants.

For this study, the data analysis involved analyzing the audio recordings from the focus groups. These files were imported into the QSR International NVivo software, and based on the previous literature on aging and bus service provision, the qualitative data were explored through a theoretical thematic analysis which was undertaken, at a semantic and realistic level which means that themes, subthemes, and codes were identified within the explicit meaning of the data, considering only what the participants said (Braun and Clarke, 2006; Robson, 2011). The analysis was conducted following the procedure described by Robson (2011). Since this study aimed to better understand the whole picture of the bus service use, and since there was no previous research and knowledge related to the use of the bus service in the studied context, the intention was to produce a rich description of the data rather than a detailed account of a particular aspect.

27.3.3 SEMI-STRUCTURED INTERVIEWS AND DOCUMENT ANALYSIS STUDY

The process of designing a service starts by gaining a clear understanding of the users' needs (Clarkson et al., 2007; Polaine et al., 2013), but understanding the context in which the service is operating is also critical to gathering insights into peoples' interactions with service touch-points (Stickdorn and Schneider, 2010; Holmlid, 2011). The use of the inclusive service design approach naturally included the users and extended this research beyond them to consider the needs of, and contributions made by, other stakeholders thereby enabling an understanding of the broader context of the service to be gleaned. To explore this broader context the research strategy used in this study, as previously stated, comprised two complementary methods; the first related to stakeholder interviews and the second to a document analysis of relevant items in the public domain.

27.3.3.1 Semi-Structured Interviews

This was a qualitative, exploratory study, which included a series of group and individual semi-structured interviews with stakeholders. Data were analyzed to give an overview of the bus service characteristics: main actors, routes, bus design, and regulation, among others. It also aimed to explore the presence of constraints in the service provision and why they were occurring. The results were interpreted in terms of how those characteristics and constraints might prevent the service from being perceived as safe, usable, and desirable from the point of view of the users.

At the beginning of this study and based on previous knowledge of the bus system, the following key stakeholders were defined:

- Users: (who took part in the previous study)
- Local authorities: those related to public transport and/or older people
- Service operators: bus companies, bus organizations, and/or bus owners
- Bus drivers

- Bus manufacturing companies: designers and managers
- Nongovernmental organizations: working in favor of public transport improvements

In total, 33 participants took part in the study either individually or as part of a group session. Initially, 11 interviews were conducted to gain an understanding of the service operation. After those interviews, five group meetings were held with people from different organizations to clarify or deepen understanding in specific subjects.

27.3.3.2 Document Analysis

To gain a better understanding of the bus system and as a supplementary method (Robson, 2011), some documents such as laws, regulations, programs and plans, and newspapers, among others were integrated into the analysis. As suggested by Bowen (2009), the procedure involved finding, selecting, assessing, and synthesizing the data contained in the documents. The following list shows some documents that were included in the analysis:

- Law on the Rights of the Elderly (INAPAM, 2011)
- General Law for the Inclusion of People with Disabilities (Sedesol, 2011)
- Federal Law to Prevent and Eliminate Discrimination (Gobierno de Mexico, 2007)
- Law of Mobility and Transport of the State of Jalisco (Gobierno de Jalisco, 2013a)
- State Development Plan—Jalisco 2013–2033 (Gobierno de Jalisco, 2013b)
- Some news reports from the digital version of the local newspapers were included in the analysis

This study comprised two different data sets which required different methods of analysis. In the first part, the data analysis method used in relation to the interviews and meetings was similar to that of the first study and was undertaken at a semantic and realistic level. However, in this case, the researchers did not have a previous theoretical perspective for analysing the information, therefore based on the themes identified from the data, an inductive analysis was used (Braun and Clarke, 2006; Robson, 2011).

The second part of the data analysis included the examination, reading, and interpretation of the documents. This process involved the combination of elements from content analysis and thematic analysis. The procedure excluded the quantification typical of conventional mass media content analysis and followed a first-pass document review to extract meaningful and relevant passages of text (Bowen, 2009). Later, as part of the thematic analysis, a more focused rereading and reviewing of the data was undertaken. The codes and themes previously defined for the interviews were used for the document analysis. However, the theme related to the current mobility situation in the city emerged mainly from the document analysis.

27.4 RESULTS AND DISCUSSION

For the sake of clarity, this section combines the results and discussion into one. The results of the focus groups study are first presented and discussed, followed by those of the second study. Following that, there is a discussion of the benefits of using the inclusive service framework in evaluating the bus service.

27.4.1 Focus Groups Results

A total of seven focus groups were conducted, four with older people and three with younger people. The majority of participants were female (65.2%) and frequent passengers (83.8%). Eleven (42.3%) older participants reported having problems going out and about; by contrast only one younger participant reported such difficulties which were due to a temporary physical impairment.

Table 27.1 shows the problematic bus service elements that were most cited by younger and older participants in the group discussions. It is noted that the bus drivers, bus design, and bus capacity

TABLE 27.1

Problematic Issues to Use the Service According to Younger and Older People

Themes Bus Service Elements	Number of Mentions	% of Participants Who Reported Each Element	
		Younger	Older
Drivers	162	76	100
Bus design	112	76	88
Crowded buses	71	94	62
Waiting time	47	64	57
Other passengers' behavior	38	29	54
Payment method	37	76	50
Bus stops	24	41	38
Lack of information	21	47	26
Distances to walk	20	12	42

were stated as problematic elements by a higher number of participants from both age groups. However, it also highlights some differences between the percentages by theme in relation to the age group. For instance, at least one concern or complaint was raised by the total of older participants related to drivers in comparison with 76% of the younger participants. Conversely, bus capacity was stated as a problem by 94% of younger participants compared to 62% of older people.

TABLE 27.2

Given Reasons of the Problematic Elements Associated with Bus Use by Age Group

Themes Bus Service Elements	Younger Participants	Older Participants
Drivers	Unfriendly drivers	Lack of consideration toward older people
	Competing for passengers	Short time to get on and off
	Inappropriate appearance or behavior	Large distance to the kerb
Bus design	Narrow aisle	Steps too high
	Uncomfortable seats	No or inappropriately placed handrails
		Steps with irregular shape
		Reduced number of priority seats
Bus capacity	Uncomfortable experience due to crowded buses	Difficulties in using the bus due to crowding
		Lack of seats
Waiting time	Unreliable service	Difficulties in standing to wait
		Unreliable service
Other passengers' behavior	Disrespectful people	Young people do not respect use of the priority seats
		Lack of consideration toward older people
Payment method	Drivers do not want to receive the half price payment from students	Drivers do not want to receive the half price payment from older people
Bus stops	Poor safety conditions	Lack of seats
		Bus stops being blocked by other roads users
Lack of information	Lack of timetables	Lack of information about route changes
	No information about routes	No information about routes
Distances to walk	Long distances	Long distances
		Bad pavement conditions
		Problems crossing roads
		Lack of pedestrian crossings

It is important to note that although younger and older users expressed concerns about the same bus service elements, each age group stated different reasons for those problematic elements (see Table 27.2) with older people experiencing additional and more serious limitations when using the bus service. This difference might be attributed to declining functionality due to the aging process (e.g., motor, visual, auditory, cognitive, or health limitations), therefore, the gap between personal abilities and environmental demands becomes wider (Rogers et al., 1998; Seidel et al., 2009). Moreover, it should be noted that the use of the bus service implies a series of tasks, such as climbing up and down stairs or moving toward a seat when the bus is moving, which are among the most challenging and hazardous types of locomotion in the daily living of older people (Startzell et al., 2000; Redfern et al., 2001).

These results suggest the existence of several elements that impose difficulties in using the service. These elements are not only physical objects; rather there are many intangible elements that need to be considered to improving the service such as driver attitude, information provision, and waiting time. The following sections describe and discuss how these problematic issues have an impact on the use of the bus service by younger and older people.

27.4.2 IMPACT OF THE PROBLEMATIC ISSUES ON USABILITY PROBLEMS

The present study aimed to identify which elements of the transport system impose greater difficulty in the use of the bus service, and their impact on users; as well as to determine if older people report different impacts from those problematic issues than younger people do. This section presents results and discussion regarding those elements of the bus system that younger and older participants pointed out as problematic in using the service.

Results of this study showed that 31 out of 43 participants reported at least one problem in using the bus service. Similar to the safety concerns, most of usability problems were reported by older participants (23), although younger participants (eight) also expressed having difficulties in using the service. However, the older participants made a higher number of comments regarding difficulties in using the bus service. Figure 27.1 shows the number of comments made by younger and older participants concerning usability problems. As discussed above, older people encounter more difficulties undertaking their daily activities (transport among them) given their decline in functionality due to the aging process (Rogers et al., 1998; Seidel et al., 2009).

Among the main problems expressed by the older participants were getting on and off the bus, traveling while standing, moving through the bus, getting in and out of seats, walking long distances, and waiting whilst standing. These are problems which have been reported in previous research (WHO, 2007; Broome et al., 2009).

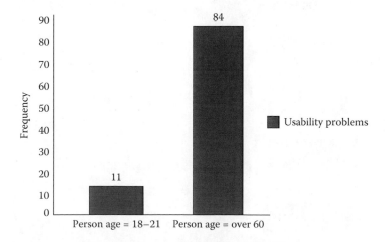

FIGURE 27.1 Number of comments relating to usability problems by age group.

In addition to difficulties in performing those activities, some older participants, especially those with severe functional limitations, stated that they avoid using the service as they consider that is impossible to do it. This situation suggests that there is a relationship between personal character-istics or limitations and usability problems (Carlsson, 2004). Such is the case of participant g4-02, who is not a frequent traveler and suffers of arthritis in her knees, who expressed:

> I do not go anywhere and do not use public transport because I cannot use it. It is impossible, because drivers do not have patience with us. They [drivers] do not wait for me to get on the bus

The analysis found that bus service use is practically affected by many issues, but it seems that issues such as drivers' behavior, bus capacity, and bus design have a higher impact for both older and younger participants, although factors such as other people's behavior, distances to walk, and lack of information were also found to have an impact. Once more, it is important to mention that the impact on passengers' experience comes through a combination of problematic issues. In the case of usability, this might be evident when considering the design of the bus, bus capacity, and driving style. Figure 27.2 graphically shows this relationship and helps to track the content of the following sections.

27.4.2.1 Drivers' Behavior, Payment Method, and Usability Problems

Most of the characteristics of drivers' behavior have already been described above. Therefore, the key point here is to explain how such behavior might cause usability problems to the passengers. In the focus groups, 23 out of 43 participants (19 older people and 4 younger) recorded having difficulties in getting on and/or off the bus, and they pointed out that these activities are further complicated: when the driver does not consider the passengers' needs, when the driver is pressurizing them to act quickly, or when there is a large horizontal distance between the bus step and the kerb. Similarly, the characteristics of a fast, jerky, or bumpy ride were highlighted by several participants as a factor making it more difficult: when traveling whilst standing; when moving through the bus; and when getting in and out of the seats.

Drivers' behavior has not been frequently reported as an issue affecting the use of public transport because the research has focused more on the physical characteristics of the vehicles or environments (Broome et al., 2009, 2010). However, results from this study are consistent with some

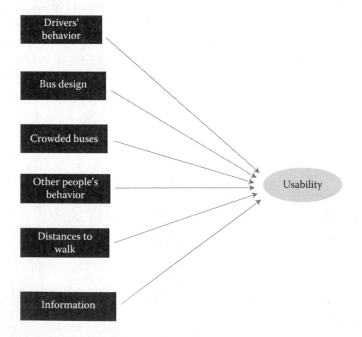

FIGURE 27.2 Bus service issues with major impact on usability.

studies that have highlighted the role of drivers as a barrier or enabler to use the bus service (WHO, 2007; Broome et al., 2010; Nickpour et al., 2012).

In addition, younger and older participants expressed the view that drivers do not want to stop and collect them due to the fact that they pay using *transvales*, a payment method which reduces the profits of the drivers. This situation causes more problems to these passengers because they then have to wait for another bus—increasing the waiting time—which could be difficult for some older users. Finally, an issue of intense concern among the older participants was their belief or the fact that drivers do not want to stop for older users. This is consistent with the findings of the WHO (2007) which pointed out that in some developing countries it was found that drivers were reluctant to pick up older people.

27.4.2.2 Crowded Buses, Peoples' Behavior, and Usability Problems

It appears that there is a strong relationship between crowded buses and difficulties in using the bus service (Coughlin, 2001; Carlsson, 2004; Katz and Garrow, 2012) as many of the participants stated that when buses are crowded all the problems increase. For instance, a younger user commented:

> Sometimes the bus is too crowded. For example, I have seen that on the route 380 sometimes there is no room even for an "ant," the bus is really full. This causes many problems, because if it is really full then it is very difficult, getting on or off, standing up, moving through the bus, and even security itself (Participant g6-02).

The data analysis of 31 out of 43 participants who referred to usability problems provided an understanding as to why and how crowding makes all aspects of bus use more difficult. For instance, when buses are crowded it increases the possibilities that drivers do not stop which therefore prolongs the waiting time and so the service provision becomes uncertain since users do not know if a driver is going to stop or not. Some participants stated that they frequently have to wait for two or three buses to go past before catching one.

According to the participants, the reasons that make it more difficult to get on and off the bus when it is crowded related to (1) the driver being in a hurry because he has stopped several times and so he now has less time to reach his destination; thus, he reduces the time he allows for passengers to get on and off in order to make up. (2) There are many people in the area of the doors which impede access to it and its use. This situation is consistent with the observations of Katz and Garrow (2012), who found that bus door crowding negatively affects the service operation.

The older participants also highlighted that getting a seat can be nearly impossible when the bus is crowded because there are no free seats; hence they are forced to travel standing. Traveling in this way is made more difficult by the driver's driving style and by the lack of opportunity to hold onto the handrails. Some participants pointed out that this situation may cause some passengers to push into each other.

27.4.2.3 Bus Design and Usability Problems

Many of the participants recorded problems relating to elements of the bus design such as high steps, number and spacing of seats, narrow aisle, inappropriately positioned handrails and bells, or poor visibility of the route number. All these elements were related to usability problems and again the highest impact was in the actions relating to: getting on and off the bus, traveling whilst standing, moving through the bus, and getting in and out of seats.

Problems in getting on and off the bus are the most common difficulties that older people report in using the bus (Rogers et al., 1998; Glasgow and Blakely, 2000; Coughlin, 2001; WHO, 2007). Step height onto the bus is generally highlighted by older people as the main cause of these difficulties (Broome et al., 2009). This can be explained by understanding that walking on stairs is a high demand task, which challenges not only the strength of the legs, but also requires balance and muscle coordination (Redfern et al., 2001).

However, in this study along with the height of steps, participants noted that it is a combination of bus design features and service provision that causes them difficulties. For instance, difficulties regarding the height of step were related to driving style, time provided to get on and off, distance between the bus and kerb, crowded buses, and other passengers' behavior. Older participants added that the shape of the steps, and the lack or inappropriate placement of handrails also have a strong relationship with difficulties in boarding and alighting. In line with this, an older participant g1-04, said:

> Regarding the design of the bus, I can tell you I have problems, especially with some buses that have curved stairs. But also they are too high; I find it very difficult to get on and off the bus. Once, a person got on one of those buses, he was more disabled than me, and even though the driver was waiting, that person had a hard time climbing up with his walker!

In addition, there were a few younger participants who also stated problems with the steps and handrail design. However, it is worth mentioning that these participants had a particular physical characteristic or a temporary disability, which did not enable them to use the bus in the same way as the other younger participants in this study. Participant f5-03 for instance, whose height is 1.47 m, commented:

> To me, the first problem is the design of the bus, because that is what causes me problems because I'm short. The first step causes the biggest problem as it is too high. But there are other buses with curved stairs, where the handrail is too far away to reach. So, I almost have to hold the second step to climb up to the first. That's even harder when I have some things in my hand.

Similarly, participant g5-05, who was recovering from a leg injury, said:

> I recently had an accident and my leg is injured. Therefore, I can hardly walk up the steps because they are very high.

Participants who recorded problems in traveling whilst standing and moving through the bus stated that along with the driving style and crowded buses, the narrow aisle and the inappropriately placed handrails in the bus are factors that increase the difficulty in its use. Concerning the handrails, two design characteristics were observed as problematic: (1) the lack of handrails in some areas of the bus and (2) handrails being placed too high for some participants to reach for use. These characteristics were mainly commented on by the older participants, who lack arm strength or who have trouble stretching their arms, or by smaller participants. On the other hand, the narrow aisle was mainly noted as problematic by the younger participants, and it was always related to crowded buses and/or when the passengers themselves were traveling carrying some belongings such as a backpack or document pocket.

Seat layout and reduced seat pitch, along with driving style and peoples' behavior, were identified as causes that make it difficult to get in and out of the seats. Participants said that usually passengers use the aisle seats because it is easier to get in and out of them, especially if the bus is moving. This situation forces a new passenger to use the windows seat which they have to access through an area confined by the aisle seat passenger.

Finally, 12 participants (six younger and six older) stated that the route number's design was a problem in identifying the correct bus and that the difficulty worsened at night-time or when there were many routes on the same road. Similarly, other participants recorded some comments relating to the bell position as being problematic when getting off the bus because generally there is only one bell, which is placed at the rear of the vehicle. This situation makes it difficult for passengers who are in the middle of the bus to ring the bell as well as for those who try to get off by the front door as is the case for older people.

Most of the bus design features noted as usability problems by the participants have been pointed out in previous studies (Broome et al., 2009). However, the analysis of focus group discussions from this study led to an understanding that bus design is not an isolated problem, but rather that its combination with service characteristics (e.g., driving style, crowded buses, and other people's behavior) make the problems even greater.

27.4.2.3.1   Distances to Walk, Bus Stop Conditions, Waiting Time, and Usability Problems

As noted earlier, walking to the bus stop may cause some safety-related problems such as risk of falling or being hit by a car when crossing roads, but also some older participants stated some problems relating to the use of the pavements which can be in a bad condition, for example, they are uneven or have holes, or there are many obstacles such as cars, shopping stalls, roots of trees, etc. Therefore, using such pavements might be dangerous or difficult, or, in a number of cases for some older people, impossible. Figure 27.3 shows an example of a pavement where the roots of a tree have destroyed the pathway. Moreover, older participants said that due to their health conditions, it is difficult to walk long distances, especially if they need to walk uphill or when the weather is very hot.

Some participants commented on usability problems relating to the bus stops. These problems were: the absence of the official bus stops and the bus stops being blocked by other vehicles thus preventing bus drivers from pulling up to the kerb which then forces the passengers to wait and board the bus from the road. The lack of seats at the bus stops was another problem which was raised and this related to the waiting time. Some older participants commented that is difficult to wait whilst standing, especially when they have to wait for a long time.

Long distances to the bus stop, uneven footpaths or footpaths with obstacles, lack of bus stops, lack of seats at the bus stops, and poor weather have been reported in previous studies as barriers in the use of the bus service by older people (Lavery et al., 1996; Broome et al., 2009; Marsden et al., 2010). It has also been established that older people generally cannot walk as fast or as far as younger adults (Burton and Mitchel, 2007).

27.4.2.4   Information and Usability Problems

Previous studies have identified that older people experience some difficulties due to incorrect information (Carlsson, 2004), or the lack of real-time information at the bus stop (Coughlin, 2001). However, in this study, some participants made complaints regarding the total lack of information.

FIGURE 27.3   Example of a pavement in bad conditions due to growth of tree roots.

They pointed out the absence of maps, timetables, and a reliable and institutional source of information. Thus, when passengers want to go to an unfamiliar place, they have to ask a friend or a relative, which often generates confusion resulting in problems in using the service or even avoiding using it altogether. For instance, an older participant (g4-05) expressed that:

> I hardly go anywhere, since I do not know the routes; if I would have to go anywhere, I would need to ask where the bus stops are, and where the bus takes me.

27.4.3 Semi-Structured Interviews and Document Analysis Results

A large volume of data was obtained from this study, from which the results were classified and summarized into a total of five themes, namely: (1) current situation of mobility in the city; (2) gaps and weaknesses in the regulation of the system; (3) the implications of the hombre-camion model (this model of service operation comprises a large number of "owner-operators" resulting in a highly fragmented industry which is only informally coordinated); (4) perception of the quality of the service; and (5) trends and commitments for improving the service. Table 27.3 shows a summary of the main results from these studies that affect the bus service operation and prevent the delivery of an inclusive service.

The table above illustrates the wide diversity of data that can be obtained from stakeholder consultations and document review, which can help in understanding the wider elements of the bus service system that negatively impact the users' experience. By extension, this data provide insights

TABLE 27.3
Themes and Issues of the Bus Service Operation

Themes	Issues Identified
Current situation of the mobility in the city	Guadalajara's mobility options comprise modes of transport which appear not to meet most of the peoples' needs. There is a high dependency on the private car. The urban investment in infrastructure has favored private car use. Conversely, there has been little investment in other mobility options. Over the last decade, public transport has lost about half of its passengers who have preferred to use the private car. The exaggerated use of cars has caused several problems such as: traffic congestion, increased travel times, air and noise pollution, and a high rate of accidents in the city. Despite of the lack of quality public transport, most of the population still uses this mode of transport.
Gaps and weaknesses in the regulation of the bus service	There is a weak regulation of the bus service, hence it presents diverse irregularities. There is a lack of information and technical studies about the service operation. The legislation has permitted the rise of the *hombre-camion* model due to its restriction of a maximum of three buses being owned by any one person. Although, there is legislation to promote social inclusion of older and disabled people, there is still a big gap related to its application.
The hombre-camion model, implications for service operation	The bus service is an informal organization comprising people who own their own bus and who work in cooperation with other. Each bus might therefore be managed differently. It is very difficult to regulate such a fragmented bus system in which there are more than 4000 people participating as bus operators. Bus drivers are paid by the number of collected passengers, and so they compete, race, and "fight" with other bus drivers. Bus operators seek to maximize profits, even if that means competing against other providers on the same route or not providing the best service for users. Drivers work shifts of up to 13 h without fixed breaks, and do not receive proper training for doing their job. Buses are designed on a truck chassis due to the lower cost. Well-designed buses are too expensive for a single owner.
Public perception of the quality of the service	Any resident in Guadalajara might say how poor the quality of the bus service is. The bus service is frequently associated with accidents. Bus drivers are habitually pointed out as being responsible for the lack of quality of the service. People perceive that bus operators only want to increase their profits and do not care about the users and the quality of service.

into the factors that may be encouraging a modal shift away from public transport to private car use and therefore provides guidance for improvements to bus service design and uptake.

27.4.4   BENEFITS OF USING THE INCLUSIVE SERVICE DESIGN APPROACH

Within this research, a number of benefits have been identified in explicitly combining both inclusive design and service design approaches in investigating the bus service, for instance, obtaining a "big picture" understanding of the inclusivity of the service through the application of a holistic service approach. The service approach led the data collection to focus on aspects relating to the broader service, that is, (1) the actors involved in the service provision (users, bus drivers, service operators, local authorities, etc.) and (2) the wider service components (information, bus design, routes, drivers, other users, etc.). As a result, a large volume of data was obtained from the study which enables an appreciation of both users' and other stakeholders' needs, for example, bus drivers. Equally, the use of this approach permitted the identification and understanding of factors that cause the failure of the service elements, for example, establishing the reasons behind the drivers' behavior, and what the implications are for service improvement. The identification of the problematic elements within all of the service is very relevant because a problem in just one part of the service might exclude users from the service as a whole.

Another advantage of this inclusive service design approach is the uniqueness of the information gained from the older users. Cassim et al. (2007) have suggested that design exclusion does not come about by chance; but through neglect, ignorance, and a lack of adequate information and data. The use of the principle of inclusion in evaluating and designing a service enables the collection of unique data from older people that comprises personal components that would be difficult for people outside of that user group to identify (or understand) the significance of. For instance, in the case of the bus service, stakeholders (who include drivers, bus operators, bus manufacturers, designers, and regulators) are usually younger than older passengers, and are unlikely to directly experience for themselves the challenges such passengers face in using the bus. In addition, most of these stakeholders (aside from the drivers) are not in direct contact with such passengers and cannot identify where aspects of the service impact specifically on older passengers.

Likewise, the use of this approach, along with the application of both studies, allowed recognition that some of the problems experienced by the users—reported in the first study—can be in part attributed to a wider range of factors identified in the second study. For instance, results from study 1 show issues with bus design (step height and shape, narrow aisle width) which relates to study 2 where due to affordability by owner-operators only truck-based vehicles are purchased for use. Equally, study 1 shows issues with long distances to walk and problems with paving and crossings which relates to study 2 where infrastructure investment has favored private car use not public transport.

A further benefit from the use of this human-centered approach is the understanding that the bus service is problematic for both users and drivers. Results from the first study suggest the presence of several issues related to drivers' attitude and behavior. Drivers are blamed for unfriendly behavior and lack of consideration toward passengers' needs, but results of the second study show that they are "forced" to work under poor working conditions. According to Polaine et al. (2013, p. 36), "when frontline staff are let down by internal systems and procedures, they become disempowered and inflexible. This is passed down the line and leads to poor customer experiences and service failures."

All in all, the benefits of using an inclusive service design approach in investigating the bus service are reflected in the abundance of information and data produced illustrating the complex interplay of service provision factors on the quality of the service experience for older passengers. As Clarkson and Coleman (2015, p. 11) state a "great product or service is typically built on a foundation of understanding the real needs of the user and other stakeholders."

27.5 CONCLUSIONS

With a view of improving bus travel for younger and older people, this research project sought to evaluate the bus service through the use of an inclusive service design approach. The specific objectives of this research included (a) the identification of the bus system elements that impose greater difficulty for younger and older people and the differential impact of those elements on them and (b) the description of the main characteristics of the bus service and the key issues that prevent delivery of an inclusive service.

Overall the results from these studies confirm the compelling necessity for a better bus service in the city. These results indicate that the current service is not characterized by being usable and desirable from the users' perspective, and since less and less people are willing to use the bus, it is becoming less profitable to service operators. More detailed results indicate that the components of the bus system that impose most difficulties to the participants are the driver's behavior, bus design, and crowded buses with factors such as "other passengers" behavior' and "distances to walk" additionally affecting the older participants. The older participants are more impacted by the difficulties they encounter with more reporting difficulties with respect to boarding/alighting due to step height as well as with respect to poor stability issues when traveling whilst standing, etc. These reported difficulties can be viewed within a broader systems context through applying a service approach. This indicated how the lack of inclusivity experienced by the participants was related to wider legislative, economic, and other factors which will need to be addressed in the future as part of a solution to the issues identified. For instance, infrastructure investment favoring car use, weak regulation, and the hombre-camion system can be linked to failing environments (poor pavements and competition for road space), lack of accessible bus designs in operation and service provision which meets the needs of the owner-drivers above the needs of the users, etc. The identification of these elements provides focus as to where improvements can be made to the bus service provision. However, it should be noted that these results cannot be generalized beyond cities exhibiting characteristics representative of those of Guadalajara.

The use of self-reporting provided a large volume of data from the users' point of view and the results from this show a range of issues identified by younger and older participants that would not be readily apparent to transport service designers, or even other groups of passengers, especially since some of these issues have a personal and temporary element. However, a potential limitation of these studies relates to the inherent self-report bias within them. Self-report measures rely on the accuracy of the participant's judgments, which can be affected by educational, cultural, language, and social differences (Johnson et al., 2010).

Therefore, performance measures and observation are also needed to achieve a more comprehensive view of the variation between these age groups and to corroborate the self-reports.

The stated aim of this chapter was to report on the use of the inclusive service design approach and based on the study's findings, the authors conclude that such an approach offers invaluable insights combining breadth and depth of understanding. The approach allows a better identification of the needs of a wider range of users (both younger and older people) as well as the wide range of service stakeholders and it enables the collection of unique data from older people that comprises personal components that would be difficult for people outside of that user group to identify (or understand) the significance of. Similarly, the use of this approach pays attention to the big picture of the service facilitating an appreciation of how wider social factors such as the extent of regulation, type of investment, etc., can impact the service provided. All in all, the benefits of using an inclusive service design approach in investigating a service are reflected on the abundance of information and data produced from a holistic and inclusive approach.

REFERENCES

Aceves-Gonzalez, C. 2014. The application and development of inclusive service design in the context of a bus service. PhD dissertation, Loughborough University. https://dspace.lboro.ac.uk/2134/16265.

445

Barrett, J. and S. Kirk. 2000. Running focus groups with elderly and disabled elderly participants. *Applied Ergonomics*, 31(6):621–29.

Bowen, G.A. 2009. Document analysis as a qualitative research method. *Qualitative Research Journal*, 9(2):27–40.

Braun, V. and V. Clarke. 2006. Using thematic analysis in psychology. *Qualitative Research in Psychology*, 3(2):77–101.

Broome, K., K. McKenna, J. Fleming, and L. Worrall. 2009. Bus use and older people: A literature review applying the person-environment-occupation model in macro practice. *Scandinavian Journal of Occupational Therapy*, 16(1):3–12.

Broome, K., E. Nalder, L. Worrall, and D. Boldy. 2010. Age-friendly buses? A comparison of reported barriers and facilitators to bus use for younger and older adults. *Australasian Journal on Ageing*, 29(1): 33–38.

BS 18477:2010. *Inclusive Service Provision – Requirements for Identifying and Responding to Consumer Vulnerability*. BSI: London, United Kingdom.

Burton, E. and L. Mitchel. 2007. *Inclusive Urban Design: Streets for Life*. Reprinted. Elsevier Ltd: Oxford, UK.

Carlsson, G. 2004. Travelling by urban public transport: Exploration of usability problems in a travel chain perspective. *Scandinavian Journal of Occupational Therapy*, 11(2):78–89.

Cassim, J. R. Coleman, J. Clarkson, and H. Dong. 2007. Why inclusive design? In: Coleman, R., J. Clarkson, H. Dong, and J. Cassim (eds.), *Design for Inclusivity. A Practical Guide to Accessible, Innovative and User-Centred Design*, Gower Publishing Limited: Hampshire, England, pp. 11–21.

Clarkson, J. and R. Coleman. 2015. History of inclusive design in the UK. *Applied Ergonomics*, 46(B): 235–247.

Clarkson, J., R. Coleman, I. Hosking, and S. Waller. 2007. *Inclusive Design Toolkit*. 1st edn. University of Cambridge: Cambridge, UK.

Clarkson, J., S. Waller, and C. Cardoso. 2015. Approaches to estimating user exclusion. *Applied Ergonomics*, 46(B): 304–310.

Coleman, R., C. Lebbon, J. Clarkson, and S. Keates. 2003. Introduction: From margins to mainstream. In: Clarkson, J., R. Coleman, S. Keates, and C. Lebbon (eds.), *Inclusive Design: Designing for the Whole Population*, Springer-Verlag: London, UK, pp. 1–29.

Coughlin, J. 2001. *Transportation and Older Persons: Perceptions and Preferences*. AARP Public Policy Institute: Washington, DC.

Design Council. 2008. *Inclusive Design Education Resource*. Design Council: London, UK. http://www.designcouncil.info/inclusivedesignresource/index.html.

Dong, H., C. McGinley, F. Nickpour, and A.S. Cifter. 2015. Designing for designers: Insights into the knowledge users of inclusive design. *Applied Ergonomics*, 46(B): 284–291.

Glasgow, N. and R.M. Blakely. 2000. Older nonmetropolitan residents' evaluations of their transportation arrangements. *Journal of Applied Gerontology*, 19(1):95–116.

Gobierno de Jalisco. 2013a. *Ley de Movilidad Y Transporte Del Estado de Jalisco*. Periodico Oficial del Estado de Jalisco: Guadalajara, Mexico.

Gobierno de Jalisco. 2013b. *Plan Estatal de Desarrollo. Jalisco 2013-2033*. Gobierno de Jalisco: Guadalajara, Mexico.

Gobierno de Mexico. 2007. *Ley Federal Para Prevenir y Eliminar La Discriminacion*. Camara de Diputados del H. Congreso de la Union: Mexico.

Holmlid, S. 2011. There is more to service than interactions. In: Meroni, A. and D. Sangiorgi (eds.), *Design for Services*. Gower: Farnham, UK, pp. 89–96.

INAPAM. 2011. *Ley de Los Derechos de Las Personas Adultas Mayores*. Instituto Nacional de las Personas Adultas Mayores: Mexico.

Johnson, D., J. Clarkson, and F. Huppert. 2010. Capability measurement for inclusive design. *Journal of Engineering Design*, 21:275–88.

Katz, D. and L.A. Garrow. 2012. The impact of bus door crowding on operations and safety. *Journal of Public Transportation*, 15(2):71–93.

Lavery, I., S. Davey, A. Woodside, and K. Ewart. 1996. The vital role of street design and management in reducing barriers to older people's mobility. *Landscape and Urban Planning*, 35:181–92.

Mager, B. and T.J. Sung. 2011. Special issue editorial: Designing for services. *International Journal of Design*, 5(2):2–4.

Marsden, G., M. Cattan, A. Jopson, and J. Woodward. 2010. Do transport planning tools reflect the needs of the older traveller? *Quality in Ageing and Older Adults*, 11(1):16–24.

Meroni, A. and D. Sangiorgi. 2011. *Design for Services*. Gower: Farnham, UK.

Metz, D.H. 2003. Transport policy for an ageing population. *Transport Reviews*, 23(4):375–86.

Mollenkopf, H., F. Marcellini, I. Rouppila, Z. Szeman, and M. Tacken. 2005. *Enhancing Mobility in Later Life: Personal Coping, Environmental Resources and Technical Support; The out-of-Home Mobility of Older Adults in Urban and Rural Regions of Five European Countries*. IOS Press: Amsterdam, The Netherlands.

Moritz, S. 2005. *Service Design: Practical Access to an Evolving Field*. KISD—Köln International School of Design: London.

Nickpour, F., P.W. Jordan, and H. Dong. 2012. Inclusive bus travel—A psychosocial approach. In: Langdon, P., J. Clarkson, P. Robinson, J. Lazar, and A. Heylighen (eds.), *Designing Inclusive Systems*. Springer: London, UK, pp. 13–23.

Parker, S. and J. Heapy. 2006. *The Journey to the Interface: How Public Service Design Can Connect Users to Reform*. Demos: London, UK. http://hm-treasury.gov.uk/d/cypreview2006_wiltshirecouncil2.pdf.

Polaine, A., L. Løvlie, and B. Reason. 2013. *Service Design: From Insight to Implementation*. Rosenfeld Media, LLC: Brooklyn, New York.

Redfern, M, R. Cham, K. Gielo-Perczak, R. Grönqvist, M. Hirvonen, H. Lanshammar, M. Marpet, C.Y. Pai, and C. Powers. 2001. Biomechanics of slips. *Ergonomics*, 44(13):1138–66.

Robson, C. 2011. *Real World Research*. 3rd edn. Wiley: Cornwall, UK.

Rogers, W., B. Meyer, N. Walker, and A.D. Fisk. 1998. Functional limitations to daily living tasks in the aged: A focus group analysis. *Human Factors: The Journal of the Human Factors and Ergonomics Society*, 40(1):111–25.

Sedesol. 2011. *Ley General Para La Inclusion de Las Personas Con Discapacidad*. Secretaria de Desarrollo Social, Gobierno de la Republica: Mexico.

Seidel, D., N. Crilly, F.E. Matthews, C. Jagger, P.J. Clarkson, and C. Brayne. 2009. Patterns of functional loss among older people: A prospective analysis. *Human Factors: The Journal of the Human Factors and Ergonomics Society*, 51(5):669–80.

Startzell, J.K., D.A. Owens, M. Lorraine, L.M. Mulfinger, and P. Cavanagh. 2000. Stair negotiation in older people: A review. *Journal of the American Geriatrics Society*, 48:567–80.

Stickdorn, M. 2010. *Definitions: Service Design as an Interdisciplinary Approach*. In: Stickdorn, M. and J. Schneider (eds.), *This is Service Design Thinking: Basics Tools Cases*, BIS Publishers: Amsterdam, The Netherlands, p. 373.

Stickdorn, M. and J. Schneider. 2010. *This is Service Design Thinking: Basics Tools Cases*. BIS Publishers: Amsterdam, The Netherlands.

United Nations Population Fund. 2012. *Ageing in the Twenty-First Century: A Celebration and a Challenge*. United Nations Population Fund: New York.

Webber, S.C., M.M. Porter, and V.H. Menec. 2010. Mobility in older adults: A comprehensive framework. *The Gerontologist*, 50(4):443–50.

WHO. 2002. *Active Ageing: A Policy Framework*. World Health Organization: Geneva. http://whqlibdoc.who.int/hq/2002/who_nmh_nph_02.8.pdf.

WHO. 2007. *Global Age-Friendly Cities: A Guide*. World Health Organization: Geneva. http://www.who.int/ageing/age_friendly_cities_guide/en/.

28 Enhancing Sustainability Embodying Human Factors in Building Design

Erminia Attaianese

CONTENTS

28.1 SUSTAINABILITY IN THE CONSTRUCTION DOMAIN

It is generally accepted that sustainable development calls for a convergence between the three pillars of economic development, social equity, and environmental protection (Drexhage and Murphy, 2010). As we know the concept was first introduced by Brundtland (UN, 1987), who defines development as "sustainable" when it "meets the needs of the present without compromising the ability of future generations to meet their own needs." It considers the long-term perspectives of the socio-economic system, to ensure that improvements occurring in the short term will not be detrimental to the future status or development potential of the system. Such kind of development implies minimizing the use of exhaustible resources, or at least, ensuring that revenues obtained from them are used to create a constant flow of income across generations, making an appropriate use of renewable resources (Bellù, 2011). Human organizations must act aiming at the same time at the effective protection of the environment; prudent use of natural resources; social progress which recognizes the needs of everyone; and maintenance of high and stable levels of economic growth employment.

Buildings and the way they are realized and operate have a fundamental role in sustainable development as they impact on the environment, consume large quantities of resources, involve large numbers of workers, and represent a large proportion of economic activity. For this reason decisions made during all stages of the construction process are vital for maximizing sustainability (Boswell and Walker, 2005). Generally the construction sector is considered to be strategic in sustainability terms, since it involves different materials and activities, affects numerous stockholders, operators and users, moving huge financial capitals. Just in the European Union (EU) the construction sector represents more than 10% of EU GDP, more than 50% of fixed capital formation, and directly employs almost 20 million people. It covers building and infrastructure design and the stages of construction, including onsite activities embracing site preparation, construction of complete buildings, building installation; manufacturing of construction materials, including building products and components; building use, maintenance and operation; and building and infrastructure reuse or disposal.

According to the assumption that often reduces sustainable development to an environmental issue (Drexhage and Murphy, 2010), the sustainability of the construction sector has been largely intended as a "green" question, considering that the energy performance of buildings and resources efficiency in manufacturing, transport, and use of construction products have a crucial impact on the environment in terms of energy and pollution issues, since it is estimated that 40%–45% of Europe's energy consumption stems from buildings with a further 5%–10% being used in processing and transport of construction products and components (FWC Sector Competitiveness, 2012).

In 1994 International Council for Building (CIB) defined sustainable construction as creating and operating a healthy built environment based on resource efficiency and ecological design, articulating the seven principles by which design and assessment decisions should be informed during each step of a building's life cycle, considering land, materials, energy, and ecosystems (Kibert, 2013).

More recently, sustainable development in construction has been intended to go over the so-called environment "box" and beyond economic viability, addressing the wide sense of the concept. As ISO TS 21929 premises in order to define sustainable indicators of buildings, sustainable construction brings about the required performance with the least unfavorable environmental impact, while encouraging economic, social, and cultural improvement at a local, regional, and global level (ISO TS, 2006). Construction is said to be sustainable when it meets environmental challenges, but responds also to social and cultural demands. The fundamental concept of sustainable construction is to deliver long term affordability, quality, and efficiency, value to clients and users, whilst decreasing negative environmental impacts and increasing economic sustainability (Bal et al., 2013).

28.2   WHEN CAN A BUILDING BE SEEN AS SUSTAINABLE?

Since it has been defined as the practice to activate a process that is environmentally responsible and resource-efficient, throughout a building's life cycle, including design, construction, operation, maintenance, renovation, and demolition (U.S. Environmental Protection Agency, 2009), building sustainability is greatly still intended as an environmental footprint both in theory and practice. Sustainable buildings are, in fact, often confused with energy-efficient buildings, as it is shown by the interchangeable use of the terms sustainable building, green building, and high-performance building (Berardi, 2013).

In scientific and technical literature, the sustainable dimension of buildings is usually defined in more ways, frequently overlapped, but mostly intended as building "greenery," in a restricted sense, and energy performance is the most used parameter to assess the sustainability of a building.

Green buildings are seen as synonymous of environmentally friendly, which design is aimed to reduce the overall impact of the built environment on human health and the natural environment.

But by an enlarged view point it has been assumed that a sustainable building has to contribute to sustainable development, through its characteristics and attributes, safeguarding and maximizing functionality and serviceability as well as aesthetic quality; minimizing life cycle and protecting and/or increasing capital; reducing land use, raw materials, and resource depletion, but also reducing malicious impacts on the environment; protecting the health, comfort, and safety of workers, occupants, users, visitors, and neighbors; and preserving cultural values and heritage (Lutzkendorf and Lorenz, 2007).

Recently high-performance building has been called on to integrate and optimize, on a life cycle basis, all major high-performance attributes, including energy conservation, environment, safety, security, durability, accessibility, cost-benefit, productivity, sustainability, functionality, and operational considerations (Clements-Croome and Jeronimidis, 2005; Fischer, 2011; Attaianese, 2012).

Even if it has been noticed that the social aspects of a sustainable building are still a rarely investigated topic (Berardi, 2013), a reinterpretation of sustainable building visions has been recently conducted and a wider consideration of social issues among sustainability principles have been addressed (CIB, 2010).

As a result a sustainable building has been defined as a healthy facility designed and built in a cradle-to-grave resource-efficient manner, using ecological principles, social equity, and life cycle quality value, and which promotes a sense of sustainable community, increasing cultural and heritage issues, traditions, human health, and social infrastructure (Berardi, 2013). Green building requires a holistic approach that would include the consideration of health and well-being of occupants in every aspects of the project, not only the individual apartment unit but the whole building, with the wider context of the environment and community around it (Chuck and Tai, 2011).

Also existing design and assessment methodologies are affected by the so-called "environment box" of sustainability, since their shared goals mainly focused on environmental issues, which involve energy and resources conservation concerns. Assessment and certification schemes that measure the sustainability of buildings have been in operation for a number of years in many countries. Some of them are members of the World Green Building Council, and dozens more are in the process of forming national councils or adopting certification standards. Rating systems have an important role because they not only provide criteria for assessing the sustainable goals of green buildings, but give also specific principles for design, operation, and construction of high performance buildings (Akadiri et al., 2012). Nevertheless even if the optimization of site potential, preservation of regional and cultural identity, minimization of energy consumption, protection and conservation of water resources, use of environmentally friendly materials and products, a healthy and convenient indoor climate, and optimized operational and maintenance practices are declared purposes that can be found in several building sustainability assessment methods (Bragança et al., 2010), the number of environmental and energy criteria strongly prevail on human and social items.

28.3 EMERGING DESIGN CHALLENGES OF SUSTAINABLE BUILDING

28.3.1 Occupants Role in Energy Goals

During the last years a general understanding about the importance of the occupants in order to achieve environmental goals of green buildings arose within energy building experts. The 2009 Passive Low Energy Architecture Conference (PLEA), for example, was themed "Architecture, Energy and Occupant's Perspective" with the ambition of positioning building inhabitants as key active determinants of energy performance in passive design through adaptive opportunities (Cole et al., 2010). Overcoming the traditional consideration of building users as passive occupants, to which indoor air quality (IAQ) and indoor comfort were usually referred (Guerin, 2012), the need of field studies considering users in term of "inhabitants" merged, capturing with this term, a more active users' engagement in building energy concerns, since "buildings don't use energy but people do" (Janda, 2011). Inhabitants are more directly involved with building systems and operation through opening and closing windows, doors, light, shading devices, thermostats, vents and other manual controls so their behaviors significantly influence building energy use. Buildings support human activities and the energy needed for doing so depends on how they are designed, mainly in relation to the operation needs and expectations of the inhabitants (Cole et al., 2010). Climate and building characteristics alone have been proven to be insufficient as determinants of energy optimization, and the roles of occupants' behaviors and socioeconomic factors have resulted in being important components (Steemers and Yun, 2009; Vale and Vale, 2010), also in terms of energy demand (Haldi and Robinson, 2011). The challenging area of investigation for building performance and evaluation has been focused on human behavior, in order to assess and improve design affordance and provide comprehensive feedback for empowering users' environment control and reduced energy use (Peffer et al., 2013); to develop maintenance and operations as dimensions to be integrated in building performance and postoccupancy evaluations (Stevenson and Leaman, 2010; Monfared and Sharples, 2011), looking at the complete life cycle of the building from initial procurement through build management process to eventual demolition (Preiser and Vischer, 2005); to better address diversities of inhabitants both in energy regulations and standards both in evaluation methods and design strategies, since the relationships between users and buildings changes over time and each situation must be studied and assessed on its own merit (Gupta and Chandiwala, 2010). To reach these goals a better understanding of users' expectations, attitudes, perceptions, and behavior by interrelated human factors directly with the physical performance of the building is required (Stevenson and Leaman, 2010).

The need to directly focus the human perspective in the energy concerns of building design and evaluation clearly emerges by user evaluation of energy-efficient buildings research outputs conducted in the last decade. The main concerns resulted in relation to the inhabitants' and occupants' perception of comfort and technical operation. In passive houses sensible differences between experienced thermal comfort and simulated indoor climate have been reported (Samuelsson and Luddeckens, 2009; Hauge et al., 2011), confirming that people perceive indoor thermal conditions differently (they change individually and vary over the time) and their perception may be influenced by several context factors (i.e., cold floor surfaces or draughts may decrease the perceived temperature, whilst the vision of fireplaces increases it). In green occupational buildings the main occupants' comfort dissatisfactions have been reported about temperature (Heerwagen and Zagreus, 2005; Leaman and Bordass, 2007), light and noise conditions (Abbaszadeh et al., 2006; Leaman and Bordass, 2007), frequently in association to the open-plan layout of offices, which characteristics have been also considered as factors inducing distractions, interruptions, and lack of concentration negatively influencing occupants working ability (Heerwagen and Zagreus, 2005). Moreover the perception of thermal comfort has often resulted in being linked to the occupants' ability and possibility to control the indoor climate, by the effective use of a manual thermostat (Peffer, 2013), opening windows (Goins, 2013), and controlling solar glare (Nicol and Roaf, 2005; Barlow and Fiala, 2007; Wagner, 2007). Occupants are more comfortable in buildings in which the amount of perceived control over temperature, ventilation, and noise is high (Boerstra et al., 2013).Therefore the question of usability of a building's controls, in terms of easiness of use and feedback, has been reported as crucial (Nicol and Roaf, 2005; Leaman and Bordass, 2007; Hauge et al., 2011). Thus research surveys conclude that energy-efficient buildings are experienced as being more satisfying than conventional ones, but the incidence of occupants that perceive them uncomfortable or that are indifferent about green buildings in which they work is relevant (Paul and Taylor, 2008). In order to overcome the gap between energy efficiency simulations and occupants' actual perceptions, longitudinal observations and the assessment of building operation and maintenance over time, together with considerations of perceived architectural and aesthetic qualities, need to be taken into account. Thus more focus on human aspects in building users' evaluation as important areas of focus for further research have been assumed (Hauge et al., 2011; Duca, 2014).

28.3.2 HIDDEN RISKS AND HUMAN SIDE EFFECTS OF SUSTAINABLE SOLUTIONS

There is strong evidence suggesting a direct correlation between some green design solutions and potentially risky building failures (Odom et al., 2007). In fact not all recurring green elements and attributes, singularly or combined, always match human needs, giving negative effects on objective and perceived health, safety, wellbeing, and task efficiency of occupants and users in general. Studies show that conflicting situations arise from maximization of energy passive performance of buildings, finally decreasing overall building usability (see Tables 28.1 through 28.4).

Fire forces studies, mostly based on reported fire incidents that are related to green issues (Tidwell and Murphy, 2010; Meacham et al., 2012), demonstrate that energy efficiency measures are critical components of green construction, worsening users' safety substantially in case of fire: fire load and

TABLE 28.1

Samples from Green Strategy and Technology for Optimizing Thermal Performance of the Exterior Envelope

Usual Elements and Attributes	Human Side Effects	References
Structural insulated panel Exterior insulation materials and finishes (rigid and spray foams, foil insulation)	Insulation increases interior temperature, increase in burning characteristics of interior and exterior materials; it can contribute to flame spread and fuel load Increased exposure to mineral fibres in demolitions Increased exposure to volatile organic compounds from, for example, paints or adhesives, and to dust, including crystalline silica for construction workers	Tidwell and Murphy (2010), Meacham et al. (2012) EASHW (2013)
High-performance glazing (i.e., insulated double glazing, triple glazing, or double pane glazing with a suspended low-e film)	Double glazing or other type of highly insulated glass is heavier than conventional glass (a double-glazed window of the same size has about twice the weight) Impact on fire force access because they can find it difficult to break for ventilation or rescue purposes While glass can be completely recycled, most high-performance glass has little recycled content The creation of glass utilizes a great deal of energy (high embodied energy) Too cold in warm weather, frequently due to big glass openings, also with the impossibility to operate window/shades control	EASHW (2013) Tidwell and Murphy (2010), Meacham et al. (2012) Heerwagen and Zagreus (2005) Guerin et al. (2012) Byrd (2012)
High-performance blocks	High-performance blocks may be heavier than conventional ones, may show inadequate grip and present possibility of injuries	Attaianese and Duca (2012b)
Double skin facade and cavity walls	Can create chimney for vertical smoke and flame spread	Tidwell and Murphy (2010), Meacham et al. (2012)
Bamboo and other cellulosic materials	Can contribute to flame spread, smoke development and fuel load	Tidwell and Murphy (2010), Meacham et al. (2012)
Recycled paper flakes and flax wool	They are impregnated with 8% boric acid (sodium tetraborate), which serves as a fire retardant and an antimicrobial agent. Boric acid has been classified as toxic to the reproductive system in the EU	EASHW (2013)
Vegetative roof systems	Can contribute to fire load, spread of fire, risk of external fire	Tidwell and Murphy (2010), Meacham et al. (2012)
Exterior vegetative covering	Can impact on fire forces access	Tidwell and Murphy (2010), Meacham et al. (2012)
Automatic windows programmed to be opened upon weather conditions	Can impact on fire forces access and rescue purposes Lack of personal control of windows opening can impact on perceived comfort influencing occupants' ability to achieve their desired conditions	Tidwell and Murphy (2010), Meacham et al. (2012) Stevens (2001) Heerwagen (1998)
Increasing vestibules and atriums in the floor plane	Vestibules used to inhibit the migration of outside air to the interior of the building will increase the degree of difficulty to deploy hose lines to the interior of the building	Tidwell and Murphy (2010), Meacham et al. (2012)

flame spread increasing due to the massive use of insulation materials and finishes; smoke development and flame spread rising due to the large open plans, the higher room volumes, and the use of solar tubes; obstructing evacuation plans and rescue purposes by the large use of high performance glasses (i.e., insulated double glazing, triple glazing) that are usually not operable windows, difficult to break for the way in and way out, and for ventilation.

TABLE 28.2

Samples from Green Strategy and Technology for Maximizing Natural Day Lighting

Usual Elements and Attributes	Human Side Effects	References
Awnings	Can impact on fire forces access also inhibiting the deployment of ladders	Tidwell and Murphy (2010), Meacham et al. (2012)
Increased glass windows	Lower sound isolating capabilities	Muehleisen (2010)
Increased internal glass walls	Reverberation can be significant in rooms with speech	
Decreased interior hard walls and opaque partitions	privacy and speech clarity issues due to large amounts of glass	
	Increased transmission of outdoor noise	
	"Acoustical" glass products highly priced and ongoing costs	
	Glass is expensive to maintain	
	While glass can be completely recycled, most high-performance glass has little recycled content	
	The creation of this kind of glass utilizes a great deal of energy (high embodied energy)	
Increasing open spaces	Too much daylight or incoming sun with problems of refection and dazzling and visual discomfort	Goins et al. (2013)
	Increased ground noise level	Heerwagen and Zagreus (2005)
	Loss of privacy	Guerin et al. (2012)
	Distraction and loss of concentration	
	Too much air movement	
	Way finding problems and lack of orientation	
Use of light shelves	The lack of window views, especially views of nature	Loftness et al. (2006)
Use of solar tubes	and proximity to windows, impact on emotional health, and occupants productivity	Edwards and Torcellini (2002)
	Tubes provide an additional means for fire transmission and smoke migration through spaces	Tidwell and Murphy (2010), Meacham et al. (2012)

A common characteristic of many "green" office buildings is a high proportion of glazing, but there is recent evidence that large areas of glazing are perceived as unsatisfactory by occupants with a potential loss in productivity (Byrd, 2012). On the other hand, green strategies to reduce heat loss in winter and sun access in summer can bring in some regions the reduction of openings on the building envelope, adopting design solutions with less windows and more light shelves and solar tubes. These solutions are reported as negative by occupants since the lack of windows views, especially views of nature and window proximity, impact on emotional health and productivity (Heerwagen, 1998; Stevens, 2001; Edwards and Torcellini, 2002; Loftness et al., 2006). Also increasing open spaces are associated to negative effects for occupants: loss of privacy with distraction and loss of concentration. Too much air movement is also reported, with way finding problems and lack of orientation (Heerwagen and Zagreus, 2005; Guerin et al., 2012). A natural tension between the benefits provided by open spaces, and the need for speech privacy to concentrate or perform confidential tasks has been reported. Complaint profiles of those dissatisfied with the acoustic quality in workspace point to problems with sound privacy, and distracting noise from people's conversation and telephone rings (Abbaszadeh, 2006).

Several postoccupancy studies report more frequent perceived discomfort associated to acoustics of green buildings, particularly due not only to open plans but to light partitions too, increasing ground noise level (Goins et al., 2013), or amplifying transmission of outdoor noise for the larger penetrations between the interior and exterior environments, with consequent loss of privacy and

TABLE 28.3

Samples from Green Strategy and Technology for Optimizing Natural Ventilation

Usual Elements and Attributes	Human Side Effects	References
Increased openings in the building enclosure	Increased transmission of outdoor noise (negative acoustic impacts due to more and larger penetrations between the interior and exterior environments) Increased outside air rates can bring unwanted outdoor pollution and humidity Production of unwanted air flows	Muehleisen (2010) Loftness et al. (2006)
Limited use of partitions so that air can flow nearly unimpeded	Increased noise transmission between rooms	Muehleisen (2010)
More open spaces horizontal More open space vertical	Increased ground noise level Loss of privacy Distraction and loss of concentration Too much air movement In case of fire, faster fire growth due to the greater air volume and the more readily available fuel sources Lack of compartmentalization to limit fire spread to smaller areas Natural ventilation can impact on ability to control smoke, can influence smoke movement depending on environmental conditions High volumes can influence sprinkler and detector performance	Heerwagen and Zagreus (2005) Tidwell and Murphy (2010), Meacham et al. (2012)

concentration of occupants. Also lower sound isolating capabilities and reverberation of glasses, frequently used for daylight maximization, bring reported negative acoustic effects (Odom et al., 2007; Muehleisen, 2010).

Most evident health concerns refer to the use of insulating and organic materials in green buildings, since they contain fibers and volatile substances, dangerous if inhaled. Furthermore open

TABLE 28.4

Samples from Green Strategy and Technology for Reducing the Use of Materials and Resources

Usual Elements and Attributes	Human Side Effects	References
Increased use of lightweight structural components (in wood, steel, or concrete)	They will rise earlier in temperature and fail more quickly than heavier components Fibers can be toxic in construction phase for workers, and for occupants if broken	Tidwell and Murphy (2010), Meacham et al. (2012)
Increased low-impact storm water technologies and other technologies that support on-site retention and ground water recharge or evapo-transpiration	The use of pervious bituminous paving and/or concrete for paving and walkways may affect pooling of flammable liquid and resulting pool fire, containment, runoff containment issues	Tidwell and Murphy (2010), Meacham et al. (2012)
Increasing use of renewable organic material (i.e., bamboo, straw, sheep wool, flax, and cork)	Might bring elevated risks of exposure to protein-based allergens, and microorganisms such as bacteria, molds and fungi, or endotoxins	EASHW (2013)
Recycled asphalt with fly ash	Fly ash contains heavy metals and may contain polycyclic aromatic hydrocarbons (PAHs), some of which are carcinogenic	EASHW (2013)

issues are under investigation for the effects of the newest and recycled materials (Odom et al., 2007; Tidwell and Murphy, 2010; Meacham et al., 2012; EASHW, 2013).

Also in relation to construction workers, some green solutions are reported as potentially unhealthy and unsafe (Rajendran et al., 2009), especially techniques for optimizing thermal performance of the exterior envelope. Structural insulated panel and exterior insulation materials and finishes (i.e., rigid and spray foams, or foil insulation) may increase exposure to volatile organic compounds from, for example, paints or adhesives, and to dust, including crystalline silica (EASHW, 2013). Moreover the need to have a high-energy envelope induces construction workers to manipulate heavy components: a double-glazed window of the same size has about twice the weight (EASHW, 2013); high-performance blocks are heavier than conventional ones, and may have inadequate grip and result in injuries (Attaianese and Duca, 2012b). On the other hand, fibers treating lightweight structural components in wood, steel, or concrete, can be toxic in the construction phase for workers, and for occupants if broken (Tidwell and Murphy, 2010; Meacham et al., 2012).

28.4 DESIGN FOR SUSTAINABLE BUILDINGS IN THE HUMAN FACTORS AND ERGONOMICS DOMAIN

Sustainability has today almost become one of the main concerns also in human factors studies (Martin et al., 2013), but the issue is not new in the ergonomics domain (Lueng, 2003). It has been observed that even if ergonomics had noticed global problems for two decades (Moray, 1993, 1995; Martin et al., 2013), sustainability has been part of the basic understanding of the human factors–ergonomic discipline for a long time, since several fundamental definitions refer ergonomics to the central principles and objectives of sustainable development (Zink and Fisher, 2013). The emphasis on the optimization of human well-being and overall system performance that IEA (International Ergonomics Association) attributes to the ergonomic–human factors focus, in order to understand the interactions among humans and other elements of a system, providing theoretical principles, data, and methods (IEA [International Ergonomics Association], 2000) is well known, for example. To break down general sustainability requirements and make the discussion about human factors and sustainable development more tangible, Zynk recently summarized some exemplary lead principles for sustainability-oriented design of human factors–ergonomic concepts and instruments: the preservation and development of human and social capital, based on the understanding that social sustainability is only one part of the three-dimensional model of sustainability; focusing on a broad systems approach including whole value creation chains; striving for a life cycle perspective in design; comprising impacts on society as well as impact on other related systems; addressing barriers for sustainable development (Zink and Fisher, 2013).

Even through the goals of sustainability and ergonomics are implicitly congruent (Martin et al., 2013), recent reported studies about the contribution of human factors to sustainability are limited among the ergonomics community. Many studies states that sustainability goals can be better achieved realizing efficient durable systems, to be used in an efficient manner (Martin et al., 2013). But even though few authors explicitly consider the supportive rule of the human factors–ergonomics approach (Steimle, 2006; Brown and Legg, 2011; Martin et al., 2013), the need to enhance design for sustainability involving systems users and their sustainable behaviors emerged (Bhramra et al., 2011). The connection between sustainability and human factor–ergonomics has been exploited through the notion of green ergonomics that focuses human factors goals in a pro-nature view. It is oriented to support the development of efficient systems, that in addition to being healthy and safe, need less energy to be used, and assist people in the comprehension of sustainable behavior change (Hedge, 2008; Hanson, 2012; Thatcher, 2012). In this view the linkage between humans and nature is bi-directional, so that green ergonomics considers both how human systems can facilitate the conservation, preservation, and restoration of natural capital, and how human interactions with nature can facilitate well-being and effectiveness. The first goal can be achieved by supporting the design of low resources systems and products, so that they are also able to favor conservative and sustainable behaviors by users (Thatcher and Milner, 2012). The second goal can be reached by designing systems and products inspired by the effectiveness of nature (Thatcher and Milner, 2012; Obiozo and Smallwood, 2013), able at the same time to stimulate human ability and positive reactions (creativity, productivity, healing effects). One of the obvious place for green ergonomics to make impact on improving individual well-being is the built environment (Thatcher, 2013), but few ergonomic studies experimented with the human component in the actual sustainability performance of buildings, in two directions: both in terms of impact of occupants' and inhabitants' actions and behaviors on building performance optimization and as an effect of this performance on human reactions and perceptions in relation to sustainability issues. It seems to be recently acknowledged that green building specifications may not automatically lead to improved physical and physiological well-being or perceived productivity gains, and the benefits of sustainable design are those also sought by ergonomists: improve well-being and productivity of all users of the design, due to improved design performance (Martin et al., 2013). It is stated that human factors can contribute to understanding how buildings are used and how people interact with their physical environment, also by identifying the needs of people who will occupy the building (Hanson, 2012), but few field studies have been carried out by ergonomists directly addressing sustainability issues (Karwowsky, 2005; Thatcher, 2012). As Haslam and Waterson (2013) recently concluded, ergonomics activity on this front appears limited and tentative.

A field strongly focused on user involvement in building performance evaluation is building usability. The starting point of this issue was in 2001 when the CIB Working Commission on the Usability of Workplaces (CIB W111, 2010), began to investigate the application of the ISO 9421 international standard on usability, previously applied in the evaluation of consumer products, to the built environment, although some studies on the issue were conducted earlier (Attaianese, 1997, 2001). It particularly included all aspects of the "user experience" in an organizational setting, encompassing the end user's interaction with an organization and its facilities and with the processes of design and management (of the built environment). According to the definition from the international standard on usability as effectiveness, efficiency and satisfaction with which a specified set of users can achieve a specified set of tasks in a particular environment (ISO, 1996), building usability, or functionality in use, is concerned with a building's ability to support the user organization's economic and professional objectives. From the user perspective, usability means that artifacts are easy and fast to learn, efficient to use, easy to remember, allow rapid recovery from errors, and offer a high degree of user satisfaction. The usability of the built environment focuses on user perceptions of the ease and efficiency with which they can use the building, considered as a facility (Jenso and Haugen, 2004). While functionality can be evaluated on the product (building), usability cannot be evaluated only analyzing the building, but looking at the context of its use, depending on users' values in culture, context, time, and situation (Alexander, 2008). Even though a partial vision of building usability, dealing with aspects of accessibility management in buildings (i.e., constructional aspects of access to buildings, to circulation within buildings, to egress from buildings in the normal course of events, and evacuation in the event of an emergency) has been recently introduced in a new international standard (ISO, 2011), some practical application of usability on specific buildings and activities show a more comprehensive consideration of the concept where usability requirements and their related markers have been formulated (Afacan and Erbug, 2009; Haruna et al., 2011; Duca, 2012). Some recurring usability areas or criteria associated with buildings are: accessibility, spatial orientation, in terms of way finding and paths efficiency, aesthetic and affective elements, comfort and well-being, flexibility and safety aspects. More recently the concept of sensory design architecture emerged (Leheman, 2011). Taking an occupant-centered approach, it is aimed at optimizing the "health" within the building, in terms of the health of an individual occupant, the health of a building's effectiveness, and the health of its ability to harmonize with surrounding environments, considering the effects of architecture on occupants, and how it can be better attuned through sensory design for healthier mind and body connection—physiologically, cognitively, emotionally, behaviorally, and spiritually (Leheman, 2011).

28.5 HUMAN FACTORS FOCAL POINTS FOR SUSTAINABLE PERFORMANCE OF BUILDINGS

As people are themselves part of the environment, green and sustainable buildings both include the human perspective in their design concerns, mainly considering the health and well-being of people involved in the design process, particularly in terms of the comfort and productivity of occupants (Miller et al., 2009). But despite frequently the inhabitants' comfort being included in an environmental-friendly building design and many standard tools for sustainable performance rating of buildings have included ergonomic related indicators (Hedge, 2008), it can be noticed that human factors are restricted to the ergonomic features of machines, equipment, and furniture, considering them as tools able to reduce discomfort and musculoskeletal disorders of occupants. Rarely are ergonomics–human factors issues included in sustainability requirements for architectural detailing of the whole-building system, and moreover, for improving the building design process (Attaianese, 2011).

In fact the ergonomic approach being aimed to optimize human interactions with systems, in order to make activities more efficient, safe, comfortable, and satisfying, architectural design and building construction and management can be enhanced by the consideration of the human factors perspective, because it gives the cultural and practical references to envisage how technical solutions and details constituting the building can be effective, efficient, and satisfying as fitting, primarily, the whole of needs derived from peoples' lives and the work activities they perform, "in or for" it (Attaianese and Duca, 2012a).

On the other hand to be sustainable a building has to offer a high performance, not only in environmental terms, as energy efficient but also through its life cycle be durable and effective, and occupant safe, secure and productive, in order to address, in a wider sense, sustainability goals.

According to the enlarged view point, above mentioned, assuming sustainable building has to contribute to sustainable development, by safeguarding and maximizing its functionality and serviceability as well as aesthetic quality; minimizing life cycle and protecting and/or increasing capital; reducing land use, raw materials, and resource depletion, but also reducing malicious impacts on the environment; protecting health, comfort, and safety of workers, occupants, users, visitors and neighbors; and preserving cultural values and heritage (Lutzkendorf and Lorenz, 2007), the following points will show how some of these aspects can be addressed by the human factors approach.

28.5.1 HUMAN FACTORS FOR ENERGY EFFICIENCY

A reduced energy footprint is the most cited element of sustainable building, and energy final use is increasingly considered a crucial element of the economic costs and environmental impacts, since the occupants' behavior influences approximately half of the total amount of energy used in buildings. The occupants' behavior is affected by more factors: politics and rules imposed by public administrators and legislators; culture, local habits, social conditioning, and lifestyles; capability to control building systems, determining reactions of adaptation or rejection of the available technologies. In fact energy consumption in buildings is linked to a general problem of adequacy, involving systems and technologies, which are asked to become more and more effective and functional, as well as occupants need to become more conscious of procedures for their use. Against the pressing necessity of lifestyles and approaches compatible with optimal resources consumption, the issue of pervasiveness of appropriate individual and collective behaviors is now emerging. This perspective contributes to enhance the role of energy end-users, which are asked to fit their needs to the conscious usage of resources, also by mean of tools and devices controlling more and more sophisticated functions (Flemming, 2008; Janda, 2011).

In this framework of ergonomics–human factors, it can be particularly helpful for the availability of methodological and operational tools able to analyze human activities, observe, and understand needs and expectancies coming from users in order to produce interfaces compatible with them

(Attaianese, 2011; Duca, 2014). It is matter of understanding and assessing the ways of human–system interaction, as well as designing devices and procedures able to improve their efficiency, assuring all stakeholders satisfaction in a balanced relation with the environment. In fact it has been considered that if green buildings are designed paying poor attention to users' preferences and needs, they can result in a sort of fragility with respect to their assumed energy performances (Leaman and Bordass, 2007). Moreover ergonomic approach to the building design process facilitates the selection of the most appropriate technologies; these are supposed to bring an optimized building functioning and, consequently, a waste reduction, thanks to the optimization of built estate management (Charytonowicz, 2007; Attaianese, 2009).

28.5.2   HUMAN FACTORS FOR FUNCTIONALITY AND SERVICEABILITY

To be sustainable a building has also to be adaptable throughout the service life and with an end-of-life strategy. It has to allow adaptation by changing performance and functionality requirements, in accordance with new constraints (CIB, 2010).

Building functionality and serviceability are both referred to as the conditions under which a building is considered useful to their occupants: the first involves building capacity to address hosted activities requirements since the early moments of its fruition by occupants; the latter concerns the conditions beyond which specified service requirements resulting from the planned use are still met. As measure of building utility they are an important element of sustainability (Fischer, 2011) that imply the protection of and/or increase in the building capital value. In fact it has been stated that the value of a product is defined by the individual needs of the customer, so that it is important to develop a design value that is adequate to specific groups of customers (Jensen, 2005). This also because a product is meant to be able to provide use value according to its efficacy (Himanen, 2003), considering how it fulfills users' needs and increases their satisfaction.

The human factors approach can implement this issue in building design, by supporting the analysis of the use context, by the gathering and survey of users' needs and expectations and by the observation and description of how all users can/could achieve their goals using the building, and also considering human differences and variability in ability and expectations (Volker and Prins, 2005; Klugengseth and Hansen, 2010; Duca, 2014). This is particularly relevant considering that sustainability is a dynamic condition, and since both users' needs and behaviors change over time, building should accommodate different changes in requirements and functions (Berardi, 2013). Human factors can support flexibility of spaces and layout (Oliveira and Elali, 2012).

28.5.3   HUMAN FACTORS FOR IMPROVING OPERATION AND MAINTENANCE

Many studies state that maintenance is necessary to achieve efficient building energy performance (Lewis, 2010). Since the main scope of maintenance is the continuity in keeping of the building and infrastructure estate capacity to perform required functions, it assumes a crucial role for building sustainable management considering the chance it gives for conscious resources utilization, as every action on a building component has the potential to improve or decrease its efficiency.

The human factors approach can contribute to increase building effectiveness, first of all, by enhancing maintainability. Maintenance efficacy is a function of processes efficacy activated by maintenance, and the human factors perspective could be crucial in the requirements design for planning and executing maintenance activities, and directly and indirectly improving environmental performances in their whole. In fact ergonomics–human factors supports a holistic approach to maintenance in which physical, cognitive, social, and organizational factors can be taken into account. These aspects have to be considered to optimize maintenance-oriented activities, by facilitating human–systems interactions.

Particular attention has to be reserved for people's interactions with technical and social systems, given that clear and effective connections bring increased autonomous and conscious usage of resources, by effective usage of built environment systems, this issue being widely considered as strategic to encourage environmentally friendly, and then sustainable, behaviors. From the design stage, when the performance and physical characteristics of the building and its durability are influenced by architectural detailing, concerning for example, the impact of design on structures and materials installed, as well as the life cycle of each component of building. At the construction and management stage, when unsatisfactory detailing, incorrect selection of building materials, components and systems, and lack of standardization have been detected as factors for the occurrence of main construction defects, also caused by incorrect assembly on site or in factory or when built environment spaces and layout create operative space often inaccessible for maintenance tasks; or when the facility management of built environment can show ineffectiveness and inefficiency due to the lack in information availability and exchange from the different skills involved. At the stage of the end-users' fruition, when incomprehensible and/or not-easy control devices of built environment systems influence resources depletion; and/or when the lack in communication between tenant and maintenance staff can affect effectiveness of building functions due to inadequate programming of the executive maintenance process (Attaianese, 2012).

28.5.4 Human Factors for Protecting Occupants Comfort

In a recent literature review it has been deduced that new approaches to indoor comfort are now emerging, all focusing on occupants rule (Cole et al., 2008). Occupants are becoming more active in shaping indoor conditions through improved means for personal control, by a perspective that assumes comfort as experienced not only in physiological terms, but also in the psychological, behavioral, and social senses. These evidences remark that indoor comfort conditions should be considered as variable and diverse rather than uniform and static, in order to simultaneously address users' needs and energy efficiency (Brown, 2009). The adaptive dimension of thermal comfort is not new (Nicol and Humphreys, 2002) such as the consideration of human factors in relation to office occupants (Erlandson, 2006), but less has been applied on related psychosocial aspects (Shove, 2008). The thermal satisfaction issue can be sustained by the ergonomic approach, since people's responses to the thermal environment are influenced by activity, clothing levels, stress, age, gender, and individual preferences. User profiling may strongly be helpful in a very early stage of the energy design of a building, for a more comprehensive analysis of needs, demands, and wishes, and a better articulation of the energy/environmental performance targets of the built environment.

Especially in homes and in occupational buildings, individuals' behaviors can be assumed as crucial starting data to collect in the sustainable design of building, in order to identify, from those actual behaviors and elements, features and components, how the energy design of a building can be improved. Moreover, user centered principles for design of energy devices are important to increase building and installation usability, considering that the most effective way to improve thermal comfort and satisfaction is by using individual controls (Peffer, 2013).

28.5.5 Human Factors for Improving Accessibility and Inclusivity

Even if mostly neglected in practice, the need for social dimension has been clearly expressed by the evolution of building sustainability principles and discussions (Berardi, 2013). Recently social and inclusion issues have been clearly included among CIB principles, since the CIB stated that sustainable building has to be healthy, comfortable, safe, and accessible for all, allowing full accessibility, and user-friendly, simple, and cost effective in the use of building facilities to everyone, and finally provide social and cultural value over time and for all the people (CIB, 2010).

Accessibility and inclusivity demands are strongly addressed by the human factors–ergonomics perspective as it focuses on cognitive, sensory, or physical abilities and impairments of all building users, setting the problem of the human variability consideration within the building design process. By analyzing human–building interactions resulting from the specific user's ability in a specific building's activities, thanks to link and layout analysis techniques, design outcomes at both whole building and detail scales can be supported, in order to better address different, variable, and sometimes conflicting human needs, not only for people with special needs, but for all. Way finding and safety performance may be improved, since green buildings are frequently conceived as open spaces with high volumes, leading people to low space control and consequent frustration with disorientation and distractions, both in ordinary use of the building and during emergencies (Attaianese and Duca, 2012a).

28.5.6 HUMAN FACTORS FOR PROTECTING HEALTH AND SAFETY OF CONSTRUCTION WORKERS

To be sustainable the building must protect all people that interact with it, regardless of the reason they have for interacting. It must allow safe working conditions during its construction and service life (CIB, 2010). This implies that at the design stage, materials, components, and construction techniques have to be previously considered not only in terms of energy performance, cost-effectiveness, or resources savings, but also in relation to their capacity to trigger risk situations for workers. The human factors approach may support this aspect giving criteria for evaluating a further aspect of resource consumption in the construction process, that is human efforts needed for set-up of any construction technique, since recent findings demonstrate that materials and techniques with similar energy performances offer different human performances according to workers perspectives, and some techniques determine working conditions that are worse than others (Duca and Attaianese, 2012). Moreover at the construction stage, ergonomic techniques for task and link analysis may increase safety performance at site, not only to protect workers but also in relation to a better management of possible interactions with inhabitants' circulation in the city centers.

28.6 CONCLUSIONS

Even if a wider vision of sustainable building has been reassumed where environmental, economic, and social values are equally identified and recognized, reviewed studies demonstrates that sustainability in construction is strongly focused on energy and environmental issues, and the so-called social pillars are actually under-considered.

The explicit reference to human and social aspects of building sustainability seems to be partially assigned to the occupants' role, since it has been generally assumed that final users are crucial in energy saving, even so this role is limited to energy questions and is far from being adequately weighted out in the assessment methods of buildings, nor effectively managed in design methodologies.

It is acknowledged that sustainable buildings have to take care of users, as part of the environment in global terms, since the sustainable design, construction, and use of buildings are based beyond the evaluation of the environmental impacts and economic aspects related to the life cycle costs, where humans are indirectly involved, but also on the direct role they have in the social aspects of sustainability. Despite that, current rating systems of buildings' sustainability show brief sections containing indicators mainly referring to users' comfort and less to social benefits.

Moreover in the last years, reported postoccupation evaluations and evidence-based studies reveal a direct correlation between some green design solutions and building failures, often eliciting conflicting situations among different building performances and triggering factors of undesirable effects on users, producing potential risks for occupants and construction workers. Therefore standard solutions to achieve building sustainability have not to be recommended, favoring instead local and case-by-case design strategies.

About the ergonomics domain, the literature shows that sustainability goals are implicitly coherent with human factors–ergonomics principles, even if a greater number of studies is aimed to understand how ergonomic intervention may be "green," rather than how ergonomics may support green building design.

To design really sustainable buildings, in a broader sense, it is crucial that solutions and details must be simultaneously selected addressing environmental, economic, and social goals. The human factors perspective can enhance building design, since ergonomic techniques are helpful to focus aspects of sustainable goals usually neglected or ignored. The focus on human–systems interactions, observation of specific users' needs and related predictable or actual tasks, let building design be informed by tailored sustainable requirements, fitting human capabilities and limitations, diversities and uniformities, variability and similarity.

To do this more ergonomics issues in whole-building design and assessment are needed.

Sustainable building assessment rating tools need to be implemented by a more comprehensive application of the sustainable principles, including a lot of the aspects of people's needs and functions, in terms of physical, emotional, as well as social dimensions. Since usability may express how building supports end-users in their activities, whole building usability could be enhanced as one of the key areas of green building design and assessment. Moreover sustainability of materials and products should be implemented considering human efforts needed for set-up of related construction techniques, since sustainable solutions with similar energy performances may offer different human performances, determining working conditions worse or better than others.

Education programs for architects need the provision of adequate knowledge about human factors and ergonomics data and methodologies, in order to enhance architectural design by the consideration of the human factors perspective and envisage how technical solutions can fit human and environmental needs derived from people's life and the work activities they perform. Designers and producers of building components need to take advantage from human factors and ergonomics issues in order to reduce the impact of green building construction on health and safety of construction workers, and to improve operability and usability of building controls for increasing end-users' comfort and well-being. The design team should frequently involve ergonomists and human factors experts.

REFERENCES

Abbaszadeh, S., Zagreus, L., Lehrer, D., and C. Huizenga. 2006. Occupant satisfaction with indoor air environmental quality in green buildings. *Proceedings of the Healthy Buildings 2006 Conference*, pp. 365–370. Lisbon, Portugal.

Afacan, Y. and C. Erbug. 2009. An interdisciplinary heuristic evaluation method for universal building design. *Applied Ergonomics* 40:731–744.

Akadiri, P.O., Chinyio, E.A., and P.O. Olomolaiye. 2012. Design of a sustainable building: A conceptual framework for implementing sustainability in the building sector. *Buildings* 2:126–152.

Alexander, K. 2008. Usability of workplaces. CIB W111 Research Report.

Attaianese, E. 2009. Ergonomics in maintenance for energy sustainable management. *Proceedings of Maintenance Management Conference—Fourth International Conference on Maintenance and Facility Management*, pp. 15–18. CNIM, Rome, Italy.

Attaianese, E. 2011. Human factors in maintenance for a sustainable management of built environment. In *Wellbeing and Innovation through Ergonomics. Proceedings of NES 2011*, eds. J. Lindfors, M. Savolainen, and S. Väyrynen, pp. 18–12. NES, Oulu, Finland.

Attaianese, E. 2012. A broader consideration of human factor to enhance sustainable building design. *Work* 41:1.

Attaianese, E. and G. Caterina. 1997. Qualità d'uso degli spazi residenziali (Use quality of residential space). *Atti del VI Congresso Nazionale SIE, Governo delle tecnologie, efficienza e creatività. Il contributo dell'ergonomia (Proceedings of VI National Congress of Italian Society of Ergonomics)*, pp. 41–45. Monduzzi, Bologna, Italy.

Attaianese, E. and G. Caterina. 2001. L'approccio ergonomico al recupero edilizio: la qualità d'uso degli ambienti di vita (Ergonomic approach to building rehabilitation: use quality of life environment). *Atti del VII Congresso Nazionale SIE: L'ergonomia nella società dell'informazione*, pp. 492–495. SIE, Firenze, Italy.

Attaianese, E. and G. Duca. 2012a. Human factors and ergonomic principles in building design for life and work activities: An applied methodology. *Theoretical Issues in Ergonomics Science* 13:2.

Attaianese, E. and G. Duca. 2012b. The human component of sustainability: A study for assessing human performances of energy efficient construction blocks. *Work* 41:1.

Bal, M., Bryde, D., Fearon, D., and E. Ochieng. 2013. Stakeholder engagement: Achieving sustainability in the construction sector. *Sustainability* 5:2.

Barlow, S. and D. Fiala. 2007. Occupant comfort in UK offices—How adaptive comfort theories might influence future low energy office refurbishment strategies. *Energy and Buildings* 39:837–846.

Bellù, L. 2011. Development Paradigms A (Reasoned) Review of Prevailing Visions. Issue Paper, EASYPol Module102. http://www.fao.org/docs/up/easypol/882/defining_development_paradigms_102en.pdf.

Berardi, U. 2013. Clarifying the new interpretations of the concept of sustainable building. *Sustainable Cities and Society* 8:72–78.

Bhramra, T., Lilley, D., and T. Tang. 2011. Design for sustainability behavior: Using product to change consumer behavior. *The Design Journal* 14:4.

Boerstra, A., Beuker, T., Loomans, M., and J. Hensen. 2013. Impact of available and perceived control on comfort and health in European offices. *Architectural Science Review* 56:1.

Boswell, P. and L. Walker. 2005. Procurement process design. *Proceedings of SB05*, Tokyo, Japan.

Bragança, L., Mateus, R., and H. Koukkari. 2010. Building sustainability assessment. *Sustainability* 2: 2010–2023.

Brown, C. and S. Legg. 2011. Human factors and ergonomics for business sustainability. In *Business and Sustainability: Concepts, Strategies and Changes, Critical Studies on Corporate Responsibility, Governance and Sustainability*, eds. G. Eweje and M. Perry, Vol. 3, pp. 61–81. Bingley, UK: Emerald Group Publishing Limited.

Brown, Z.B. 2009. Occupant comfort and engagement in green buildings: Examining the effects of knowledge, feedback and workplace culture. PhD thesis, The University of British Columbia Vancouver, Canada.

Byrd, H. 2012. Post-occupancy evaluation of green buildings: The measured impact of over-glazing. *Architectural Science Review* 55(3):206–212.

Charytonowicz, J. 2007. Reconsumption and recycling in the ergonomic design of architecture. Universal Access. *Lecture Notes in Computer Science. Human-Computer Interaction*. Ambient Interaction, Vol. 4555.

Chuck, W.F. and J. Tai. 2011. Building environmental assessment schemes for rating of IAQ. In *Indoor and Built Environment* 20:15–15.

CIB W111. 2010. Usability of workplaces. Phase 3, Research Report. In *CIB Report 330 International Council for Research and Innovation in Building and Construction*, ed. K. Alexander. Rotterdam.

Clements-Croome, D. and G.Jeronimidis. 2005. Sustainable building solutions: A review of lessons from the natural world. *Building and Environment* 40:319–328.

Cole, R.J., Brown, Z. and S. McKay. 2010. Building human agency: A timely manifesto. *Building Research & Information* 38:3.

Cole, R.J., Robinson, J., Brown, Z., and M. O'shea. 2008. Re-contestualizing the notion of comfort. *Building Research & Information* 36:4.

Drexhage, J. and D. Murphy. 2010. Sustainable development: From Brundtland to Rio 2012. Background Paper, High Level Panel on Global Sustainability, United Nations Headquarters, New York http://www.un.org/wcm/webdav/site/climatechange/shared/gsp/docs/GSP16_Background%20on%20Sustainable%20Devt.pdf.

Duca, G. 2012. Usability requirements for buildings: A case study on primary schools. *Work* 41:1.

Duca, G. 2014. From energy-efficient buildings to energy-efficient users and back: Ergonomic issues in intelligent buildings design. *Intelligent Buildings International* 6:4.

Duca, G. and E. Attaianese. 2012. From usability requirement to technical specifications for hand-held tools and materials: An applied research in the construction field. *Work* 41:1.

EASHW. 2013. Occupational safety and health issues associated with green building. E-fact 70 https://osha.europa.eu/en/publications/e-facts/e-fact-70-occupational-safety-and-health-issues-associated-with-green-building/view.

Edwards, L. and P. Torcellini. 2002. A literature review of the effects of natural light on building occupants. *Technical Report of National Renewable Energy Laboratory*, Colorado.

Erlandson, T., Cena, K., De Dear R., and G. Havenith. 2006. Environmental and human factors influencing thermal comfort of office occupants in hot-humid and hot-arid climates. *Ergonomics* 46:6.

Fischer, E.A. 2011. *Issues in Green Building and the Federal Response: An Introduction*. Washington DC: Congressional Research Service, 2010.7-5700, R40147. Retrieved on line August 28, 2011 www.crs.gov.

Flemming, S., Hilliardand, A., and G.A. Jamieson. 2008. The need of human factors in the sustainability domain. In *Proceedings of the Human Factors and Ergonomics Society 52nd Annual Meeting*. Thousand Oaks, California: SAGE Publications, USA.

FWC Sector Competitiveness. 2012. Studies N° B1/ENTR/06/054—Sustainable competitiveness of the construction sector, Final Report ECORYS SCS Group, Rotterdam.

Goins, J., Chun, C., and H. Zhang. 2013. User perspectives on outdoor noise in open-plan offices with operable windows. *Architectural Science Review* 56(1): 42–47.

Guerin, D.A., Brigham, J.K., Kim, H.Y., Choi, S.M., and A. Scott. 2012. Post-occupancy evaluation of employees 'work performance and satisfaction as related to sustainable design criteria and workstation TYPE. *Journal of Green Building* 7(4): 85–99.

Gupta, R. and S. Chandiwala. 2010. Understanding occupants: Feedback techniques for large-scale low-carbon domestic refurbishments. *Building Research & Information* 38:5.

Haldi, F. and D. Robinson. 2011. The impact of occupants' behavior on building energy demand. *Journal of Building Performance Simulation* 4:4.

Hanson, M.A. 2012. Green ergonomics: Challenges and opportunity. *Ergonomics* 56:3.

Haruna, S.N., Hamid, M.Y., Talib, A., and Z.A. Rahim. 2011. Usability evaluation: Criteria for quality architecture in-use. In *The 2nd International Building Control Conference 2011*, ed. Tawil, N.M. *Procedia Engineering* 20(2011): 135–146.

Haslam, R. and P. Waterson. 2013. Ergonomics and sustainability. *Ergonomics* 56:3.

Hauge, A.L., Thomsen, J., and T. Berker. 2011. User evaluations of energy efficient buildings: Literature review and further research. *Advance in Building Energy Research* 5:1.

Hedge, A. 2008. The sprouting of "Green" ergonomics. *HFES Bulletin* 51:12.

Heerwagen, J. and L. Zagreus. 2005. *The Human Factors of Sustainability: A Post Occupancy Evaluation of the Philip Merrill Environmental Center*. Summary Report for U.S. Department of Energy, Center for the Built Environment. Berkeley, California: University of California.

Heerwagen, J.H. 1998. Design, productivity and well being: What are the Links? *The American Institute of Architects Conference on Highly Effective Facilities*, Cincinnati, Ohio.

Himanen, M. 2003. The intelligence of intelligent buildings. The feasibility of the intelligent building concept in office buildings. PhD thesis. Espoo, Finland: VTT Building Technology.

IEA (International Ergonomics Association). 2000. *Ergonomics International News and Information*. London: Marshall Associates.

ISO 9421:1996. 1996. Ergonomic requirements for office work with visual display terminals.

ISO 21542:2011. 2011. Building construction—Accessibility and usability of the built environment.

ISO/TS 21929-1. 2006. Sustainability in building construction—Sustainability indicators—Part 1: Framework for development of indicators for buildings.

Janda, K.B. 2011. Buildings don't use energy: People do. *Architectural Science Review* 54:15–22.

Jensen, P.A. 2005. Value concepts and value based collaboration in building projects. In *Proceedings of the CIB W096 Architectural Management*, eds. S. Emmitt and M. Prins, pp. 3–10. Rotterdam: CIB.

Jenso, M. and T. Haugen. 2004. Usability of hospital buildings. Is patient focus leading to usability in hospital buildings. *Paper CIB W70 Hong Kong International Symposium, Facility Management & Asset Maintenance, The human Element in Facilities Management*, Hong Kong.

Karwowsky, W. 2005. Ergonomics and human factors: The paradigms for science, engineering, design, technology and management of human–compatible systems. *Ergonomics* 48:5.

Kibert, C.J. 2013. *Sustainable Construction: Green Building Design and Delivery*. Hoboken, NJ: John Wiley & Sons.

Klugengseth, N.J. and G.K. Hansen. 2010. What is building's usability? Retrieved online May 5, 2010 http://www.metamorfose.ntnu.no/Artikler/What%20is%20a%20buildings%20usability_EuroFM09.pdf.

Leaman, A. and B. Bordass. 2007. Are users more tolerant of 'green' buildings? *Building Research and Information* 35:6.

Leheman, M.V. 2011. How sensory design brings value to buildings and their occupants. *Intelligent Buildings International* 3:1.

Lewis, A.P.E. 2010. Designing For Energy-Efficient Operations and Maintenance. http://www.esmagazine.com/Articles/Feature_Article/BNP_GUID_9-5-2006_A_10000000000000873588.

Loftness, V., Hartkopf, V., and L.K. Poh. 2006. Sustainability and health are integral goals for the building environment. In *HB 2006 Healthy Buildings. Proceedings,* eds. E. De Oliveira Fernandes, M. Gameiro da Silva, and J. Rosado Pinto. Lisbon, Portugal: Universidade do Porto.

Lueng, M.H. 2003. Ergonomics and sustainable development. *Ergonomics in the Digital Age. Proceedings of the XVth Triennial Congress of the International Ergonomics Association & the 7th Joint Conference of the Ergonomics Society of Korea & the Japan Ergonomics Society.* Ergonomics Society of Korea, Seoul, Korea.

Lutzkendorf, T. and D. Lorenz. 2007. Integrating sustainability into property risk for market transformation. *Building Research & Information* 35:6.

Martin, K., Legg, S., and C. Brown. 2013. Designing for sustainability: Ergonomics—Carpe diem. *Ergonomics* 56:3.

Meacham, B., Poole, B., Echeverria, J., and R. Cheng. 2012. *Fire Safety Challenges of Green Buildings. Final Report.* Fire Protection Research Foundation. Quincy, Massachusetts.

Miller, N.G., Pogue, D., Gough, Q.D., and S.M. Davis. 2009. Green buildings and productivity. *Journal of Sustainable Real Estate* 1:1.

Monfared, I.G. and S. Sharples. 2011. Occupants' perceptions and expectations of a green office building: A longitudinal case study. *Architectural Science Review* 54:344–355.

Moray, N. 1993. Technosophy and humane factors. *Ergonomics in Design* 1:4.

Moray, N. 1995. Ergonomics and the global problems of the twenty-first century. *Ergonomics* 38:8.

Muehleisen, R.T. 2010. Acoustics of green buildings. *InformeDesign* 8:1.

Nicol, F. and S. Roaf. 2005. Post-occupancy evaluation and field studies of thermal comfort. *Building Research and Information* 33(4):338–346.

Nicol, J.J. and M. Humphreys. 2002. Adaptive thermal comfort and sustainable thermal standards for buildings. *Energy and Buildings* 34(6):563–572.

Obiozo, R. and R. Smallwood. 2013. The role of 'greening' and an ecosystem approach to enhancing construction ergonomics. In *Proceedings of 29th Annual ARCOM Conference*, eds. S.D. Smith and D.D. Ahiaga-Dagbui. Reading, UK: Association of Researchers in Construction Management.

Odom, J.D., Scott, R., and G.H. DuBose. 2007. *The Hidden Risks of Green Buildings: Avoiding Moisture and Mold Problems.* Washington, DC: National Council of Architectural Registration Boards (NCARB).

Oliveira R.C. and G.A. Elali. 2012. Minimum housing spaces, flexibility and sustainability: A reflection on the basis of ergonomics intervention. *Work* 41:1409–1416.

Paul, W.L. and P.A. Taylor. 2008. A comparison of occupant comfort and satisfaction between a green building and a conventional building. *Building and Environment* 43:1858–1870.

Peffer, T., Perry, D., Pritoni, M., Aragon, C., and A. Meier. 2013. Facilitating energy savings with programmable thermostat: Evaluation and guidelines for the thermostat user interface. *Ergonomics* 56:9.

Preiser, W. and J. Vischer (eds.). 2005. *Assessing Building Performance.* Oxford: Elsevier.

Rajendran, S., Gambatese, J., and M. Behm. 2009. Impact of green building design and construction on worker safety and health. *Journal of Construction Engineering and Management* 135:10.

Samuelsson, M. and T. Luddeckens. 2009. *Passivhus ur en brukares perspektiv.* Växjö, Sweden: Växjö Universitet.

Shove, E., Chappells, H., Lutzenhiser, L., and B. Hackett. 2008. Comfort in a lower carbon society. *Building Research & Information* 36:4.

Steemers, K. and G.Y. Yun. 2009. Household energy consumption: A study of the role of occupants. *Building Research & Information* 37:5–6.

Steimle, U. 2006. Sustainable development and human factors. In *International Encyclopedia of Ergonomics & Human Factors.* 2nd edition, Vol. 2. Boca Raton, FL: CRC Press.

Stevens, S. 2001. Intelligent facades: Occupant control and satisfaction. *International Journal of Solar Energy* 21:147–160.

Stevenson, F. and A. Leaman 2010. Evaluating housing performance in relation to human behavior: New challenges. *Building Research & Information* 38:5.

Thatcher, A. 2013. Green ergonomics: Definition and scope. *Ergonomics* 56:3.

Thatcher, A. and K. Milner. 2012. The impact of a 'green' building on employees' physical and psychological wellbeing. *Work* 41:3816–3823.

Tidwell, J. and J.J. Murphy. 2010. *Bridging the Gap-Fire Safety and Green Buildings.* USA: National Association for Fire Marshals.

UN. 1987. Report of the World Commission on Environment and Development: Our Common Future. United Nations. p. 6.

U.S. Environmental Protection Agency. 2009. Green Building Basic Information. Retrieved on line December 10, 2009, from http://www.epa.gov/greenbuilding/pubs/about.htm.

Vale, B. and R. Vale. 2010. Domestic energy use, lifestyles and POE: Past lessons for current problems. *Building Research & Information* 38:5.

Volker, L. and M. Prins. 2005. Exploring the possibilities of correlating management with value in architectural design. In *Proceedings of the CIB W096 Architectural Management,* eds. S. Emmitt and M. Prins. Rotterdam: CIB.

Wagner, A., Gossauer, E., Moosmann, C., Gropp, T., and R. Leonhart. 2007. Thermal comfort and workplace occupant satisfaction—Results of field studies in Germany low energy office buildings. *Energy and Buildings* 39:758–769.

Zink, K.J. and K. Fisher. 2013. Do we need sustainability as a new approach in human factors and ergonomics? *Ergonomics* 56:3.

29 Dressing Autonomy for Frozen Shoulder Users
Analysis of Five Different Tops

Letícia Schiehll, Fernando Moreira da Silva, and Inês Simões

CONTENTS

29.1 INTRODUCTION

Independence and autonomy are essential to maintain the quality of life of an individual. Determined by the body's anatomy, its movements, and functional capacities, the lack of independence or autonomy has a direct effect in the way activities are performed throughout the day by an individual, being reflected in all the following systems: the physical, psychological, and social. Since all these systems represent a cyclical systemic relationship, when one fails, the others collapse.

We took the terms "activity limitations" and "body functions," as defined by CIF (2003), to refer to functional limitations. While the first concept is defined as "the difficulties that an individual may encounter in executing activities" (CIF, 2003, p. 11) and the second as "the physiological functions of organic systems (including the psychological functions)" (CIF, 2003, p. 11), body functional limitations are understood in this paper as the difficulties perceived by the individual in the performance of mechanical, physical, biochemical, or psychological functions. As discussed by Andrade (2009), the deficit in functional capacity, which causes an individual to be dependent on others to perform daily activities, has different degrees.

For a proper contextualization of the terms covered in this study, we must expand on the meaning of independence and autonomy. Accordingly, independence is defined as "the ability to perform daily life functions and the ability to live independently in community, with few or no help from others" (WHO, cited apud Almeida, 2009). Although independence is usually used as a synonym of autonomy, it is important to mention that some people with reduced functional and physical abilities have the capacity to choose and control part of their environment (Garcia, 2009). Thus, dependence is a complex phenomenon, as it is usually defined as a person's clinical condition that leads to the need of being helped in performing basic daily activities, or the inability of a person to act satisfactorily without someone's aid or equipment.

In the strict sense, dependence conveys the need for assistance in performing daily activities that a person is no longer capable of performing or does not want to perform. Therefore, it is the support one requires to perform survival tasks, such as having a bath, dressing, eating, etc. (Andrade, 2009; Garcia, 2009; Pavarini and Neri, 2000).

As for autonomy, it means "the ability to control, act or make personal decisions about how to live every day, according to one's own rules or preferences" (WHO, apud Almeida, 2009).

Autonomy means self-sufficiency, the ability of a person to choose the rules for his/her own functioning, the orientation of his/her actions, and the risks he/she is willing to take, which make up one's own behavior, choices, and values. Freedom of choice and self-determination are also associated with autonomy. Thus, it involves independence of action, speech, and thought, although independence at the motor level is the most felt by an individual, as it often interferes with other levels of independence (Academia das Ciências de Lisboa, 2001; Agich, 2008; Garcia, 2009). Although behaviors of autonomy do not only require social skills, perceptual, dexterity, and cognitive skills are very important for greater autonomy (Ferland, 2006).

Regarding elderly people, the maintenance of autonomy and independence is closely related to the quality of life. It is the perception of an individual about his/her life in the context of the culture and value system existing in society and guiding his/hers goals, expectations, standards, and concerns (WHO, s.d.). Thus, the perception of a lesser quality life, depression, or isolation states reduce the individual's autonomy and independence, just as the maintenance of these capacities contributes to the quality of life.

On the other hand, Lemos and Medeiros (2002) argue that autonomy and independence are great indicators of health among older adults, and stress the need of every human being not to be dependent on others.

This acceptation of autonomy is extended by Soutinho (2006) who realized that older women prefer to wear trousers and dresses which are more malleable, softer and looser, easier to don, doff, and move with, and suggests that this preference is possibly due to the joint stiffness combined with decreased strength and motor coordination. Martins (2008, p. 325) points out that, "if children's clothing was in the past the miniaturization of adults clothes, the latter is nowadays designed under the same concept as the clothing for young people, even though the elderly do not have the same mobility, reach, biotype and needs." The author's words clearly illustrate the contemporary emphasis on youth, beauty, autonomy, independence, and ability to be productive and reproductive, which leaves aside the individuals with any kind of limitations.

29.2   DRESSING PROCESS WITH AUTONOMY

The process of dressing is among the activities that relate human beings and clothes although it is the interaction (dress) between the body and clothes that gives meaning or purpose to the latter. Clothes are designed to facilitate being donned and doffed by oneself or with the aid of others. According to Katz (Lessa et al., 1994), along with other activities—such as, having a bath/shower (1), getting dressed (2), performing cleansing and grooming activities (3), lying down and getting up from a bed/chair (4), eating (5), and urinating and evacuating (6), donning and doffing are necessary actions to maintain independence and autonomy, as they are part of daily life.

These daily activities are used to evaluate functional capacity and thus to measure the performance of an individual's autonomy and independence. The inefficiency in carrying out any of these activities corresponds to a decline in functional capacity, which means the degree of dependence caused by functional aging. The ability to perform daily activities depends thus on the body, being the vehicle that provides us expression, be it in motion, or as a support (Falcão, 2011; Lessa et al., 1994).

These are actions related to one's ease of handling combined with anatomical, anthropometric, and biomechanical aspects (Martins, 2008), that is, they are actions that depend on the body and its movements.

Regarding clothing design, Menezes and Spaine (2010) underline the need for a designer to understand the body, its movements, joints, human proportions, structures, and among others, before understanding patterns, so as to include the actions of the body in the designed patterns.

29.2.1 USUAL DRESSING PROCESS

After studying the dressing process, we identified seven moments or steps.

1. The "selection of clothes" is the moment when someone identifies or chooses the piece or pieces of clothing that he/she wants to wear among his/hers wardrobe. This step involves only the issues of cognitive ability of the user (Figure 29.1).
2. "Getting the clothes" from the closet or drawer. This step involves the issues of fine and gross motor skills ability of the user (Figure 29.2).
3. "Preparation" of the clothes is the moment when the user places the back of the garment on his/her lap. Here, besides the cognitive ability, the user needs to perform fine motor skills—such as, pincer grip—as well as gross motor skills—such as, the extension and flexion of the arms, among others (Figure 29.3).
4. "Donning" or dressing oneself is when the user puts the selected garment or garments on the body, which normally involves covering each segment of the body with the corresponding part of clothing. To don a T-shirt, for example, one has to put the head into the neckline from the inside, both arms into the armholes—also from the inside—and to pull it down so that the hemline gets positioned around the hip region. For a good donning performance, it is necessary to have good cognitive ability as well as to have good fine and gross motor skills (Figure 29.4).
5. After getting into the clothes properly, it is necessary to "fasten" them. The user needs to handle one type or several types of closures (e.g., buttons, zippers, hooks, ties, elastic bands, etc.), for which fine motor skills are crucial.

FIGURE 29.1 Selection of clothes (Schiehll).

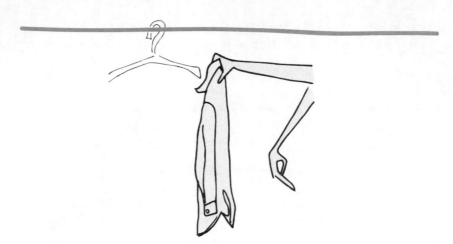

FIGURE 29.2   Getting the clothes (Schiehll).

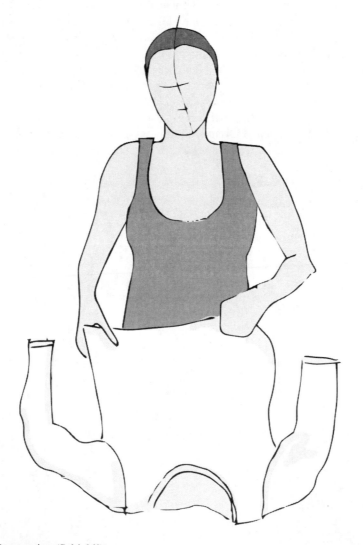

FIGURE 29.3   Preparation (Schiehll).

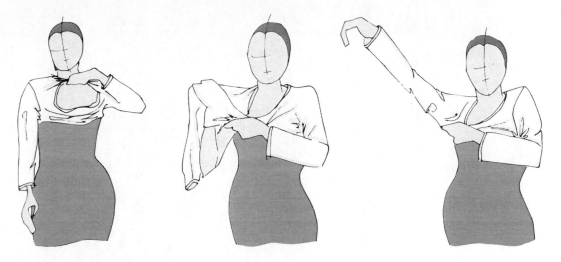

FIGURE 29.4 Donning or dressing (Schiehll).

6. At this point, the user "adjusts the donned clothes" by properly positioning them along the body segments. This step also requires having fine motor skills—so that the clothes serve their proper function and do not look awkward on the body—although gross motor skills also play an important role here (Figure 29.5).

7. Doffing is the last step of this process, which also involves, though in reverse, performing fine and gross motor activities while the individual takes off the pieces of clothing that were donned and worn that day (Figure 29.6).

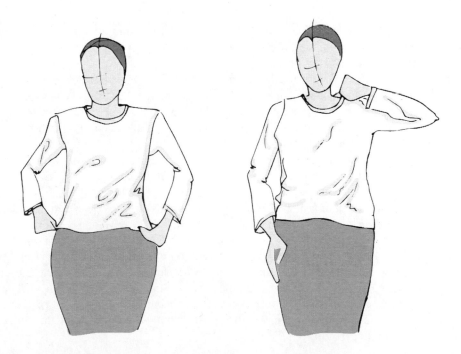

FIGURE 29.5 Adjusting the donned clothes (Schiehll).

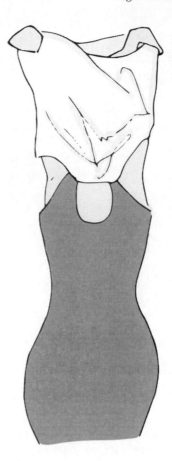

FIGURE 29.6   Doffing or undressing (Schiehll).

We realized that people must have sufficient levels of flexibility, joint mobility, and coordination of the body segments in question to perform the rather common activities involved in the dressing process in its entirety (Vale et al., 2006).

The development of the skills involved in donning and doffing by children can be used to help us understand the body and to obtain spatial orientation (Pfeifer, s.d.). Children around 2 years of age try to exercise part of their autonomy as their bodies have already acquired some control over their fine and gross motor skills (Nucci et al., 1996). In fact, a 2-year-old child is already capable of taking clothes off the body. From the age of three, a child can be encouraged to put clothes on all by her/himself, but only around the age of five or six a child is able to don and doff without any help, for instance, to button and unbutton skirts, pants, and tops and to tie and untie shoe laces (Drescher, 1999; Early, 2013; Papalia, 2010).

Ferland (2006) points out that fastening clothing with buttons or snaps placed in the back, neck, and wrists are activities that take more time to develop. Although the development of fine and gross motor skills—that is, the coordination between what an individual wants to do and what an individual is able to do in terms of donning and doffing—is accomplished between the ages of three and six (Papalia, 2010). In other words, these skills acquired in childhood can be affected throughout life by a series of circumstances: accidents, trauma, disease, and aging, among others. Therefore, any person of any age can experience a limitation of movement at some point of his/her life, an occurrence, that is, in many cases associated with the individual's secondary aging that reduces his/her functional capacity to perform.

29.2.2 WEAKENED DRESSING PROCESS

Diseases of the musculoskeletal system are the main causes for the weakening of the dressing process, particularly in the donning and doffing steps. The skeletal system includes all of the bones and joints in the body, the latter being the location at which bones connect. Joints are categorized into two types: structural and functional. The structural category refers to how bones are connected to each other, while the functional category is related to the degree of movement between the articulating bones.

The skeletal system also provides attachment points for the muscles. The muscular system is composed of about 600 muscles, having particular characteristics determined by their type, action, or location. Together, the skeletal system and the muscular system form the musculoskeletal system, which is responsible for movement of the body. Controlled through the nervous system, muscles enable body segments to shorten the distance between the fixed ends (contraction), granting the myriad positions and postures assumed by the body, as well as its dynamism and stillness. So, muscles are the active elements of movement, while bones are the passive elements (Grave, 2004).

The mechanics of muscle movement involves the contraction of the muscle, resulting in displacement, since the muscles are linked to two bones and cross one or more joints. During contraction, the muscle shortens down to one-sixth of its initial length in order to dislocate the bone. This activity depends on the range and force of muscle contraction. The musculature presents a fixed origin point, tied to the bone, which is not moveable (Grave, 2004).

Locomotor system pathologies—or musculoskeletal pathologies—culminate in lower movement, loss of force and strength, the frozen shoulder being the most frequent pathology in patients with relapsing difficulties in the dressing process.

The term "frozen shoulder" is commonly used instead of *adhesive capsulitis*—the technical term—and other conditions associated with loss of range of motion at the joint. This particular condition is characterized by the decrease of amplitude of the active and passive range of motion—that is, the range of motion when someone else moves the patient's shoulder—which causes pain in a first stage followed by severe stiffness—or mechanical lock. It occurs because the capsule becomes inflamed, thickened, and contracted. The frozen shoulder is characterized by an insidious onset and in most cases is a form of reflex algoneurodystrophy located at the shoulder that causes a progressive loss of movement. It affects the nondominant arm in most patients, mostly women between 40 and 65, and in patients with clinical depression (Snider, 2000; SOS Med, 2014; Xhardez, 1990).

After the installation—or freezing—stage and stiffness—or frozen/adhesive—stage, the complete return to normal or close to normal strength and range of motion typically takes from 6 months to 2 years without treatment. During this final stage—also known as the thawing stage—joint mobility returns gradually, external rotation being the last motion to be recovered. Although during the 4–6 months period of the freezing stage, daily activities are very difficult but possible to handle by oneself, during the stiffness stage through the final stage, one may depend on others throughout the dressing process (AAOS, 2003; Xhardez, 1990). Adhesive capsulitis is a syndrome defined as idiopathic restriction of shoulder movement, that is, usually painful at onset. Secondary causes include alteration of the supporting structures of and around the shoulder resulting from an underlying condition, such as diabetes mellitus, rotator cuff tendinopathy or tear, subacromial bursitis, biceps tendinopathy, recent shoulder surgery or trauma, and inflammatory diseases (Ewald, 2011).

The incidence of adhesive capsulitis is approximately 3% in the general population, and is rare in children. Patients with frozen shoulders are typically between 40 and 70 years of age, mostly women. Furthermore, persons with a history of adhesive capsulitis are at increased risk of developing the condition on the contralateral side (Ewald, 2011; Snider, 2000; UW Medicine, 2013).

In order to analyze which the main difficulties are in patients with a frozen shoulder regarding donning, adjusting, and doffing procedures, we must understand, first, the structure of the shoulder joint (Figure 29.7) so as to identify the movements involved.

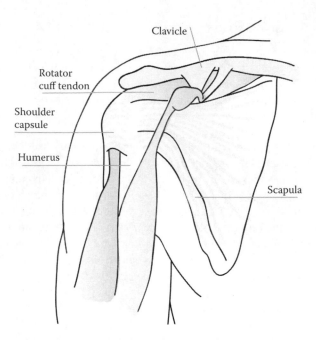

FIGURE 29.7   Shoulder joint structure (Schiehll).

Formed by the articulation of the head of the humerus with the glenoid cavity of the scapula, the shoulder joint is a ball and socket synovial joint—or spheroidal joint—that allows a wide range of movement (ROM), such as extension, flexion, abduction, adduction, medial rotation, and lateral rotation. The shoulder girdle, which consists primarily of the scapula bone and the clavicle bone which move together as a unit, allows the following ROM: elevation, depression, retraction, and protraction (adduction and abduction). Organized in pairs of opposite motions, the referred movements consist of

- Scapular elevation is the raising movement of the scapula; the opposite motion is scapular depression, which is when the scapula is lowered from elevation.
- Scapular retraction is when the scapula is moved posteriorly and medially along the back, moving the arm and shoulder joint posteriorly; the opposite motion is scapular protraction, in which the scapula is moved anteriorly and laterally along the back, moving the arm and shoulder joint anteriorly.
- Arm extension is the backwards movement of the arm; the opposite motion is arm flexion, which is the forward movement of the arm (in sagittal plane).
- Arm abduction is the movement away from midline; the opposite motion is arm adduction, which is the movement toward midline (in coronal plane).
- Medial rotation of the arm is the rotation toward the midline, so that the thumb is pointing medially. Lateral rotation of the arm is the rotation away from the midline, so that the thumb is pointing laterally.

According to UW Medicine (2013), the normal shoulder is the most moveable joint in the body, as "it enables us to put our hand in a wide range of positions, for example, reaching overhead reaching cross the body reaching up the back and rotating out to the side." UW Medicine (2013) also adds, "these motions are accomplished by motion between the humerus—arm bone—and scapula—shoulder blade—as well as between the scapula and the chest wall. These motions are called humeroscapular and scapulothoracic motions."

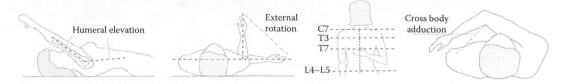

FIGURE 29.8 Forward flexion; external rotation; internal rotation; and cross body (Schiehll).

The generally accepted values for normal ROM in the shoulder joint as measured in degrees are flexion 0°–180°, extension 0°–50°, abduction 0°–90°, adduction 90°–0°, lateral rotation 0°–90°, and medial rotation 0°–90°.

Loss of motion with forward flexion, abduction, and external and internal rotation should rise for adhesive capsulitis. Because adhesive capsulitis does not affect the dynamic stabilizers of the shoulder (i.e., rotator cuff, biceps tendon, and deltoid muscle), strength should theoretically be preserved in all planes. However, patients with adhesive capsulitis may not have enough range of motion to perform strength testing. In Figure 29.8, it is possible to see the forward flexion at 170°, an external rotation at 80°; an internal rotation at the T5 segment, and the cross body adduction, with the elbow 14 cm away from the opposite shoulder (Ewald, 2011; UW Medicine, 2013).

Apart from the described movements, the section of this paper reporting the results of the study informs about other basic movements involved in the dressing process.

29.3 METHODOLOGY

29.3.1 USER-CENTERED DESIGN AND INCLUSIVE DESIGN

The reported research used the user-centered design (UCD) process. UCD is the term that describes the design processes in which the end-users influence how a design takes shape. This is a very common methodology in ergonomics project development. Therefore, in UCD, users are given extensive attention at each stage of the design process in order to make inclusive products and/or services. The principle behind UCD is that the designed products and/or services need to facilitate the people's needs, limitations, capabilities, desires, and motivations, as they are the persons who will use the product or system being built.

By definition, inclusive design is the design of mainstream products and/or services that are accessible to, and usable by, as many people as reasonably possible, regardless of age, sex, or ability and without the need for special adaptation or specialized design. The main goal of inclusive design is to make life easier for all people. In this context, the designer may have a critical world vision, and adopts a holistic and sustainable approach toward the product being designed (Moreira da Silva, 2013).

In this way, UCD can be characterized as a multi-phase process to solve problems that involves the designer and real end-users. The designer's duties comprise analyzing and forecasting how users will use a particular product. The users' duties comprise testing the validity of the designer's assumptions regarding their behavior in the real world. The extent to which a product can be used is specified by users so as to achieve the specified goals with effectiveness, efficiency, and satisfaction in a specified context of use (Vredenburg et al., 2013).

Along with the described principles, we selected a group of 10 women to observe and get their feedback about five tops. The selected group of women is composed of semi dependent or independent women evidencing some difficulties in the dressing process: they are women with the musculoskeletal pathology known as frozen shoulder. For the project's empirical phase, we have chosen two sample groups: one in Portugal and another one in Brazil. Each of the sample groups is composed of five women. The Portuguese group of women with frozen shoulders is composed of residents of "*Casa do Artista*" in Lisbon—a retirement home for actors; the Brazilian sample group is composed of nonresident patients of a physiotherapy clinic in Rio Grande do Sul. In both groups, all women are undergoing treatment for the stated condition at different stages.

29.3.2 FIVE TOPS

To identify which the most difficult movements are for an individual with a frozen shoulder, we observed 10 women during the dressing process, in particular, the steps of donning, doffing, and adjusting, for which we used five generic garments for the upper body with different characteristics—namely a poncho, a tank top, a long sleeve T-shirt, a batwing sleeve top, and a top with a front zipper—that we viewed as relevant for the evaluation of movement and to identify ROM limitations.

For a better understanding of the used garments (Figure 29.9), we consider relevant to describe them:

1. The top with a front zipper is a garment with long sleeves. Its main characteristic is a full-length opening at the center front, a feature that differentiates the process of donning it from the process of putting on the four garments described below, as the user has to put first one arm into one sleeve, then the other arm into the other sleeve, and only after this, she buttons down the font placket while adjusting the top along the torso. The main difficulty to don this type of garment is to grab the second sleeve from the back.
2. The poncho is a sleeveless garment with an A-line silhouette and slits for the arms. To don this piece, the user only has to put the head into the neckline from the inside and adjust, if necessary, its surface along the torso. Because it does not have sleeves and darts, this type of garment hangs straight from the neckline to the hemline, while resting on the shoulders.
3. The batwing sleeve top is a type of garment that instead of having a set-in sleeve—that is, a sleeve joined to the torso by a seam starting at the edge of the shoulder and continuing around the armhole—it has the front and back pieces of the torso combined with the front and back parts of the sleeve, respectively. The donning and adjusting steps also involve the movements of the arms, head, and torso.
4. The long sleeve T-shirt is a tight fitting garment. It has a defined armhole where long and narrow sleeves are set. The stages required to don this type of garment are very similar to those of the tank top, as they also involve the arms, head, and torso. In most cases, the user is required to adjust it along the torso, neck, forearms, and arms as the shirt can get slightly displaced while it is being donned. Because it was designed with a sleeve—the main characteristic of this type of garment—some extra degree of difficulty may occur while donning it.
5. The tank top is a sleeveless garment with a tight fitting neckline and armholes. To don this piece, the user has to put the head into the neckline from the inside, the arms into the armholes also from the inside, and pull the front and back down. In most cases, the user is required to adjust it around the neckline and armholes as they can get slightly displaced while it is being donned. This type of garment involves the extension of the arms.

FIGURE 29.9   Top with front zip, poncho, batwing sleeve top, long sleeve T-shirt and tank top (Schiehll).

29.3.3 Movements Analysis

Using direct observation methodology for the actions of donning, doffing, and adjusting the five tops, the researchers identified the kind of difficulties undergone by the sample groups. The identified movements for evaluation pertaining to the mobility of the upper limbs and torso are as such:

1. Alternate arm extension and flexion
2. Neck flexion
3. Arm bending extending from the forearm
4. Arm bending with flexion of the forearm
5. Arm extension with forearm flexion (behind the back)
6. Wrist rotation
7. Forearm extension
8. Shoulder rotation and elevation
9. Forearm flexion
10. Handle with force
11. Thumb pressure
12. Pincer grip

By definition, flexion is a bending movement around a joint in a limb that decreases the angle between the bones of the limb at the joint, and extension is an unbending movement around a joint in a limb that increases the angle between the bones of the limb at the joint (Grave, 2004; Kendall and Kendall, 1980).

The assessment was made using the research methodology developed by Foddy (1994) and the answers were translated into a numerical scale from 1 to 5, where 1 stands for "high difficulty" while performing the task and 5 for "no difficulty."

29.4 DRESSING PROCESS WITH A FROZEN SHOULDER: RESULTS

29.4.1 Movements Performed during the Dressing Process of the Five Tops

The assessment considers the characteristics of each top in relation to the movements of the upper limbs and torso performed during the dressing process (Table 29.1). The movements were broken down using the insight of kinesiologists such as Rasch and Burke (1977) and Falcão (2011) because their analyses helped to observe and define the difficulties in the dressing process for each top.

Using direct observation, we identified which movements are required to don each top as well as the movements that patients with a frozen shoulder have greater difficulty to perform, particularly the rotation and elevation of the shoulder, and the flexion and extension of the forearm, as these actions require the displacement of the humerus at the scapula cavity. In addition, activities that require strength are difficult to perform on the affected side.

By analyzing this table, we infer that the poncho and the top with a front zipper are the easiest pieces to don and doff, as their dressing process exclude some movements involved in the dressing process of the tank top, the long sleeve T-shirt, and the batwing sleeve top. We also realize that the batwing sleeve top requires less range of motion, for which we ranked it as the third garment that provides greater ease in donning and doffing considering the amount of implicated movements, but the second one—coming right after the poncho—if we think in terms of ROM.

29.4.2 Difficulties of Donning and Doffing Each Top

From the analysis of the implicated movements, we can define the difficulty of donning and doffing each top, the difficulty of adjusting each top after donning it, and listing the tops by order of difficulty as felt by the focus group during the dressing process.

TABLE 29.1
Movements Performed to Don and Doff the Five Tops (Authors)

Top with Front Zipper	Poncho	Batwing Sleeve Top	Long Sleeve T-Shirt	Tank Top
Arm extension followed by arm flexion	Arm extension followed by arm flexion	Arm extension followed by arm flexion	Arm extension followed by arm flexion	Arm extension followed by arm flexion
	Neck flexion	Neck flexion	Neck flexion	Neck flexion
Arm bending extending from the forearm[a]	Arm bending extending from the forearm[a]	Arm bending extending from the forearm[a]	Arm bending extending from the forearm[a]	Arm bending extending from the forearm[a]
Arm bending with forearm flexion[a]	Arm bending with forearm flexion[a]	Arm bending with forearm flexion[a]	Arm bending with forearm flexion[a]	Arm bending with forearm flexion[a]
Arm extension with forearm flexion (behind the back)[a]		Arm extension with forearm flexion (behind the back)[a]	Arm extension with forearm flexion (behind the back)[a]	Arm extension with forearm flexion (behind the back)[a]
Wrist rotation	Wrist rotation	Wrist rotation	Wrist rotation	Wrist rotation
Forearm extension		Forearm extension	Forearm extension	Forearm extension
Shoulder rotation and elevation[a]	Shoulder rotation and elevation[a]	Shoulder rotation and elevation[a]	Shoulder rotation and elevation[a]	Shoulder rotation and elevation[a]
Forearm flexion[a]	Forearm flexion[a]	Forearm flexion[a]	Forearm flexion[a]	Forearm flexion[a]
			Handle with force[a]	Handle with force[a]
Thumb pressure[a]		Thumb pressure[a]	Thumb pressure[a]	
Pincer grip	Pincer grip	Pincer grip	Pincer grip	Pincer grip

[a] Movements that a frozen shoulder is not able to do.

29.4.2.1 Donning and Doffing Clothes: Identified by the Researchers

Regarding the donning and doffing procedures performed by the focus group, we infer that the batwing sleeve top was the easiest top to don, as nine out of the 10 women showed medium difficulty or no difficulty at all. However, one of the women had great difficulty, primarily because of this type of garment being unfamiliar to her. Apart from this occurrence, the batwing sleeve top was the only utilized garment to be donned and doffed without any difficulty, and this fact must be highlighted (Chart 29.1).

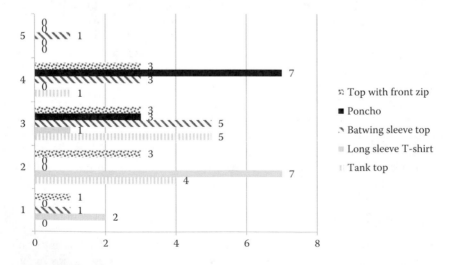

CHART 29.1   Difficulties identified by the researchers (Authors).

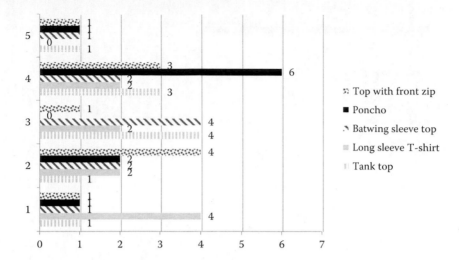

CHART 29.2 Difficulties to adjust the garments to the body after donning (Authors).

We also emphasize that the poncho was ranked between presenting medium and little difficulty to don. The biggest difficulties presented by the two groups of women was with respect to the long sleeve T-shirt—as nine of them had either little difficulty or much difficulty to don it, and one had medium difficulty.

Regarding the tank top, we observed that it presented little or medium difficulty. Similarly, the top with a front zipper presented little or medium difficulty: three women had medium and little difficulty to don it, respectively; one woman had much difficulty due to being unable to perform arm extension with forearm flexion behind the back.

29.4.2.2 Adjusting the Clothes: Identified by the Researchers

We noticed that most of the women had a higher degree of difficulty to adjust the donned tops than to don them. The long sleeve T-shirt was the garment that presented the highest difficulty; the poncho presented little difficulty; the batwing sleeve top and the tank top presented medium difficulty (Chart 29.2).

The fact that the difficulty to adjust the donned tops is greater than the difficulty to don them reflects, once again, the problem of performing arm extension with forearm flexion behind the back displayed by people with a frozen shoulder. The rotation and elevation of the shoulder combined with the strength to position the garments correctly after donning is also problematic for people with a frozen shoulder.

29.4.2.3 Donning and Doffing the Clothes: Identified by the Users

The long sleeve T-shirt was identified by the sample group as the most difficult top to don, as nine of them pointed out. For those users, the armhole and the sleeve were the main difficulties. In contrast, eight ladies considered the poncho the easiest garment to don (Chart 29.3).

The batwing sleeve top was ranked in the middle followed by the top with a front zipper, as both presented little, medium, and no difficulties. According to the users, the tank top was also one of the hardest tops to don.

29.5   DISCUSSION AND CONCLUSIONS

The main conclusion formed from the researchers' and users' evaluation was that the poncho, batwing sleeve top, and top with a front zipper are more appropriate for people with a frozen

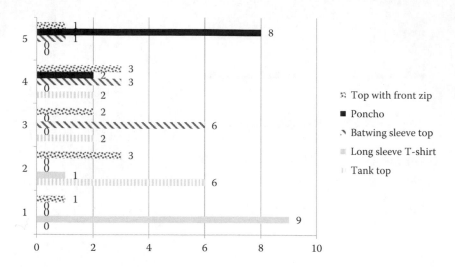

CHART 29.3 Difficulty identified by the user (Authors).

shoulder—the targeted users—as they provided greater autonomy and independence. The main reason is due to the styles themselves, as garments that have fewer seams and/or components—particularly the poncho and the batwing sleeve top—require fewer steps for donning and doffing and thus decrease the amount of performed movements as well as the range of motion.

We emphasize the results of the batwing sleeve top and the top with a front zipper, as they present similar ease in putting the arms into the sleeve, both styles allowing the armhole area to spread out more and, therefore, requiring less range of motion, particularly with reference to the elevation of the shoulder, arm, and forearm.

However, the top with a front zipper requires arm extension with forearm flexion behind the back and such movement present difficulty in the donning and doffing process. Besides, it requires the involvement of another person or equipment to facilitate putting on the second sleeve in most cases.

On the other hand, the long sleeve T-shirt and the top with a front zipper, both with tight armholes, were considered as the most difficult garments to don by the users.

We conclude, thus, that the easier tops to don are the ones that require a smaller amount of movements as well as a smaller range of motion, and emphasize the role played by the armholes in the donning and doffing procedure: tops with set-in sleeves or sleeveless tops are difficult to don as people with a frozen shoulder are not able to flex, extend, and rotate the affected arm and consequently the shoulder (Figure 29.10).

We also emphasize that sleeves also increase the time to don a garment. In this category, batwing sleeve tops may be easier to don than tops with front zippers because the process includes putting one arm into the sleeve, next the head into the neckline, and only then the other arm into the other sleeve, whereas in the case of tops with front zippers, the second sleeve has to be reached from the back after the first sleeve was donned, a combination of movements that we identified as being one of the most difficult for people with a frozen shoulder to accomplish.

The choice of including two sample groups, added to this research not only a geographic quality, but also quality in quantity and perspective. This last variable is mainly due to the fact that the Brazilian women, who live in their own homes, thus having a better quality of life than the residents of *Casa do Artista*, have shown greater interest to take part in this research. From the collected data, we infer that the inclusion of users before and during the design process by means of the UCD process not only provides a valuable interaction between the designer and the targeted user throughout the whole process, but also increases the chance of success of the product in meeting the real needs of the targeted user. Likewise, the inclusion of kinesiological analyses in the design of clothes

FIGURE 29.10 Donning and doffing procedure difficulties (Schiehll).

increases the efficiency of the donning and doffing procedures, thus ensuring greater autonomy to the users.

Finally, we infer that the problems related to lack of strength were also critical, which means that it is necessary to consider the combination of use of force and ease of motion and to exclude critical movements difficult for a frozen shoulder in the design of clothing.

REFERENCES

AAOS. 2003. Frozen shoulder. OrthoInfo. http://orthoinfo.aaos.org/topic.cfm?topic=a00071 (accessed October 2, 2013).

Academia das Ciências de Lisboa. 2001. *Dicionário da Língua Portuguesa*. Lisboa: Academia das Ciências de Lisboa.

Agich, G.J. 2008. *Dependência e Autonomia na Velhice: um modelo e´tico para o cuidado de longo prazo*. São Paulo: Loyola.

Almeida, M. 2009. Promoção da Saúde depois dos 65 anos: Elementos para uma política integrada de envelhecimento. Tese de doutorado não publicada. Escola Nacional de Saúde Pública, Universidade Nova de Lisboa, Lisboa.

Andrade, F. 2009. O Cuidado Informal à Pessoa Idosa Dependente em Contexto Domiciliário: Necessidades Educativas do Cuidador Principal. Dissertação de mestrado não publicada. Universidade do Minho, Instituto de Educação e Psicologia, Guimarães.

CIF. 2003. Classificação Detalhada com definições: Todas as categorias com as suas definições, inclusões e exclusões. Organização Mundial da Saúde (WHO), Direcção-Geral da Saúde. 2003. 222. http://arquivo.ese.ips.pt/ese/cursos/edespecial/CIFIS.pdf (accessed October 23, 2010).

Drescher, J.M. 1999. *Sete necessidades básicas da criança*. São Paulo: Mundo cristão.

Early, M.B. 2013. *Physical Dysfunction Practice Skills for the Occupational Therapy Assistant*. St.Louis, Missouri: Mosby/Elsevier.

Ewald, A.M.D. 2011. Adhesive capsulitis: A review. *American Family Physician*, 83(4): 417–422. Columbus, Ohio: Grant Medical Center.

Falcão, S. 2011. Autonomia e Movimento do Corpo Idoso. Dissertação de mestrado não publicada. ULisboa I FMH, Lisboa.

Ferland, F. 2006. *O desenvolvimento da criança no dia-a-dia. Do berço até à escola primária.* Lisboa: Climepsi Editores.

Foddy, W. 1994. *Constructing Questions for Interviews and Questionnaires.* Cambridge, Massachusetts: Cambridge University Press.

Garcia, A.M. 2009. A satisfação dos idosos em relação ao apoio prestado por uma IPSS. Monografia não publicada. Faculdade de Ciências da Saúde, Universidade Fernando Pessoa, Porto.

Grave, M.F. 2004. *A Modelagem sob a ótica da Ergonomia.* São Paulo: Zenneex Piblishing.

Kendall, H.O., Kendall, F.P. 1980. *Músculos, provas e funções.* São Paulo: Editora Manole.

Lemos, N., Medeiros, S.L. 2002. Suporte social ao idoso dependente. In Freitas, E.V., Py, L., Neri, A.L., Cançado, F.A.C., Gorzoni, M.L. and Rocha, S.M. (eds.), *Tratado de geriatria e gerontologia,* pp. 892–897. Rio de Janeiro: Guanabara Koogan.

Lessa, A., Rendas, A., Samouco, M.H., Botelho, M.A., Ramilo, M.T. 1994. *Imagem e capacidade funcional da pessoa idosa: o envelhecimento nas periferias urbanas: um estudo no concelho de Oeiras.* Universidade internacional. Centro de estudos de ecologia humana. Centro de estudos de gerontologia social. Lisboa: E.I Editora Internacional, 1994, p. 275. CBN 1157.

Martins, S. 2008. Ergonomia e moda: repensando a segunda pele. In Pires, D.B. (ed.), *Design de Moda: Olhares Diversos,* pp. 319–336. São Paulo: Estação das letras e cores.

Menezes, M.S., Spaine, P.A.A. 2010. Modelagem Plana Industrial do Vestuário: diretrizes para a indústria do vestuário e o ensino & aprendizado. *Projética, Londrina,* 1(1), 82–100.

Moreira da Silva, F. 2013. *Colour and Inclusivity: A Visual Communication Design Project with Older People.* Lisboa: Caleidoscópio.

Nucci, L.P., Killen, M., Smetana, J.G. 1996. Autonomy and the personal: Negotiation and social reciprocity. In Killen, M. (ed.), *Children's Autonomy, Social Competence and Interactions with Adults and Other Children: Exploring Connections and Consequences,* pp. 7–24. New York: Joussey-Bass.

Papalia, D.E. 2010. *O mundo da criança: da infância à adolescência.* Porto Alegre: AMGH.

Pavarini, S.C.I., Neri, A.L. 2000. Compreendendo dependência, independência e autonomia no context domiciliar: conceitos, atitude e comportamentos. In Duarte, Y.A.O., Diogo, M.J.D. (eds.), *Atendimento domiciliar: um enfoque gerontológico,* pp. 49–70. São Paulo: Atheneu.

Pfeifer, L.I. s.d. AVD: em busca da qualidade de vida. http://www.profala.com/artto3.htm (accessed March 12, 2013).

Rasch, P., Burke, R. 1977. *Cinesiologia e Anatomia Aplicada: A ciência do movimento humano.* Rio de Janeiro: Guanabara Koogan.

Snider, R.K. 2000. *Tratamento das Doenças do Sistema Musculoesquelético.* São Paulo: Editora Manole.

SOS Med. 2014. Frozen shoulder syndrome (adhesive capsulitis). http://www.sosmed.org/specialties/shoulder-elbow/frozen-shoulder-syndrome-adhesive-capsulitis/ (accessed October 2, 2013).

Soutinho, H.F.C. 2006. Vestuário desportivo: novos desenvolvimentos e novas funcionalidades. Dissertação de Mestrado de Design e Marketing (MSc Thesis). DET/EE/UM. Universidade do Minho.

UW Medicine. 2013. Orthopaedics and sports medicine: Evaluation of the stiff shoulder. Seattle. http://www.orthop.washington.edu/?q=patient-care/articles/shoulder/evaluation-of-the-stiff-shoulder.html (accessed October 2, 2013).

Vale, R.G.S., Pernambuco, C.S., Novaes, J.S., Dantas, E.H.M. 2006. Teste de autonomia functional: vestir e tirar uma camisa (VTC). In: Silva, F.M., Matsudo, V.K.R., Matsudo, S.M.M., Araujo, T.L., Melo, G.F., Prestes, J. (eds.), *Brazilian Journal of Science and Movement,* pp. 71–78. Brasilia: Universa.

Vredenburg, K., Isensee, S., Righi, C. 2013. *User-Centred Design: An Integrated Approach.* Upper Saddle River, New Jersey: Prentice Hall PTR.

WHO. s.d. Ageing and life course: Our ageing world. http://www.who.int/ageing/en/ (accessed February 21, 2011).

Xhardez, Y. 1990. *Manual de cinesioterapia: técnicas, patologia, indicações, tratamento.* Rio de Janeiro: Atheneu.

30 Neurodesign
Applications of Neuroscience in Design and Human–System Interactions

*Tareq Ahram, Christianne Falcão, Rafaela Q. Barros,
Marcelo M. Soares, and Waldemar Karwowski*

CONTENTS

30.1 INTRODUCTION

The human world is complex and exhibits inherent uncertainty and vagueness. Therefore, in order to study the behavior of human systems it is necessary to develop dynamic and approximate modeling approaches. From the systems design viewpoint, the main issue in design is to determine the requisite, and absolutely essential compatibility of the artifact–human system, and use it as a reference point for system improvements. In this chapter we introduce the term neurodesign, which is the merger between neuroscience knowledge and design theory and practices to assure such requisite compatibility in the functioning of the artifact–human systems with respect to complex and uncertain (cognitive and perceptual) interrelationships between key design principles and users, machines, and the environment. Neurodesign must account for the dynamics of human cognitive and perceptual processes. This chapter examines some of the critical issues and methodological needs in neuroscience, design principles, and best practices.

The science of neuroergonomics, which is derived from neuroscience human factors, and ergonomics, aims to model and explain how people interact with their environments and perceive the surroundings, given the specific human characteristics, limitations, and performance capabilities. In doing so, designers face the problem of increased system's complexity and the related human and system-based dynamics. Neural dynamics and fuzziness is not only a state of human mind, but the essence of human development and existence, and a necessary condition for human learning, growth, and survival. The application of neuroscience in design offers methods and tools that facilitate the creative design processes (Hevner et al., 2014). Neuroscience-based knowledge and applications help evaluate the artifact from the time it evolves in the laboratory environment till its release. With regard to the product user experience, neuroscience principles in combination with cognitive science have given the rise to the neurodesign research field (Kirkland, 2012). This approach allows designers to better understand the user experience from the brain activity measurement, and assists in product design decision making. In this context, neuroscience makes it possible to explain why a product or service experience is fundamentally good or bad, allowing members of the project team get true insights about product design features (Kirkland, 2012).

Understanding the neural and cognitive models of users allows designers to develop intuitive and supportive solutions while achieving customer satisfaction. Using what the scientific community have learned in neuroscience in the field of design will empower designers to understand how solutions evolve and the user interaction patterns (Kirkland, 2012). On the other side, understanding neural-system dynamics methodologies allow accounting for complex and natural human cognitive fuzziness, and provide the necessary framework for successful modeling efforts in the human perception and ergonomics discipline. In order to develop the proper relationship between people and the outside surroundings (natural and artificial, i.e., technology-based), the intrinsic dynamics and fuzziness of human perception must be treated by system designers and engineers as natural requirements of their everyday activities.

30.2   BACKGROUND

The human world is complex, therefore, in order to study user behavior it is necessary to develop dynamic and approximate modeling approaches. The human world-based uncertainty and vagueness can be conceptualized and modeled through the notion of fuzziness and dynamics. Karwowski (1991, 1992) suggested that perceptual fuzziness should be looked upon as the natural model of people at work and those human-made systems that interact with people, embedded into traditional descriptions of human sensory, information processing, and communication, or physiological functioning processes. In this context, complexity and fuzziness is not just a product of the human mind that can be described and comprehended only through formal mathematical theories. Fuzziness, as a basic quality of human understanding, is also the essence of human development and existence, and a necessary condition for human learning and growth.

While the contemporary discipline of design with respect to human factors and ergonomics attempts to model and describe how people interact with their outside environment, the common understanding is that human beings are too complex to be fully understood or described in all their characteristics, limits, tolerances, performance, and perceptual capabilities, and that no unified or comprehensive mathematical models are available to describe and integrate all the abovementioned measures and findings about human behavior. The mathematical models currently used in modeling human perceptual capabilities suffer from the measuring problem, which includes difficulties in encoding and describing the varying task load and human workload, the state of social and physical environments, and design and measurement of information flow in people and machines (Karwowski et al., 1999). The unified methodology of human perceptual systems dynamics is viewed as a powerful design modeling tool that can allow the overcoming of some of the above problems.

30.2.1   Neuroscience: From Cognitive Science to Human Perception

According to Kirkland (2012) neurodesign allow better communication with designers to convey creative direction. This approach investigates the brain's cognitive triggers, explaining the reasoning for good or bad customer experience and using them to help designers make better-informed design decisions that are reflected in user behavior, product use, and user interactions. This approach helps designers and usability professionals understand user interaction and explain user experiences that result in optimal solutions. Neurodesign supports "behavioral and contextual" research before making decisions about design key priorities. By conducting interviews at the beginning of a design project, team members can get customer insights. Cognitive ergonomics can be defined as the science that aims to assure compatibility in the artifact–human functioning with respect to complex and uncertain interrelationships between system users, machines, and environments (Guastello, 1995; Karwowski et al., 1999; Ahram and Karwowski, 2011a). Such analysis must account for natural nonlinear dynamics (chaos) and fuzziness of human cognitive processes.

Uncertainty due to vagueness of human decision making and thoughts is inherent to any complex system, and to human perceptual processes. In order to study complex artifact–human systems it is necessary to use modeling approaches that are approximate in nature. Systems dynamics and fuzziness is a useful model for human language and categorizing processes. Fuzziness describes an event ambiguity and measures the degree to which an event occurs, not whether it occurs or not (Zadeh, 1973; Kosko, 1992). Systems dynamics modeling (SDM) provides a useful framework for modeling a variety of complex tasks, situations, artifacts and environments, and their interactions with people. Since this type of complexity occurs at all levels of human interactions with the outside environments, ranging from physical to cognitive tasks, it can be used as the natural model of human sensory, information processing, communication, or physiological functioning. The potential advantages for SDM applications in design are as follows:

1. Model user behavior and categorize interrelated perceptual processes
2. Augment conventional statistical techniques in analysis of user complex and fuzzy decision making
3. Supplement statistical and design data collection techniques such as reliability analysis and regressions, and structurally oriented methods such as hierarchical clustering and multidimensional scaling (Ahram et al., 2010)

In view of the above, neurodesign and systems dynamics methodology can provide a useful modeling framework for applied human perception, and especially for modeling a variety of complex task situations, systems, artifacts, and environments, and their interactions with people.

30.2.2 User Perceptual Processes and Memory

User perception is a dynamic nonlinear process under control of the observer who is consciously aware of the observed phenomena. The process of conscious perception is a fuzzy one, with ambiguity due to both physiological thresholds for physical stimuli and pattern recognition and judgment requirements. For example, a question of "at what frequency will an intermittent light be perceived as just flickering" is an ambiguous one, and, hence, the eye flicker frequency a fuzzy and dynamic with environmental conditions, not random phenomenon, as it refers to the degree of noticeable flickering rather than the question of whether it occurs (Karwowski, 1992). It should be noted here that the psychophysical laws, which relate the magnitude of change in physical stimuli (ΔI) that will just be noticed by an observer, are also fuzzy laws. According to the Weber–Fechner law, the magnitude of change in a physical stimulus that will be just noticed by an observer is a constant proportion of the stimulus. Although defined in terms of the magnitude of ΔI that will result in a judgment of a difference in the levels of physical stimuli 50% of the time (so-called just noticeable difference), this law refers not to the probabilistic statement of whether the perception of change will occur, but at what level of the difference in physical stimuli such a change will be observed (Karwowski, 1992). The degree to which many sources of object stimuli and different dimensions of the same objects can be processed by the human brain exemplifies the fuzzy dynamics nature of

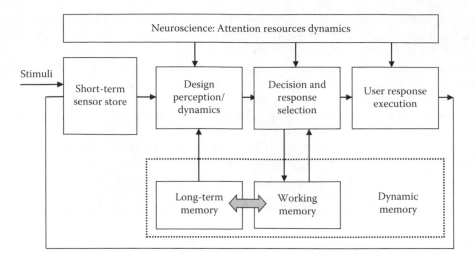

FIGURE 30.1 System dynamics model of human information processing. (After Karwowski, W., Salvendy G. 1992. *An Introduction to Fuzzy Logic Applications in Intelligent Systems*, Kluwer Academic Publishers, Boston, pp. 201–220, 1992.)

human perceptual processes (see Figure 30.1). The human ability to process information from different stimulus objects at one time is limited. However, several dimensions of a single object can be processed in parallel (Wickens, 1987).

30.2.3 DESIGN FOR FUNCTIONALITY AND DYNAMICS IN PERCEPTION AND MEMORY

The user perceptual system is limited in its ability to process information from different stimulus objects at one time. However, several dimensions of a single design functionality and object can be processed in parallel. Therefore, designers should ask the most important question: *To what degree can several sources of design stimuli and different dimensions or functionality of the same design be processed by the user brain?* and the answer to this question indicates the need to consider the fuzzy and dynamics nature of user perceptual processes. For example, in a human–machine system, the displayed information can be arranged along a continuum which defines the degree to which that information is spatial-analog in nature (i.e., information about relative locations, transformations, or continuous motion), linguistic–symbolic or verbal (i.e., a set of instructions, alphanumeric codes, directions or logical operations). The border between these two systems (i.e., the verbal–spatial), however, is ambiguous and subject to user perception dynamics and fuzzy interpretation. Although the distinction between the short-term (working) memory and long-term memory has been universally accepted, the degree to which the memory is short or long is also a fuzzy and dynamic category (see Figure 30.1). The same applies, and even more so, to the classification of spatial and verbal perceptual memory systems.

30.3 DYNAMICS IN HUMAN–MACHINE RESEARCH

Human–machine studies aim to optimize work systems or product design options with respect to the physical and psychological characteristics of the users, and investigate complex and ill-defined relationships among people, machines, and physical environments. The main goal of such investigation is to remove the incompatibilities between the user and the product or system task, and to make the design or workplace healthy, productive, and comfortable. Such user-centered systems are very complex and difficult to analyze due to the complexity of the relationships between people and their

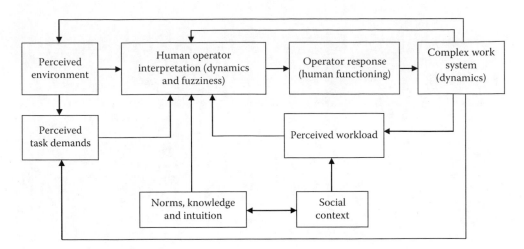

FIGURE 30.2   Fuzzy systems dynamics in human-machine interfacing. (After Karwowski, W., Salvendy G. 1992. *An Introduction to Fuzzy Logic Applications in Intelligent Systems*, Kluwer Academic Publishers, Boston, pp. 201–220.)

working environments, an inherent nonlinear dynamics and fuzziness in human thinking processes, and subjective perception of the outside world. Figure 30.2 illustrates types of dynamics fuzziness that should be accounted for in modeling of any user–machine systems based on research study by Karwowski and Salvendy (1992).

30.3.1   MODELING USER SENSATIONS

Considering the uncertainty and complexity in the process of user information interpretation, dynamic fuzzy systems based on neurodesign principles can be a useful modeling approach for analysis and design of better user–machine interactions. Shimizu and Jindo (1995) proposed a framework for dealing with ambiguities and nonlinearity of the human information processing relevant to the modeling of human sensitivity. The conventional methods to quantify the relationships between human sensations and the physical characteristics which influence them are typically the multivariate analysis techniques such as multiple regression analysis and quantification theory. However, when higher order data are involved, it is much more difficult to find a model formula that suitably represents the nonlinearity factor. Moreover, many conventional methods have traditionally excluded the ambiguities that can arise in the process of recognizing and making subjective evaluations of the physical characteristics. The main advantage of applying neurodesign knowledge to understand user behavior via dynamic fuzzy systems is that the ambiguities and nonlinearity (e.g., of user sensation) can be taken into account and quantified to derive correlations with the considered physical characteristics of the product.

A dynamic fuzzy regression method can be applied for the evaluation of user perceptual sensitivity, for example, the perceived thermal sensation in a car interior. The nonlinearity of human sensation modeled through the traditional method of multiple regression analysis is illustrated in Figure 30.3. While it is possible to treat this marked nonlinearity (at least to some extent) by transforming the variables, it is difficult to determine how the variables should be transformed into nonfuzzy regression. It was shown that 85% of the data items obtained through the subjective evaluations of the car temperature fell within the range of the predicted values obtained from the fuzzy regression analysis method. The results also showed that dynamic fuzzy logic supported making predictions which take into account the natural ambiguity of human perceptual sensations. Furthermore, the application of fuzzy regression analysis made it is relatively easy to obtain results which were closer to the true subjective evaluations made by the people.

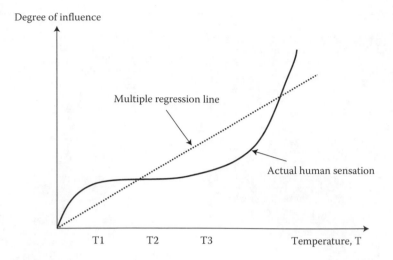

FIGURE 30.3 Example of nonlinearity of human sensation and perception. (After Shimizu, Y., Jindo, T., *International Journal of Industrial Ergonomics*, 15, 39–47, 1995.)

30.3.2 User Response Processes

Decisions are typically followed by responses. The human response processes are intrinsically dynamic, even though the prevalent definition of the relationship between the choice reaction time and the degree of choice (the Hick–Hyman law) is based upon the information content of a stimulus (S) in bits as follows: $RT = a + b(S)$. The very visible presence of human fuzziness cannot be overlooked in the paradigm of stimulus–response compatibility. This paradigm relates to the physical relationship (compatibility) between a set of stimuli and a set of responses, as it affects the speed of human response. As pointed out by Karwowski (1992), the spatial relations between arrangements of signals and response devices in human–machine systems with respect to direction of movement and adjustments are often ambiguous, with a high degree of uncertainty regarding the effects of intended control actions.

30.4 APPLICATION OF NEUROSCIENCE TO DESIGN VIA MODELING USER PERCEPTUAL CAPACITIES

Systems Modeling Language (SysML) is a modeling language that aids in visualizing complex software or systems of systems (SoS), ideally suited to modeling and visualizing neuroergonomic and cognitive constructs. The brain and its capabilities may be simulated on one part of the model, while a computer system on another (Ahram and Karwowski, 2009; Ahram et al., 2009). The respective dynamics, strengths, and capabilities combined with the shortcomings of each system can be shown graphically, and task performance or decision making can be simulated, analyzed, and optimized.

Using real data on user perceptual and cognitive performance combined with the current understanding of brain structure and function (see Figure 30.4), SysML along with systems dynamics can create cognitive models predicting perceptual task performance by implementing models of known structures, their interactions, and perhaps implement the resulting fuzzy logic to create meaningful, visual, and quantitative models. Figure 30.4 shows an example model of the dorsal/ventral streams in visual perception.

The SysML model breaks down brain perceptual and decision-making structures and their functions. It can be used for training or educational purposes in addition to simulation of dynamic decision making. Simulations can be quickly made from this and related models (Ahram et al., 2009).

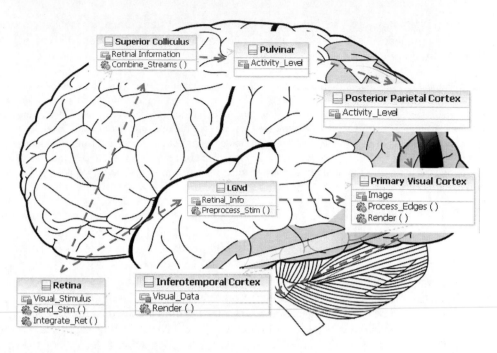

FIGURE 30.4 Example SysML Model of Dorsal/Ventral stream in visual perception. (Adapted from Original by Karwowski, W. et al. 2012. *Neuroadaptive Systems: Theory and Applications*, October 29, 2012, CRC Press, Boca Raton.)

Systems dynamics along with SysML will prove to be an exciting addition to the growing neuroergonomics set of toolbox capabilities.

30.4.1 Modeling Complex User Perception in Design Features

Neurodesign based on SDM of human perception is a novel graphical approach to make the modeling of dynamical perceptual systems accessible by combining the relations we perceive in such systems. According to Fuchs (2006, 2010) the simple ideas behind SD (system dynamics) models correspond to a basic form of human thought. It makes use of a very few structures which are projected onto virtually any type of dynamical system and its processes, that is, it makes strong use of analogical reasoning (Fuchs, 2002a,b). For example, Figure 30.4 shows an example systems dynamics model in the context of a warfare technology concept.

System dynamics applies to dynamic problems arising in complex social, managerial, economic, or ecological systems (i.e., any dynamic system characterized by interdependence, mutual interaction, information feedback, and circular causality). The field developed initially from the work of Forrester (1961, 1969) in his seminal book *Industrial Dynamics* which is still a significant statement of philosophy and methodology in the field. Within 10 years of its publication, the span of applications grew from corporate and industrial problems to include the engineering, science, research and development, and the dynamics of growth in a finite world. SDM is now applied in defence and theory building in social science (Ahram and Karwowski, 2011a), and other areas, as well as decision making (Ahram and Karwowski, 2011a,b; Systems Dynamics Society, 2011) and service engineering (Karwowski et al., 2010).

The name industrial dynamics no longer does justice to the breadth of the field, so it has become generalized to system dynamics. According to Morecroft (2007) the modern name suggests links to other systems engineering methodologies. System dynamics emerges out of servomechanisms engineering, not general systems theory or cybernetics (Richardson, 1991/1999).

Forrester's (1969) provided the following framework for system dynamics structure: closed boundary, which includes feedback loops, levels, rates, goal, observed condition, discrepancy, and desired action. According to the Systems Dynamics Society (2011):

> The importance of levels and rates appears most clearly when one takes a continuous view of structure and dynamics. Although a discrete view, focusing on separate events and decisions, is entirely compatible with an endogenous feedback perspective, the system dynamics approach emphasizes a continuous view. The continuous view strives to look beyond events to see the dynamic patterns underlying them. Moreover, the continuous view focuses not on discrete decisions but on the policy structure underlying decisions. Events and decisions are seen as surface phenomena that ride on an underlying tide of system structure and behavior. It is that underlying tide of policy structure and continuous behavior that is the system dynamicist's focus.

Complex systems develop because of the natural tendency of components and subsystems to resonate and synchronize (Robertson-Dunn, 2009). This form of complex assembly develops because of existing nonlinearities such as those aspects related to growth needs and increased interaction between various human perceptual and cognitive components. As energy levels rise, structures that require high energy to form can be created and become very stable because the energy levels needed to destroy them are no longer present (Robertson-Dunn, 2009).

Research into complex systems has examined the understanding of specific systems rather than looking at the phenomenon as a whole (Norling et al., 2008). The modeling paradigm described in this chapter is based upon a wide range of evidence and an approach that focuses on the behavior of complex systems in general. Emergent, self-organizing, complex systems are nonstationary, nonequilibrium, nonhomogeneous, and nonlinear. They are formed via processes that are episodic, during which structures are created and energy stored in those structures. Structures can be formed by removing or applying energy. All complex systems including human cognitive and perceptual processes are dynamic and are capable of displaying some form of oscillation (Sterman, 2000; Robertson-Dunn, 2009). The oscillation may be because of an external, driving force, or be a characteristic of the system itself.

In the case of falling energy levels for user perception dynamics, neural structures can arise and develop more connections that reflect particular relationships between components and subsystems or a new acquired knowledge based on human perceptual capacities. When the energy levels cycle above and below particular values, structures already formed can be transformed and/or participate in other more complex structures increasing the nonlinearity and dynamics of such a human perceptual system.

It is a characteristic of complex systems that the perceptual components of a system can be influenced and changed by the system itself. This happens when system designed behavior leads to the accumulation of high levels of energy which result in the components being impacted.

The self-organizing nature of complex systems can be the result of one or more phenomena such as energy wells and resonance due to the human perceptual and learning process itself and can be assisted by the presence of existing structures which can facilitate the development of specific, more likely, or more efficient structures including those responsible for decision making (Robertson-Dunn, 2009). While this might look like the phenomena of "design" in general human designed systems, in reality it is simply a catalytic effect or a preference for that which already exists. A summary of the analysis of complex systems is in Table 30.1.

As detailed in Section 30.2 discussion on human perception and memory, the fuzzy systems dynamics model has three parts:

1. Neuroscience-based complex systems arise in an environment of cyclic energy levels
2. Components critically dependent on nonlinear structures, and
3. They are facilitated by the presence of preexisting patterns in human perception and unique seeds

TABLE 30.1
Design Dimension and Principles of Complex Systems

Design Dimension	Impact
Nonlinearities	Stable complex system structures are dependent on the nonlinear nature of the components and their interactions
Emergence	Properties that emerge from a complex perceptual and cognitive system are consequences of its structure and development over a long period of time
Hierarchies	Complex systems can act as components of systems that are more complex
Creation	Complex systems are created via episodic energy interchanges with their environment or through the learning process where individuals acquire new knowledge
Stability	Complex systems have structures that are stable within certain limits. They are created as energy levels vary—sometimes by the application of large amounts of perceptual and environmental needs, energy, and sometimes by "setting" or "annealing"
Structures and complexity	A variety of structures, energy storage, and/or resonant frequencies are required for systems of greater complexity
Interactions	Complex systems interact and synchronize with other systems via resonant and nonlinear mechanisms
"Human" energies	Complex human systems can form based upon conceptual energy storage mechanisms and resonance mechanisms

Source: Robertson-Dunn, B. 2009. Meta modelling self organising, emergent and complex systems. Retrieved from: http://www.drbrd.com/docs/MetaModellingComplexSystems.doc.

Human cognitive and perceptual complex systems are the new scientific frontier which emerged in the past decades with advances of modern computing technology and the study of new parametric domains in natural systems (Robertson-Dunn, 2009). An important challenge involves unprecedented difficulty in predicting human behavior by their structure and the strength of interactions between perceptual and cognitive system components. Complex perceptual and cognitive systems are shielded completely by their specific individual features. So these fuzzy and dynamic systems are a counter example to reductionism, which has been influential in science with the Cartesian method that is only valid for complicated systems.

Whether complex perceptual and cognitive systems are obeying strict laws like classical systems is still unclear, it is however possible today to develop methods using systems dynamics and fuzzy nonlinear modeling which allow the handling of some dynamical properties of such system. They should comply with representing system self-organization when passing from complicated to complex, which rests upon the new paradigm of passing from classical trajectory space to more abstract trajectory manifolds associated to natural system invariants characterizing complex system dynamics and similar to complex information systems (Ahram et al., 2010). So they are basically of qualitative nature, independent of system state space dimension and, because of generic impreciseness, privileging robustness to compensate for not well-known system parameter and functional variations (Robertson-Dunn, 2009).

30.5   NEURODESIGN APPLICATIONS

One of the main principles of neuroscience is mirror neurons. Mirror neuron are important to understand user behavior and actions or how users develop in learning new skills by imitation, for example, or maximizing the rewards associated with user actions to enhance the user learning process. Neuroscientists discovered that experiences actually change the brain structure and work to help create a perception of experience. This approach helps designers understand user social and behavioral trends and how this evolves. For example the Amazon web store utilizes neurodesign to prioritize and fulfill customer needs by capturing social and trend behavior by offering a list of related products other customers have bought, or through the concept of relatedness and other people's reviews and likes. One of the interesting facts learned from the online store experience is that the more reviews the product have the better it is even if the reviews are negative experiences, since customers tend to like the product that attracts the most attention and has the best price showing as a deal. Starbucks utilized a different neurodesign approach by fulfilling customer senses like colors or aroma to bring customer enjoyment memory of the previous purchase.

30.5.1 APPLICATION OF ELECTROENCEPHALOGRAPHY TO DESIGN AND EVALUATION OF USER SATISFACTION

Parasuraman and Rizzo (2007) stressed the importance of electroencephalography (EEG) as a performance evaluation tool for measuring workload, mental stress, mental fatigue, and excitement level. They found that changes in the performance of the tasks can be automatically detected and measured using algorithms that combine EEG power spectrum parameters in multivariate functions, corresponding to an effective method for measuring changes in cognitive load.

During usability evaluation, participants can do much more than simply completing a task and a questionnaire. Gestures and movements, for example, can reflect user frustrations or satisfaction. These correspond to some of the gestures that represent user behavior and provide important data (Tullis and Albert, 2008). Many resources are used to verify human satisfaction from emotional behavior, such as those expressed by verbal and nonverbal gestures. The shape of the vocal tract, the change of tone and behavior in the speech process, such as duration and pause, are some of the features that have been used to measure human emotion from auditory signals (Scherer, 2003). Facial expressions correspond to a nonverbal behavior and have much to reveal about the participant's experience (Tullis and Albert, 2008). The researchers highlighted the correlation between emotion and facial expression (Pantic and Rothkrantz, 2000). However, Esfahani and Sundararajan (2011), indicated that the identification of facial expression is limited to face-to-face interaction and that it may not always be noticed by the researcher. Other research also interprets satisfaction from physiological signals such as heart rate and peripheral temperature (Kim et al., 2004; Anttonen and Surakka, 2005; Tullis and Albert, 2008). Despite the possibility of the use of those resources, the evaluation of user satisfaction still represents a challenge for the fact that it is an internal state, which cannot be completely measured through behavior. This factor initiated the idea of using EEG resources to measure user satisfaction in various situations, including evaluation of the usability of products.

Mental tasks activate certain parts of the brain and brain computer interfaces aim to detect these activities using EEG patterns and relate them to the emotional state (Esfahani and Sundararajan, 2011). Coan and Allen (2004) from a literature review showed that the asymmetry in the frontal EEG activity has been linked consistently to constructs related to motivation and emotion.

Esfahani and Sundararajan (2011) conducted experiments using the collection of brain activity from a brain computer interface to estimate the level of satisfaction. In their experiment, users used the EPOC headset to control the movements of a robot from the mental imagery, with the same command was not met by the robot in response to the emotional level of user satisfaction was affected and then measured. As a result, the experiment proved an accuracy of 79.2% detection level of user satisfaction based on the EEG.

A number of studies have also been conducted in marketing in order to investigate changes in brain activity while participants watched TV commercials. These studies indicated that the level of cortical activity of the frontal and parietal areas was high for TV commercials that were remembered or recalled by the viewer as compared with the activity caused by that which were forgotten (Astolfi et al., 2008; Ohme et al., 2010). The activity of the alpha band was observed in the occipital regions, as was also observed theta activity in the midline and the frontal cortical regions for the most memorable commercials. To design tests using EEG, especially for websites, eye-tracking devices are also used in a complementary way. The information provided by an eye-tracking system can be quite useful in usability testing because it allows the researcher to know where the participant is looking in real time. In studies by Khushaba et al. (2013), the EEG headset system by EMOTIV was used together with an eye-tracking tool to analyze the EEG spectral changes in a context of simple choice (decision), designed to measure specific features in the participant's choice by products that most pleased him.

As researchers pointed, research has been conducted applying neuroscience to design, but has not provided significant studies in the application of EEG in the evaluation of usability. In neuromarketing, most research has collected EEG data on cognition, emotion, and preferences, while research in neuroergonomics have been collecting data on job consequences.

In view of the above, a great opportunity to research appears in the use of methods and brain computer interface technologies for assessing the experience and the emotional aspects of the user, especially studies focusing on the usability of the product with the EEG approach.

30.5.2 EMOTIONAL DESIGN

Emotions seem to dictate rules in people's everyday lives, sometimes we make decisions based on emotional state or mood which are based on cognitive neurological impairments state of mind, whether it is happiness, sadness, irritation, annoyance, or frustration. Emotions are very often intertwined with a range of psychological phenomena (neurological impairments): humor, temperament, personality, disposition, and motivation. Shin and Wang (2015) argue that, according to some theories, cognition and mental processes are important aspects of emotions and in the psychology of decision making. Emotions are usually defined as a complex state of feelings that responds by way of physical and psychological changes, which can influence an individual's thinking and behavior toward a product.

An experience is the learning resulting from an interaction between an individual and the components that make up the environment at a given moment. The user's interaction with the product can be a source of stimulation and thus the unleashing of emotions in people's everyday lives. Each interaction of the user with a product is said to lead to an experience of using it.

Shin and Wang (2015) comment on the importance of experience in the user's interaction with the product because alongside this experience, it is possible to include the perception and identification of a product; to make associations; to perceive memories that are related to it; feelings and emotions that can be provoked; and to make judgments. Thus, people tend to provoke various emotions simultaneously based on their experience of a given product. These emotions are not only prompted by the aesthetic characteristics of the product, but are also evoked by other aspects of its composition, such as its function, ergonomics, brand, and events associated with product interaction.

In this perspective, Norman (2004) coined the term "emotional design" so as to study some issues on design and emotion. In this area, he created three types of emotional designs which function based on three levels of processing information and which are totally dependent on each other. The three levels are visceral design, behavioral design, and reflective design.

Overbeeke and Hekkert (1999) argue that studies on the effect of the emotions on the use of products was first discussed at an event, held in 1999 in Delft, The Netherlands, the theme of which tackled the effects of the relationship between emotion and design. This became known as "The First Conference on Design and Emotion" and brought together researchers from different areas of knowledge, including designers.

Therefore, emotional design should be present to improve the user's experience and hold a central position in the process of product development. This process is of vital importance to deal with the relationship between people and products and is the key to evaluating the product from the point of view of the users.

In addition, the study of emotional design not only helps the connection with users but also evokes positive emotions. According to Norman (2004), the essential factor in the emotional design process is to create "emotional ties" between user and product.

According to Person (2003), some research studies in the area of design and emotion are investigating methods to assess the individual's emotional interaction with the product. These studies can aid decision making during the project and even create decision-aid tools (e.g., identification, measurement in the design process). In this perspective, some studies on notions such as pleasure (Jordan, 2000; Green and Jordan, 2002) and perception and emotion (Seva et al., 2011) analyze user satisfaction in accordance with the sensations of experience with the product.

Damasio (2004) points out that the supporters of this new approach understand that industrial artifacts have emotional competence and unleash all kinds of emotions in their integration. With a holistic view to understand the emotional aspects of user–product interaction, in recent years, studies in human neurophysiology, design, cognitive sciences, and artificial intelligence have been undergoing several enhancements (Marar, 2007). Therefore, the application of neuroscience to emotional design is presented in an extensive range of possibilities for evaluating users' behavior during the decision-making process for the handling of products and artifacts.

Esperidião-Antonio et al. (2008) declares that studies that use the technique of neuroimaging seek to broaden knowledge about the neural bases of and the processes related to the emotions by undertaking in-depth studies of the limbic system. Therefore, what is of interest for these studies is to understand the relationship of emotional, cognitive, and homeostatic and neural impairments processes, and the individual's respective physiological responses. In other words, the discussion deals with the emotions and the relationship of brain circuits in different situations that the individual experiences.

30.5.3 NEURODESIGN DYNAMICS AND USER COGNITION

Many of the nonlinear features and behaviors of complex systems are typically disregarded in exchange for a simple system understanding (Gleick, 1987; Karwowski, 1991). In the traditional view, simple systems were thought to behave in simple ways, while only complex behavior implied complex causes. Recently, chaos theory (Kaplan and Glass, 1995) has shown that relatively simple systems can give rise to complex behaviors, while complex systems can give rise to simple behaviors. Chaos theory provides evidence that rather simple deterministic perceptual models could produce unstable behaviors with exquisite structures that look like, but are not, random behaviors. It was also shown that even though system "periodicity" is the most complicated orderly behavior, deterministic systems can produce much more than just periodic behavior. Furthermore, the systems that are locally unpredictable could also be globally very stable (Karwowski, 1991).

It should be noted here that complex human perceptual and cognitive systems are the rule rather than an exception in ergonomic and neuroscience studies (Karwowski, 1991). This is due to the interaction and continuous exchange of information with the surrounding environment, such systems could experience turbulence and coherence at the same time, with abrupt changes in behavior. It is plausible that many such systems exhibit the so-called "Feigenbaum phenomenon" (Feigenbaum, 1980), that is, under certain circumstances these perceptual and cognitive decision making systems develop chaotic behavior, which cannot be explained in the framework of traditional research approaches used in the cognitive or neuroscience ergonomics of today.

30.6 CONCLUSIONS

The study of the relation of product design and user perceptual complexity and human factors and ergonomics faces the problems of fuzzy dynamics and work system's complexity and related human and system-based fuzziness that increase the ever-present incompatibility between users and their living and working environments. As pointed out in this chapter, fuzziness and dynamics are not just a product of the human mind, but are the essence of human development and existence in the ever more complex world we live in, and a necessary condition for human learning, growth, and survival in a challenging environment rich in computing technologies and fast processing mobile services. In order to develop the cohesive and durable relationship between users and the design and outside surroundings (both natural and artificial), the intrinsic dynamics and fuzziness of the user perceptual and cognitive processing must be treated by designers and engineers as a natural system design requirement with complex interaction with the surroundings. Fuzziness can also be viewed as an expression of human entropy, with models of fuzziness and dynamics as representations of such entropy. The neuroscience science implemented to study user behavior and user–product dynamics represents the natural model of people at work, and those man-made systems that interact with people of which there is a continuous exchange of information in data rich environments. Systems engineering and systems dynamics methodologies allow accounting for natural human fuzziness and human perceptual complexity, and provide the necessary framework for successful modeling efforts in the human factors engineering discipline. In addition, the study of emotional design through neuroscience appears to open a wide and promising application for future user interaction and design research.

REFERENCES

Ahram, T., Karwowski, W., Amaba, B. 2011. Collaborative systems engineering and social networking approach to design and modeling of smarter products. *Behaviour and Information Technology* 30 (1): 13–26.

Ahram, T. Z., Karwowski, W. 2009. Human systems integration modeling. *Human Factors and Ergonomics Society Annual Meeting Proceedings* 53 (24), 1849–1853(5). *53rd Annual Meeting of the Human Factors and Ergonomics Society (HFES 2009)*, October 19–23, San Antonio, Texas.

Ahram, T. Z., Karwowski, W. 2011a. Social networking applications: Smarter product design for complex human behavior modeling, human centered design. In *Proceedings of the 14th International Conference on Human-Computer Interaction (HCII 2011)*, Orlando, FL, July 9–14, vol. 6776, pp. 471–480.

Ahram, T. Z., Karwowski, W. 2011b. Developing human social, cultural, behaviour (HSCB) ontologies: Visualizing & modeling complex human interactions, *Presented at the Office of Secretary of Defense Human Social, Culture Behavior Modeling (HSCB Focus 2011)*, Chantilly, Virginia, February 8–10, pp. 269–316.

Ahram, T. Z., McCauley-Bush, P., Karwowski, W. 2010. Multicriteria weighted model to estimate document collections intrinsic dimensionality and enhance information retrieval performance. *Information Sciences* 180 (15): 2845–2855.

Ahram, T. Z., Karwowski, W., Amaba, B., Obeid, P. 2009. Human systems integration: Development based on SysML and the rational systems platform, *Proceedings of the 2009 Industrial Engineering Research Conference*, Miami, Florida, pp. 2333–2338 (Chair, Systems Architecting Research Sessions).

Anttonen, J., Surakka, V. 2005. Emotions and heart rate while sitting on a chair. *Proceedings of ACM SIGCHI Conference on Human Factors in Computer System*. Portland, Oregon, pp. 491–499.

Astolfi, L. et al. 2008. Neural basis for brain responses to TV commercials: A high-resolution EEG study. *IEEE Transactions on Neural Systems and Rehabilitation Engineering* 27 (6): 522–531.

Coan, J. A., Allen, J. 2004. Frontal EEG asymmetry as a moderator and mediator of emotion. *Biological Psychology* 67: 7–49.

Damasio, A. 2004. Emotions and feelings: A neurological perspective. In Manstead, A., Fridja, N., and Fischer, A., (Eds.), *Feelings and Emotions*, Chapter 4, pp. 49–57. Cambridge University Press: Cambridge, UK.

Esfahani, E., Sundararajan, V. 2011. Using brain-computer interfaces to detect human satisfaction in human-robot interaction. *International Journal of Humanoid Robotics* 08 (01): 87–101. Doi: 10.1142/S0219843611002356.

Esperidião-Antonio, V. et al. 2008. *Revista de Psiquiatria Clínica* 35 (2): 55–65.

Feigenbaum, M. J. 1980. Universal behavior in nonlinear systems. *Los Alamos Science* 1, 4.

Forrester, J. W. 1961. *Industrial Dynamics*. The MIT Press: Cambridge, Massachusetts. Reprinted by Pegasus Communications, Waltham, Massachusetts.

Forrester, J. W. 1969. *Urban Dynamics*. The MIT Press: Cambridge, Massachusetts. Reprinted by Pegasus Communications, Waltham, Massachusetts.

Fuchs, H. U. 2002a. *Modeling of Uniform Dynamical Systems*. Orell Füssli Verlag: Zürich.

Fuchs, H. U. 2002b. A simple continuum model leading to the reciprocity relation for thermoelectric effects. Zurich University of Applied Sciences at Winterthur.

Fuchs, H. U. 2006. System dynamics modeling. In *Science and Engineering*, Retrieved from: http://136.145.236.36/isdweb/Congreso-ISD/pres%20H.%20Fuchs.pdf.

Fuchs, H. U. 2010. *The Dynamics of Heat*, 2nd Edition. Springer-Verlag: New York.

Gleick, J. 1987. *Chaos: Making a New Science*. Penguin Books: New York.

Green, W., Jordan, P. (Eds.). 2002. *Pleasure with Products: Beyond Usability*. Taylor & Francis: London.

Guastello, S. J. 1995. *Chaos, Catastrophe, and Human Affairs*. Lawrence Erlbaum Associates: Mahwah, New Jersey.

Hevner, A. R., Davis, C., Collins, R. W., Gill, T. G. 2014. A neurodesign model for IS research. *Informing Science: The International Journal of an Emerging Transdiscipline* 17: 103–132.

Jordan, P. 2000. *Designing Pleasurable Products: An Introduction to the New Human Factors*. Taylor & Francis: London.

Kahman, R., Henze, L. 2002. Mapping the user-product relationship (in product sesign). In W. S. Green and P. W. Jordan (Eds.), *Pleasure with Products: Beyond Usability*, Taylor & Francis, London.

Kaplan, D., Glass, L. 1995. *Understanding Nonlinear Dynamics*. Springer-Verlag: New York.

Karwowski, W. 1992. The human world of fuzziness, human entropy, and the need for general fuzzy systems theory. *Journal of Japan Society for Fuzzy Theory and Systems* 4(5): 591–609.

Karwowski, W., Ahram, T. Z., Andrzejczak, C., Fafrowicz, M., Marek, T. 2012. Potential applications of systems modeling language (SysML) and systems dynamics to simulate and model complex human brain functions, in: M. Fafrowicz, T. Marek, W. Karwowski, and D. Schmorrow (Eds.), *Neuroadaptive Systems: Theory and Applications*, October 29, 2012, CRC Press: Boca Raton

Karwowski, W., Salvendy, G., Ahram, T. 2010. A human-centered approach to design and modeling of service systems, in: G. Salvendy and W. Karwowski (Eds.), *Introduction to Service Engineering*, John Wiley & Sons, pp. 179–206.

Karwowski, W., J. Grobelny, Y. Yang, W. G. Lee. 1999. Applications of fuzzy systems in human factors, in H. Zimmermman (Ed.), *Handbook of Fuzzy Sets and Possibility Theory*, Kluwer Academic Publishers, Boston, 589–620.

Karwowski, W. 1991. Complexity, fuzziness and ergonomic incompatibility issues in the control of dynamic work environments. *Ergonomics* 34: 671–686.

Karwowski, W., Salvendy, G. 1992. Fuzzy-set-theoretic applications in modeling of man-machine interactions, in: R. R Yager and L. A Zadeh (Eds.), *An Introduction to Fuzzy Logic Applications in Intelligent Systems*, Kluwer Academic Publishers: Boston, pp. 201–220.

Khushaba, R., Wise, C., Kodagoda, S., Louviere, J., Kahn, B., Townsend, C. 2013. Consumer neuroscience: Assessing the brain response to marketing stimuli using electroencephalogram (EEG) and eye tracking. *Expert Systems with Applications* 40: 3803–3812.

Kim, K., Bang, S., Kim, S. 2004. Emotion recognition system using short-term monitoring of physiological signals. *Medical & Biological Engineering & Computing* 42: 419–427.

Kirkland, L. 2012. *Using Neuroscience to Inform Your UX Strategy and Design*, Published July 9, 2012. Last accessed September 22, 2015: http://www.uxmatters.com/mt/archives/2012/07/using-neuroscience-to-inform-your-ux-strategy-and-design.php#top.

Kosko, B. 1992. *Neural Networks and Fuzzy Systems: A Dynamical Systems Approach to Machine Intelligence*. Prentice–Hall: Englewood Cliffs.

Marar, J. F. 2007. Research and project—Design and artificial Intelligence: An integrated system of smart selection of materials and its impacts in product design (in Portuguese).

Morecroft, J. 2007. *Strategic Modeling and Business Dynamics: A Feedback Systems Approach*. John Wiley & Sons. ISBN 0470012862.

Norling, E., Powell, C. R., Edmonds, B. 2008. Cross-disciplinary views on modelling complex systems, *Proceedings of the 9th Workshop on Multiagent-Based Simulation*, Estoril, Portugal (May 12–13), pp. 573–575.

Norman, D. A. 2004. *Emotional Design: Why We Love (or Hate) Everyday Things*. Basic Books: NewYork.

Ohme, R., Reykowska, D., Wiener, D., Choromanska, A. 2010. Application of frontal EEG asymmetry to advertising research. *Journal of Economic Psychology* 31 (5): 785–793.

Overbeeke, C. J., Hekkert, P. (Eds.) 1999. *Proceedings of the 1st International Conference on Design and Emotion*. Delft: Delft University of Technology.

Pantic, M., Rothkrantz, L. 2000. Automatic analysis of facial expressions: The state of the art. *Pattern Analysis and Machine Intelligence, IEEE Transactions on* 22 (12): 1424–1445.

Parasuraman, R., Rizzo, M. 2007. *Neuroergonomics: The Brain at Work*. Oxford University Press.

Person, O. 2003. *Usability is not Enough: The First Underline of a Functional Model for Describing Emotional Response Towards Products*. The Norwegian University of Science and Technology (NTNU).

Richardson, G. P. 1991/1999. *Feedback Thought in Social Science and Systems Theory*. University of Pennsylvania Press: Philadelphia. Reprinted by Pegasus Communications, Waltham, Massachusetts.

Robertson-Dunn, B. 2009. Meta modelling self organising, emergent and complex systems. Retrieved from: http://www.drbrd.com/docs/MetaModellingComplexSystems.doc.

Seva, R., Gosiaco, K., Santos, Ma, C., Pangilinan, D. 2011. Product design enhancement using apparent usability and affective quality. *Applied Ergonomics* 42: 511–517.

Scherer, K. 2003. Vocal communication of emotion: A review of research paradigms. *Speech Communication* 40: 227–256.

Shimizu, Y., Jindo, T. 1995. A fuzzy logic analysis method for evaluating human sensitivities. *International Journal of Industrial Ergonomics* 15: 39–47.

Shin, D., Wang, Z. 2015. The experimentation of matrix for product emotion, *Procedia Manufacturing* 3: 2295–2302. Doi: 10.1016/j.promfg.2015.07.375.

Sterman, J. 2000. *Business Dynamics: Systems Thinking and Modeling for a Complex World*. McGraw-Hill: New York.

Systems Dynamics Society 2011: http://www.systemdynamics.org.

Tullis, T., Albert, B. 2008. *Measuring the User Experience: Collecting, Analyzing and Presenting Usability Metrics*. Elsevier Inc.

Wickens, C. D. 1987. Information processing, decision-making, and cognition, in: G. Salvendy (Ed.), *Handbook of Human Factors*, John Wiley & Sons, New York, pp. 72–107.

Zadeh, L. A. 1973. Outline of a new approach to the analysis of complex systems and decision processes. *IEEE Transactions on Systems, Man and Cybernetics* SMC-3, 28–44.

Index